RSME Springer Series

Volume 10

As of 2015, RSME - Real Sociedad Matemática Española - and Springer cooperate in order to publish works by authors and volume editors under the auspices of a co-branded series of publications including advanced textbooks, Lecture Notes, collections of surveys resulting from international workshops and Summer Schools, SpringerBriefs, monographs as well as contributed volumes and conference proceedings. The works in the series are written in English only, aiming to offer high level research results in the fields of pure and applied mathematics to a global readership of students, researchers, professionals, and policymakers.

Antonio Alarcón • Vicente Palmer • César Rosales
Editors

New Trends in Geometric Analysis

Spanish Network of Geometric Analysis 2007-2021

Real Sociedad Matemática Española

Springer

Editors
Antonio Alarcón
Departamento de Geometría y Topología
Universidad de Granada
Granada, Spain

Vicente Palmer
Departament de Matemàtiques
Universitat Jaume I
Castellón, Spain

César Rosales
Departamento de Geometría y Topología
Universidad de Granada
Granada, Spain

ISSN 2509-8888 ISSN 2509-8896 (electronic)
RSME Springer Series
ISBN 978-3-031-39915-2 ISBN 978-3-031-39916-9 (eBook)
https://doi.org/10.1007/978-3-031-39916-9

This Springer imprint is published by the registered company Springer Nature Switzerland AG
The registered company address is: Gewerbestrasse 11, 6330 Cham, Switzerland

Preface

Geometric analysis lies in the interplay between geometry and partial differential equations and has applications to several branches of mathematics such as Riemannian geometry, topology, calculus of variations, potential theory and complex variables, as well as to general relativity, crystallography, material science, architecture, and other fields.

In this sense, function theory on manifolds equipped with a metric structure and its relation with potential theory and the theory of PDEs have been revealed as a sharpen analysis instrument of the geometric structure of the manifold. This analysis takes as a starting point the study of the analytic properties of some distinguished functions defined on a manifold, namely, the solution of partial differential equations formulated using differential operators deeply related with the given metric structure (such as, for instance, the Laplace operator in the case of Riemannian manifolds), and deals with the connection among these functional properties and the geometric properties of the manifold.

Among many other achievements, the solutions of the positive mass conjecture in general relativity, the Poincaré conjecture in topology, the Lawson, Willmore and Yau conjectures in differential geometry, or the Kobayashi conjecture in complex geometry are some highlights which show the current relevance and usefulness of geometric analysis in pure mathematics.

As suggested by the title, the aim of this book is to provide an overview of some of the progress made by the Spanish Network of Geometric Analysis (REAG, by its Spanish acronym). REAG was created in 2007 with the objective of enabling the interchange of ideas and the knowledge transfer between several Spanish groups having Geometric Analysis as a common research line. This currently includes nine groups at Universidad Autónoma de Barcelona, Universidad Autónoma de Madrid, Universidad de Granada, Universidad Jaume I de Castellón, Universidad de Murcia, Universidad de Santiago de Compostela, and Universidad de Valencia. The success of REAG has been substantiated with regular meetings and the publication of research papers obtained in collaboration between the members of different nodes.

REAG is a Research Network that has been supported by the successive Spanish Research Programs via the following grants.

• MTM2006-27480-E, Ministerio de Ciencia e Innovación.
Coordinated by Olga Gil-Medrano (Valencia), May 2007–April 2008.
• MTM2008-01013-E, Ministerio de Ciencia e Innovación.
Coordinated by José A. Gálvez (Granada), October 2008–September 2009.
• MTM2009-06054-E/MTM, Ministerio de Ciencia e Innovación.
Coordinated by Joan Porti (Barcelona), October 2009–September 2010.
• MTM2010-09693-E/MTM, Ministerio de Ciencia e Innovación.
Coordinated by Luis Guijarro (Madrid), January 2011–June 2012.
• MTM2011-15848-E/MTM, Ministerio de Economía y Competitividad.
Coordinated by Manuel Ritoré (Granada), October 2012–October 2013.
• MTM2014-57309-REDT, Ministerio de Economía y Competitividad.
Coordinated by José A. Gálvez (Granada), January 2015–December 2016.
• MTM2016-81938-REDT, Ministerio de Economía y Competitividad
Coordinated by Luis J. Alías (Murcia), July 2017–June 2019.
• RED2018-102361-T, Ministerio de Ciencia, Innovación y Universidades.
Coordinated by Antonio Alarcón (Granada), January 2020–April 2022.

On the occasion of the 15th anniversary of REAG, this book aims to collect some old and new contributions of this network to geometric analysis. The book consists of 13 independent chapters, all of them authored by current members of REAG. The topics under study cover geometric flows, constant mean curvature surfaces in Riemannian and sub-Riemannian spaces, integral geometry, potential theory, and Riemannian geometry, among others. Some of these chapters have been written in collaboration between members of different nodes of the network, and show the fruitfulness of the common research atmosphere provided by REAG. The rest of the chapters survey a research line or present recent progresses within a group of those forming REAG.

We would like to conclude this preface by warmly thanking the authors of the chapters for submitting their valuable works. Likewise, we wish to take this opportunity to express our gratitude to all colleagues who in one or the other way have contributed with their effort to the development and success of REAG during this 15 years.

Granada, Spain — Antonio Alarcón
Castellón, Spain — Vicente Palmer
Granada, Spain — César Rosales
November 2021

Contents

Snapshots of Non-local Constrained Mean Curvature-Type Flows

Esther Cabezas-Rivas

Abstract The *mean curvature flow* is the most natural way to deform a hypersurface according to its curvature, since it evolves the parametrization by means of the heat equation. In 1987, G. Huisken (J. Reine Angew. Math. 382:35–48 (1987)) introduced a variant that keeps the enclosed volume constant while the area decreases. For this modification, a global term is added to the speed of the original flow, which makes the usual methods in geometric flows (e.g. maximum principles) either fail or become more intricate. Moreover, the resulting evolution is not equivalent to the unconstrained flow because, for instance, an initially embedded curve may develop self-intersections (cf. Mayer and Simonett, Differ. Integr. Equ. 13(7–9):1189–1199 (2000)). Accordingly, constrained mean curvature-type flows are more challenging; indeed, there was no extension of Huisken's result to a non-Euclidean ambient space until 2007 (Cabezas-Rivas and Miquel, Indiana Univ. Math. J. 56(5):2061–2086 (2007)). Here, we will give a sketchy portrait of the development of the theory from the early stages to the recent applications, including the proof of several geometric inequalities (Andrews et al., J. Eur. Math. Soc. (JEMS) 23(7):2467–2509 (2021)) and new classification results for solitons (Cabezas-Rivas and Scheuer, The quermassintegral preserving mean curvature flow in the sphere. Preprint arxiv.org/abs/2211.17040 (2022)).

Keywords Volume-preserving · Mean curvature flow · Area-preserving · Curve shortening flow

E. Cabezas-Rivas (✉)
Departament de Matemàtiques, Universitat de València, Burjassot, Spain
e-mail: Esther.Cabezas-Rivas@uv.es

A. Alarcón et al. (eds.), *New Trends in Geometric Analysis*,
RSME Springer Series 10, https://doi.org/10.1007/978-3-031-39916-9_1

1 Non-local Versus Unconstrained Mean Curvature Flow

1.1 The Classical Mean Curvature Flow

Consider $M^n \subset \mathbb{R}^{n+1}$ a closed hypersurface (i.e. compact and without boundary), which is smooth and embedded. The idea is to move each point $x \in M$ in the normal direction and with velocity proportional to the way that the hypersurface is curved within the ambient space. More precisely,

Definition 1 Let $F_0 : M^n \to \mathbb{R}^{n+1}$ be a smooth immersion. The evolution of $M_0 = F_0(M)$ by the mean curvature flow is a one-parameter family of immersions $F(\cdot, t) : M \to \mathbb{R}^{n+1}$ satisfying

$$\partial_t F(p,t) = \mathbf{H}_{F(p,t)} = -(H\nu)(p,t) \tag{1}$$

for all $p \in M$ and $t \geq 0$, with initial condition $F(\cdot, 0) = F_0$.

Here, $H = \kappa_1 + \cdots + \kappa_n$ is the mean curvature, $\kappa_1 \leq \cdots \leq \kappa_n$ are the principal curvatures, and ν denotes the outward unit normal of the evolving hypersurface $M_t = F(M,t)$. The sign conventions are taken so that convex hypersurfaces (i.e. with $\kappa_1 \geq 0$) move inwards.

The first instances of this evolution equation appeared in materials science (cf. [50, 63]) to model the growth of cell, grain and bubble structures. Brakke [9] gave an early mathematical treatment from the viewpoint of geometric measure theory. The approach by means of differential geometry and classical PDE comes from Huisken in [31] for $n \geq 2$ and Gage [25] for $n = 1$ (the version of (1) for curves in the plane is known as *curve shortening flow*).

Between all the possible ways to deform a hypersurface according to its curvature, (1) is unquestionably the most natural motion, as it turns out to be the heat equation for the parametrization:

$$\partial_t F = \Delta_{M_t} F.$$

Unfortunately, because the non-linear terms hidden within the time dependence of the Laplacian, we actually have a reaction-diffusion equation, and hence there is a competition between two opposed effects. In fact, (1) tends to regularize the curvature and make it constant (as in the classical heat diffusion, understanding the curvature as a sort of geometric temperature) and simultaneously pushes the curvature to blow up in finite time.

Indeed, the diffusion term cannot prevent the formation of a singularity, but it is strong enough to keep embeddedness of the evolving shapes and produce a *very symmetric singularity*, as shown by the following:

Theorem 1 (Gage and Hamilton [27] for $n = 1$, Huisken [31] for $n \geq 2$) *If $M_0 \subset \mathbb{R}^{n+1}$ is a closed convex and embedded hypersurface, then the solution M_t of* (1) *starting at M_0 exists on a finite time interval* $[0, T)$*; stays smooth, embedded* *and convex; and converges as $t \to T$ to a round point. This means that if $\widetilde{M}_t$ is obtained by rescaling M_t so that the area is constant in t, then $\widetilde{M}_t$ converges smoothly and exponentially fast to a round sphere as $t \to \infty$.*

In the one-dimensional case, we can even drop the convexity assumption:

Theorem 2 (Grayson [28]) *Under the curve shortening flow, a general embedded curve becomes convex and thus by Theorem 1 eventually shrinks to a round point.*

1.2 The Volume-Preserving Version of the Flow

The aim now is to look for an ad hoc evolution process to find solutions of the isoperimetric problem. Accordingly, the new flow should keep the enclosed volume constant while decreasing the area. As (1) turns out to be the gradient flow of the area functional, it is already the most efficient way to reduce the area. Therefore, the idea is to modify the flow velocity adding an extra term as follows:

$$\partial_t F = (h(t) - H)\nu,$$

and then define the global function $h(t)$ so that the volume of the domain Ω_t enclosed by M_t is constant along the flow.

As the changes of area $|M_t|$ and enclosed volume $|\Omega_t|$ under an extrinsic evolution are well-known [55], we get that

$$\frac{d}{dt}|\Omega_t| = \int_M (h - H)\, d\mu_t = 0 \quad \text{implies} \quad h(t) = \frac{1}{|M_t|}\int_M H\, d\mu_t = \overline{H}(t),$$

where $d\mu_t$ denotes the volume element on M_t. Next, we can check that by choosing the global term as the averaged mean curvature $\overline{H}$, the area is actually decreasing under the flow:

$$\frac{d}{dt}|M_t| = \int_{M_t} (\overline{H} - H)H = \int_{M_t} (\overline{H} - H)H - \int_{M_t} (\overline{H} - H)\overline{H} = -\int_{M_t} (\overline{H} - H)^2 \leq 0.$$

This leads to the definition of the *volume-preserving mean curvature flow* introduced by Huisken in [33]:

$$\partial_t F = (\overline{H} - H)\nu. \tag{2}$$

Physically, this can be regarded as the deformation of a super elastic rubber membrane that surrounds an incompressible fluid.

From the mathematical viewpoint, the resultant evolution problem is particularly *appealing*, since it is specially well suited for applications to the isoperimetric problem, and *challenging*, because the presence of a non-local term $\overline{H}$ in all the relevant evolution equations causes a plethora of extra complications; e.g. a basic principle for (1) (*disjoint surfaces remain disjoint, and embedded curves remain embedded*) fails in general for (2).

Indeed, Gage [26] gave an early sketch of an initially embedded curve that may develop self-intersections. This was confirmed by Mayer [47] using numerical simulations, until a rigorous treatment extending the example to higher dimensions was done by Mayer and Simonett in [48].

To study the behaviour of singularities is still the major open problem in the work with (2). Certainly, this task is really harder than the corresponding one for (1), since there is no hope to apply in this setting the methods by Huisken and Sinestrari (see [36, 37]) for hypersurfaces of positive mean curvature under (1). The reasons are the counterexamples to the preservation of mean convexity ($H > 0$) under (2) found by Miquel and the author in [13]. Here, we also confirm that the next natural curvature condition (i.e. positive scalar curvature) is neither invariant under (2).

In short, the knowledge of this flow is considerably poorer than that of the unconstrained version, as the presence of global terms in the evolution equations makes difficult to reach a priori estimates, unless we have extra hypotheses that guarantee some control on $\overline{H}$, for instance, that it has a sign or is uniformly bounded. As these properties are granted for the area-preserving evolution of curves, we will use the one-dimensional situation as a model case to discuss the main features and obstacles of constrained flows.

2 The Area-Preserving Curve Shortening Flow

It was introduced by Gage in [26]: let $\gamma : \mathbb{S}^1 \times [0, T) \to \mathbb{R}^2$ be a time-dependent family of plane curves which is a solution of

$$\partial_t \gamma = (\bar{\kappa} - \kappa)\nu = \partial_s^2 \kappa + \bar{\kappa}\nu, \tag{3}$$

where

$$\bar{\kappa} = \frac{1}{L_t} \int_{\mathbb{S}^1} \kappa \, ds.$$

Here, $s = s(t)$ denotes the arc-length parameter of the evolving curve $\Gamma_t = \gamma(\mathbb{S}^1, t)$, whose length is $L_t = L(t)$. Notice that (3) can be regarded as a non-linear heat equation with a non-local forcing term.

If the evolving curves remain embedded, the flow has three basic features:

1. $\bar{\kappa}(t) = \dfrac{2\pi}{L_t} > 0$.
2. *The flow is area-preserving and length-decreasing*. More precisely, if $A = A(t)$ is the area of the domain Ω_t enclosed by Γ_t, we have

$$A'(t) = 0 \qquad \text{and} \qquad L'(t) = -\int \kappa^2\, ds + \frac{4\pi^2}{L_t} \leq 0, \tag{4}$$

where the latter follows by Cauchy-Schwarz inequality.
3. *The global term is non-decreasing and uniformly bounded*. In particular, by item 2 above, we have

$$0 < \bar{\kappa}(0) = \frac{2\pi}{L_0} \leq \frac{2\pi}{L_t} = \bar{\kappa}(t) \leq \sqrt{\frac{\pi}{A_0}}, \tag{5}$$

where we have applied the classical isoperimetric inequality

$$L^2 \geq 4\pi A. \tag{6}$$

Notice that equality in (5) holds for round circles.

Hereafter, we will usually drop the explicit time dependence of quantities like L_t unless it is unclear from the context.

2.1 Evolution of Convex Curves

In this case, the flow works exactly as expected from the corresponding unconstrained results in [27]:

Theorem 3 (Gage [26]) *Let Γ_0 be a closed embedded and convex curve in $\mathbb{R}^2$. Then the solution Γ_t of* (3) *starting at Γ_0 exists for all times $t > 0$; stays smooth, embedded and convex; and converges smoothly and exponentially fast as $t \to \infty$ to a round circle of radius $\sqrt{A_0/\pi}$.*

The convex one-dimensional case of the flow takes advantage of the availability of isoperimetric inequalities that are stronger than (6):

Lemma 1 *Let Γ be a closed convex and embedded curve in the plane of length L that encloses an area A.*

- ***Gage's isoperimetric inequality [24]:***

$$\int \kappa^2 ds \geq \frac{\pi L}{A}. \tag{7}$$

- ***Bonnesen inequality [51]:***

$$\frac{L^2}{A} - 4\pi \geq \frac{\pi^2}{A}(r_{out} - r_{in})^2, \tag{8}$$

where r_{out} and r_{in} represent the inradius and outradius, respectively, of the domain Ω enclosed by Γ.

Remark 1

(a) Cauchy-Schwarz for simple curves leads to $\int \kappa^2 ds \geq \frac{4\pi^2}{L}$. Notice that (6) implies that Gage's inequality (7) is stronger than Cauchy-Schwarz in this setting. Moreover, equality holds in (7) if and only if Γ is a round circle (cf. [52]).

(b) An example of Jacobowitz (see [24]) suggests that (7) fails for non-convex curves. The idea is to construct a dumbbell-formed curve by means of straight segments and circle arcs of radius r, with the neck long enough so that A and L of the ends are negligible. Then it holds

$$\int \kappa^2 ds \approx \int \frac{1}{r^2} ds \propto \frac{2\pi}{r}.$$

But on the other hand we also have $\frac{L}{A} \approx \frac{L}{2\varepsilon L/2} = \frac{1}{\varepsilon}$, and hence we can take ε small enough so that (7) breaks down.

The following four key observations suggest that Theorem 3 should be true, modulo technicalities:

1. *A convex solution becomes instantaneously strictly convex under* (3).
This follows by inspection of the evolution equation

$$\partial_t \kappa = \partial_s^2 \kappa + (\kappa - \bar{\kappa})\kappa^2.$$

In fact, if one can find $(p, \tau) \in \mathbb{S}^1 \times (0, T)$ with $\kappa(p, \tau) = 0$, then the strong maximum principle implies that $\kappa \equiv 0$ on $\mathbb{S}^1 \times [0, \tau]$, which is impossible since Γ_t is a compact curve.

In the ordinary curve shortening flow (where it holds $(\partial_t - \partial_s^2)\kappa = \kappa^3$), notice that the cubic term pushes the curvature to blow up in finite time. For the current case, this effect becomes softened for convex curves but gets worse if we have points of negative curvature.

2. *If Γ_0 is embedded and convex, then Γ_t is embedded for each t.*
The statement comes from a delicate two-point maximum principle argument applied to the evolution of the Euclidean distance

$$d : \mathbb{S}^1 \times \mathbb{S}^1 \times [0, T) \longrightarrow \mathbb{R}, \qquad d(p, q, t) = \|\gamma(p, t) - \gamma(q, t)\|^2$$

between any two points in the evolving curve (see [27, Theorem 3.2.1]).

3. *The isoperimetric deficit is decreasing with time:*

$$\omega(t) := \frac{L^2}{A} - 4\pi \le Ce^{-2\pi t/A_0}. \tag{9}$$

This is established by integration of the inequality

$$\omega'(t) \underset{(4)}{=} -\frac{2}{A}\Big(L\int \kappa^2\,ds - \pi\frac{L^2}{A} + \pi\omega\Big) \underset{(7)}{\le} -\frac{2\pi}{A_0}\omega.$$

4. *If $T = \infty$, then Γ_t approaches a round circle in the Hausdorff distance.*

This claim holds by taking limits as $t \to \infty$ in

$$Ce^{-2\pi t/A_0} \underset{(9)}{\ge} \omega(t) \underset{(8)}{\ge} (r_{\text{out}} - r_{\text{in}})^2.$$

The preservation of convexity can be exported to higher dimensions (see [33, Theorem 1.3]) by means of a version of the maximum principle for symmetric two tensors. But let us stress that the last two proofs are very specific for the convex curve case, due to the application of the strong inequalities from Theorem 1.

2.2 Basic Features for the Flow of Embedded Curves

Let $\gamma(\cdot, t)$ be a solution of (3) defined on a maximal time interval $[0, T)$, for which we use the notation $|\kappa|_{\max}(t) = \max_{\mathbb{S}^1} |\kappa(\cdot, t)|$. To strengthen the type of convergence, that is, to jump from Hausdorff to C^∞, we need the following properties of the flow that hold for non-necessarily convex curves, provided that embeddedness is preserved under the flow:

1. *If the curvature is bounded, then all its space and time derivatives are also controlled*:

$$|\kappa| \le C_0 \text{ on } \mathbb{S}^1 \times [0, T) \quad \Rightarrow \quad |\partial_t^m \partial_s^n \kappa| \le C_{n,m} \text{ on } \mathbb{S}^1 \times [0, T). \tag{10}$$

This is established arguing by induction and applying maximum principles to the corresponding evolution equations.

2. *The only reason for singularities is curvature blow-up*, that is,

$$T < \infty \quad \to \quad \lim_{t \nearrow T} |\kappa|_{\max} = \infty.$$

The idea for the proof is a contradiction argument: suppose that $|\kappa|_{\max} \leq C_0$ for some constant. Then we can take a sequence of times $t_i \in [0, T)$ with $t_i \nearrow T$ as $i \to \infty$ so that

$$\|\gamma(\cdot, t_j) - \gamma(\cdot, t_i)\| \leq \int_{t_i}^{t_j} \|\partial_t \gamma\| \, dt \underset{(3)}{=} \int_{t_i}^{t_j} \|\bar{\kappa} - \kappa\| \, dt \leq 2C_0(t_j - t_i). \tag{11}$$

In particular, the same computation, but integrating between 0 and T, shows that Γ_t stays in a bounded region of $\mathbb{R}^2$. Accordingly, $\{\gamma(\cdot, t_i)\}_{i \in \mathbb{N}}$ is a bounded uniform Cauchy sequence of smooth functions on $\mathbb{S}^1$ and hence by standard arguments admits a continuous limit $\gamma_T : \mathbb{S}^1 \to \mathbb{R}^2$.

By redoing the computation (11) with derivatives of γ and using the bounds in (10), one can argue that the limit is a smooth embedded curve, which by short time existence allows to extend the solution beyond T. The contradiction to the maximality of T settles the statement.

3. *Lower blow-up rate for the curvature:*

$$T < \infty \qquad \Rightarrow \qquad 2\,|\kappa|_{\max} \geq 1/\sqrt{T - t}.$$

The proof is a straightforward adaptation of [34, Lemma 1.2]. This property allows to distinguish between type I singularities, if there is also an upper bound $2\,|\kappa|_{\max}\sqrt{T - t} \leq C$, and type II otherwise.

Let us highlight that all the above properties have their corresponding versions for higher dimensions, but we have chosen the curve case for simplicity of the exposition. Notice that, to guarantee long time existence ($T = \infty$), the only remaining task is to ensure that $|\kappa|_{\max}$ is uniformly bounded.

2.3 Towards a Grayson-Type Theorem

Expanding curve flows like $\partial_t \gamma = -\frac{1}{\kappa}\nu$ already had to face the difficulty of non-preservation of embeddedness for general non-convex curves. To overcome this, Chow, Liou and Tsai [18] considered a class of curves satisfying an extra condition for the turning angle:

$$\theta_{\min}(0) := \min_{p,q \in \mathbb{S}^1} \angle(T_p, T_q) = \int_p^q \kappa \, ds \geq -\pi, \tag{12}$$

where T_p denotes the unit tangent vector of $\gamma(p, 0)$. Notice that this condition is weaker than star-shapedness.

In the setting of the flow (3), it turns out that this condition is sharp, in the sense that for any initial embedded curve with $\theta_{\min}(0) < -\pi$ the evolving Γ_t develops self-intersections (see [20]). The key fact is that the angle between tangent vectors at any two points of the curve (modulo the local winding number), that is, the function

$$\theta : \mathbb{S}^1 \times \mathbb{S}^1 \times [0, T) \longrightarrow \mathbb{R} \quad \text{given by} \quad \theta(p, q, t) = \int_p^q \kappa(\cdot, t)\, ds,$$

satisfies a heat-type equation $\partial_t \theta = \Delta_{\Gamma_t} \theta$, and hence condition (12) is preserved under the flow. This allows to adapt the ideas from [35] with a good control of the extra terms that arise in the evolution equations due to the presence of $\bar{\kappa}$. Therefore, the following Grayson-type theorem holds:

Theorem 4 (Dittberner [20]) *Let Γ_0 be a closed embedded curve in $\mathbb{R}^2$ that satisfies* (12). *Then the solution Γ_t of* (3) *starting at Γ_0 exists for all times $t > 0$, stays smooth and embedded and converges smoothly and exponentially fast as $t \to \infty$ to a round circle of radius $\sqrt{A_0/\pi}$.*

2.4 Further Results

The flow (3) can be generalized in many different directions, for instance, one can allow self-intersections. Indeed, for closed immersed curves, there is a notion of signed area which is preserved under (3) with global term

$$\bar{\kappa}(t) = \frac{2\pi m}{L_t},$$

where $m \in \mathbb{Z}$ is the winding number. Here are the known results for the evolution of non-simple, locally convex, closed curves:

- If $A_0 < 0$ or $L_0^2 < 4\pi m A_0$, then a singularity forms in $T < \infty$ [22].
- One can find a class of highly symmetric curves and a family of Abresh-Langer-type curves (both with $A_0 > 0$) for which Γ_t converges smoothly as $t \to \infty$ to an m-fold circle [64].
- Assume $T = \infty$. Then we have convergence to an m-fold circle if $A_0 > 0$ and shrinkage to a point if $A_0 = 0$ [65].

As Bonnesen inequalities may fail for general non-simple curves, the above results use an energy method which amounts to bound under the flow the quantity $E(t) = \int (\kappa - \bar{\kappa})^2\, ds$. These statements were further generalized in [59] to area-preserving flows of the type $\partial_t \gamma = (\bar{\kappa} - \kappa^\alpha)\nu$ for any $\alpha > 0$, where the global term is defined ad hoc so that the area is kept constant.

On the other hand, there are related Neumann free boundary-type problems: take a convex support curve Σ and an initial Γ_0 embedded and strictly convex, whose interior points are contained in the outer domain of Σ and the endpoints meet Σ perpendicularly. Then evolve Γ_0 by (3) under the additional request that the endpoints of Γ_t also touch Σ orthogonally.

In this framework, Mäder-Baumdicker [45] found an upper bound for the length of Γ_0 depending on A_0 and $|\kappa|_{\max}(0)$, which guarantees smooth convergence of Γ_t to an arc of a round circle as $t \to \infty$. Moreover, [46] gives suitable conditions on Γ_0 ensuring that a singularity happens in finite time, and it should be of type II.

Finally, there is a wide literature of global flows for various choices of $\bar{\kappa}$:

- If $\bar{\kappa} = \frac{1}{2\pi}\int \kappa^2\, ds$, then (3) is length-preserving and area-decreasing. The results corresponding Theorem 3 and Theorem 4 are obtained in [44] and [20], respectively.
- For $\bar{\kappa} = \frac{L}{2A}$, the flow becomes area-increasing and length-decreasing; in fact, this is the gradient flow of the isoperimetric ratio $\frac{L^2}{4\pi A}$ [39].
- For expanding flows $\partial_t \gamma = (\frac{1}{\kappa} - \bar{\kappa})\nu$, the choices $\bar{\kappa} = \frac{1}{L}\int \frac{1}{\kappa}\, ds$ in [43] and $\bar{\kappa} = -\frac{L}{2\pi}$ in [53] give area- and length-preserving flows, respectively. For $\bar{\kappa} = -2\frac{A}{L}$, the flow is both area- and length-increasing (see [54]).

3 Volume-Preserving Mean Curvature-Type Flows

3.1 Convex Evolution and Stability Results

The higher-dimensional version of Theorem 3 works as expected:

Theorem 5 (Huisken [33]) *If M_0 is a closed embedded and convex hypersurface in $\mathbb{R}^{n+1}$, then the solution M_t of (2) starting at M_0 exists for all $t > 0$, stays smooth, embedded and convex; and converges smoothly and exponentially fast as $t \to \infty$ to a round sphere enclosing the same volume as M_0.*

Li [42] replaces the convexity assumption by requiring $\overline{H}(0) > 0$ and that the traceless second fundamental form is small enough in an integral sense. By means of viscosity solutions, Kim and Kwon [40] prove the corresponding result for initial hypersurfaces that satisfy a stronger version of star-shapedness.

Huisken even gave an intuitive counterexample stressing how hard it would be to export Theorem 5 to a non-flat ambient space. In fact, unlike (1) (where such an extension [32] was done 2 years after the Euclidean version [31]), this remained an open question for (2) from 1987 to 2007, when Miquel and the author found a Riemannian manifold with a suitable notion of convexity invariant under (2). More precisely,

Theorem 6 ([10]) *Let M_0 be a closed, embedded hypersurface in the hyperbolic space $\mathbb{H}^{n+1}$:*

(1) *If M_0 is h-convex (i.e. $\kappa_1 \geq 1$), then under* (2) ***(a)*** *the evolving hypersurfaces M_t remain h-convex,* ***(b)*** *the flow exists for all positive time, and* ***(c)*** *the solution converges exponentially (in the C^k topology, for each $k \in \mathbb{N}$) to a geodesic sphere in $\mathbb{H}^{n+1}$.*

(2) *If M_0 is not h-convex but close enough (in a $C^{1+\alpha}$-Hölder sense) to a geodesic sphere of $\mathbb{H}^{n+1}$, then **(b)** and **(c)** remain true.*

The proofs combine methods from geometric flows but applied in a more sophisticated way due to the non-local nature of (2): the geometric properties of h-convexity in $\mathbb{H}^{n+1}$ [8] and a technique based on the maximal regularity theory for certain PDEs (see [23] for the earlier Euclidean version of **(2)**).

The condition of h-convexity has been recently relaxed to positive intrinsic sectional curvature ($\kappa_1\kappa_2 > 0$) in [2] for a wider class of velocities and constrained flows that preserve the *quermassintegrals*, which are given by

$$W_k(\Omega) = \frac{(n+1-k)\,\omega_{k-1}\cdots\omega_0}{(n+1)\,\omega_{n-1}\cdots\omega_{n-k}} \int_{\mathcal{L}_k} \chi(L_k \cap \Omega)\,dL_k, \quad k = 1, \ldots, n.$$

Here, χ is the characteristic function of $L_k \cap \Omega$, $\mathcal{L}_k$ represents the space of k-dimensional totally geodesic subspaces L_k in a space form $\mathbb{M}^{n+1}_K$ of constant sectional curvature K, and ω_n is the area of a unit sphere $\mathbb{S}^n \subset \mathbb{R}^{n+1}$. This notion includes the area of the hypersurface $|M| = (n+1)W_1(\Omega)$ and the enclosed volume $W_0(\Omega) = |\Omega|$. Thus, a flow $\partial_t F = (h - H)\nu$ that preserves $W_0(\Omega_t)$ coincides with (2).

In $\mathbb{R}^{n+1}$, the quermassintegrals coincide, up to a constant factor, with the *curvature integrals* or *mixed volumes*, defined as follows:

$$V_{n-\ell}(\Omega) = \int_M H_\ell \, d\mu, \qquad \text{for} \quad \ell = 1, \ldots, n,$$

where

$$H_\ell = \binom{n}{\ell}^{-1} \sum_{1 \le i_1 < \cdots < i_\ell \le n} \kappa_{i_1} \cdots \kappa_{i_n}. \tag{13}$$

Indeed, mixed volumes and quermassintegrals are related [61, Proposition 7] in a space of constant curvature $\mathbb{M}^{n+1}_K$ by means of

$$\frac{1}{n+1} V_{n-\ell}(\Omega) = W_{\ell+1}(\Omega) - K \frac{\ell}{n+2-\ell} W_{\ell-1}(\Omega), \qquad \ell = 1, \ldots, n, \tag{14}$$
$$V_n(\Omega) = (n+1)W_1(\Omega) = |M|.$$

The study of constrained flows that preserve the mixed volumes in $\mathbb{R}^{n+1}$ was started by McCoy [49].

While $\mathbb{H}^{n+1}$ is the model space with constant negative sectional curvature, the positive counterpart is the sphere $\mathbb{S}^{n+1}$. However, Huisken [33] gave convincing reasons that convexity of hypersurfaces in $\mathbb{S}^{n+1}$ is not preserved under the flow (2) in general. If r denotes the radial distance to a fixed origin O in $\mathbb{S}^{n+1}$, Scheuer and the author introduce the first example of a non-local mean curvature flow which preserves convexity in the sphere.

Theorem 7 ([15]) *Let $n \geq 2$ and M_0 be a smooth, embedded and strictly convex hypersurface of $\mathbb{S}^{n+1}$. Then there exists a finite system of origins $(O_i)_{0 \leq i \leq m}$ and numbers $0 = t_0 < t_1 < \cdots < t_m < t_{m+1} = \infty$, such that the curvature flow defined by*

$$\begin{cases} \partial_t F = (h_i(t)\cos(r_i) - H)\nu, \\ F(M, 0) = M_0 \\ F(M, t_i) = \lim\limits_{t \nearrow t_i} M_t, \quad 1 \leq i \leq m, \end{cases} \qquad \text{with} \quad h_i(t) = \frac{\int_M H H_\ell \, d\mu_t}{\int_M \cos(r_i) H_\ell \, d\mu_t}, \tag{15}$$

where r_i is the distance to O_i, preserves the quermassintegral $W_\ell(\Omega_t)$ and has a solution $F : M \times [0, \infty) \to \mathbb{S}^{n+1}$ with strictly convex evolving hypersurfaces. Furthermore, the restriction $F : M \times [t_m, \infty) \to \mathbb{S}^{n+1}$ is smooth and converges for $t \to \infty$ in C^∞ to a geodesic sphere around O_m with radius determined by $W_\ell(B_r) = W_\ell(\Omega_0)$.

3.2 Free Boundary Problems with Rotational Symmetry

To replace convexity, a natural geometric condition, invariant under (2) and which still softens the problems caused by the global term, is to take the initial M_0 to be a revolution hypersurface generated by the graph of a function. In an ambient space with rotational symmetry around an axis $\mathcal{A}$, Miquel and the author study the evolution by (2) of such M_0 whose boundary intersects orthogonally two totally geodesic hypersurfaces π_{tg} orthogonal to $\mathcal{A}$.

Theorem 8 ([11]) *Asking that the evolving hypersurface M_t meets π_{tg} orthogonally at each time and under suitable hypotheses on the negativity of some ambient sectional curvatures, it holds the following:*

(1) *While M_t does not touch $\mathcal{A}$, **(a)** the flow exists, **(b)** the generating curve is a graph over $\mathcal{A}$, and **(c)** the global term is bounded above and below by positive constants.*
(2) *Under a hypothesis relating the enclosed volume to the area of M_0, we achieve **(a)** long time existence and **(b)** convergence to a revolution hypersurface of constant mean curvature.*
(3) *The singularity set (if not empty) is discrete along $\mathcal{A}$.*

This completes some results from [4, 5] for $\mathbb{R}^n$ and extends them to a broader class of ambient manifolds. In such ambients, unlike $\mathbb{R}^n$, the hypersurfaces π_{tg} are not at constant distance from each other. Hence, in [12], we address the same problem but considering regions limited by equidistant hypersurfaces π_{eq}; as a by-product, we also cover a new situation in $\mathbb{R}^n$: the case where π_{eq} are spheres instead of hyperplanes. Moreover, the claims analogous to **(1a)**, **(1b)** and **(2)** allow (for the first time in the literature on (2)) positive ambient curvatures and M_t with non-necessarily positive global term. Notice that the situation in [12] is much harder, since the geometry does not allow to repeat M_t periodically, so each step has the further complication of analysing what happens at the boundary.

In the same spirit, [30] replaces the rotational symmetry by asking that $M_0 \subset \mathbb{R}^{n+1}$ is close enough (in a suitable little Hölder space) to a cylinder of sufficiently large radius, which ensures that an immortal solution exists and converges exponentially fast to a cylinder.

Finally, [6] studies axially symmetric hypersurfaces in $\mathbb{R}^{n+1}$ under the assumption that they do not develop singularities along $\mathcal{A}$ during the flow. In this framework, M_t converges to a hemisphere when M_0 has a free boundary and meets $\mathcal{A}$ orthogonally, and to a sphere if M_0 is closed.

3.3 Evolution with Speeds Different from the Mean Curvature

The constrained evolution of convex hypersurfaces was studied for a large class of speeds of homogeneous degree one in the principal curvatures (cf. [49]). For higher degree, the analysis was restricted to local flows either in dimension two (e.g. [57]) or for specific choices of the speed, requiring also a *pinching condition* stronger than convexity (cf. [17, 58]).

In the same line, Sinestrari and the author [14] initiated the use of volume-preserving-type flows to analyse the evolution of closed hypersurfaces in $\mathbb{R}^{n+1}$ by means of velocities with degree of homogeneity > 1.

Theorem 9 ([14]) *Let M_0 be a hypersurface of $\mathbb{R}^{n+1}$ satisfying the following relation between the Gauss and the mean curvature: there is a universal constant $c > 0$ such that*

$$K(p) > cH^n(p) > 0, \qquad \textit{for all} \qquad p \in M_0. \tag{16}$$

Then deform M_0 under the flow

$$\partial_t F = (h(t) - H_m^k)\nu, \quad \textit{for any} \quad k \geq 1/n,$$

where the global term h is chosen so that the flow is volume-preserving and H_m is the mth mean curvature (13). *In this setting,* (16) *is preserved, the solution exists for all times, and we get exponential and smooth convergence to a round sphere.*

Here, the *speeds* are (i) *not concave/convex in the second derivatives* (hence, the standard regularity theory [41] for fully non-linear parabolic PDEs cannot be applied), and (ii) *not uniformly elliptic,* so the equation may a priori degenerate at infinite time. To overcome (i), we prove (inspired by [62]) space regularity at a fixed time by using a result for elliptic equations [16]. We attacked (ii) by getting just sufficient control on how fast the curvature can decay to ensure that the limit (if exists) is totally umbilic; secondly, we rewrote the evolution of H_m as a divergence porous medium equation, which admits Hölder interior estimates for degenerate parabolic PDEs [19].

Theorem 9 was extended to the hyperbolic space later in [29]. Back to the Euclidean setting, one can exploit that the flow improves the isoperimetric ratio to remove the pinching condition (16) and prove the corresponding version of (9) for strictly convex hypersurfaces (see [60] for $m = 1$ and [3, 7] for the general case).

The latter is a remarkable example where constrained flows work nicer than their non-local counterparts, thanks to the monotonicity of the isoperimetric quotient. This was already observed earlier by Andrews [1] in his anisotropic version of (2).

3.4 Applications

Despite the technical difficulties and the poorer development compared with the local unconstrained situation, the flow (2) and its relatives already provided powerful applications in different contexts. Indeed, they are used as a tool to

- Construct spacelike hypersurfaces of prescribed mean curvature in cosmological spacetimes [21].
- Produce a stable constant mean curvature foliation in the exterior region of an asymptotically flat manifold of positive mass [38]. This foliation is used to define the centre of mass for an observer located infinitely far.
- Build up a foliation of the end of an asymptotically hyperbolic three-manifold of positive Bondi mass by constant mean curvature surfaces, provided that the metric approaches the hyperbolic metric fast enough [56].
- Get an alternative proof of the Minkowski inequalities for smooth convex domains in $\mathbb{R}^{n+1}$ using a mixed volume-preserving flow [49].
- Obtain novel Alexandrov-Fenchel-type inequalities for h-convex hypersurfaces in $\mathbb{H}^{n+1}$:

$$W_k(\Omega) \geq f_k \circ f_\ell^{-1}\big(W_\ell(\Omega)\big)$$

  for any $0 \leq \ell < k \leq n$ with equality for geodesic balls. Here, $f_k(r) = W_k(B_r)$ is the quermassintegral of a geodesic ball of radius r. This relies on the definition of a curvature flow under which one quermassintegral is preserved while others change monotonically (see [66]). An improved version of these inequalities can be found in [2] by using some modified quermassintegrals and an ad hoc preserving flow. Moreover, for $\ell = 0$ and any $k = 1, \ldots n$, the h-convexity assumption can be weakened to positive sectional curvature [2].

- Deduce new classification results for some constant curvature-type equations in $\mathbb{S}^{n+1}$, as well as for convex solitons both in the sphere and De Sitter space (see [15]).

Acknowledgments This work is partially supported by the AEI (Spain) and FEDER project PID2019-105019GB-C21 and by the GVA project AICO 2021 21/378.01/1. The author was also partially supported by project PID2022-136589NB-I00 as well as the network RED2022-134077-T, both funded by MCIN/AEI/10.13039/501100011033.

References

1. Andrews, B.: Volume-preserving anisotropic mean curvature flow. Indiana Univ. Math. J. **50**(2), 783–827 (2001)
2. Andrews, B., Xuzhong, C., Wei, Y.: Volume preserving flow and Alexandrov-Fenchel type inequalities in hyperbolic space. J. Eur. Math. Soc. (JEMS) **23**(7), 2467–2509 (2021)
3. Andrews B., Wei, Y.: Volume preserving flow by powers of the k-th mean curvature. J. Differ. Geom. **117**(2), 193–222 (2021)
4. Athanassenas, M.: Volume-preserving mean curvature flow of rotationally symmetric surfaces. Comment. Math. Helv. **72**(1), 52–66 (1997)
5. Athanassenas, M.: Behaviour of singularities of the rotationally symmetric, volume-preserving mean curvature flow. Calc. Var. Part. Differ. Equ. **17**(1), 1–16 (2003)
6. Athanassenas, M., Kandanaarachchi, S.: Convergence of axially symmetric volume-preserving mean curvature flow. Pac. J. Math. **259**(1), 41–54 (2012)
7. Bertini, M.C., Sinestrari, S.: Volume preserving flow by powers of symmetric polynomials in the principal curvatures. Math. Z. **289**(3–4), 1219–1236 (2018)
8. Borisenko, A.A., Miquel, V.: Total curvatures of convex hypersurfaces in hyperbolic space. Ill. J. Math. **43**(1), 61–78 (1999)
9. Brakke, K.: The motion of a surface by its mean curvature. Mathematical Notes. Princeton University Press, Princeton, NJ (1978)
10. Cabezas-Rivas, E., Miquel, V.: Volume preserving mean curvature flow in the hyperbolic space. Indiana Univ. Math. J. **56**(5), 2061–2086 (2007)
11. Cabezas-Rivas, E., Miquel, V.: Volume-preserving mean curvature flow of revolution hypersurfaces in a rotationally symmetric space. Math. Z. **261**(3), 489–510 (2009)
12. Cabezas-Rivas, E., Miquel, V.: Volume preserving mean curvature flow of revolution hypersurfaces between two equidistants. Calc. Var. Part. Differ. Equ. **43**(1–2), 185–210 (2012)
13. Cabezas-Rivas, E., Miquel, V.: Non-preserved curvature conditions under constrained mean curvature flows. Differ. Geom. Appl. **49**, 287–300 (2016)
14. Cabezas-Rivas, E., Sinestrari, C.: Volume-preserving flow by powers of the mth mean curvature. Calc. Var. Part. Differ. Equ. **38**(3-4), 441–469 (2010)
15. Cabezas-Rivas, E., Scheuer, J.: The quermassintegral preserving mean curvature flow in the sphere. Preprint arxiv.org/abs/2211.17040 (2022)
16. Caffarelli, L.A.: Interior a priori estimates for solutions of fully nonlinear equations. Ann. Math. (2) **130**(1), 189–213 (1989)
17. Chow, B.: Deforming convex hypersurfaces by the nth root of the Gaussian curvature. J. Differ. Geom. **22**(1), 117–138 (1985)
18. Chow, B., Liou L.-P., Tsai, D.-H.: Expansion of embedded curves with turning angle greater than $-\pi$. Invent. Math. **123**(3), 415–429 (1996)
19. DiBenedetto, E., Friedman, A: Hölder estimates for nonlinear degenerate parabolic systems. J. Reine Angew. Math. **357**, 1–22 (1985)

20. Dittberner, F: Curve flows with a global forcing term. J. Geom. Anal. **31**(8), 8414–8459 (2021)
21. Ecker, K., Huisken, G.: Parabolic methods for the construction of spacelike slices of prescribed mean curvature in cosmological spacetimes. Commun. Math. Phys. **135**(3), 595–613 (1991)
22. Escher, J., Ito, K.: Some dynamic properties of volume preserving curvature driven flows. Math. Ann. **333**(1), 213–230 (2005)
23. Escher, J., Simonett, G.: The volume preserving mean curvature flow near spheres. Proc. Am. Math. Soc. **126**(9), 2789–2796 (1998)
24. Gage, M.E.: An isoperimetric inequality with applications to curve shortening. Duke Math. J. **50**(4), 1225–1229 (1983)
25. Gage, M.E.: Curve shortening makes convex curves circular. Invent. Math. **76**(2), 357–364 (1984)
26. Gage, M.E.: On an area-preserving evolution equation for plane curves. In Nonlinear problems in geometry (Mobile, Ala., 1985), volume 51 of Contemp. Math., 51–62. Amer. Math. Soc., Providence, RI (1986)
27. Gage, M.E., Hamilton, R. S.: The heat equation shrinking convex plane curves. J. Differ. Geom. **23**(1), 69–96 (1986)
28. Grayson, M.A.: The heat equation shrinks embedded plane curves to round points. J. Differ. Geom. **26**(2), 285–314 (1987)
29. Guo, S., Li, G., Wu, C.: Volume-preserving flow by powers of the m-th mean curvature in the hyperbolic space. Commun. Anal. Geom. **25**(2), 321–372 (2017)
30. Hartley, D.: Motion by volume preserving mean curvature flow near cylinders. Commun. Anal. Geom. **21**(5), 873–889 (2013)
31. Huisken, G.: Flow by mean curvature of convex surfaces into spheres. J. Differ. Geom. **20**(1), 237–266 (1984)
32. Huisken, G.: Contracting convex hypersurfaces in Riemannian manifolds by their mean curvature. Invent. Math. **84**(3), 463–480 (1986)
33. Huisken, G.: The volume preserving mean curvature flow. J. Reine Angew. Math. **382**, 35–48 (1987)
34. Huisken, G.: Asymptotic behavior for singularities of the mean curvature flow. J. Differ. Geom. **31**(1), 285–299 (1990)
35. Huisken, G.: A distance comparison principle for evolving curves. Asian J. Math. **2**(1), 127–133 (1998)
36. Huisken, G., Sinestrari, C.: Convexity estimates for mean curvature flow and singularities of mean convex surfaces. Acta Math. **183**(1), 45–70 (1999)
37. Huisken, G., Sinestrari, C.: Mean curvature flow singularities for mean convex surfaces. Calc. Var. Part. Differ. Equ. **8**(1), 1–14 (1999)
38. Huisken, G., Yau, S-T.: Definition of center of mass for isolated physical systems and unique foliations by stable spheres with constant mean curvature. Invent. Math. **124**(1–3), 281–311 (1996)
39. Jiang, L., Pan, S.: On a non-local curve evolution problem in the plane. Commun. Anal. Geom. **16**(1), 1–26 (2008)
40. Kim, I., Kwon, D.: Volume preserving mean curvature flow for star-shaped sets. Calc. Var. Part. Differ. Equ. **59**(2), Paper No. 81, 40 pp. (2020)
41. Krylov, N.V., Safonov, M.V.: A property of the solutions of parabolic equations with measurable coefficients. Izv. Akad. Nauk SSSR Ser. Mat. **44**(1), 161–175, 239 (1980)
42. Li, H.: The volume-preserving mean curvature flow in Euclidean space. Pac. J. Math. **243**(2), 331–355 (2009)
43. Ma, L., Cheng, L.: A non-local area preserving curve flow. Geom. Dedicata **171**, 231–247 (2014)
44. Ma, L., Zhu, A.: On a length preserving curve flow. Monatsh. Math. **165**(1), 57–78 (2012)
45. Mäder-Baumdicker, E.: The area preserving curve shortening flow with Neumann free boundary conditions. Geom. Flows **1**(1), 34–79 (2015)
46. Mäder-Baumdicker, E.: Singularities of the area preserving curve shortening flow with a free boundary condition. Math. Ann. **371**(3–4), 1429–1448 (2018)

47. Mayer, U.F.: A numerical scheme for moving boundary problems that are gradient flows for the area functional. Eur. J. Appl. Math. **11**(1), 61–80 (2000)
48. Mayer U.F., Simonett, G.: Self-intersections for the surface diffusion and the volume-preserving mean curvature flow. Differ. Integr. Equ. **13**(7–9), 1189–1199 (2000)
49. McCoy, J.A.: Mixed volume preserving curvature flows. Calc. Var. Part. Differ. Equ. **24**(2), 131–154 (2005)
50. Mullins, W.W.: Two-dimensional motion of idealized grain boundaries. J. Appl. Phys. **27**, 900–904 (1956)
51. Osserman, R.: Bonnesen-style isoperimetric inequalities. Am. Math. Month. **86**(1), 1–29 (1979)
52. Pan, S.L., Tang, X.Y., Wang, X.Y.: A strengthened form of Gage's isoperimetric inequality. Chin. Ann. Math. Ser. A **29**(3), 301–306 (2008)
53. Pan, S., Yang, J.: On a non-local perimeter-preserving curve evolution problem for convex plane curves. Manuscripta Math. **127**(4), 469–484 (2008)
54. Pan, S., Zhang, H.: On a curve expanding flow with a non-local term. Commun. Contemp. Math. **12**(5), 815–829 (2010)
55. Reilly, R.C.: Variational properties of functions of the mean curvatures for hypersurfaces in space forms. J. Differ. Geom. **8**, 465–477 (1973)
56. Rigger, R.: The foliation of asymptotically hyperbolic manifolds by surfaces of constant mean curvature (including the evolution equations and estimates). Manuscripta Math. **113**(4), 403–421 (2004)
57. Schnürer, O.C.: Surfaces contracting with speed $|A|^2$. J. Differ. Geom. **71**(3), 347–363 (2005)
58. Schulze, F.: Convexity estimates for flows by powers of the mean curvature. Ann. Sc. Norm. Super. Pisa Cl. Sci. (5) **5**(2), 261–277 (2006)
59. Sesum, N., Tsai, D-H., Wang, X-L.: Evolution of locally convex closed curves in the area-preserving and length-preserving curvature flows. Commun. Anal. Geom. **28**(8), 1863–1894 (2020)
60. Sinestrari, C.: Convex hypersurfaces evolving by volume preserving curvature flows. Calc. Var. Part. Differ. Equ. **54**(2), 1985–1993 (2015)
61. Solanes, G.: Integral geometry and the Gauss-Bonnet theorem in constant curvature spaces. Trans. Am. Math. Soc. **358**(3), 1105–1115 (2006)
62. Tsai, D.-H.: $C^{2,\alpha}$ estimate of a parabolic Monge-Ampère equation on S^n. Proc. Am. Math. Soc. **131**(10), 3067–3074 (2003)
63. von Neumann, J.: Discussion remark concerning paper of C. S. Smith, "Grain shapes and other metallurgical applications of topology". In: Proceedings of the Metal Interfaces, Lake Geneva, WI, USA, September 1952, pp. 108–110
64. Wang, X-L., Kong, L-H.: Area-preserving evolution of nonsimple symmetric plane curves. J. Evol. Equ. **14**(2), 387–401 (2014)
65. Wang, X-L., Wo, W.F., Yang, M.: Evolution of non-simple closed curves in the area-preserving curvature flow. Proc. Roy. Soc. Edinburgh Sect. A **148**(3), 659–668 (2018)
66. Wang, G., Xia, C.: Isoperimetric type problems and Alexandrov- Fenchel type inequalities in the hyperbolic space. Adv. Math. **259**, 532–556 (2014)

Spherical Curves Whose Curvature Depends on Distance to a Great Circle

Ildefonso Castro, Ildefonso Castro-Infantes, and Jesús Castro-Infantes

Dedicated to Professor Óscar Garay, in memoriam

Abstract Motivated by a problem posed by David A. Singer in 1999 and by the elastic spherical curves, we study the spherical curves whose curvature is expressed in terms of the distance to a great circle (or from a point). By introducing the notion of *spherical angular momentum*, we provide new characterizations of some well-known curves, like the mentioned elastic curves, the spherical catenaries, the loxodromic-type spherical curves, the Viviani's curve, and the spherical Archimedean spiral curves. Furthermore, we show that they may be obtained as critical points of some energy curvature functionals. We also find out several new families of spherical curves whose intrinsic equations are expressed in terms of elementary functions or Jacobi elliptic functions, and we are able to get arc length parametrizations of them.

Keywords Spherical curves · Spherical angular momentum · Elastic curves · Catenaries · Loxodromic curves

I. Castro (✉)
University of Jaén, Jaén, Spain
e-mail: icastro@ujaen.es

I. Castro-Infantes
University of Alicante, Alicante, Spain
e-mail: ildefonso.castro@ua.es

J. Castro-Infantes
Polytechnic University of Madrid, Madrid, Spain
e-mail: jesus.castro@upm.es

A. Alarcón et al. (eds.), *New Trends in Geometric Analysis*,
RSME Springer Series 10, https://doi.org/10.1007/978-3-031-39916-9_2

1 Introduction

The plane curves are uniquely determined up to rigid motion by its intrinsic equation giving its curvature κ as a function of its arc length. However, such curves are impossible to find explicitly in practice in most cases, due to the difficulty in solving the three quadratures appearing in the integration process. In [16], David A. Singer considered a different sort of problem: *Can a plane curve be determined if its curvature is given in terms of its position?*

Probably, the most interesting solved problem in this setting corresponds to the Euler elastic curves, whose curvature is proportional to one of the coordinate functions, e.g., $\kappa(x, y) = c\,y$. Motivated by the above question and by the classical elasticae, the authors studied in [7] the plane curves whose curvature depends on the distance to a line (say the x-axis, and so $\kappa = \kappa(y)$) and in [8] the plane curves whose curvature depends on the distance from a point (say the origin, and so $\kappa = \kappa(r)$, $r = \sqrt{x^2 + y^2}$) requiring in both cases the computation of three quadratures too. But the simple case $\kappa(r) = r$, where elliptic integrals appear, illustrated that the fact that the corresponding differential equation is integrable by quadratures does not mean that it is easy to perform the integrations. In [16], only the very pleasant special case of the classical Bernoulli lemniscate, $r^2 = 3\cos 2\theta$ in polar coordinates, was solved explicitly, where the corresponding elliptic integral becomes elementary.

In this paper, we pay our attention to the Singer's problem version for curves lying in a sphere:

Can a spherical curve be determined when its curvature is given in terms of its position?

The geodesic curvature κ of a spherical curve ξ given as a function of its arc length s determines the curve (up to isometries of the sphere) by integration of its Frenet equations. However, it is expectable that if the curvature κ of $\xi = (x, y, z)$ is given by a function of its position, i.e., $\kappa = \kappa(x, y, z)$, the situation becomes quite complicated since the general form of this problem is equivalent to solving the nonlinear differential equation

$$\begin{vmatrix} x(s) & y(s) & z(s) \\ \dot{x}(s) & \dot{y}(s) & \dot{z}(s) \\ \ddot{x}(s) & \ddot{y}(s) & \ddot{z}(s) \end{vmatrix} = \kappa(x(s), y(s), z(s))$$

with the constraints

$$x(s)^2 + y(s)^2 + z(s)^2 = 1 \quad \text{and} \quad \dot{x}(s)^2 + \dot{y}(s)^2 + \dot{z}(s)^2 = 1.$$

The purpose of this article is to study the aforementioned cases of the classical Singer's problem in the setting of spherical curves, considering geodesics in the role of lines. Concretely, we consider a curve $\xi = (x, y, z)$ lying in the unit sphere $\mathbb{S}^2$ centered at the origin and write $z = \sin\varphi$ (φ being the latitude of ξ); we aim to control those curves ξ whose geodesic curvature κ satisfies the condition $\kappa =$

$\kappa(\varphi) \Leftrightarrow \kappa = \kappa(z)$. We point out that this condition includes both types of problems involving curvature and distance, since φ is the distance to the equator (the great circle $\varphi = 0 \Leftrightarrow z = 0$) and the colatitude $\pi/2 - \varphi$ is the distance to the North pole (the point (0,0,1)).

As in the Euclidean case, it will be necessary again the computation of three quadratures when $\kappa(z)$ is a continuous function (see Theorem 1), and the key tool will be the notion of *spherical angular momentum*, which completely determines a spherical curve (up to a family of distinguished isometries) in relation with its relative position with respect to a fixed geodesic.

In this way, we find out several interesting new families of spherical curves whose intrinsic equations can be expressed in terms of elementary or Jacobi elliptic functions. We also provide new characterizations of some well-known curves, like elastic-type spherical curves, spherical catenaries, loxodromic-type spherical curves, Viviani's curve, and spherical Archimedean spiral curves. In addition, we show that they may be obtained as critical points of some energy curvature functionals.

2 Spherical Curves such that $\kappa = \kappa(z)$ and the Spherical Angular Momentum

We introduce a smooth function associated with any spherical curve, which completely determines it (up to a family of distinguished isometries) in relation with its relative position with respect to a fixed geodesic.

Indeed, let $\xi = \xi(s): I \subseteq \mathbb{R} \to \mathbb{S}^2$ be an immersed curve parametrized by the arc length, i.e., $|\xi(s)| = |\dot{\xi}(s)| = 1$, for any $s \in I$, where I is some interval in $\mathbb{R}$. Along the paper, $\dot{}$ will denote derivative with respect to s and $\langle \cdot, \cdot \rangle$ and $\times$ the Euclidean inner product and the cross product in $\mathbb{R}^3$, respectively. Let $T = \dot{\xi}$ be the unit tangent vector and $N = \xi \times \dot{\xi}$ the unit normal vector of ξ. If ∇ is the connection in $\mathbb{S}^2$, the oriented geodesic curvature κ of ξ is given by the Frenet equation $\nabla_T T = \kappa N$, which implies

$$\ddot{\xi} = -\xi + \kappa N, \quad \dot{N} = -\kappa \dot{\xi} \tag{1}$$

and so $\kappa = \det(\xi, \dot{\xi}, \ddot{\xi})$.

We are interested in the geometric condition that the curvature of ξ depends on the distance to a geodesic of $\mathbb{S}^2$. If $\mathbf{e} \in \mathbb{R}^3$ is a unit length vector, then $\langle \xi, \mathbf{e} \rangle$ is the signed distance to the orthogonal plane to $\mathbf{e}$ passing through the origin. Without restriction, we consider $\mathbf{e} := (0, 0, 1)$ and write $\xi = (x, y, z)$ with $x^2 + y^2 + z^2 = 1$. So we can pay our attention to study the condition $\kappa = \kappa(z)$ since $z = \langle \xi, \mathbf{e} \rangle$ represents the signed distance to the great circle $\mathbb{S}^2 \cap \{z = 0\}$. Concretely, we use geographical coordinates in $\mathbb{S}^2$ and write

$$\xi = (\cos\varphi \cos\lambda, \cos\varphi \sin\lambda, \sin\varphi), \; -\pi/2 \leq \varphi \leq \pi/2, \; -\pi < \lambda \leq \pi.$$

Then it is interesting to notice that the latitude φ is the signed distance to the equator $\mathbb{S}^2 \cap \{z = 0\} \equiv \varphi = 0$ and, in addition, the colatitude $\pi/2 - \varphi$ gives the distance to the North pole $(0, 0, 1)$.

At a given point $\xi(s)$ on the curve, we introduce the *spherical angular momentum* (with respect to the z-axis) $\mathcal{K}(s)$ as the (signed) volume of the parallelepiped spanned by the position $\xi(s)$, the unit tangent $T(s)$, and the vector $\mathbf{e} := (0, 0, 1)$. Concretely, we define

$$\mathcal{K}(s) := -\det(\xi(s), T(s), \mathbf{e}) = -\langle N(s), \mathbf{e}\rangle = \dot{x}(s)y(s) - x(s)\dot{y}(s). \tag{2}$$

In physical terms, as a consequence of Noether's theorem (cf. [1]), $\mathcal{K}$ may be described as the angular momentum of a particle of unit mass with unit-speed and spherical trajectory $\xi(s)$. We point out that $\mathcal{K}$ assumes values in $[-1, 1]$ and it is well defined, up to the sign, depending on the orientation of the normal to ξ. In geographical coordinates, $\mathcal{K}$ is given by

$$\mathcal{K} = -\dot{\lambda}\cos^2\varphi. \tag{3}$$

The unit-speed condition on ξ implies that $\dot{\varphi}^2 + \dot{\lambda}^2\cos^2\varphi = 1$, and assuming φ is nonconstant and using (3), we deduce that

$$ds = \frac{d\varphi}{\sqrt{1 - \dot{\lambda}^2\cos^2\varphi}} = \frac{\cos\varphi\, d\varphi}{\sqrt{\cos^2\varphi - \mathcal{K}^2}} = \frac{dz}{\sqrt{1 - z^2 - \mathcal{K}^2}} \tag{4}$$

and

$$d\lambda = -\frac{\mathcal{K}ds}{\cos^2\varphi} = \frac{\mathcal{K}ds}{z^2 - 1}. \tag{5}$$

Hence, given $\mathcal{K} = \mathcal{K}(z)$ as an explicit function, looking at (4) and (5), one may attempt to compute $z(s)$ (and so $\varphi(s)$) and $\lambda(s)$ in three steps: integrate (4) to get $s = s(z)$, invert to get $z = z(s)$, and integrate (5) to get $\lambda = \lambda(s)$. We remark that the integration constants appearing in (4) and (5) simply mean a translation of the arc parameter and a rotation around the z-axis, respectively.

In addition, using (1) and (2), we have that $\dot{\mathcal{K}} = -\langle \dot{N}, \mathbf{e}\rangle = \kappa\langle\dot{\xi}, \mathbf{e}\rangle = \kappa\dot{z}$, and if we take into account the assumption $\kappa = \kappa(z)$ (being z nonconstant), we finally arrive at

$$d\mathcal{K} = \kappa(z)dz, \tag{6}$$

that is, $\mathcal{K}(z)$ can be interpreted as an antiderivative of $\kappa(z)$.

As a summary, we can determine by quadratures in a constructive explicit way the spherical curves such that $\kappa = \kappa(z)$ in the spirit of [16, Theorem 3.1].

Theorem 1 *Let* $\kappa = \kappa(z)$ *be a continuous function. Then the problem of determining locally a spherical curve whose curvature is* $\kappa(z)$*—*z *representing the (nonconstant) signed distance to the great circle* $z = 0$*—with spherical angular momentum* $\mathcal{K}(z)$ *satisfying* (6) *is solvable by quadratures considering the unit-speed curve* $\xi(s) = (x(s), y(s), z(s))$*, with* $x(s) = \cos\varphi(s)\cos\lambda(s)$, $y(s) = \cos\varphi(s)\sin\lambda(s)$, $z(s) = \sin\varphi(s)$*, where* $\varphi(s)$ *and* $\lambda(s)$ *are obtained through* (4) *and* (5) *after inverting* $s = s(z)$*. Such a curve is uniquely determined by* $\mathcal{K}(z)$ *up to a rotation around the* z*-axis (and a translation of the arc parameter* s*).*

In other words, any spherical curve $\xi = (x, y, z) : I \subseteq \mathbb{R} \to \mathbb{S}^2$, with z nonconstant, is uniquely determined by its spherical angular momentum $\mathcal{K}$ as a function of its coordinate z, that is, by $\mathcal{K} = \mathcal{K}(z)$. The uniqueness is modulo rotations around the z-axis. Moreover, the curvature of ξ is given by $\kappa(z) = \mathcal{K}'(z)$.

Remark 2 If we prescribe a continuous function $\kappa = \kappa(z)$ as curvature, the proof of Theorem 1 clearly implies the computation of three quadratures, following the sequence:

(i) A one-parameter family of antiderivatives of $\kappa(z)$:

$$\int \kappa(z)dz = \mathcal{K}(z).$$

(ii) Arc length parameter s of $\xi = (x, y, z)$ in terms of z, defined—up to translations of the parameter—by the integral

$$s = s(z) = \int \frac{dz}{\sqrt{1 - z^2 - \mathcal{K}(z)^2}},$$

where $-\sqrt{1 - z^2} < \mathcal{K}(z) < \sqrt{1 - z^2}$, and inverting $s = s(z)$ to get $z = z(s)$ and so the latitude of ξ is $\varphi(s) = \arcsin z(s)$.

(iii) Longitude of $\xi = (\cos\varphi\cos\lambda, \cos\varphi\sin\lambda, \sin\varphi)$ in terms of s, defined—up to a rotation around the z-axis—by the integral

$$\lambda(s) = \int \frac{\mathcal{K}(z(s))}{z(s)^2 - 1} ds,$$

where $|z(s)| < 1$.

We note that we get a one-parameter family of spherical curves satisfying $\kappa = \kappa(z)$ according to the spherical angular momentum $\mathcal{K}(z)$ chosen in (i) and verifying $\mathcal{K}(z)^2 + z^2 < 1$. It will distinguish geometrically the curves inside a same family by their relative position with respect to the equator (or the z-axis).

We now show two illustrative examples applying steps (i)–(iii) of the algorithm described in Remark 2:

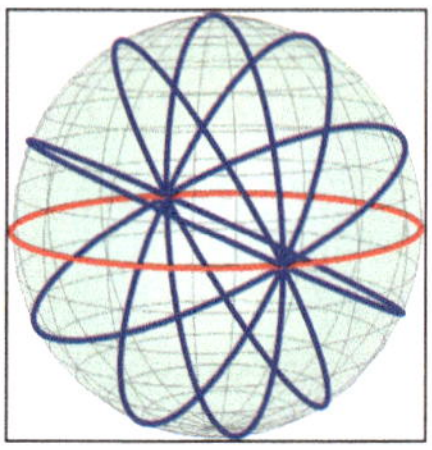

Fig. 1 Great circles $\mathbb{S}^2 \cap \{\sqrt{1-c^2}\, y + c\, z = 0\}$: $\mathcal{K} \equiv c \in (-1, 1)$

Example 3 ($\kappa \equiv 0$) Then $\mathcal{K} \equiv c \in \mathbb{R}$, and $s = \int \frac{dz}{\sqrt{1-c^2-z^2}} = \arcsin \frac{z}{\sqrt{1-c^2}}$, with $|c| < 1$. Therefore, $z(s) = \sqrt{1-c^2} \sin s$. This gives that $\lambda(s) = -\arctan(c \tan s)$ and finally $\xi(s) = (\cos s, -c \sin s, \sqrt{1-c^2} \sin s)$. It corresponds to the great circle $\mathbb{S}^2 \cap \{\sqrt{1-c^2}\, y + c\, z = 0\}$. Up to rotations around the z-axis, they provide arbitrary great circles in $\mathbb{S}^2$, except the equator. As a consequence of Theorem 1, we deduce that the great circle $\mathbb{S}^2 \cap \{\sqrt{1-c^2}\, y + c\, z = 0\}$ is the only spherical curve (up to rotations around the z-axis) with constant spherical angular momentum $\mathcal{K} \equiv c$ (see Fig. 1).

For example, taking $c = 0$, we get the meridian $\mathbb{S}^2 \cap \{y = 0\}$, and hence the meridians are the only spherical curves with null spherical angular momentum. We notice that the limiting cases $c = \pm 1$ lead to the equator $\mathbb{S}^2 \cap \{z = 0\}$.

Example 4 ($\kappa \equiv k_0 > 0$) Now, $\mathcal{K}(z) = k_0 z + c$, $c \in \mathbb{R}$. In this case, it is not difficult to get that

$$z(s) = \frac{1}{1+k_0^2} \left(\sqrt{1-c^2+k_0^2}\, \sin\left(\sqrt{1+k_0^2}\, s\right) - c\, k_0 \right)$$

with $|c| < \sqrt{1+k_0^2}$. But the expression of λ is far from trivial and depends on the values of c. After a long computation, we deduce the following:

- If $|c| \neq k_0$: $\lambda(s) = \arctan\left(\frac{\sqrt{1-c^2+k_0^2} + (1-ck_0+k_0^2)\tan(\frac{1}{2}\sqrt{1+k_0^2}s)}{(k_0-c)\sqrt{1+k_0^2}} \right) + \arctan\left(\frac{\sqrt{1-c^2+k_0^2} + (1+ck_0+k_0^2)\tan(\frac{1}{2}\sqrt{1+k_0^2}s)}{(k_0+c)\sqrt{1+k_0^2}} \right)$.
- If $c = k_0$: $\lambda(s) = \arctan\left(\frac{1-(1+2k_0^2)\tan(\frac{1}{2}\sqrt{1+k_0^2}s)}{2k_0\sqrt{1+k_0^2}} \right)$.
- If $c = -k_0$: $\lambda(s) = \arctan\left(\frac{1+(1+2k_0^2)\tan(\frac{1}{2}\sqrt{1+k_0^2}s)}{2k_0\sqrt{1+k_0^2}} \right)$.

Of course, up to rotations around the z-axis, we get all the nonparallel small circles of $\mathbb{S}^2$. The parameter c distinguishes the position of the circle with respect to the equator. If $0 \leq |c| < 1$, the circles intersect the equator transversely; in particular, when $c = 0$, we obtain the orthogonal circles to the equator. If $c = \pm 1$, the circles

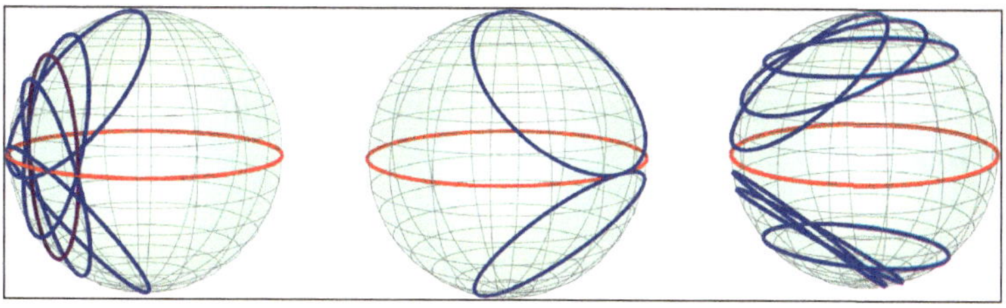

Fig. 2 Small circles: $\mathcal{K}(z) = k_0\, z + c,\ k_0 > 0$;
$0 \leq |c| < 1$ (left), $c = \pm 1$ (center), $1 < |c| < \sqrt{1+k_0^2}$ (right)

are tangent to the equator. Finally, if $1 < |c| < \sqrt{1+k_0^2}$, the circles do not intersect the equator (see Fig. 2).

For example, if $c = 0$, we arrive (up to rotations around the z-axis) at $\lambda(s) = \arctan\left(\cos(\sqrt{1+k_0^2}\, s)/k_0\right)$ and finally

$$\xi(s) = \frac{1}{\sqrt{1+k_0^2}}\left(k_0, \cos\left(\sqrt{1+k_0^2}\, s\right), \sin\left(\sqrt{1+k_0^2}\, s\right)\right),$$

that corresponds to the small circle $\mathbb{S}^2 \cap \{x = k_0/\sqrt{1+k_0^2}\}$. This is the only spherical curve, up to rotations around the z-axis, with spherical angular momentum $\mathcal{K}(z) = k_0 z$ (see Fig. 2).

Remark 5 The main difficulties one can find carrying on the strategy described in Remark 2 (or in Theorem 1) to determine a spherical curve whose curvature is $\kappa = \kappa(z)$ are the following:

(1) The integration of $s = s(z)$: Even in the case that $\mathcal{K}(z)$ was polynomial, the corresponding integral is not necessarily elementary. For example, when $\mathcal{K}(z)$ is a quadratic polynomial, it can be solved using Jacobian elliptic functions (see, for example, [6]). This is equivalent to $\kappa(z)$ be linear, i.e., $\kappa(z) = 2az + b$, $a, b \in \mathbb{R}$. We will study such spherical curves in Sect. 3.
(2) The previous integration gives us $s = s(z)$; it is not always possible to obtain explicitly $z = z(s)$, what is necessary to determine the curve.
(3) Even knowing explicitly $z = z(s)$, the integration to get $\lambda(s)$ may be impossible to perform using elementary or known functions.

Nevertheless, along the paper, we will study different families where we are successful with the procedure described in Remark 2, and we will recover some known curves and find out new spherical curves characterized by their spherical angular momentum.

3 Elastic-Type Curves on the Sphere

3.1 A New Characterization and a Generalization of Elastic Curves

A unit-speed spherical curve ξ is said to be an *elastica under tension* σ (see [14]) if its curvature κ satisfies the differential equation

$$2\ddot{\kappa} + \kappa^3 + (2 - \sigma)\kappa = 0 \tag{7}$$

for some $\sigma \in \mathbb{R}$. They are the critical points of the elastic energy functional

$$\int_\xi (\kappa^2 + \sigma) ds$$

acting on spherical curves with suitable boundary conditions. If $\sigma = 0$, then the constraint on arc length is removed and ξ is called a *free elastica*.

A possible generalization of free elasticae was considered in [4], where the authors studied the elastic curves in $\mathbb{S}^2$ which are circular at rest. They are called λ*-elastic curves*. These curves are critical points of the functional

$$\int_\xi (\kappa + \lambda)^2 ds, \ \lambda \in \mathbb{R}$$

and are characterized by the Euler-Lagrange equation

$$2\ddot{\kappa} + \kappa^3 + (2 - \lambda^2)\kappa + 2\lambda = 0. \tag{8}$$

It is obvious that the 0-elastic curves are the free elasticae. The main result of this section deals with the spherical curves which are the critical points of the bending energy for variations with constant length (including both previous types of elastic curves) relating them with the case commented in part (1) of Remark 5.

Theorem 6 *Let ξ be a spherical curve whose curvature κ satisfies*

$$\kappa = 2a\langle \xi, \mathbf{e} \rangle + b, \ a \neq 0, \ b \in \mathbb{R}, \tag{9}$$

for some $\mathbf{e} \in \mathbb{R}^3$, $|\mathbf{e}| = 1$. Then

(i) *The spherical angular momentum $\mathcal{K}$ of ξ is given by*

$$\mathcal{K} = a\langle \xi, \mathbf{e} \rangle^2 + b\langle \xi, \mathbf{e} \rangle + c,$$

for some $c \in \mathbb{R}$.

(ii) ξ *is a critical point of the functional*

$$\int_\xi (\kappa^2 - 2b\kappa + b^2 - 4ac)\, ds$$

and so κ *satisfies the corresponding Euler-Lagrange equation*

$$2\ddot{\kappa} + \kappa^3 + \left(2 - (b^2 - 4ac)\right)\kappa - 2b = 0. \tag{10}$$

If $b = 0$, ξ *is an elastica under tension* $\sigma = -4ac$, *and if* $c = 0$, ξ *is a* λ*-elastic curve with* $\lambda = -b$.

Conversely, if ξ *is a critical point of the functional*

$$\mathcal{F}_\sigma^\lambda(\xi) := \int_\xi \left((\kappa + \lambda)^2 + \sigma\right) ds,\ \lambda, \sigma \in \mathbb{R},$$

then the curvature of ξ *can be written as in* (9).

Remark 7 It is remarkable the similitude of condition (9) characterizing the generalized elasticae considered in Theorem 6 with the geometric property satisfied by the classical Euler elastic curves in the plane: their curvature is proportional to one of the coordinate functions, say $\kappa(x, y) = 2\lambda y + \mu$, $\lambda \neq 0$, $\mu \in \mathbb{R}$ (see Section 3 in [7] and Section 1 in [16]). Even something similar happens to spacelike and timelike elastic curves in Lorentz-Minkowski plane (see [9] and [10]).

Proof of Theorem 6 From (2), (1), and (9), we get that

$$\frac{d}{ds}\left(-\mathcal{K} + a\langle \xi, \mathbf{e}\rangle^2 + b\langle \xi, \mathbf{e}\rangle\right) = 0.$$

This proves part (i).

We also have from (9), (1), and (2), that $\dot{\kappa} = 2a\langle \dot{\xi}, \mathbf{e}\rangle$ and $\ddot{\kappa} = 2a(-\langle \xi, \mathbf{e}\rangle - \kappa\mathcal{K})$. Now, we can easily check that κ given by (9) satisfies (10) since, after a straightforward computation using (i) and putting $\langle \xi, \mathbf{e}\rangle = (\kappa - b)/2a$, we arrive at (10). We observe that if $b = 0$, then ξ satisfies (7) with $\sigma = -4ac$ and if $c = 0$, then ξ satisfies (8) with $\lambda = -b$. Moreover, in [2], it is shown that for a given differentiable function $P(\kappa)$, the critical points of the functional $\int_\xi P(\kappa) ds$ are characterized by the Euler-Lagrange equation

$$(\kappa^2 + 1)P'(\kappa) + \frac{d^2(P'(\kappa))}{ds^2} = \kappa P(\kappa). \tag{11}$$

It is an exercise to check that putting $P(\kappa) = \kappa^2 - 2b\kappa + b^2 - 4ac$ in (11) we obtain (10). This finishes the proof of part (ii).

Multiplying (10) by $\dot{k}$, we obtain a first integral

$$\dot{\kappa}^2 + \kappa^4/4 + \left(1 - (b^2 - 4ac)/2\right)\kappa^2 - 2b\kappa = E,$$

where E is a real constant. After a long computation involving (9), and part (i), (2) and using that $\langle \dot{\xi}, \mathbf{e}\rangle^2 = 1 - \langle \xi, \mathbf{e}\rangle^2 - \langle N, \mathbf{e}\rangle^2$, since $\{\xi, \dot{\xi}, N\}$ is an orthonormal basis in $\mathbb{R}^3$, we get that

$$E = 4a^2 - b^2 - (b^2 - 4ac)^2/4. \tag{12}$$

On the other hand, suppose now that ξ is a critical point of $\mathcal{F}_\sigma^\lambda$. Taking into account (11), we deduce that κ verifies the differential equation

$$2\ddot{\kappa} + \kappa^3 + (2 - \lambda^2 - \sigma)\kappa + 2\lambda = 0. \tag{13}$$

Multiplying (13) by $\dot{\kappa}$ and integration allow us to deduce the energy $E \in \mathbb{R}$ of ξ:

$$E := \dot{\kappa}^2 + \frac{\kappa^4}{4} + \left(1 - \frac{\lambda^2 + \sigma}{2}\right)\kappa^2 + 2\lambda\kappa. \tag{14}$$

We want to prove that κ has an expression like in (9). For this purpose, we first observe that if $\lambda = 0$, then $\kappa \equiv 0$ is a trivial solution to (13). For example, we can take $\mathbf{e} \in \mathbb{R}^3$, $|\mathbf{e}| = 1$ the unit normal vector orthogonal to the vectorial plane containing the corresponding great circle in $\mathbb{S}^2$. Now, we must look for $a \neq 0$ and $b \in \mathbb{R}$ satisfying (9). Comparing (10) and (13), we take $b = -\lambda$ and observe that σ must satisfy $\sigma = -4ac$, with $a \neq 0$, $c \in \mathbb{R}$. Using (14), we have that $4E + 4\lambda^2 + (\lambda^2 + \sigma)^2 = 4\dot{\kappa}^2 + 4(\kappa + \lambda)^2 + \left(\kappa^2 - (\lambda^2 + \sigma)\right)^2 > 0$.

But looking at (12), E must satisfy $E = 4a^2 - b^2 - (b^2 - 4ac)^2/4$, and eliminating c, we finally arrive at $0 < 4E + 4\lambda^2 + (\lambda^2 + \sigma)^2 = 16a^2$, which allows us to obtain the searched value for a.

The study of (free) elastic curves on the sphere has been considered under different approaches (see, for example, [2, 5, 13, 14], or [17]), paying special attention to the closed ones. The closed λ-elastic spherical curves were studied in [4]. All the mentioned articles are based on the study of the differential equation for the geodesic curvature of the spherical curve, being sometimes integrated directly in terms of Jacobi elliptic functions.

In our approach of Theorem 6, we can choose $\mathbf{e} = (0, 0, 1)$ without loss of generality and so $\langle \xi, \mathbf{e}\rangle = z$. In this way, we arrive at the conditions

$$\kappa(z) = 2az + b,\ \mathcal{K}(z) = az^2 + bz + c,\ a \neq 0, b, c \in \mathbb{R}, \tag{15}$$

in the notation of Theorem 1. Thus, using suitable coordinates in the sphere, we can conclude the following uniqueness result for the spherical elastic curves considered in literature:

Corollary 8

(a) *The elasticae under tension σ are the only spherical curves (up to rotations around z-axis) with spherical angular momentum $\mathcal{K}(z) = az^2 + c$, $a \neq 0, c \in \mathbb{R}$, with $\sigma = -4ac$. In particular, the free elasticae are characterized by the spherical angular momentum $\mathcal{K}(z) = az^2$, $a \neq 0$.*
(b) *The λ-elastic curves are the only spherical curves (up to rotations around z-axis) with spherical angular momentum $\mathcal{K}(z) = az^2 + bz$, $a \neq 0, b \in \mathbb{R}$, with $\lambda = -b$.*

Remark 9 Inspired by Langer and Singer's work on the Kirchhoff elastic rod [15], in [3], it is obtained by geometrical means the first integrals of the Euler-Lagrange equations of curvature energy functionals $\int_\xi P(\kappa)ds$, where P is a smooth function and κ denotes the curvature of the spherical curve ξ in $\mathbb{S}^3$. Assuming that $P''(\kappa) \neq 0$, the critical points of the functional $\int_\xi P(\kappa)ds$ are characterized by a couple of differential equations naturally related to a system of cylindrical coordinates in the three-sphere adapted to the curve ξ.

In the case that the curve ξ lies in $\mathbb{S}^2$, from Section 3 of [3], using concretely equation (33) with $\theta = 0$ ($b = 0$), then we have that $\cos^2 \psi = P'(k)^2/a^2$, $a \neq 0$, where $\psi = \pi/2 - \varphi$ is the colatitude of ξ.

As a consequence, *if a spherical curve ξ is a critical point of $\int_\xi P(\kappa)ds$, then there exist geographical coordinates (φ, λ) adapted to ξ such that*

$$\sin \varphi = \delta\, P'(\kappa),\ \delta \neq 0.$$

This result is consistent with Theorem 6, since if $P(\kappa) = (\kappa + \lambda)^2 + \sigma$, then $\sin \varphi = z = \frac{\kappa - b}{2a} = \delta\, P'(\kappa)$, taking $\delta = \frac{1}{4a}$ and $\lambda = -b$.

If one follows the strategy described in Remark 2 to determine the generalized elasticae satisfying (15), it is necessary to perform the integral

$$s = s(z) = \int \frac{dz}{\sqrt{P(z)}},\quad P(z) = 1 - z^2 - (az^2 + bz + c)^2. \tag{16}$$

Since $P(z)$ is a fourth-order polynomial, (16) can be solved in terms of Jacobi elliptic functions once the nature and multiplicity of the roots of $P(z)$ are determined. After inverting $s = s(z)$ to get $z = z(s)$, we would arrive at the expressions of $\kappa = \kappa(s) = 2az(s) + b$ compatible with the ones given in [2, 5, 13, 14] or [17], and [4], at least when $b = 0$ or $c = 0$. Besides helices and circles, the borderline, orbitlike, and wavelike elasticae appear as well as a more general case according to the expressions of the geodesic curvature in terms of the arc parameter (see [17]).

In order to get the explicit expression of the generalized elasticae, we also have to compute

$$\lambda(s) = \int \frac{az(s)^2 + bz(s) + c}{z(s)^2 - 1}\, ds, \tag{17}$$

which, in general, leads to complicated elliptic integrals. In the next two sections, we are going to illustrate the aforementioned computations in two interesting and attractive cases.

3.2 *Seiffert's Spherical Spirals*

The Seiffert's spirals are defined as those spherical curves obtained when one moves along the surface of a sphere with constant speed while maintaining a constant angular velocity with respect to a fixed diameter (cf. [11]). These curves are given in cylindrical coordinates (r, θ, z), $r^2 + z^2 = 1$, by the parametric equations

$$r = \mathrm{sn}(s, p),\ \theta = p\, s,\ z = \mathrm{cn}(s, p),\ (p > 0), \tag{18}$$

where p is a positive constant and sn and cn are the elementary Jacobi elliptic functions (cf. [6], for instance). Erdös provided in [11] a derivation of the equations of this curve, as well as an analysis of its properties, including conditions for obtaining periodic orbits. When $p > 1$, the spiral is located entirely in the northern hemisphere.

Now, we prove that these curves are elastic curves with positive tension corresponding to the conditions $\kappa(z) = 2az$, $\mathcal{K}(z) = az^2 - a$, i.e., $b = 0$, $a + c = 0$ in (15). Then $\sigma = 4a^2 > 0$, and there is no restriction if we consider $a > 0$. So (16) can be written as

$$\left(\frac{dz}{ds}\right)^2 = (1 - z^2)\left(1 - a^2 + a^2 z^2\right), \quad \frac{a^2 - 1}{a^2} < z^2 < 1,$$

which implies that $z(s) = \mathrm{cn}(s, a)$ (cf. [6], for instance) and thus $r(s) = \mathrm{sn}(s, a)$. Using that $a + c = 0$ in (17), we get that $\lambda(s) = a\, s$, and so we arrive at the Seiffert's spirals (see Fig. 3). As a summary

Corollary 10 *The Seiffert's spirals* (18) *are the only spherical curves (up to rotations around z-axis) with spherical angular momentum* $\mathcal{K}(z) = pz^2 - p,\ p > 0$.

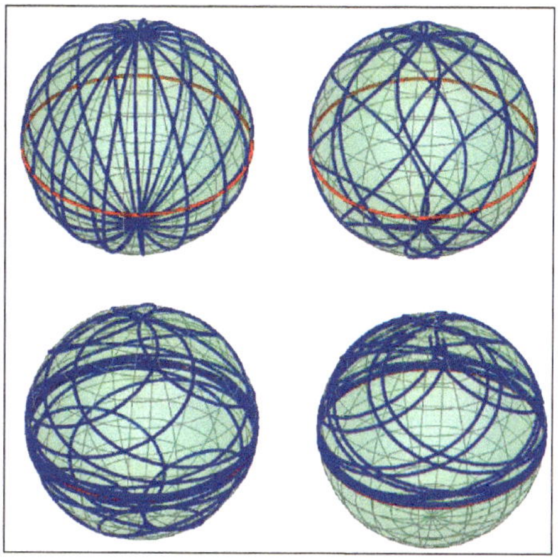

Fig. 3 Seiffert's spirals ($p \approx 0$, $0 < p < 1$, $p \approx 1$, $p > 1$): $\mathcal{K}(z) = pz^2 - p$, $p > 0$

3.3 Borderline Spherical Elastic Curves

We study in this section elastic curves with null energy producing elasticae under positive tension $\sigma > 0$. This corresponds to the conditions $\kappa(z) = 2az$, $\mathcal{K}(z) = az^2 - 1$, i.e., $b = 0$, $c = -1$ in (15). Then $E = 0$ according to (12), and we can take $a > 0$ in order to $\sigma = 4a > 0$. Now, (16) leads to

$$s = s(z) = \int \frac{dz}{z\sqrt{2a - 1 - a^2z^2}}, \quad 2a - 1 - a^2z^2 > 0,$$

which implies that $a > 1/2$. The above integral becomes elementary, and after inverting $s = s(z)$ and up to a translation on the parameter s, we get

$$z(s) = \frac{2a - 1}{a} \operatorname{sech}(\sqrt{2a - 1}\, s). \tag{19}$$

Looking at (17), we get that if $a = 1$ then $\lambda(s) = s$ (and $z(s) = \operatorname{sech} s$), and when $a > 1/2$ with $a \neq 1$, we obtain

$$\lambda(s) = s + \arctan\left(\frac{\sqrt{2a - 1}}{1 - a} \tanh(\sqrt{2a - 1} s)\right). \tag{20}$$

This family corresponds to the "borderline elasticae" described in [17] which are asymptotic to the equator. We show some pictures of them in Fig. 4. In conclusion, we deduce the following uniqueness result:

Corollary 11 *The borderline elasticae given by* (19) *and* (20) *are the only spherical curves (up to rotations around z-axis) with spherical angular momentum* $\mathcal{K}(z) = az^2 - 1$, $a > 1/2$.

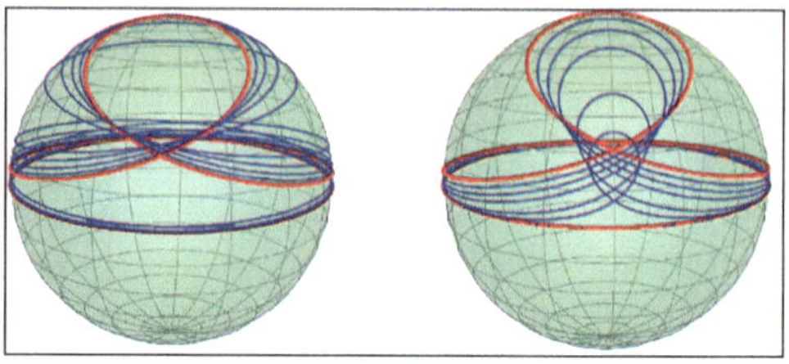

Fig. 4 Borderline elastic curves: $\mathcal{K}(z) = az^2 - 1$, $a > 1/2$ (left, $1/2 < a \leq 1$; right, $a \geq 1$)

We remark that when $a = 1$, we recover the Seiffert's spiral corresponding to $p = 1$, since $\mathrm{cn}(s, 1) = \mathrm{sech}\, s$ (see [6]).

4 Loxodromic-Type Curves on the Sphere

In this section, we are interested in critical points of the functional

$$\mathcal{F}^{\mu}(\xi) := \int_{\xi} \sqrt{\kappa^2 + \mu^2}\, ds, \ \mu > 0.$$

This functional was considered in Section 6 of [2] acting on the space of immersed closed curves in $\mathbb{S}^2$, motivated by total $\mathbb{R}^3$ curvature-type functionals. We aim to connect the critical points of $\mathcal{F}^{\mu}$ with the classical loxodrome curves in $\mathbb{S}^2$ and others spherical curves with similar characteristics.

Taking into account Remark 9, we are devoted to study the spherical curves with curvature

$$\kappa(z) = \frac{\mu\, z}{\sqrt{\delta^2 - z^2}}, \ \mu > 0, \delta \neq 0. \tag{21}$$

We will distinguish cases according to $0 < \mu < 1$, $\mu = 1$, and $\mu > 1$.

4.1 *Case* $0 < \mu < 1$*: Spherical Loxodromes*

The loxodromes are interesting curves in the sphere studied, among others, by Pedro Nunes in 1537, Simon Stevin in 1608, or Maupertuis in 1744. They are also known as rhumb lines because they make a constant angle $\alpha \in (0, \pi/2)$ with the meridians (cf. [12]). Analytically, using geographical coordinates (φ, λ), they are defined by equation

$$d\lambda = \cot\alpha \frac{d\varphi}{\cos\varphi}. \tag{22}$$

The aim of this section is the study of the spherical curves satisfying

$$\kappa(\varphi) = a \tan\varphi, \ 0 < a < 1, \tag{23}$$

or, equivalently,

$$\kappa(z) = \frac{az}{\sqrt{1-z^2}}, \ 0 < a < 1. \tag{24}$$

So they correspond in (21) to the election $\mu = a \in (0, 1)$ and $\delta = 1$.

The trivial solution of (24) is given by the equator $z = 0$. We follow the method described in Theorem 1 and Remark 2, considering the spherical angular momentum

$$\mathcal{K}(z) = -a\sqrt{1-z^2}, \ 0 < a < 1.$$

Then, we have

$$s = \int \frac{dz}{\sqrt{1-z^2-(-a\sqrt{1-z^2})^2}} = \int \frac{dz}{\sqrt{(1-a^2)(1-z^2)}} = \frac{1}{\sqrt{1-a^2}} \arcsin z,$$

and so $z(s) = \sin(\sqrt{1-a^2}\, s)$ and therefore

$$\varphi(s) = \sqrt{1-a^2}\, s. \tag{25}$$

On the other hand, from (25), we get

$$\lambda(s) = \int \frac{a\, ds}{\cos\varphi(s)} = \frac{a}{\sqrt{1-a^2}} \log\left(\sec(\sqrt{1-a^2}\, s) + \tan(\sqrt{1-a^2}\, s)\right), \tag{26}$$

where $|s| < \dfrac{\pi}{2\sqrt{1-a^2}}$.

We deduce from (25) and (26) that

$$d\lambda = \frac{a}{\sqrt{1-a^2}} \frac{d\varphi}{\cos\varphi}$$

and, taking into account (22) and that $\mathcal{K}(\varphi) = -a\cos\varphi$, we conclude the following characterization of the loxodromes:

Corollary 12 *The loxodromes given by $d\lambda = \cot\alpha\, d\varphi/\cos\varphi$ are the only spherical curves (up to rotations around z-axis) with spherical angular momentum $\mathcal{K}(\varphi) = -\cos\alpha\cos\varphi$, $\alpha \in (0, \pi/2)$.*

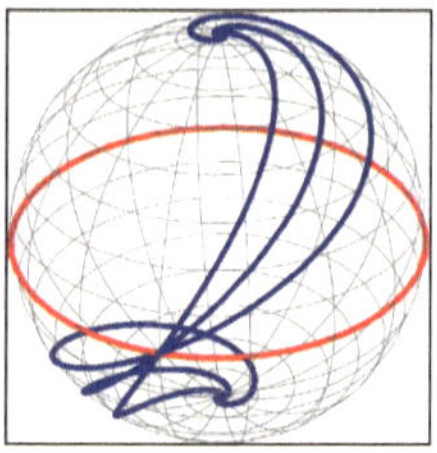

Fig. 5 Loxodromes: $\mathcal{K}(\varphi) = -\cos\alpha\cos\varphi$, $\alpha \in (0, \pi/2)$

From (23) and (25), we arrive at the intrinsic equation of the loxodromes (see Fig. 5), given by

$$\kappa(s) = \cos\alpha\ \tan(\sin\alpha\, s),\ \alpha \in (0, \pi/2).$$

4.2 Case $\mu = 1$

We now want to determine the spherical curves whose curvature is given by

$$\kappa(z) = \frac{z}{\sqrt{a - z^2}},\ 0 < a < 1. \tag{27}$$

Looking at (21), we are now considering $\mu = 1$ and $\delta = \sqrt{a}$.

The trivial solution is given by the equator $z = 0$. We follow the method described in Theorem 1 or Remark 2 for the nontrivial case, considering the spherical angular momentum

$$\mathcal{K}(z) = -\sqrt{a - z^2}.$$

In this way, we get

$$s = s(z) = \int \frac{dz}{\sqrt{1 - z^2 - (-\sqrt{a - z^2})^2}} = \int \frac{dz}{\sqrt{1-a}} = \frac{z}{\sqrt{1-a}},$$

and thus $z(s) = \sqrt{1-a}\, s$.

We write $a = \sin^2\alpha$, $0 < \alpha < \pi/2$, and abbreviate $c_\alpha = \cos\alpha, s_\alpha = \sin\alpha$. Hence, we have

$$\lambda(s) = \int \frac{\sqrt{s_\alpha^2 - c_\alpha^2 s^2}}{1 - c_\alpha^2 s^2}\, ds,$$

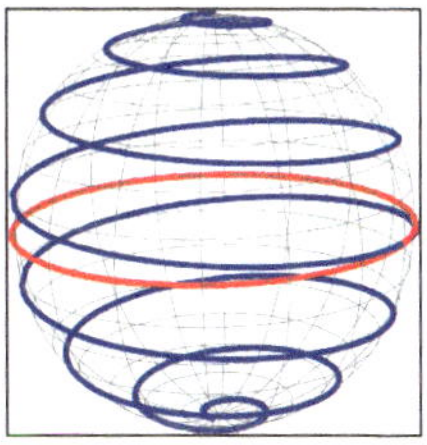

Fig. 6 Spherical curve with $\mathcal{K}(z) = -\sqrt{a - z^2}$, $0 < a < 1$

which gives

$$\lambda(s) = \frac{1}{c_\alpha} \arctan\left(\frac{c_\alpha s}{\sqrt{s_\alpha^2 - c_\alpha^2 s^2}}\right)$$
$$-\frac{1}{2}\arctan\left(\frac{c_\alpha s + s_\alpha^2}{c_\alpha \sqrt{s_\alpha^2 - c_\alpha^2 s^2}}\right) - \frac{1}{2}\arctan\left(\frac{c_\alpha s - s_\alpha^2}{c_\alpha \sqrt{s_\alpha^2 - c_\alpha^2 s^2}}\right),$$

where $|s| < \tan\alpha$.

We observe that $|z(s)| < \sin\alpha$. Using (27) and $z(s) = c_\alpha s$, we get the intrinsic equation

$$\kappa(s) = \frac{c_\alpha s}{\sqrt{s_\alpha^2 - c_\alpha^2 s^2}}, \quad |s| < \tan\alpha,$$

of this family of spherical curves of loxodromic-type (see Fig. 6) characterized by the geometric angular momentum $\mathcal{K}(z) = -\sqrt{\sin^2\alpha - z^2}$, $0 < \alpha < \pi/2$.

4.3 Case $\mu > 1$

Finally, we wish to study the spherical curves whose curvature is given by

$$\kappa(z) = \frac{az}{\sqrt{1 - az^2}}, \; a > 1. \tag{28}$$

We observe that corresponds in (21) with the elections $\mu = \sqrt{a}$ and $\delta = 1/\sqrt{a}$. The trivial solution is given by the equator $z = 0$. Otherwise, we make use of the method proposed in Theorem 1 or Remark 2, considering the spherical angular momentum

$$\mathcal{K}(z) = -\sqrt{1 - az^2}.$$

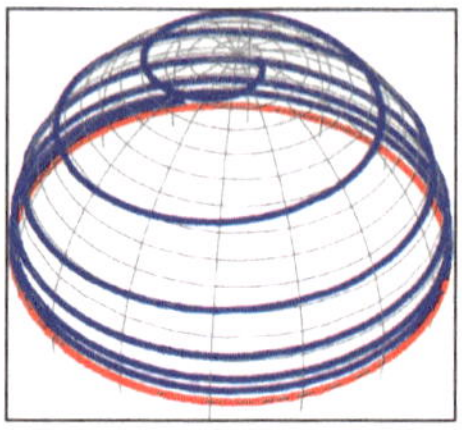
Fig. 7 Spherical curve with $\mathcal{K}(z) = -\sqrt{1-az^2}$, $a > 1$

In this way, we obtain

$$s = s(z) = \int \frac{dz}{\sqrt{1 - z^2 - (-\sqrt{1-az^2})^2}} = \int \frac{dz}{\sqrt{a-1}\,|z|} = \frac{\log|z|}{\sqrt{a-1}},$$

and so $|z(s)| = e^{\sqrt{a-1}\,s} > 0$.

If we write $a = \cosh^2\delta$, $\delta > 0$, then $z(s) = \pm e^{\sinh\delta\, s}$ and

$$\lambda(s) = \int \frac{\sqrt{1-\cosh^2\delta\, e^{2\sinh\delta\, s}}}{1-e^{2\sinh\delta\, s}} ds =$$

$$= -\frac{\operatorname{arctanh}\left(\sqrt{1-\cosh^2\delta\, e^{2\sinh\delta\, s}}\right)}{\sinh\delta} + \arctan\left(\frac{\sqrt{1-\cosh^2\delta\, e^{2\sinh\delta\, s}}}{\sinh\delta}\right),$$

where $s < -\log\cosh\delta/\sinh\delta$.

Using (28), we get the intrinsic equation

$$|\kappa(s)| = \frac{\cosh^2\delta\, e^{\sinh\delta\, s}}{\sqrt{1-\cosh^2\delta\, e^{2\sinh\delta\, s}}}, \quad s < -\log\cosh\delta/\sinh\delta,$$

of this family of spherical curves of loxodromic-type (see Fig. 7) characterized by the geometric angular momentum $\mathcal{K}(z) = -\sqrt{1-\cosh^2\delta z^2}$.

5 Spherical Catenaries

In this section, we are interested in critical points of the functional

$$\mathcal{F}(\xi) := \int_\xi \sqrt{\kappa}\, ds.$$

The above functional $\int_\xi \kappa^{1/2}\, ds$ was considered in Section 5 of [2] acting on the space of convex ($\kappa > 0$) closed curves in $\mathbb{S}^2$, motivated by the study of ($r = 1/2$)-

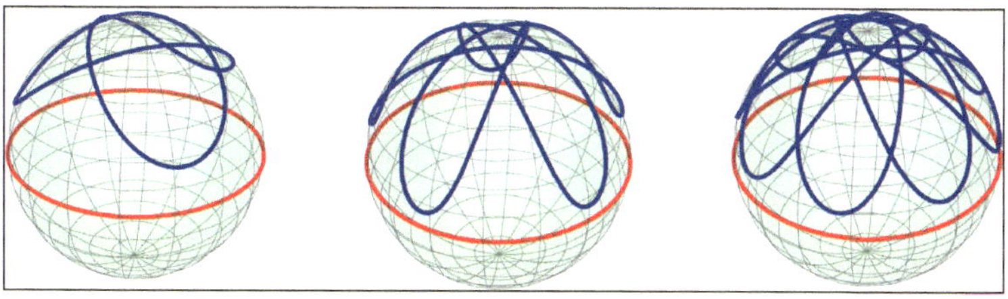

Fig. 8 Closed spherical catenaries

generalized elastic curves in $\mathbb{S}^2$. We aim to connect the critical points of $\mathcal{F}$ with the classical catenary curves in $\mathbb{S}^2$.

Taking into account Remark 9, we are devoted to study the spherical curves with curvature

$$\kappa(z) = \frac{\delta^2}{4z^2}, \ \delta \neq 0. \tag{29}$$

The spherical catenaries are the equilibrium lines of an inelastic flexible homogeneous infinitely thin massive wire included in a sphere, placed in a uniform gravitational field. Like any catenaries, their centers of gravity have the minimal altitude among all the curves with given length passing by two given points. They were studied by Bobillier in 1829 and by Gudermann in 1846 (cf. [12]). See Fig. 8.

Using cylindrical coordinates (r, θ, z) in $\mathbb{R}^3$, they can be described analytically by the following first integral of the corresponding ordinary differential equation:

$$(z - z_0)\, r^2 \frac{d\theta}{ds} = \text{constant}. \tag{30}$$

We study in this section spherical curves satisfying the condition

$$\kappa(z) = a/z^2, \ a > 0. \tag{31}$$

So they correspond in (29) to the election $\delta = 2\sqrt{a}$. For any $a > 0$, it is easy to prove that there exists a unique angle $\varphi_0 \in (0, \pi/2)$ such that $a = \tan\varphi_0 \sin^2\varphi_0$. Thus, the parallel $z = \sin\varphi_0$ is a constant solution to (31).

We now apply Theorem 1 and Remark 2 considering (31) and

$$\mathcal{K}(z) = -\frac{a}{z}, \ a > 0.$$

Then, we have

$$s = s(z) = \int \frac{z\,dz}{\sqrt{z^2(1 - z^2) - a^2}},$$

which implies that $a < 1/2$ and $1 - \sqrt{1-4a^2} < 2z^2 < 1 + \sqrt{1-4a^2}$, and it is not difficult to get

$$z(s) = \sqrt{\frac{1 + \sqrt{1-4a^2}\sin 2s}{2}}. \tag{32}$$

In addition, we have that

$$d\lambda = \frac{a}{z(1-z^2)}\, ds \tag{33}$$

Looking at (30), taking into account that $r^2 + z^2 = 1$ and $\theta = \lambda$, we deduce from (33) that we get a spherical catenary (with $z_0 = 0$ and constant $a \in (0, 1/2)$). However, the explicit computation of λ in terms of the arc parameter s requires elliptic integrals of the first and third kind. As a summary, we have proved the following uniqueness result:

Corollary 13 *The spherical catenaries* (33) *are the only spherical curves (up to rotations around the z-axis) with spherical angular momentum* $\mathcal{K}(z) = -a/z$ *(and curvature* $\kappa(z) = a/z^2$*),* $0 < a < 1/2$.

Combining (31) and (32), we have that the intrinsic equation of the spherical catenaries is given by

$$\kappa(s) = \frac{2a}{1 + \sqrt{1-4a^2}\sin 2s}, \quad 0 < a < 1/2.$$

6 New and Classical Spherical Curves

The purpose of this section is to find out new curves $\xi = (x, y, z)$ in $\mathbb{S}^2$, expressed in terms of elementary functions or in terms of Jacobi elliptic functions, prescribing their curvature as a function of the distance from the equator in such a way we can avoid the difficulties described in Remark 5. In addition, we provide uniqueness results for some well-known spherical curves in terms of the spherical angular momentum introduced in Sect. 2.

6.1 Spherical Curves Such That $\kappa(\varphi) = p\cos 2\varphi/\cos\varphi$, $0 < p < 1$

The purpose of this section is to find out new curves in $\mathbb{S}^2$ expressed in terms of Jacobi elliptic functions prescribing in a suitable way their curvature in terms of their latitude. Concretely, we aim to study the spherical curves whose curvature is given by

$$\kappa(\varphi) = \frac{p\cos 2\varphi}{\cos\varphi}, \quad 0 < p < 1. \tag{34}$$

Recalling that $z = \sin\varphi$, (34) is equivalent to

$$\kappa(z) = \frac{p(1-2z^2)}{\sqrt{1-z^2}}, \quad 0 < p < 1. \tag{35}$$

We follow the strategy proposed by Theorem 1 or Remark 2, considering the spherical angular momentum

$$\mathcal{K}(z) = p\,z\sqrt{1-z^2}.$$

Then, we get

$$s = \int \frac{dz}{\sqrt{(1-z^2)(1-p^2z^2)}} = \int \frac{d\varphi}{\sqrt{1-p^2\sin^2\varphi}} = F(\varphi, p) = F(\arcsin z, p),$$

where $F(\cdot, p)$ denotes the elliptic integral of first class of modulus p (see, for example, [6]). Hence, $\varphi(s) = \mathrm{am}(s, p)$ and $z(s) = \mathrm{sn}(s, p)$, where $\mathrm{am}(\cdot, p)$ is the Jacobi amplitude and $\mathrm{sn}(\cdot, p)$ is the Jacobi sine amplitude of modulus p (see, for example, [6]). In addition,

$$\lambda(s) = -p\int \frac{\mathrm{sn}(s,p)}{\mathrm{cn}(s,p)}\,ds$$

where $\mathrm{cn}(\cdot, p)$ is the Jacobi cosine amplitude of modulus p. Using formula 316.01 of [6], we finally arrive at the following expression for the longitude:

$$\lambda(s) = -\frac{p}{2p'}\log\left(\frac{\mathrm{dn}(s,p)+p'}{\mathrm{dn}(s,p)-p'}\right),$$

where $\mathrm{dn}(\cdot, p)$ is the Jacobi delta amplitude of modulus p and $p' = \sqrt{1-p^2}$ is the complementary modulus.

Using (35) and that $z(s) = \mathrm{sn}(s, p)$, joint to formula 124.02 of [6], we get the intrinsic equation

$$\kappa(s) = p\,(2\,\mathrm{cn}(s,p) \quad 1/\mathrm{cn}(s,p)), \quad 0 < p < 1,$$

of the only spherical curves (up to rotations around z-axis) with spherical angular momentum $\mathcal{K}(z) = p\,z\sqrt{1-z^2}$ or, equivalently, $\mathcal{K}(\varphi) = (p/2)\sin 2\varphi$, $0 < p < 1$.

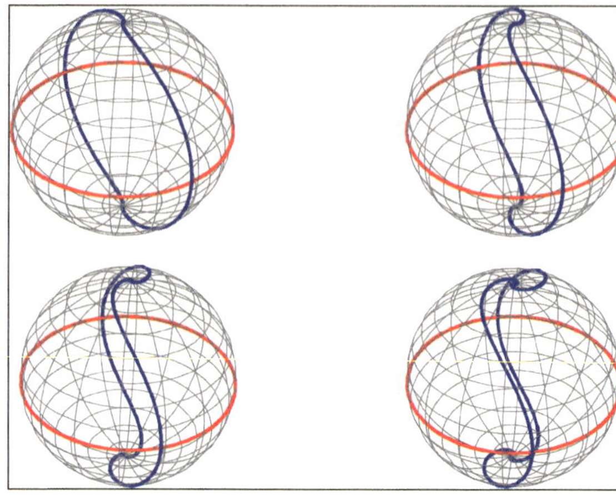

Fig. 9 Spherical curves with $\mathcal{K}(z) = p\,z\sqrt{1-z^2}$, $0 < p < 1$

These curves are embedded and closed since $\xi(s + 4K(p)) = \xi(s)$, where $K(p)$ is the complete elliptic integral of first class of modulus p (see Fig. 9).

6.2 Viviani's Curve and Spherical Archimedean Spirals

The Viviani's curve is the intersection between a sphere of radius R and a cylinder of revolution with diameter R such that a generatrix passes by the center of the sphere; so this curve is at the same time spherical and cylindrical. We can obtain a Viviani's curve by sticking the tip of a compass inside a cylinder of revolution and tracing on this cylinder a "circle" with radius equal to the cylinder diameter. It was studied by Vincenzo Viviani in 1692 (cf. [12]). In geographical coordinates of $\mathbb{S}^2$, the Viviani's curve can be simply described as $\varphi = \lambda$ (see Fig. 10).

We study in this section spherical curves satisfying the condition

$$\kappa(z) = \frac{z(3-z^2)}{(2-z^2)^{3/2}}, \tag{36}$$

applying Theorem 1 and Remark 2, considering the spherical angular momentum

$$\mathcal{K}(z) = \frac{z^2-1}{\sqrt{2-z^2}}.$$

Then, we have

$$s = s(z) = \int \sqrt{\frac{2-z^2}{1-z^2}}\,dz = E(\arcsin z, 1/2) \tag{37}$$

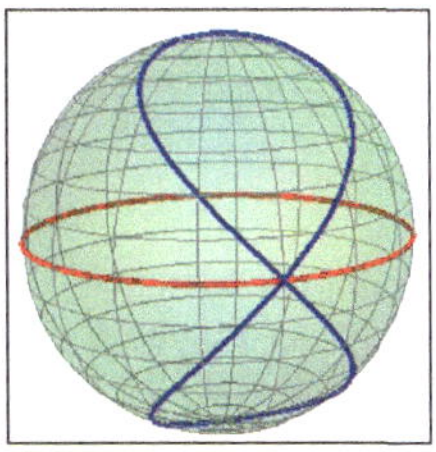

Fig. 10 Viviani's curve

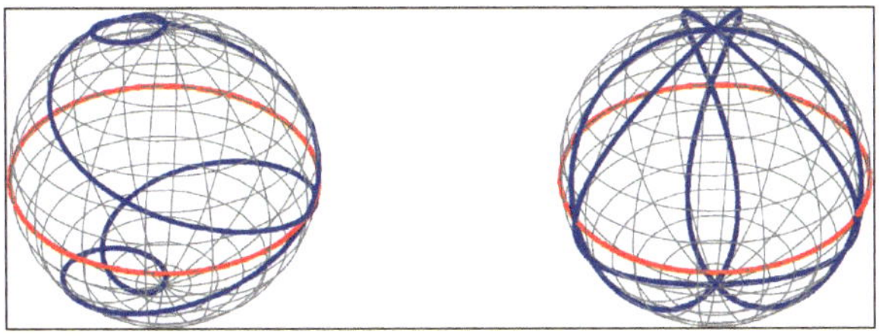

Fig. 11 Spherical Archimedean spiral curves: $\varphi = n\lambda$, $n = 1/3$, and $n = 3$

which involves an elliptic integral E of second kind (see, for example, [6]). In addition, we get

$$d\lambda = \frac{ds}{\sqrt{2 - z^2}} \tag{38}$$

Using that $z = \sin\varphi$, (37), and (38), we get easily that $d\lambda = d\varphi$. Hence, we have proved the following characterization of the Viviani's curve:

Corollary 14 *The Viviani's curve $\varphi = \lambda$ is the only spherical curve (up to rotations around the z-axis) with spherical angular momentum $\mathcal{K}(\varphi) = -\cos^2\varphi/\sqrt{1+\cos^2\varphi}$.*

The spherical Archimedean spiral curves are natural generalizations of the Viviani's curve, since they are described in geographical coordinates by $\varphi = n\lambda$, $n > 0$. A spherical Archimedean spiral is algebraic if and only if $n \in \mathbb{Q}$. They were studied by Guido Grandi in 1728, also called *clelias*. They are the loci of a point P on a meridian of a sphere rotating at constant speed ω around the polar axis, the point P also moving at constant speed $n\,\omega$ along this meridian (see Fig. 11). Therefore, physically, we obtain a clelia when peeling an orange or when rewinding regularly a spherical wool ball.

A similar argument used in the preceding section, considering now

$$\kappa(z) = \frac{z(2n^2 + 1 - z^2)}{(n^2 + 1 - z^2)^{3/2}}, \tag{39}$$

gives us the following uniqueness result, whose proof we will omit.

Corollary 15 *The spherical spiral curves $\varphi = n\lambda$, $n > 0$, are the only spherical curves (up to rotations around the z-axis) with spherical angular momentum $\mathcal{K}(\varphi) = -\cos^2\varphi/\sqrt{n^2+\cos^2\varphi}$.*

Acknowledgments The authors were partially supported by State Research Agency and European Regional Development Fund via the Grant No. PID 2020-117868GB-I00 supported by MCIN/AEI/10.13039/501100011033: The first named author was supported also by Maria de Maeztu Excellence Unit IMAG CEX2020-001105-M funded by MCIN/AEI/10.13039/501100011033/ and by the EBM/FEDER UJA 2020 project 1380860. The second named author was supported also by project PGC2018-097046-B-I00, supported by MCIN/AEI/10.13039/501100011033/ FEDER "Una manera de hacer Europa" and by Ayudas a proyectos para el desarrollo de investigación científica y técnica por grupos competitivos, included in the Programa Regional de Fomento de la Investigación Científica y Técnica (Plan de Actuación 2022) of the Fundación Séneca-Agencia de Ciencia y Tecnología de la Región de Murcia, REF. 21899/PI/22. The third named author was also supported by MCIN/AEI project PID-2019-111531GA-I00.

We would also like to thank Álvaro Pámpano for helpful discussion about the topic, especially for comments concerning Remark 9.

References

1. Arnold, V.I.: Mathematical Methods of Classical Mechanics. Springer, New York (1978)
2. Arroyo, J., Garay, O.J., Mencía, J.J.: Closed generalized elastic curves in $\mathbb{S}^2(1)$. J. Geom. Phys. **48**, 339–353 (2003)
3. Arroyo, J., Garay, O.J., Mencía, J.J.: Extremals of curvature energy actions on spherical closed curves J. Geom. Phys. **51**, 101–125 (2004)
4. Arroyo, J., Garay, O.J., Mencía, J.J.: Elastic circles in 2-spheres. J. Phys. A Math. Gen. **39**, 2307–2324 (2006)
5. Brunnett, G., Crouch, P.E.: Elastic curves on the sphere. Adv. Comput. Math. **2**, 23–40 (1994)
6. Byrd, P.F., Friedman, M.D.: Handbook of Elliptic Integrals for Engineers and Physicists. 2nd edn. Springer, New York (1971)
7. Castro, I., Castro-Infantes, I.: Plane curves with curvature depending on distance to a line. Diff. Geom. Appl. **44**, 77–97 (2016)
8. Castro, I., Castro-Infantes, I., Castro-Infantes, J.: New plane curves with curvature depending on distance from the origin. Mediterr. J. Math. **14**, 108 (2017)
9. Castro, I., Castro-Infantes, I., Castro-Infantes, J.: Curves in Lorentz-Minkowski plane: elasticae, catenaries and grim-reapers. Open Math. **16**, 747–766 (2018)
10. Castro, I., Castro-Infantes, I., Castro-Infantes, J.: Curves in Lorentz- Minkowski plane with curvature depending on their position. Open Math. **18**, 749–770 (2020)
11. Erdös, P.: Spiraling the Earth with C. G. J. Jacobi. Amer. J. Phys. **68**, 888–895 (2000)
12. Ferréol, R.: Encyclopédie des formes mathématiques remarquables. www.mathcurve.com
13. Jurdjevic, V.: Non-Euclidean elastica. Am. J. Math. **117**, 93–124 (1995)
14. Langer, J., Singer, D.: The total squared curvature of closed curves. J. Diff. Geom. **20**, 1–22 (1984)
15. Langer, J., Singer, D.: Lagrangian aspects of the Kirchhoff elastic rod. SIAM Rev. **38**, 1–17 (1996)
16. Singer, D.: Curves whose curvature depends on distance from the origin. Am. Math. Month. **106**, 835–841 (1999)
17. Singer, D.: Lectures on elastic curves and rods. Curvature and variational modelling in physics and biophysics. AIP Conf. Proc. vol. **1002**, 3–32 (2008)

Conjugate Plateau Constructions in Product Spaces

Jesús Castro-Infantes, José M. Manzano, and Francisco Torralbo

Abstract This survey paper investigates, from a purely geometric point of view, Daniel's isometric conjugation between minimal and constant mean curvature surfaces immersed in homogeneous Riemannian three-manifolds with isometry group of dimension four. On the one hand, we collect the results and strategies in the literature that have been developed so far to deal with the analysis of conjugate surfaces and their embeddedness. On the other hand, we revisit some constructions of constant mean curvature surfaces in the homogeneous product spaces $\mathbb{S}^2 \times \mathbb{R}$, $\mathbb{H}^2 \times \mathbb{R}$, and $\mathbb{R}^3$ having different topologies and geometric properties depending on the value of the mean curvature. Finally, we also provide some numerical pictures using Surface Evolver.

Keywords Constant mean curvature · Compact surfaces · Homogeneous three-manifolds · Product spaces · Conjugate constructions

1 Introduction

The study of minimal and constant mean curvature surfaces (H-surfaces in the sequel) represents a central topic in surface theory with a long trajectory dating back to works of Euler and Lagrange. Minimal surfaces were popularized by Plateau's experiments on soap films, which gave rise to the so-called Plateau problem of

J. Castro-Infantes
Universidad Politécnica de Madrid, Dpto. Matemática Aplicada a las Tecnologías de la Información y las Comunicaciones, Madrid, Spain
e-mail: jesus.castro@upm.es

J. M. Manzano (✉)
Universidad de Jaén, Dpto. Matemáticas, Jaén, Spain
e-mail: jmprego@ujaen.es

F. Torralbo
Universidad de Granada, Dpto. Geometría y Topología, Granada, Spain
e-mail: ftorralbo@ugr.es

A. Alarcón et al. (eds.), *New Trends in Geometric Analysis*,
RSME Springer Series 10, https://doi.org/10.1007/978-3-031-39916-9_3

finding the least-area surface spanning a given contour. Likewise, H-surfaces with $H > 0$ were first investigated in connection with the isoperimetric problem of finding the least-area surface enclosing a given volume. A geometrically appealing question that has caught the attention of many geometers during the last century is to produce nontrivial examples of H-surfaces with a high number of symmetries. This is well exemplified by the vast number of triply periodic minimal surfaces in Euclidean space $\mathbb{R}^3$ that appear in the literature, e.g., see Schoen's report [101] and Weber's Minimal Surface Archive [107].

However, in the case $H > 0$, very few examples of H-surfaces were known prior to Lawson's discovery [58] of an isometric duality or conjugation for H-surfaces in space forms $\mathbb{M}^3(c)$ of constant sectional curvature $c \in \mathbb{R}$. It allowed him to produce the first examples of doubly periodic embedded 1-surfaces in $\mathbb{R}^3 = \mathbb{M}^3(0)$ by means of minimal surfaces in the round sphere $\mathbb{S}^3 = \mathbb{M}^3(1)$. The minimal surfaces he employed were indeed solutions to appropriate Plateau problems over certain spherical geodesic polygons whose sides become planar lines of curvature in the conjugate 1-surface. Furthermore, he noticed that an extension of the minimal surface by axial symmetries about the boundary components agrees with an extension of the conjugate *one*-surface by mirror symmetry about the planes containing the conjugate boundary. This technique is known as the *conjugate Plateau construction* and has been used to obtain many examples of complete 1-surfaces in $\mathbb{R}^3$ as we explain below. It is important to say that Lawson's result actually yields a two-parameter isometric deformation in which we can change not only the ambient curvature but also a phase parameter which rotates the second fundamental form. In particular, Lawson's correspondence generalizes the classical notion of associate Bonnet family of minimal surfaces in $\mathbb{R}^3$ (see also [56]). However, when we speak of *conjugation*, we are prescribing the phase angle equal to $\frac{\pi}{2}$, which gives a more precise control of the corresponding surfaces, as we will explain throughout this work.

As for the conjugation from minimal surfaces in $\mathbb{S}^3$ to 1-surfaces in $\mathbb{R}^3$, Lawson's constructions were resumed by Karcher [51], who realized that many of Schoen's triply periodic minimal surfaces in [101] admit many planes of mirror symmetry and their fundamental pieces can be obtained as conjugate Plateau constructions. This allowed him to deform Schoen's minimal examples into H-surfaces by considering geodesic polygons in the $\mathbb{S}^3$ similar to those needed in $\mathbb{R}^3$. In his paper, Karcher also improved some of Lawson's ideas about conjugate curves and developed their connection with the different Hopf fibrations in $\mathbb{S}^3$, making it clear that the conjugate technique had the capability to produce beautiful highly symmetric H-surfaces (see also [53]). Große-Brauckmann [28] took it one step further and produced many interesting examples of 1-surfaces in $\mathbb{R}^3$, including k-unduloids, whose $k \geq 2$ ends are asymptotic to the classical Delaunay unduloids. In fact, triunduloids (with $k = 3$) were later proved to be properly embedded 1-surfaces in $\mathbb{R}^3$ by Große-Brauckmann, Kusner, and Sullivan [30]. Große-Brauckmann and Wohlgemuth [32] also used the conjugation to prove that the triply periodic minimal surface known as *gyroid* is embedded and can be deformed into an H-surface (see also [29] for numerical experiments). The conjugation has also been used in different ambient spaces, e.g., Karcher, Pinkall, and Sterling [55] constructed minimal surfaces in $\mathbb{S}^3$ by conjugating minimal surfaces in $\mathbb{S}^3$; Karcher [52] and Rossman [97] obtained 1-surfaces in hyperbolic space $\mathbb{H}^3 = \mathbb{M}^3(-1)$ from minimal surfaces in $\mathbb{R}^3$; and Karcher and Polthier [56] also used related techniques to revisit conjugate minimal surfaces in $\mathbb{R}^3$.

There have been other approaches to produce H-surfaces in space forms, three of which will be highlighted here concerning compact examples: First, using implicit methods, Karcher, Pinkall, and Sterling [55] obtained compact embedded minimal surfaces in $\mathbb{S}^3$ with arbitrary genus, and Kapouleas [46, 47] found compact immersed H-surfaces in $\mathbb{R}^3$ with arbitrary genus. The second technique is often referred to as the Dorfmeister-Pedit-Wu (DPW) method [21], which uses integrable systems and can be thought of as a global version of the Weierstraß representation. The DPW method was pioneered by Pinkall and Sterling [88] and Hitchin [41] to study H-tori. Heller, Heller, and Traizet [39] (see also [5] for numerically constructed surfaces) have recently shown the existence, for large genus g, of a complete and smooth family of compact H-surfaces in the round 3-sphere deforming the Lawson surface $\xi_{1,g}$ (see [58, §6]) into a doubly covered geodesic 2-sphere. Third and last, we have gluing methods, which can be illustrated by Kapouleas' constructions of compact minimal surfaces in $\mathbb{S}^3$ by connecting two parallel Clifford tori [49] or equatorial 2-spheres [48] by means of catenoidal bridges.

Back to the conjugate techniques, they have been extended recently to the case of simply connected homogeneous three-manifolds with isometry group of dimension 4, which are the most symmetric spaces after the space forms. These spaces form a two-parameter family $\mathbb{E}(\kappa,\tau)$, $\kappa \neq 4\tau^2$, containing the product spaces $\mathbb{M}^2(\kappa) \times \mathbb{R} = \mathbb{E}(\kappa, 0)$ as well as the Lie groups $\widetilde{\mathrm{SL}}_2(\mathbb{R})$, $\mathrm{SU}(2)$, and Nil_3 with some left-invariant metrics (see Table 2). Their geometry will be discussed in Sect. 2. The cornerstone of this extension is the work of Daniel [14], who found a Lawson-type correspondence within this family that connects, for a phase angle of $\frac{\pi}{2}$, minimal surfaces in $\mathbb{E}(4H^2+\kappa, H)$ with H-surfaces in $\mathbb{M}^2(\kappa) \times \mathbb{R}$ for all $\kappa, H \in \mathbb{R}$ (see Table 3). Hauswirth, Sa Earp, and Toubiana [38] also found this correspondence as a particular case of the associate family for $H = 0$ and arbitrary $\kappa \in \mathbb{R}$. We also remark that Daniel's correspondence contains other cases with phase angle $\frac{\pi}{2}$, but mirror symmetries only exist in the case of product spaces, which makes this case the most tractable one (see Lemmas 2 and 3). Observe that in the case $\kappa = 0$, we have that $\mathbb{E}(4H^2, H) = \mathbb{M}^3(H^2)$ and Daniel's conjugation reduces to the classical Lawson's conjugation between minimal surfaces in $\mathbb{S}^3(H^2)$ and H-surfaces in $\mathbb{R}^3$.

The starting tools in a conjugate construction, i.e., the Plateau problem (and its improper version known as the Jenkins-Serrin problem) in $\mathbb{E}(\kappa,\tau)$ have been solved under quite general conditions [13, 76, 79, 81, 84, 108] (see Sect. 4.1 and Sect. 4.2). They provide us with plenty of surfaces with boundary a geodesic polygon (with possibly some components at infinity in the Jenkins-Serrin case) that are at our disposal to act as initial minimal surfaces in the conjugate construction. Also, the extension by axial and mirror symmetries and the absence of singularities rely on general results that also apply (see Proposition 3).

It is necessary to point out that the family $\mathbb{E}(\kappa, \tau)$ has fewer isometries than space forms, which forces us to consider only initial polygons consisting of horizontal and vertical geodesics (with respect to the Killing submersion $\pi : \mathbb{E}(\kappa, \tau) \to \mathbb{M}^2(\kappa)$, see Sect. 2). This reduces both the number of possible configurations and the directions in which we are able to control the involved surfaces.

In Sect. 3, we will collect and present different features of Daniel's correspondence that are essential in the conjugate constructions, specially a detailed study of conjugate curves, some of them in more generality than those in the literature. We will pay special attention in Sect. 3.2 to some classes of surfaces in which the correspondence is well understood, such as equivariant surfaces, ideal Scherk graphs, or ruled minimal surfaces. In this respect, it is worth mentioning that Daniel's correspondence has been a formidable tool to analyze surfaces satisfying preserved geometric conditions and has played a key role in their classification, e.g., H-surfaces with zero Abresch-Rosenberg quadratic differential [1, 19, 23], H-surfaces with certain bounds on the intrinsic or extrinsic area growth [67], or H-surfaces with constant Gauss curvature [15].

We will now give a brief overview of constructions of H-surfaces in $\mathbb{E}(\kappa, \tau)$-spaces that use conjugation, and we will begin with the minimal case. Morabito and Rodríguez [83] used a conjugate Jenkins-Serrin construction to obtain minimal k-noids in $\mathbb{H}^2 \times \mathbb{R}$ with genus 0 and k-ends asymptotic to vertical planes, as well as minimal saddle towers in $\mathbb{H}^2 \times \mathbb{R}$ similar to those in $\mathbb{R}^3$ obtained by Karcher [50]. Pyo [92] also found the minimal k-noids independently assuming additionally that the vertical planes are disposed symmetrically. Rodríguez [94] extended this construction to give minimal examples in $\mathbb{H}^2 \times \mathbb{R}$ with infinitely many ends and an arbitrary (finite or countable) number of limit ends. Martín and Rodríguez [74] have also used a conjugate Jenkins-Serrin construction to produce minimal embeddings of any non-simply connected planar domain in $\mathbb{H}^2 \times \mathbb{R}$. Mazet, Rodríguez, and Rosenberg [77] used a conjugate Jenkins-Serrin construction to produce examples for their classification of doubly periodic embedded minimal surfaces in $\mathbb{H}^2 \times \mathbb{R}$. The first two authors [10] have obtained minimal k-noids in $\mathbb{H}^2 \times \mathbb{R}$ with genus 1 and $k \geq 3$ ends by another conjugate Jenkins-Serrin construction, inspired by a construction of Plehnert [89] of similar surfaces for $H = \frac{1}{2}$. Plehnert and the second and third authors [68] have also obtained embedded minimal surfaces of type Schwarz P in $\mathbb{S}^2 \times \mathbb{R}$ by a conjugate Plateau construction.

In the non-minimal case, the study of conjugate constructions was initiated independently by Plehnert [90] (to obtain k-noids with genus 0 in $\mathbb{H}^2 \times \mathbb{R}$ and $0 < H \leq \frac{1}{2}$) and by the second and third authors [70] (to obtain horizontal unduloid-type H-surfaces in $\mathbb{H}^2 \times \mathbb{R}$ and $\mathbb{S}^2 \times \mathbb{R}$). The latter was subsequently improved in [72] to obtain horizontal nodoid-type H-surfaces in $\mathbb{H}^2 \times \mathbb{R}$ and $\mathbb{S}^2 \times \mathbb{R}$ and to determine which of these examples are embedded. The second and third authors have also provided compact orientable embedded H-surfaces in $\mathbb{S}^2 \times \mathbb{R}$ with arbitrary genus by means of a different conjugate Plateau construction. In the case of $\mathbb{H}^2 \times \mathbb{R}$ and $0 < H \leq \frac{1}{2}$, Rodríguez and the first and second authors [11] have produced k-noids and saddle towers that extend Plehnert's, plus another family of H-surfaces, called

k-nodoids, with genus 0 and $k \geq 2$ ends that approach the asymptotic vertical planes from the convex side (unlike the k-noids, which lie in their concave side).

We will illustrate how the technique works by sketching five of the above constructions (see Table 1). A deeper motivation for each of them will be given in the corresponding sections, but we will say here that we have chosen them because we would like to cover all ranges of the mean curvature in both $\mathbb{S}^2 \times \mathbb{R}$ and $\mathbb{H}^2 \times \mathbb{R}$ in order to visualize the dissimilarities between the critical, supercritical, and subcritical cases (see Sect. 2.5). These constructions are also part of some ongoing open research lines to which the authors have contributed:

- On the one hand, we are interested in the classification of compact embedded H-surfaces in $\mathbb{S}^2 \times \mathbb{R}$ attending to their genus. The only non-equivariant known examples are the families of H-tori given by Theorem 8 with $H > \frac{1}{2}$ and the arbitrary genus H-surfaces given by Theorem 9 with $H < \frac{1}{2}$. We expect that the former are the only embedded H-tori. Since they are not equivariant, their characterization should be more involved than in the case of H-tori in $\mathbb{S}^3$ (see Andrews and Li [3]).
- On the other hand, we are also interested in H-surfaces with finite total curvature in $\mathbb{H}^2 \times \mathbb{R}$ displaying different topologies and asymptotic behaviors. In Sect. 6.2, we discuss and collect some properties of this class of surfaces in the minimal case, which is probably the most studied class of minimal surfaces in $\mathbb{H}^2 \times \mathbb{R}$. It is widely believed that H-surfaces in $\mathbb{H}^2 \times \mathbb{R}$ with $0 < H < \frac{1}{2}$ should behave similarly, but this is still an unexplored area of research. In particular, we expect that the (H, k)-noids and (H, k)-nodoids given by Theorem 11 have finite total curvature when the mean curvature is not critical.

In conjugate constructions, one easily reaches the central but tough question of embeddedness, which probably has not been well understood yet. We will treat it carefully throughout the constructions in this survey by emphasizing the different approaches to answer this question. In the case of horizontal Delaunay H-surfaces, embeddedness follows from spotting a geometric function in the common stability operator of the conjugate immersions and finding one-parameter groups of isometries that induce this function (see Sect. 5.1.5). In the case of arbitrary genus H-surfaces, we rely on the estimates on the curvature of the boundary of an H-bigraph given by the second author in [62]. In the case of minimal Schwarz P-surfaces, convexity of some boundary curves come in handy along with some isoperimetric inequalities. In the case of (H, k)-noids and (H, k)-nodoids, we are able to find some embedded limits, and then we can ensure that some of the examples are embedded (and some of them are not) by continuity (see Proposition 2). Finally, in the case of the minimal k-noids of genus 1, we use the Krust-type property given by Hauswirth, Sa Earp, and Toubiana [38] (see Proposition 5) to also show that the surfaces are embedded provided that the parameters are controlled. The Krust property is an essential tool that facilitates the conjugation of minimal surfaces in $\mathbb{H}^2 \times \mathbb{R}$, but it is not true in general for other values of H (see Sect. 6.1.3).

Table 1 Constructions sketched in the document. Note that the special value $H = \frac{1}{2}$ is the critical mean curvature in $\mathbb{H}^2 \times \mathbb{R}$ but it is also relevant for the embeddedness of compact H-surfaces in $\mathbb{S}^2 \times \mathbb{R}$; see Sects. 5.1 and 5.2

<table>
<tr><th></th><th>$H = 0$</th><th>$0 < H < \frac{1}{2}$</th><th>$H = \frac{1}{2}$</th><th>$H > \frac{1}{2}$</th></tr>
<tr><td rowspan="2">$\mathbb{S}^2 \times \mathbb{R}$</td><td rowspan="2">Theorem 10: Schwarz P-surfaces</td><td>Theorem 9: compact embedded surfaces of arbitrary genus</td><td></td><td>Theorem 8: embedded tori</td></tr>
<tr><td colspan="3">Theorem 7: Horizontal Delaunay surfaces</td></tr>
<tr><td>$\mathbb{H}^2 \times \mathbb{R}$</td><td>Theorem 12: Genus-1 k-noids</td><td colspan="2">Theorem 11: Saddle towers, (H, k)-noids, and (H, k)-nodoids</td><td>Theorem 7: Horizontal Delaunay surfaces</td></tr>
</table>

Numerical methods are an invaluable tool in the visualization of H-surfaces and prove useful to forecast theoretical results. We will highlight three different approaches in this respect: First, Weierstraß representation along with numerical solutions of the period problems has been used to represent many families of minimal surfaces in $\mathbb{R}^3$ (see the Mathematica notebooks given by Weber in [107] for details). Second, the use of the DPW method along with a numerical adjustment of parameters has been developed to get high genus H-surfaces in $\mathbb{S}^3$ by Bobenko, Heller, and Schmitt [5, 40], where the figures in the first paper were done with the software XLAB developed by Schmitt). This approach presents the drawback that particular Weierstraß data might be difficult to derive in more complicated examples. A last and more *variational* approach was developed by Brakke in his software Surface Evolver [7] by minimizing energies (e.g., area functional) of triangulated surfaces under certain constraints (e.g., volume and boundary constrains, a.k.a. free boundary problem) to get H-surfaces, although the method may display some issues when approximating the solution since fundamental domains might not be stable minima. In Sect. 7, we present some numerical experiments using Surface Evolver in order to visualize the minimal examples in $\mathbb{S}^2(\kappa) \times \mathbb{R}$ constructed in Sect. 5.3. We also remark that to overcome some of the issues in the above methods, Pinkall and Polthier [87] used discrete differential techniques to implement the conjugation directly. The philosophy of this procedure is that the Plateau solution is usually stable and easier to obtain by minimization than the free boundary solution which is more likely unstable. Examples of surfaces produced with this technique can be found in [31, 87], where the software GRAPE [91] is used. We would like to say that, unfortunately and as far as we know, neither XLAB nor GRAPE is publicly available.

As a final comment, in the present work, we have made an effort to keep the notation homogeneous by writing: a tilde for the elements of the initial minimal surface in $\mathbb{E}(4H^2 + \kappa, H)$, no tilde for the corresponding elements of the conjugate target H-surface in $\mathbb{M}^2(\kappa) \times \mathbb{R}$, and an asterisk * to indicate the completion of the surface after successive reflections about its boundary components. Also, we have indicated as subscripts the parameters that are part of the construction (e.g., $\Sigma_{a,b}$) and in functional notation those which are auxiliary and will likely disappear after a continuity argument (e.g., $\Sigma(a, b)$). Regarding the figures, we have colored the boundary curves in red and blue for the horizontal and vertical components of the initial polygon, respectively. We hope this will help the reader to easily follow some of the geometric discussions in the document. Finally, we have deliberately not normalized the spaces by homotheties, so that we will do our constructions in $\mathbb{H}^2(\kappa) \times \mathbb{R}$ and $\mathbb{S}^2(\kappa) \times \mathbb{R}$. This is because the limit case $\kappa = 0$ always gives some insight by comparing with the Euclidean counterparts.

2 The Geometry of $\mathbb{E}(\kappa, \tau)$-spaces

Simply connected oriented homogeneous Riemannian three-manifolds with four-dimensional isometry group can be arranged in a two-parameter family $\mathbb{E}(\kappa, \tau)$, where $\kappa, \tau \in \mathbb{R}$ and $\kappa - 4\tau^2 \neq 0$. These parameters are geometrically characterized by the existence of a Riemannian submersion $\pi : \mathbb{E}(\kappa, \tau) \to \mathbb{M}^2(\kappa)$ whose fibers are the integral curves of a unit Killing vector field ξ (also called *Killing submersion*) and such that the bundle curvature is constant and equal to τ, that is,

$$\overline{\nabla}_X \xi = \tau X \times \xi \tag{1}$$

holds true for all vector fields $X \in \mathfrak{X}(\mathbb{E}(\kappa, \tau))$ (see [14, 64]). The cross product $\times$ reflects the orientation of the ambient space, being $\{u, v, u \times v\}$ a positively oriented basis for all linearly independent tangent vectors u and v. As a matter of fact, $\mathbb{E}(\kappa, \tau)$ and $\mathbb{E}(\kappa, -\tau)$ are the same space for all κ and τ, but opposite orientations have been chosen. More generally, $\mathbb{E}(a^2\kappa, a\tau)$ is homothetic to $\mathbb{E}(\kappa, \tau)$ with a conformal factor a^2 for any constant $a \neq 0$.

In the case $\kappa = 4\tau^2$, the above description is also valid, though the space $\mathbb{E}(\kappa, \tau)$ has constant sectional curvature (it is isometric to the space form $\mathbb{M}^3(\tau^2)$), whence its isometry group is six-dimensional. We remark that hyperbolic spaces $\mathbb{H}^3(c)$ are not $\mathbb{E}(\kappa, \tau)$-spaces for any sectional curvature $c < 0$ because they do not admit Killing fields of constant length.

All $\mathbb{E}(\kappa, \tau)$-spaces are isometric to Lie groups endowed with left-invariant metrics, except for $\mathbb{E}(\kappa, 0)$ with $\kappa > 0$ (see [78, Thm. 2.4]). Indeed, the condition $\tau = 0$ indicates that the distribution orthogonal to ξ is integrable, so $\mathbb{E}(\kappa, 0)$ is better thought of as the Riemannian product space $\mathbb{M}^2(\kappa) \times \mathbb{R}$ (observe that there is no Lie group with underlying manifold $\mathbb{S}^2 \times \mathbb{R}$). As shown in Table 2, in the case $\tau \neq 0$, we encounter the universal cover of the special linear group $\widetilde{\mathrm{SL}}_2(\mathbb{R})$, the Heisenberg group Nil_3, and the special unitary group $\mathrm{SU}(2)$, endowed with left-invariant metrics in which ξ is a biinvariant vector field and hence Killing. In the case of $\mathrm{SU}(2)$, these spaces are known as Berger spheres and will be denoted by $\mathbb{S}^3_{\mathrm{B}}(\kappa, \tau)$ in the sequel. Observe that the Lie groups $\mathrm{SU}(2)$ and $\widetilde{\mathrm{SL}}_2(\mathbb{R})$ also admit left-invariant metrics with three-dimensional isometry groups that will not be considered here.

As Killing submersions, the $\mathbb{E}(\kappa, \tau)$-spaces admit natural notions of vertical and horizontal directions, defined as those tangent and orthogonal to the unit Killing vector field ξ, respectively. Also, the isometries spanned by ξ, called *vertical translations*, form a one-parameter group of isometries $\{\Phi_t\}_{t \in \mathbb{R}}$ that plays a fundamental role in the geometry of these spaces.

Table 2 Different geometries in $\mathbb{E}(\kappa, \tau)$-spaces

	$\kappa > 0$	$\kappa = 0$	$\kappa < 0$
$\tau = 0$	$\mathbb{S}^2 \times \mathbb{R}$	$\mathbb{R}^3$	$\mathbb{H}^2 \times \mathbb{R}$
$\tau \neq 0$	$\mathrm{SU}(2)$	Nil_3	$\widetilde{\mathrm{SL}}_2(\mathbb{R})$

2.1 Geodesics

There are two distinguished types of geodesics in $\mathbb{E}(\kappa,\tau)$, namely, the vertical ones (fibers of the submersion over $\mathbb{M}^2(\kappa)$) and horizontal ones (horizontal lifts of geodesics of $\mathbb{M}^2(\kappa)$). More generally, if $\tau\neq 0$, then any non-vertical geodesic γ of $\mathbb{E}(\kappa,\tau)$ with unit speed projects onto a curve $\pi\circ\gamma$ of $\mathbb{M}^2(\kappa)$ of constant curvature, say c, and meets the Killing direction with constant angle $\langle\gamma',\xi\rangle = \frac{c}{\sqrt{4\tau^2+c^2}}$ (see [64, Prop. 3.6] and Fig. 1). In the case $\tau = 0$, any non-vertical geodesic of $\mathbb{E}(\kappa,\tau)=\mathbb{M}^2(\kappa)\times\mathbb{R}$ factors as the product of geodesics of each factor, so it always projects onto a geodesic of $\mathbb{M}^2(\kappa)$.

Horizontal and vertical geodesics have infinite length if $\kappa\leq 0$. However, if $\kappa>0$ and $\tau\neq 0$, the length of all vertical geodesics is $\frac{8\tau\pi}{\kappa}$, whereas the length of all horizontal geodesics is $\frac{4\pi}{\sqrt{\kappa}}$. Note that this is twice the length of a great circle of $\mathbb{S}^2(\kappa)$ because horizontal geodesics in Berger spheres project two-to-one onto great circles. This contrasts the case $\kappa>0$ and $\tau=0$ because horizontal geodesics of $\mathbb{S}^2(\kappa)\times\mathbb{R}$ project one-to-one onto great circles of $\mathbb{S}^2(\kappa)$.

Explicit parametrizations of all geodesics of $\mathbb{E}(\kappa,\tau)$, $\kappa\leq 0$, as well as some discussion of their minimization properties can be found in [67, §2]. The case $\kappa>0$ and $\tau\neq 0$ were studied by Rakotoniaina in [93]. A discussion of Jacobi fields along any geodesic of any $\mathbb{E}(\kappa,\tau)$ can be found in [19, §4].

2.2 Isometries

An isometry $f\in\mathrm{Iso}(\mathbb{E}(\kappa,\tau))$ induces another isometry $h\in\mathrm{Iso}(\mathbb{M}^2(\kappa))$ such that $\pi\circ f = h\circ\pi$. Conversely, given $h\in\mathrm{Iso}(\mathbb{M}^2(\kappa))$, there is an orientation-preserving

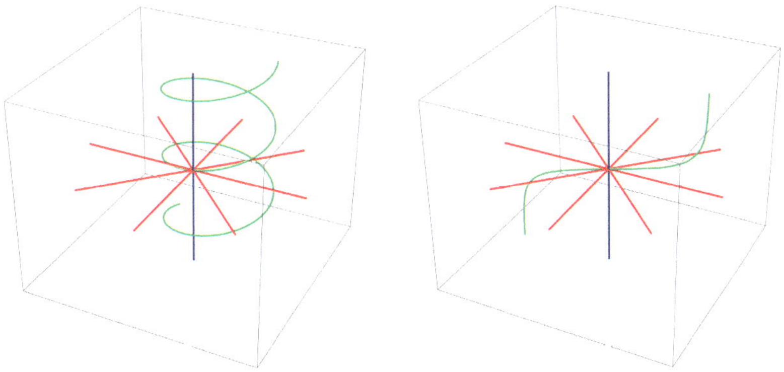

Fig. 1 A vertical geodesic in blue, four horizontal geodesics in red, and an oblique geodesic in green. They are represented in Nil_3 (left) and in $\mathbb{H}^2\times\mathbb{R}$ (right) in the Cartan model (see Sect. 2.3)

isometry $f_+ \in \mathrm{Iso}(\mathbb{E}(\kappa,\tau))$ such that $\pi \circ f_+ = h \circ \pi$. If $\tau = 0$, then there is also an orientation-reversing isometry $f_- \in \mathrm{Iso}(\mathbb{E}(\kappa,\tau))$ such that $\pi \circ f_- = h \circ \pi$. Both f_+ and f_- are unique up to composition with vertical translations. Moreover, all the isometries of $\mathbb{E}(\kappa,\tau)$ are of these types provided that $\kappa - 4\tau^2 \neq 0$; in particular, there are no orientation-reversing isometries in $\mathbb{E}(\kappa,\tau)$ if $\kappa - 4\tau^2 \neq 0$ and $\tau \neq 0$. We refer to [64, Thm. 2.8] for a classification of the isometries that preserve the Killing direction in a general Killing submersion.

In $\mathbb{E}(\kappa,\tau)$, there are rotations of any angle about vertical geodesics, which are recovered as orientation-preserving lifts of rotations in $\mathbb{M}^2(\kappa)$ with center the point onto which the vertical geodesic projects. In the case of horizontal geodesics, there are axial symmetries (i.e., rotations of angle π) about them, recovered as orientation-preserving lifts of axial symmetries with respect to their projections, but there are no rotations of arbitrary angle if $\kappa - 4\tau^2 \neq 0$. These are the only orientation-preserving isometries that keep all points of a certain geodesic *fixed*.

Lemma 1 *Given a non-vertical geodesic $\gamma : \mathbb{R} \to \mathbb{E}(\kappa,\tau)$, there is a unique one-parameter group $\{T_t\}_{t\in\mathbb{R}}$ of isometries of $\mathbb{E}(\kappa,\tau)$ such that $T_t(\gamma(s)) = T_{t+s}(\gamma(0))$ for all $t, s \in \mathbb{R}$.*

Proof Since $\pi \circ \gamma$ is a curve of constant curvature parametrized with constant speed, we can consider a one-parameter group $\{h_t\}_{t\in\mathbb{R}}$ of isometries of $\mathbb{M}^2(\kappa)$ such that $h_t(\pi(\gamma(0))) = \pi(\gamma(t))$. Therefore, we can find a unique orientation-preserving lift $T_t \in \mathrm{Iso}(\mathbb{E}(\kappa,\tau))$ of h_t such that $T_t(\gamma(0)) = \gamma(t)$. By the uniqueness of the lift and the uniqueness of a geodesic given initial conditions, it easily follows that $\{T_t\}_{t\in\mathbb{R}}$ is a group satisfying the desired condition.

The elements of $\{T_t\}_{t\in\mathbb{R}}$ will be called translations along γ; if γ is vertical, then such a group is not unique, but translations along γ are defined as the vertical translations. In general, a one-parameter group $\{f_t\}_{t\in\mathbb{R}}$ of isometries is not necessarily the group of translations along a geodesic, but translations will allow us to interpret the rest of them as their screw motions (i.e., a composition with vertical translations):

- If $\tau \neq 0$ and the induced group $\{h_t\}_{t\in\mathbb{R}}$ of isometries of $\mathbb{M}^2(\kappa)$ has a fixed point p, then $\{f_t\}_{t\in\mathbb{R}}$ is a group of rotations about the vertical geodesic $\pi^{-1}(p)$ or their screw motions. If $\{h_t\}_{t\in\mathbb{R}}$ has no fixed points, then $\kappa \leq 0$, and we have two possibilities:
  - If $\{h_t\}_{t\in\mathbb{R}}$ consists of translations along a geodesic α of $\mathbb{R}^2$ or $\mathbb{H}^2(\kappa)$, then $\{f_t\}_{t\in\mathbb{R}}$ is a group of (hyperbolic) translations along a horizontal geodesic of $\mathbb{E}(\kappa,\tau)$ projecting to α or their screw motions.
  - The second possibility only occurs if $\kappa < 0$ and $\{h_t\}_{t\in\mathbb{R}}$ are parabolic translations. In this case, we can choose an oblique geodesic γ projecting to an orbit of $\{h_t\}_{t\in\mathbb{R}}$, which is a horocycle of $\mathbb{H}^2(\kappa)$. Then, $\{f_t\}_{t\in\mathbb{R}}$ is a group of (parabolic) translations along γ or their screw motions.
- If $\tau = 0$, then one-parameter groups of isometries of $\mathbb{E}(\kappa,\tau)$ factor as the product of one-parameter groups of isometries of $\mathbb{M}^2(\kappa)$ and $\mathbb{R}$. This makes it

easier to classify such a one-parameter group as rotations, translations, parabolic translations, or hyperbolic translations (when they leave the factor $\mathbb{R}$ fixed) and their screw motions (otherwise).

If $\tau = 0$, then there are also mirror symmetries with respect to vertical planes (i.e., preimages of geodesics of $\mathbb{M}^2(\kappa)$ under π) or with respect to horizontal slices (i.e., surfaces of the form $\mathbb{M}^2(\kappa) \times \{t_0\}$ for some $t_0 \in \mathbb{R}$). These symmetries are orientation-reversing lifts of axial symmetries and of the identity in $\mathbb{M}^2(\kappa)$, respectively. In particular, vertical planes and horizontal slices are totally geodesic surfaces in $\mathbb{E}(\kappa, 0)$, yet there are no totally umbilical surfaces in $\mathbb{E}(\kappa, \tau)$ if $\tau \neq 0$, as shown by Souam and Toubiana [102].

2.3 *Working in Coordinates*

The following common framework for all $\mathbb{E}(\kappa, \tau)$-spaces was originally introduced by Cartan [8, §296]. We consider

$$\Omega_\kappa = \{(x, y) \in \mathbb{R}^2 : \lambda_\kappa(x, y) > 0\}, \text{ where } \lambda_\kappa(x, y) = \frac{1}{1 + \frac{\kappa}{4}(x^2 + y^2)}.$$

Therefore, Ω_κ is a disk of radius $\frac{2}{\sqrt{-\kappa}}$ if $\kappa < 0$ or the whole $\mathbb{R}^2$ if $\kappa \geq 0$. Then, we define $M(\kappa, \tau)$ as $\Omega_\kappa \times \mathbb{R} \subseteq \mathbb{R}^3$ endowed with the Riemannian metric

$$\mathrm{d}s^2 = \lambda_\kappa^2(\mathrm{d}x^2 + \mathrm{d}y^2) + (\mathrm{d}z + \lambda_\kappa \tau(y\mathrm{d}x - x\mathrm{d}y))^2.$$

Moreover, the orientation is chosen such that

$$E_1 = \tfrac{1}{\lambda_\kappa}\partial_x - \tau y\, \partial_z, \quad E_2 = \tfrac{1}{\lambda_\kappa}\partial_y + \tau x\, \partial_z, \quad E_3 = \partial_z \tag{2}$$

is a global positively oriented orthonormal frame. It easily follows that $\pi(x, y, z) = (x, y)$ is a Killing submersion from $M(\kappa, \tau)$ to $(\Omega_\kappa, \mathrm{d}s_\kappa^2)$ with constant bundle curvature τ and unit Killing vector field $\xi = E_3$, where the Riemannian metric $\mathrm{d}s_\kappa^2 = \lambda_\kappa^2(\mathrm{d}x^2 + \mathrm{d}y^2)$ has constant curvature κ. Therefore, $M(\kappa, \tau)$ is a global model of $\mathbb{E}(\kappa, \tau)$ if $\kappa \leq 0$ but fails to be complete otherwise. More precisely, if $\kappa > 0$, then $M(\kappa, \tau)$ is isometric to the universal cover of $\mathbb{E}(\kappa, \tau)$ minus a vertical fiber, as we will discuss shortly.

In the frame (2), the Levi-Civita connection $\overline{\nabla}$ of $\mathbb{E}(\kappa, \tau)$ reads

$$\begin{array}{lll}
\overline{\nabla}_{E_1} E_1 = \frac{\kappa y}{2} E_2, & \overline{\nabla}_{E_1} E_2 = -\frac{\kappa y}{2} E_1 + \tau E_3, & \overline{\nabla}_{E_1} E_3 = -\tau E_2, \\
\overline{\nabla}_{E_2} E_1 = -\frac{\kappa x}{2} E_2 - \tau E_3, & \overline{\nabla}_{E_2} E_2 = \frac{\kappa x}{2} E_1, & \overline{\nabla}_{E_2} E_3 = \tau E_1, \\
\overline{\nabla}_{E_3} E_1 = -\tau E_2, & \overline{\nabla}_{E_3} E_2 = \tau E_1, & \overline{\nabla}_{E_3} E_3 = 0.
\end{array} \tag{3}$$

By evaluating at the frame, it is not difficult to check that the Riemann curvature tensor $\overline{R}$ of $\mathbb{E}(\kappa,\tau)$ is given by

$$\begin{aligned}\overline{R}(X,Y,Z,W) &= \langle \overline{\nabla}_X\overline{\nabla}_Y Z - \overline{\nabla}_X\overline{\nabla}_Y Z - \overline{\nabla}_{[X,Y]}Z, W\rangle \\ &= -\tau^2\langle X\times Y, Z\times W\rangle - (\kappa-4\tau^2)\langle X\times Y,\xi\rangle\langle Z\times W,\xi\rangle.\end{aligned} \tag{4}$$

Note that our sign convention for $\overline{R}$ is the opposite to Daniel's in [14, Prop. 2.1]. Equation (4) implies that the sectional curvature in $\mathbb{E}(\kappa,\tau)$ equals τ^2 for vertical planes and $\kappa-3\tau^2$ for horizontal planes. It follows that the sectional curvature is constant if and only if $\kappa-4\tau^2=0$ as already discussed. On the other hand, it also follows that $\mathbb{E}(\kappa,\tau)$ has constant scalar curvature $2(\kappa-\tau^2)$.

Remark 1 Using the model $M(\kappa,\tau)$, we can understand why the geometry of $\mathbb{E}(\kappa,\tau)$ twists in the presence of bundle curvature, which leads to a very different behavior with respect to product spaces ($\tau=0$). Assume that $\alpha:[0,\ell]\to\Omega_\kappa$ is a piecewise-C^1 Jordan curve enclosing a relatively compact region U with the interior of U to the left when traveling on α. Let $\overline{\alpha}:[0,\ell]\to M(\kappa,\tau)$ be a horizontal lift of α (i.e., $\pi\circ\overline{\alpha}=\alpha$ and $\overline{\alpha}$ is everywhere orthogonal to ξ), which is unique up to vertical translations. The signed vertical distance from $\overline{\alpha}(0)$ to $\overline{\alpha}(\ell)$ is given by $2\tau\mathrm{Area}(U)$ (see [18, Prop. 1.6.2] and [64, Prop. 3.3]).

2.3.1 A Global Model for Berger Spheres

If $\kappa>0$ and $\tau\neq 0$, then $\mathbb{E}(\kappa,\tau)$ or $\mathbb{S}^3_{\mathrm{B}}(\kappa,\tau)$ is a Berger sphere modeled in complex coordinates as the usual 3-sphere $\mathbb{S}^3=\{(z,w)\in\mathbb{C}^2:|z|^2+|w|^2=1\}$ equipped with the Riemannian metric

$$\mathrm{d}s^2(X,Y)=\tfrac{4}{\kappa}\left[\langle X,Y\rangle+\tfrac{16\tau^2}{\kappa^2}\left(\tfrac{4\tau^2}{\kappa}-1\right)\langle X,\xi\rangle\,\langle Y,\xi\rangle\right],$$

being $\langle\cdot,\cdot\rangle$ the usual inner product in $\mathbb{C}^2\equiv\mathbb{R}^4$. The vector field ξ is defined by $\xi_{(z,w)}=\frac{\kappa}{4\tau}(iz,iw)$, and the Killing submersion is the Hopf fibration:

$$\pi:\mathbb{S}^3_{\mathrm{B}}(\kappa,\tau)\to\mathbb{S}^2(\kappa)\subset\mathbb{C}\times\mathbb{R}\equiv\mathbb{R}^3,\qquad \pi(z,w)=\tfrac{2}{\sqrt{\kappa}}\left(z\bar{w},\tfrac{1}{2}(|z|^2-|w|^2)\right),$$

(see [105, §2]). A local isometry between $\mathbb{S}^3_{\mathrm{B}}(\kappa,\tau)$ and $M(\kappa,\tau)$ is given by the Riemannian covering map $\Theta: M(\kappa,\tau)\to\mathbb{S}^3_{\mathrm{B}}(\kappa,\tau)-\{(e^{i\theta},0):\theta\in\mathbb{R}\}$, where

$$\Theta(x,y,z)=\frac{1}{\sqrt{1+\frac{\kappa}{4}(x^2+y^2))}}\left(\tfrac{\sqrt{\kappa}}{2}(y+ix)\exp(i\tfrac{\kappa}{4\tau}z),\exp(i\tfrac{\kappa}{4\tau}z)\right). \tag{5}$$

2.3.2 The Half-Space Model

In the case $\kappa < 0$, another model of $\mathbb{E}(\kappa, \tau)$ is the half-space model given by $\{(x, y, z) \in \mathbb{R}^2 : y > 0\}$ endowed with the metric

$$\frac{dx^2 + dy^2}{-\kappa y^2} + \left(dz + \frac{2\tau}{\kappa y} dx\right)^2. \tag{6}$$

Notice that the conformal factor $\frac{1}{y\sqrt{-\kappa}}$ defines a metric of constant curvature κ in the upper half-plane and the orientation is chosen such that

$$E_1 = y\sqrt{-\kappa}\,\partial_x + \tfrac{2\tau}{\sqrt{-\kappa}}\partial_z, \quad E_2 = y\sqrt{-\kappa}\,\partial_y, \quad E_3 = \partial_z, \tag{7}$$

is a positively oriented orthonormal frame. The Killing submersion is again $\pi(x, y, z) = (x, z)$ with unit Killing vector field $\xi = E_3$. A global isometry from $M(\kappa, \tau)$ to the half-space model is given by

$$\Theta(x, y, z) = \left(\frac{\frac{4}{\sqrt{-\kappa}} y}{\left(\frac{2}{\sqrt{-\kappa}} + x\right)^2 + y^2}, \frac{-\frac{4}{\kappa} - x^2 - y^2}{\left(\frac{2}{\sqrt{-\kappa}} + x\right)^2 + y^2}, z + \frac{4\tau}{\kappa} \arccos\left(\frac{y}{\sqrt{\left(\frac{2}{\sqrt{-\kappa}} + x\right)^2 + y^2}} \right) \right).$$

The following expression for the Levi-Civita connection in the global frame (7) can be deduced directly from [64, Eq. (5–1)] for $\lambda = \frac{1}{y\sqrt{-\kappa}}$:

$$\begin{aligned}
&\overline{\nabla}_{E_1} E_1 = \sqrt{-\kappa} E_2, && \overline{\nabla}_{E_1} E_2 = -\sqrt{-\kappa} E_1 + \tau E_3, && \overline{\nabla}_{E_1} E_3 = -\tau E_2,\\
&\overline{\nabla}_{E_2} E_1 = -\tau E_3, && \overline{\nabla}_{E_2} E_2 = 0, && \overline{\nabla}_{E_2} E_3 = \tau E_1,\\
&\overline{\nabla}_{E_3} E_1 = -\tau E_2, && \overline{\nabla}_{E_3} E_2 = \tau E_1, && \overline{\nabla}_{E_3} E_3 = 0.
\end{aligned} \tag{8}$$

2.4 *Fundamental Data*

Let $\phi : \Sigma \to \mathbb{E}(\kappa, \tau)$ be an isometric immersion of an orientable Riemannian surface Σ with global smooth unit normal N. The shape operator of ϕ with respect to N can be seen as a symmetric $(1, 1)$-tensor A identified with the smooth field of self-adjoint linear operators:

$$A_p : T_p\Sigma \to T_p\Sigma, \qquad A_p(v) = -\overline{\nabla}_{d\phi_p(v)} N_p, \quad \text{for all } v \in T_p\Sigma.$$

We can also consider the angle function $\nu \in C^\infty(\Sigma)$ and the tangent part of the Killing field $T \in \mathfrak{X}(\Sigma)$ defined by $\nu(p) = \langle \xi_{\phi(p)}, N_p \rangle$ and $d\phi_p(T_p) = \xi_{\phi(p)} - \nu(p)N_p$, respectively, for all $p \in \Sigma$. The orientation in $\mathbb{E}(\kappa, \tau)$ and the choice of

N induce an orientation in Σ expressed in terms of a $\frac{\pi}{2}$-rotation J in the tangent bundle (i.e., J is a smooth field of linear operators such that $J^2 = -\mathrm{id}$) defined by assuming that $\{\mathrm{d}\phi_p(u), \mathrm{d}\phi_p(Ju), N_p\}$ is positively oriented or equivalently

$$\mathrm{d}\phi_p(Ju) = N_p \times \mathrm{d}\phi_p(u), \tag{9}$$

for all nonzero $u \in T_p\Sigma$ and $p \in \Sigma$.

Daniel [14] showed that the quadruplet (A, T, J, ν), also called the *fundamental data* of the immersion, satisfies the following equations for all $X, Y \in \mathfrak{X}(\Sigma)$:

(i) $K = \det(A) + \tau^2 + (\kappa - 4\tau^2)\nu^2$,
(ii) $\nabla_X AY - \nabla_Y AX - A[X, Y] = (\kappa - 4\tau^2)(\langle Y, T\rangle X - \langle X, T\rangle Y)\nu$,
(iii) $\nabla_X T = (AX - \tau JX)\nu$
(iv) $\nabla\nu = -AT - \tau JT$,
(v) $\|T\|^2 + \nu^2 = 1$,

where K is the Gauß curvature of Σ and ∇ is its Levi-Civita connection. The identities (i) and (ii) are nothing but the Gauß and Codazzi equations. Observe that we have considered J as a fundamental datum to state explicitly that the orientation plays a key role; equivalently, Daniel assumes that Σ is oriented.

Conversely, any simply connected Riemannian surface Σ carrying a quadruplet (A, T, J, ν) that satisfies the above conditions (i)–(v) can be isometrically immersed in $\mathbb{E}(\kappa, \tau)$ with shape operator A, tangent part of the Killing T, and angle function ν, being A and ν defined with respect to the normal N compatible with J, in the sense that (9) holds true. Such an immersion is determined up orientation-preserving isometries that also preserve the orientation of vertical fibers (see [14, Thm. 4.3]). A similar discussion of the fundamental equations in terms of a conformal parameter on Σ is given in [24].

Remark 2 To understand the uniqueness, it is important to mention how other geometric transformations affect the fundamental data:

- A change of the sign of N results in $(A, T, J, \nu) \mapsto (-A, T, -J, -\nu)$.
- A composition with an orientation-preserving isometry that reverses the orientation of the fibers gives $(A, T, J, \nu) \mapsto (A, -T, J, -\nu)$.
- If $\tau = 0$, a composition with an orientation-reversing isometry that preserves the orientation of the fibers gives $(A, T, J, \nu) \mapsto (A, T, -J, \nu)$.

These transformations are also discussed in [14, Rmk. 4.12] and [27, Rmk. 3.4].

2.5 Cylinders and Multigraphs

A vertical cylinder in $\mathbb{E}(\kappa, \tau)$ is a surface Σ invariant under vertical translations, so it is foliated by vertical geodesics, and its angle function vanishes identically. This amounts to say that Σ is the preimage $\pi^{-1}(\beta)$ of a curve $\beta \subset \mathbb{M}^2(\kappa)$, from where it easily follows that the mean curvature of Σ is half of the curvature of β as a curve of $\mathbb{M}^2(\kappa)$. In particular, we define the H-cylinders as the preimages under π of complete curves of constant curvature $2H$ in $\mathbb{M}^2(\kappa)$. This implies a different behavior if they project onto circles ($4H^2+\kappa>0$), straight lines ($H=\kappa=0$), horocycles ($4H^2+\kappa=0$ and $\kappa<0$), or curves of $\mathbb{H}^2(\kappa)$ equidistant to a geodesic ($4H^2+\kappa<0$).

On the opposite side of vertical cylinders, another distinguished class of surfaces in $\mathbb{E}(\kappa,\tau)$ are the so-called vertical multigraphs, which are surfaces everywhere transverse to ξ. The angle function of a multigraph Σ never vanishes, whence the projection $\pi_{|\Sigma}$ is a local diffeomorphism. If $\pi_{|\Sigma}$ is also one-to-one, the surface is called a *vertical graph* over the domain $\pi(\Sigma)\subseteq\mathbb{M}^2(\kappa)$. Such a graph is said to be *entire* if $\pi_{|\Sigma}$ is a global diffeomorphism onto $\mathbb{M}^2(\kappa)$.

Smooth graphs over $\Omega\subseteq\mathbb{M}^2(\kappa)$ can be parametrized by smooth functions $u\in C^\infty(\Omega)$ as $F_u:\Omega\to\mathbb{E}(\kappa,\tau)$ given by $F_u(q)=\Phi_{u(q)}F_0(q)$ for all $q\in\Omega$, where $F_0:\Omega\to\mathbb{E}(\kappa,\tau)$ is a zero section and $\{\Phi_t\}_{t\in\mathbb{R}}$ is the group of vertical translations. If $\tau=0$, then $\mathbb{M}^2(\kappa)\times\{0\}$ is the natural zero section, but it is important to remark that there is no natural zero section whenever $\tau\neq 0$. Working in the model $M(\kappa,\tau)$, it is common to consider $F_0(x,y)=(x,y,0)$ as zero section and then parametrize any graph over $\Omega\subseteq\Omega_\kappa$ as $F_u(x,y)=(x,y,u(x,y))$. The mean curvature of this parametrization is given in divergence form as

$$H=\frac{1}{2}\operatorname{div}\left(\frac{Gu}{\sqrt{1+\|Gu\|^2}}\right), \tag{10}$$

where the divergence and norm are computed in the geometry of $\mathbb{M}^2(\kappa)$. The vector field $Gu\in\mathfrak{X}(\Omega)$, given by $Gu=(\frac{u_x}{\lambda_\kappa}+\tau y)\frac{\partial_x}{\lambda_\kappa}+(\frac{u_y}{\lambda_\kappa}-\tau x)\frac{\partial_y}{\lambda_\kappa}$, is also known as the *generalized gradient* of u. It does not depend on the choice of the zero section, and $\sqrt{1+\|Gu\|^2}$ is the area element of the surface (see [60, §3]).

Any complete H-multigraph in $\mathbb{E}(\kappa,\tau)$ is either a horizontal slice in $\mathbb{S}^2(\kappa)\times\mathbb{R}$ (with $H=0$) or a graph (with $4H^2+\kappa\leq 0$) over a simply connected domain of $\mathbb{M}^2(\kappa)$ bounded by curves of constant curvature $\pm 2H$ where the function defining the graph tends to $\pm\infty$ (see [69]). Such a complete H-graph must be entire if $\kappa+4H^2=0$ (see [18, Cor. 4.6.3] and the references therein). Note that by the Fernández and Mira's solution to the Bernstein problem [25] (see also [65]), there are plenty of entire graphs with critical mean curvature, generically a two-parameter family of them for each choice of a holomorphic quadratic differential on the complex plane $\mathbb{C}$ (not identically zero) or on the unit disk $\mathbb{D}\subseteq\mathbb{C}$.

If $\kappa+4H^2\leq 0$, by a standard application of the maximum principle for H-surfaces, the existence of entire H-graphs prevents the existence of compact immersed H-surfaces. However, if $\kappa+4H^2>0$, then there do exist compact H-surfaces in $\mathbb{E}(\kappa,\tau)$, e.g., the rotationally invariant H-spheres. As a matter of fact, Abresch and Rosenberg [1] solved the Hopf problem in $\mathbb{E}(\kappa,\tau)$ by showing that these are the only immersed H-spheres in $\mathbb{E}(\kappa,\tau)$. Note that some of these H-spheres are non-embedded [104, Thm. 1]. We will show in Sect. 5 that there are plenty of compact embedded H-surfaces in $\mathbb{S}^2(\kappa) \times \mathbb{R}$ which are not equivariant. It is important to remark that the classical Alexandrov problem concerning the classification of compact embedded H-surfaces has not been hitherto settled in $\mathbb{E}(\kappa, \tau)$-spaces other than $\mathbb{H}^2(\kappa) \times \mathbb{R}$ (see [43]) and $\mathbb{R}^3$, where Alexandrov's moving plane technique applies. In $\mathbb{S}^2(\kappa) \times \mathbb{R}$ and $\mathbb{S}^3_B(\kappa, \tau)$, there do exist nonspherical H-surfaces, e.g., the rotational embedded H-tori (see [86, 104] or the non-equivariant examples in Sect. 5.1 and Sect. 5.2).

All the above discussions reveal that the value of H, if any, such that $4H^2 + \kappa = 0$ is geometrically relevant, and it is called the *critical mean curvature* in $\mathbb{E}(\kappa, \tau)$. Accordingly, H-surfaces with $4H^2 + \kappa > 0$ and $4H^2 + \kappa < 0$ will be said to have *supercritical* and *subcritical* mean curvature, respectively.

3 The Conjugate Construction

Let $\kappa, \tau, H, \widetilde{\kappa}, \widetilde{\tau}, \widetilde{H} \in \mathbb{R}$ be constants such that $\kappa - 4\tau^2 = \widetilde{\kappa} - 4\widetilde{\tau}^2$ and $\tau + iH = e^{i\theta}(\widetilde{\tau} + i\widetilde{H})$ for some $\theta \in [0, 2\pi)$. Given a simply connected Riemannian surface Σ, there is an isometric correspondence between $\widetilde{H}$-immersions $\widetilde{\phi} : \Sigma \to \mathbb{E}(\widetilde{\kappa}, \widetilde{\tau})$ and H-immersions $\phi : \Sigma \to \mathbb{E}(\kappa, \tau)$ by means of the following transformation of fundamental data:

$$(A, T, J, \nu) = \big(\mathrm{Rot}_\theta \circ (\widetilde{A} - \widetilde{H}\,\mathrm{id}) + H\,\mathrm{id}, \mathrm{Rot}_\theta(\widetilde{T}), \widetilde{J}, \widetilde{\nu}\big), \tag{11}$$

where $\mathrm{Rot}_\theta = \cos(\theta)\mathrm{id} + \sin(\theta)J$ is a rotation of angle θ in the tangent bundle of Σ. The immersions $\widetilde{\phi}$ and ϕ are called *sister immersions* and determine each other up to orientation-preserving isometries that also preserve the orientation of the fibers (see [14, Prop. 4.1]). We would like to remark that the simple connectedness of Σ is not an essential assumption here, for the correspondence can be applied after considering the Riemannian universal cover of Σ.

Remark 3 (Lawson Correspondence) If $\kappa - 4\tau^2 = 0$, then Daniel correspondence reduces to a correspondence between $\widetilde{H}$-immersions in $\mathbb{M}^3(\widetilde{\tau}^2)$ and H-immersions in $\mathbb{M}^3(\tau^2)$ such that $\widetilde{H}^2 + \widetilde{\tau}^2 = H^2 + \tau^2$ in which the shape operator is rotated by an angle θ. This is a particular case of the two-parameter correspondence given by Lawson [58, Thm. 8] (see also [28, 51] for a more geometric description). If $\widetilde{\kappa} = \widetilde{\tau} = \widetilde{H} = 0$, then $\kappa = \tau = H = 0$, and Daniel correspondence reduces to the classical Bonnet associate family of minimal surfaces in $\mathbb{R}^3$.

Using the uniqueness, we can prove a result similar to [58, Prop. 2.12]:

Proposition 1 *Let $\widetilde{\phi} : \Sigma \to \mathbb{E}(\widetilde{\kappa}, \widetilde{\tau})$ and $\phi : \Sigma \to \mathbb{E}(\kappa, \tau)$ be sister immersions. Then ϕ is equivariant if and only if $\widetilde{\phi}$ is equivariant.*

Proof Assume that $\{S_t\}_{t\in\mathbb{R}}$ is a one-parameter group of isometries of $\mathbb{E}(\kappa,\tau)$ under which ϕ is invariant. Since this is a continuous group, it must preserve both the orientation and the orientation of the fibers and induces a one-parameter group $\{R_t\}_{t\in\mathbb{R}}$ of isometries of Σ such that $S_t\circ\phi=\phi\circ R_t$. The fundamental data of the immersions $\phi\circ R_t:\Sigma\to\mathbb{E}(\kappa,\tau)$ do not depend on t, whence their sister immersions have the same fundamental data as $\widetilde{\phi}\circ R_t:\Sigma\to\mathbb{E}(\widetilde{\kappa},\widetilde{\tau})$ (which does not depend on t either) in view of (11). By the uniqueness of the correspondence, we can find isometries $\widetilde{S}_t$ of $\mathbb{E}(\widetilde{\kappa},\widetilde{\tau})$ preserving the orientation and the orientation of the fibers such that $\widetilde{S}_t\circ\widetilde{\phi}=\widetilde{S}_t\circ\widetilde{\phi}\circ R_0=\widetilde{\phi}\circ R_t$ for all $t\in\mathbb{R}$. It follows that

$$\widetilde{S}_{s+t}\circ\widetilde{\phi}=\widetilde{\phi}\circ R_{s+t}=\widetilde{\phi}\circ R_s\circ R_t=\widetilde{S}_s\circ\widetilde{\phi}\circ R_t=\widetilde{S}_s\circ\widetilde{S}_t\circ\widetilde{\phi}.$$

This implies that $\widetilde{S}_{s+t}=\widetilde{S}_s\circ\widetilde{S}_t$ unless $\widetilde{\phi}$ is part of vertical cylinder or a horizontal slice. However, were it the case, then $\nu\equiv 0$ or $\nu\equiv\pm1$, and it follows that ϕ is also part of a vertical cylinder or a horizontal slice, so the statement holds true.

The continuity of the sister correspondence also follows from its uniqueness. The proof is a direct generalization of the particular case in [11, Prop. 2.3].

Proposition 2 (Continuity) *Let Σ be a smooth surface, and let $\widetilde{\phi}_n:(\Sigma,\mathrm{d}s_n^2)\to\mathbb{E}(\widetilde{\kappa},\widetilde{\tau})$ be a sequence of isometric $\widetilde{H}$-immersions that converge on the C^m-topology on compact subsets (for all $m\geq 0$) to an isometric $\widetilde{H}$-immersion $\widetilde{\phi}_\infty:(\Sigma,\mathrm{d}s_\infty^2)\to\mathbb{E}(\widetilde{\kappa},\widetilde{\tau})$. Given $\theta\in\mathbb{R}$ not depending on n, the sister H-immersions ϕ_n (up to suitable isometries of $\mathbb{E}(\kappa,\tau)$) converge in the same mode to the sister H-immersion ϕ_∞.*

Observe that $\mathbb{E}(\kappa,\tau)$ has bounded geometry in view of (4). If a sequence of H-surfaces in $\mathbb{E}(\kappa,\tau)$ has uniformly bounded second fundamental form in a neighborhood (of uniform intrinsic radius) around an accumulation point, then there is a subsequence that converges in the topology C^m for all m to some H-surface. The condition on the accumulation point amounts to translate the surfaces appropriately using the homogeneity, whereas the uniform bound of the second fundamental form usually follows from stability, as shown by Rosenberg, Souam, and Toubiana [96]. If the surfaces in the sequence are complete and stable, then their limit is also complete.

Remark 4 (Rescaling) Consider homotheties $\widetilde{\rho}_\mu:\mathbb{E}(\widetilde{\kappa},\widetilde{\tau})\to\mathbb{E}(\mu^2\widetilde{\kappa},\mu\widetilde{\tau})$ and $\rho_\mu:\mathbb{E}(\kappa,\tau)\to\mathbb{E}(\mu^2\kappa,\mu\tau)$ that multiply lengths by a constant factor μ^{-1}. In the Cartan model, these transformations are nothing but $(x,y,z)\mapsto(\frac{x}{\mu},\frac{y}{\mu},\frac{z}{\mu^2})$.

Given an isometric $\widetilde{H}$-immersion $\widetilde{\phi}:(\Sigma,\mathrm{d}s_{\widetilde{\Sigma}}^2)\to\mathbb{E}(\widetilde{\kappa},\widetilde{\tau})$ with fundamental data $(\widetilde{A},\widetilde{T},J,\nu)$, we have that $\widetilde{\rho}_\mu\circ\widetilde{\phi}:(\Sigma,\mu^{-2}\mathrm{d}s_{\widetilde{\Sigma}}^2)\to\mathbb{E}(\mu^2\widetilde{\kappa},\mu\widetilde{\tau})$ is an isometric $(\mu\widetilde{H})$-immersion with fundamental data $(\mu\widetilde{A},\mu\widetilde{T},J,\nu)$. Given a sister H-immersion $\phi:\Sigma\to\mathbb{E}(\kappa,\tau)$ for some phase angle $\theta\in[0,2\pi)$ and reasoning likewise, the fundamental data of $\rho_\mu\circ\phi$ is $(\mu A,\mu T,J,\nu)$. By looking at (11), we deduce that $\widetilde{\rho}_\mu\circ\widetilde{\phi}$ is the sister of $\rho_\mu\circ\phi$ for the same value of θ, that is, the sister correspondence commutes with rescaling the metrics.

Other geometric objects in the theory of H-surfaces in $\mathbb{E}(\kappa, \tau)$ also behave nicely with respect to the correspondence. For instance, at the conformal level, the Abresch-Rosenberg holomorphic quadratic differentials [1] of sister immersions are related by $\widetilde{Q} = e^{-2i\theta} Q$. Also, the harmonic Gauß maps into hyperbolic space spanning these differentials in the case of multigraphs of critical mean curvature are associate for sister immersions (see [16, Prop. 5.6]). We also point out that sister immersions define the same stability operator given by

$$L = \Delta - 2K + 4H^2 + \kappa + (\kappa - 4\tau^2)\nu^2, \tag{12}$$

where Δ stands for the Laplacian on Σ (see [14, Prop. 5.11]).

As for the construction of H-surfaces, we will focus on the case $\theta = \frac{\pi}{2}$ and $\widetilde{H} = 0$, which gives $\widetilde{\kappa} = 4H^2 + \kappa$ and $\widetilde{\tau} = H$, i.e., we can associate an H-immersion $\phi : \Sigma \to \mathbb{M}^2(\kappa) \times \mathbb{R}$ to any minimal immersion $\widetilde{\phi} : \Sigma \to \mathbb{E}(4H^2 + \kappa, H)$. Under these assumptions, we will call ϕ and $\widetilde{\phi}$ *conjugate* immersions in the sequel. They fulfill some additional properties that will allow us to control the geometry of ϕ in terms of the geometry of $\widetilde{\phi}$ and will be discussed in Sect. 3.1. Recall that we will use the tilde notation for the minimal surface (the minimal surface we will begin with), because ϕ (without tilde) will stand for our target surface in a product space. Conjugation is characterized in terms of the fundamental data $(A, T, J, \nu) = (J \circ \widetilde{A} + H\,\mathrm{id}, J\widetilde{T}, J, \widetilde{\nu})$, plus the orientations satisfy the compatibility relations $\mathrm{d}\phi_p(Ju) = N \times \mathrm{d}\phi_p(u) = N^* \times \mathrm{d}\phi^*_p(u)$ for all $u \in T_p\Sigma$.

Table 3 shows the different possible configurations and explains why H-surfaces with critical, supercritical, and subcritical mean curvature in $\mathbb{M}^2(\kappa) \times \mathbb{R}$ display qualitatively different behaviors, as discussed in Sect. 2.5. Their minimal isometric conjugate surfaces lie in Berger spheres (supercritical case), in the Heisenberg group (critical case), in $\widetilde{\mathrm{SL}}_2(\mathbb{R})$ (subcritical case), or in $\mathbb{H}^2 \times \mathbb{R}$ and $\mathbb{S}^2 \times \mathbb{R}$ (minimal case). In this last case, there is a notion of associate family of minimal immersions, discovered by Hauswirth, Sa Earp, and Toubiana [38, Cor. 10]. They found that conjugate surfaces come from conjugate harmonic maps as in the aforesaid case of critical mean curvature. This justifies the use of the term *conjugation* in the general context.

Table 3 Possible configurations for conjugate surfaces in product spaces

	Produces an H-surface in		
A minimal surface in	$\mathbb{S}^2(\kappa) \times \mathbb{R}$	$\mathbb{H}^2(\kappa) \times \mathbb{R}$	$\mathbb{R}^3$
$\mathbb{S}^3_{\mathrm{B}}(4H^2 + \kappa, H)$	$H > 0$	$H > \sqrt{-\kappa}/2$	$H > 0$
Nil_3	–	$H = \sqrt{-\kappa}/2$	–
$\widetilde{\mathrm{SL}}_2(\mathbb{R})(4H^2 + \kappa, H)$	–	$0 < H < \sqrt{-\kappa}/2$	–
$\mathbb{H}^2(\kappa) \times \mathbb{R}$	–	$H = 0$	–
$\mathbb{S}^2(\kappa) \times \mathbb{R}$	$H = 0$	–	–
$\mathbb{R}^3$	–	–	$H = 0$

Remark 5 We could also have chosen $\theta = -\frac{\pi}{2}$, but it would lead to an isometric surface in $\mathbb{E}(\kappa, -\tau)$. Note that composition with an isometry from $\mathbb{E}(\kappa, \tau)$ to $\mathbb{E}(\kappa, -\tau)$ that preserves the orientation of the fibers produces the change of fundamental data $(A, T, J, \nu) \mapsto (-A, -T, J, \nu)$ and also the change of parameters $(\kappa, \tau, H) \mapsto (\kappa, -\tau, -H)$. This is equivalent to adding $\pm\pi$ to θ in (11).

3.1 Conjugate Curves

Let $\widetilde{\phi} : \Sigma \to \mathbb{E}(4H^2+\kappa, H)$ be a minimal immersion, and let $\phi : \Sigma \to \mathbb{M}^2(\kappa)\times\mathbb{R}$ be its conjugate H-immersion. We recall that the shape operators A and $\widetilde{A}$ of ϕ and $\widetilde{\phi}$ are related by $A = J\widetilde{A} + H\,\mathrm{id}$ (see (11)).

3.1.1 Vertical and Horizontal Geodesics

Given $\alpha : [0, \ell] \to \Sigma$ a unit-speed regular curve, we shall investigate the relation between conjugate curves $\widetilde{\gamma} = \widetilde{\phi}\circ\alpha$ and $\gamma = \phi\circ\alpha$. More precisely, we will collect some results to understand the geometry of γ whenever $\widetilde{\gamma}$ is a vertical or a horizontal geodesic. The proof of the following two lemmas, sometimes in particular cases, is scattered across the references [10, 11, 66, 68, 70, 90, 105]:

Lemma 2 (Horizontal Geodesics) *If $\widetilde{\gamma}$ is a horizontal geodesic segment, then γ lies in a totally geodesic vertical plane P that the immersion ϕ meets orthogonally:*

(a) *If $\gamma = (\beta, z) \in \mathbb{M}^2(\kappa)\times\mathbb{R}$ is decomposed component-wise, then $|z'| = \sqrt{1-\nu^2}$ and $\|\beta'\| = |\nu|$. In particular, z and β might fail to be locally one-to-one only around points where $\nu = \pm 1$ and $\nu = 0$, respectively (see Fig. 2).*
(b) *Write $\widetilde{N}_{\gamma(t)} = \cos(\theta(t))X(t) + \sin(\theta(t))\widetilde{\xi}_{\widetilde{\gamma}(t)}$ for some $\theta \in C^\infty([0, \ell])$, where $\{\widetilde{\gamma}', \widetilde{\xi}, X\}$ is a positively oriented orthonormal frame along $\widetilde{\gamma}$. The curve γ has geodesic curvature $\kappa_g^P = \theta'$ as a curve of P with respect to N as conormal.*

Proof We will provide the proof of item (b), which is not in the literature. Since $\overline{\nabla}_{\widetilde{\gamma}'}\widetilde{\xi} = H\widetilde{\gamma}'\times\widetilde{\xi} = HX$ and $\overline{\nabla}_{\widetilde{\gamma}'}X = \overline{\nabla}_{\widetilde{\gamma}'}(\widetilde{\gamma}'\times\widetilde{\xi}) = -H\widetilde{\xi}$ (see (1)), we easily compute $\overline{\nabla}_{\widetilde{\gamma}'}\widetilde{N} = (\theta' - H)\widetilde{N}\times\widetilde{\gamma}'$. This implies that

$$\begin{aligned}\theta' - H &= \langle\overline{\nabla}_{\widetilde{\gamma}'}\widetilde{N}, \widetilde{N}\times\widetilde{\gamma}'\rangle = -\langle\widetilde{A}\alpha', J\alpha'\rangle = \langle JA\alpha' - HJ\alpha', J\alpha'\rangle \\ &= \langle A\alpha', \alpha'\rangle - H = -\langle\overline{\nabla}_{\gamma'}N, \gamma'\rangle - H = \langle\overline{\nabla}_{\gamma'}\gamma', N\rangle - H = \kappa_g^P - H.\end{aligned}$$

Lemma 3 (Vertical Geodesics) *If $\widetilde{\gamma}$ is a vertical segment, then γ is contained in a totally geodesic horizontal slice $P = \mathbb{M}^2(\kappa)\times\{t_0\}$ that ϕ meets orthogonally.*

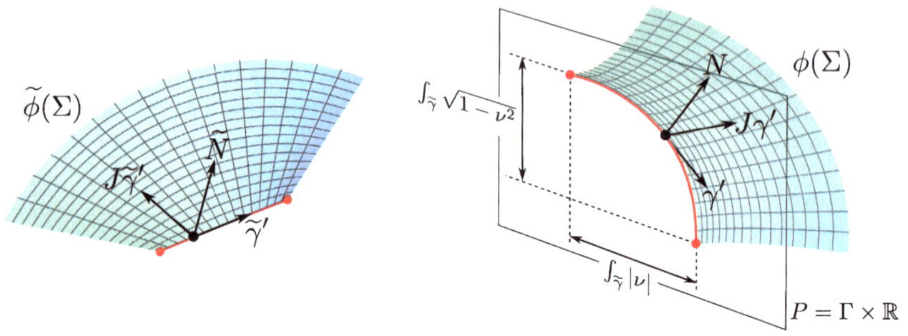

Fig. 2 Conjugate surfaces $\widetilde{\phi}(\Sigma)$ and $\phi(\Sigma)$ in a neighborhood of a horizontal geodesic. The quantities $\int_{\widetilde{\gamma}} |\nu|$ and $\int_{\widetilde{\gamma}} \sqrt{1-\nu^2}$ indicate the lengths of the projections of γ to the factors $\mathbb{M}^2(\kappa)$ and $\mathbb{R}$, respectively (with multiplicity)

Assume that $\widetilde{\gamma}' = \widetilde{\xi}$, and write $\widetilde{N}_{\widetilde{\gamma}(t)} = \cos(\theta(t))X_1(t) + \sin(\theta(t))X_2(t)$ for some $\theta \in C^\infty([0, \ell])$, where $\{X_1, X_2, \xi\}$ is a positively oriented orthonormal frame along $\widetilde{\gamma}$ such that X_1 and X_2 project to constant tangent vectors of $\mathbb{M}^2(4H^2+\kappa)$:

(a) *The conjugate curve γ has geodesic curvature $\kappa_g^P = 2H - \theta'$ as a curve of P with respect to N as normal.*
(b) *Assume that $\alpha \subset \partial R$ for some region $R \subseteq \Sigma$ on which $\nu > 0$, and let $\widetilde{\Omega} \subset \mathbb{M}^2(4H^2 + \kappa)$ and $\Omega \subset \mathbb{M}^2(\kappa)$ be the (possibly non-embedded) domains over which $\widetilde{\phi}(R)$ and $\phi(R)$ are multigraphs, respectively:*

- *If $\theta' > 0$, then $J\widetilde{\gamma}'$ (resp. $J\gamma' = \xi$) is a unit outer conormal to $\widetilde{\phi}$ (resp. ϕ) along $\widetilde{\gamma}$ (resp. γ), N points to the interior of Ω along γ, and $\phi(R)$ lies in $\mathbb{M}^2(\kappa) \times (-\infty, t_0]$ locally around γ (see Fig. 3, top).*
- *If $\theta' < 0$, then $J\widetilde{\gamma}'$ (resp. $J\gamma' = \xi$) is a unit inner conormal to $\partial\widetilde{\phi}$ (resp. ϕ) along $\widetilde{\gamma}$ (resp. γ), N points to the exterior of Ω along γ, and $\phi(R)$ lies in $\mathbb{M}^2(\kappa) \times [t_0, +\infty)$ locally around γ (see Fig. 3, bottom).*

The function $\theta \in C^\infty([0, \ell])$ in Lemmas 2 and 3 will be called the *angle of rotation* of $\widetilde{N}$ along $\widetilde{\gamma}$. In general, it is impossible to obtain θ explicitly, but $\theta(\ell) - \theta(0) = \int_0^\ell \theta'(t)\mathrm{d}t$ gives useful information because it is related to $\int_\gamma \kappa_g^P$, the total geodesic curvature of γ by means of the formulas in the above statements. It is also important to remark that if P is a vertical plane (Lemma 2), then P is flat, and $\int_\gamma \kappa_g^P$ is the total rotation of the normal of γ as a curve of P. However, if P is a horizontal slice (Lemma 3), then P has constant curvature κ, and there is no clear geometric interpretation of $\int_\gamma \kappa_g^P$ if $\kappa \neq 0$. Gauß-Bonnet theorem offers some information, as we will discuss in the constructions.

Remark 6 If $\widetilde{\gamma}$ is vertical, the curvature κ_g^P can be related to the rotation angle of γ with respect to foliations of $\mathbb{M}^2(\kappa)$ by curves. This idea was devised by Plehnert [89, Lem. 4.7] for a foliation of $\mathbb{H}^2$ by horocycles, but alike formulas show up for other foliations of $\mathbb{M}^2(\kappa)$ by curves of constant curvature.

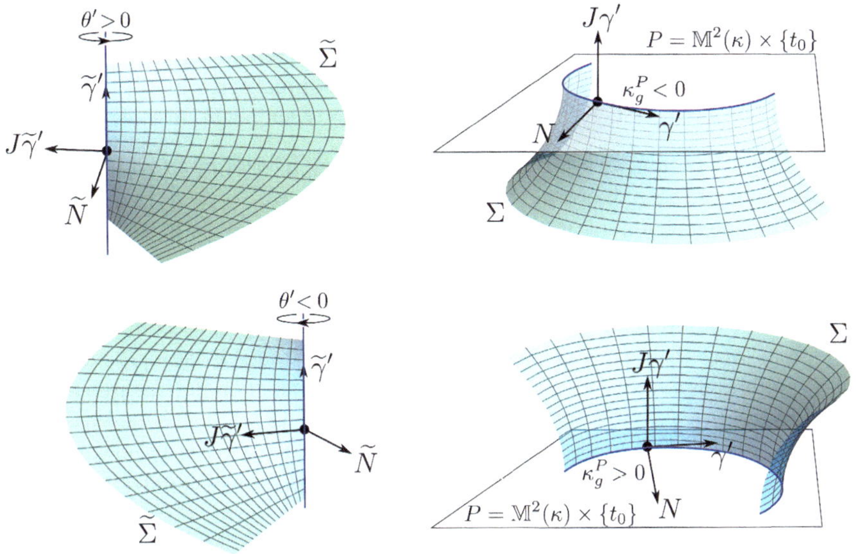

Fig. 3 Orientation of conjugate surfaces $\widetilde{\phi}(\Sigma)$ and $\phi(\Sigma)$ according to the direction of rotation of $\widetilde{N}$ along a vertical geodesic $\widetilde{\gamma}$ such that $\widetilde{\gamma}' = \widetilde{\xi}$. Recall that this is true on the boundary of a region where $\nu > 0$, but we can obtain a similar picture if $\nu < 0$ after changing some signs

In the halfspace model of $\mathbb{H}^2(\kappa) \times \mathbb{R}$, we can use the frame $\{E_1, E_2, E_3\}$ given by (7) to express

$$\gamma'(t) = \cos(\psi(t))E_1 + \sin(\psi(t))E_2, \tag{13}$$

where we call $\psi \in C^\infty([0, \ell])$ the *angle of rotation* of γ with respect to a foliation by horocycles (these horocycles are the integral curves of E_1). Using (8) and taking derivatives in (13), we get $\overline{\nabla}_{\gamma'}\gamma' = (\psi' + \sqrt{-\kappa}\cos(\psi))(-\sin(\psi)E_1 + \cos(\psi)E_2)$. On the other hand, as the surface meets P orthogonally, we infer that $N = \gamma' \times J\gamma' = \gamma' \times E_3 = \sin(\psi)E_1 - \cos(\psi)E_2$, so we reach the desired formula

$$\kappa_g^P = \langle \overline{\nabla}_{\gamma'}\gamma', N\rangle = -\psi' - \sqrt{-\kappa}\cos(\psi).$$

This also makes sense in the limit case $\kappa = 0$, in which ψ becomes the angle rotation of the normal of γ as a plane curve, whose derivative is well known to agree with the curvature of γ.

3.1.2 The Control of the Angle Function

Lemma 2 exposes the convenience of knowing at which points the angle function takes values 0 or ± 1 in order to obtain geometric information about the conjugate curves. We collect here some ideas that prove useful when dealing with this type of analysis in different constructions. They are complementary to the maximum principles (in the interior or in the boundary) for minimal surfaces. To simplify the notation, we will denote by $\widetilde{\Sigma}$ and Σ the (immersed) conjugate surfaces:

- **Strategy 1** (points where $\nu = 0$). If $p \in \widetilde{\Sigma}$ is an interior point with $\nu(p) = 0$, then compare with the vertical cylinder T_p tangent to Σ at p.

  The intersection of minimal surfaces consists of a set of regular curves meeting transversally at some isolated points, with at least four rays emanating from tangency points. More generally, there are $2n + 2$ rays emanating from a point with order of contact n, i.e., where all derivatives of both surfaces (as graphs over their tangent planes) coincide up to order n (see [79, Lem. 2]). Therefore, this strategy consists in reaching a contradiction if there are too many rays emanating from p and ending in $\partial\widetilde{\Sigma} \cap T_p$, which generically consists of very few points. However, we must ensure that they cannot enclose regions either of $\widetilde{\Sigma}$ or in T_p by some maximum principle at the interior or at infinity. Let us also comment about two variants of this strategy:

  - Comparing with T_p also gives information if $p \in \partial\widetilde{\Sigma}$ by applying the boundary maximum principle or by considering p as an interior point after extending the surface across that boundary.
  - In the case there are two or more curves where $\nu = 0$ meeting at p, then $\nabla\nu(p) = 0$, and there are at least six rays in $\widetilde{\Sigma} \cap T_p$ emanating from p (see [72, Lem. 3.5]). Note that ν lies in the kernel of the stability operator given by (12), which is a second-order linear elliptic operator, and hence its zeros also form set of regular curves intersecting transversally at some isolated points (see [4] and the references therein).

- **Strategy 2** (points where $\nu = \pm 1$). The is pretty similar to Strategy 1, but we compare with the umbrella $\mathcal{U}_p$ centered at an interior or boundary point $p \in \widetilde{\Sigma}$ where $\nu(p) = \pm 1$ (see Example 3). Since $\mathcal{U}_p$ is tangent to $\widetilde{\Sigma}$, at least four rays in $\mathcal{U}_p \cap \widetilde{\Sigma}$ arise from p, and we have to figure out where they end.

  If p belongs to a horizontal boundary component $\widetilde{h} \subset \partial\widetilde{\Sigma}$, then it is often useful to compare with the invariant surface $\mathcal{I}_{\widetilde{h}}$ defined in Example 4, whose angle is constant 1 along $\widetilde{h}$. In particular, some interior arc in $\mathcal{I}_{\widetilde{h}} \cap \widetilde{\Sigma}$ emanates from each point of $\widetilde{h}$ where $\nu = \pm 1$.
- **Strategy 3** (comparison along the boundary). Sometimes, it is possible to find a minimal surface B (also called a *barrier*) that contains a horizontal component $\widetilde{h} \subset \partial\widetilde{\Sigma}$ and stays (locally around $\widetilde{h}$) at one side of $\widetilde{\Sigma}$. The boundary maximum principle imposes a restriction on the normal $\widetilde{N}$ of $\widetilde{\Sigma}$ along $\widetilde{h}$. If we are able to control the points where $\nu = 0$ or $\nu = \pm 1$, the existence of such a barrier often translates into estimates for the angle function ν.

  This might become a bit subtle. For instance, if we know that $\widetilde{\Sigma}$ and B are multigraphs with angle functions $\nu > 0$ and $\nu_B > 0$, respectively, and Σ stays above B along $\widetilde{h}$, then this does not necessarily imply that $\nu_B < \nu$ or $\nu < \nu_B$. However, if we know additionally that the invariant surface $\mathcal{I}_{\widetilde{h}}$ stays below B (resp. above $\widetilde{\Sigma}$), then it follows that $0 < \nu < \nu_B$ (resp. $\nu_B < \nu < 1$).

Note that this strategy also applies to vertical geodesics, in which case we obtain an estimate for the angle of rotation of the normal (instead of the angle function), which is bounded by the angle of rotation of the barrier.

3.1.3 Completion and Embeddedness

A minimal surface in $\mathbb{E}(\kappa, \tau)$ containing a horizontal or a vertical geodesic segment in the interior must be axially symmetric with respect to that segment, which follows from the work of Leung [61]. Sa Earp and Toubiana [99] have also proved that a minimal surface which has a vertical or horizontal geodesic segment in the boundary and is of class C^1 up to that boundary can be analytically extended by axial symmetry. By comparing the fundamental data of axially symmetric surfaces and mirror symmetric surfaces (see Remark 2), it is not difficult to show that a minimal surface in $\mathbb{E}(4H^2 + \kappa, H)$ axially symmetric with respect to a geodesic segment γ has a conjugate H-surface in $\mathbb{M}^2(\kappa) \times \mathbb{R}$ mirror symmetric with respect to the totally geodesic surface P containing the conjugate curve γ. Also, if two boundary components of $\partial\Sigma$ meet in a vertex with angle $\frac{\pi}{k}$ for some $k \geq 2$, then we can apply successive reflections, and the possible singularity at the vertex can be removed by means of a result of Choi and Schoen [12, Prop. 1]. All in all, we can state the following result:

Proposition 3 *Let $\widetilde{\phi} : \Sigma \to \mathbb{E}(4H^2+\kappa, H)$ be a minimal immersion with boundary a geodesic polygon consisting of vertical and horizontal geodesic arcs (of finite or infinite length). Assume that $\widetilde{\phi}$ is of class C^1 up to the interior of each boundary component and the interior angle at each vertex is an integer divisor of π:*

(a) *The immersion $\widetilde{\phi}$ can be extended to a complete minimal immersion $\widetilde{\phi}^*$ by successive axial symmetries about the boundary components.*
(b) *The conjugate immersion $\phi : \Sigma \to \mathbb{M}^2(\kappa) \times \mathbb{R}$ can be extended to a complete H-immersion ϕ^* by successive mirror symmetries about the boundary components.*

Once a complete surface is obtained by successive mirror symmetries, it is natural to investigate whether it is embedded or not. This is one of the toughest problems in conjugate constructions, because it must be solved just in terms of the prescribed geodesic polygon, since none of the immersions $\widetilde{\phi}$ and ϕ are explicitly known. It is often convenient to relax this assumption and consider *Alexandrov embeddedness* instead, which means that the immersed surface can be recovered as the boundary of a three-manifold immersed (but possibly non-embedded) in $\mathbb{E}(\kappa, \tau)$. To simplify future discussions, we will distinguish the two possible situations where non-embeddedness occurs in conjugate constructions:

- **Type I self-intersections:** The immersion ϕ is not an embedding, i.e., the fundamental piece has self-intersections.
- **Type II self-intersections:** The immersion ϕ is an embedding, but the completion ϕ^* has self-intersections (see Figs. 10 and 19).

On the one hand, self-intersections of type I can be prevented if the boundary of ϕ projects one-to-one to $\mathbb{M}^2(\kappa)$ by a standard application of the maximum principle (see also Proposition 6). Convexity has also something to say in this respect; e.g., the following result for multigraphs follows from the maximum principle and from Gauß-Bonnet theorem and assumes that the angle at one vertex is convex to prove that the conjugate vertical geodesic is embedded.

Proposition 4 ([11, Lem. 2.1]) *Let $\widetilde{\phi} : \Sigma \to \mathbb{E}(4H^2 + \kappa, H)$ and $\phi : \Sigma \to \mathbb{H}^2(\kappa) \times \mathbb{R}$ be conjugate multigraphs, where $\kappa < 0$ and $4H^2 + \kappa \leq 0$. Assume that $\nu > 0$ and $\widetilde{\gamma}$ is a vertical geodesic such that $\widetilde{\gamma}' = \xi$. If $\theta' > 0$ and $\int_{\widetilde{\gamma}} \theta' \leq \pi$, then γ is embedded.*

In the case $H = 0$, Hauswirth, Sa Earp, and Toubiana [38, Thm. 14] extended a theorem of Krust in $\mathbb{R}^3$ (unpublished, see [54, Thm. 2.4.1]) exploiting the convexity of the domain. Notice that a similar property does not hold true in general if $H > 0$, as we shall discuss in Sect. 6.1.

Proposition 5 (Krust Property) *Let $\widetilde{\phi} : \Sigma \to \mathbb{M}^2(\kappa) \times \mathbb{R}$ and $\phi : \Sigma \to \mathbb{M}^2(\kappa) \times \mathbb{R}$ be sister minimal immersions. If $\kappa \leq 0$ and $\widetilde{\phi}$ is a graph over some convex domain $\Omega \subset \mathbb{M}^2(\kappa)$, then ϕ is a graph (and hence embedded).*

On the other hand, self-intersections of type II are usually treated by an a priori subdivision of the target space $\mathbb{M}^2(\kappa) \times \mathbb{R}$ in disjoint congruent regions bounded by horizontal and vertical planes, so that the fundamental piece fits in one of those regions with its boundary lying in the boundary planes. If the desired symmetries and the tessellation can be chosen properly, then there will be one copy of the fundamental piece on each region, so that no self-intersections will be produced by reflection. The constructions in Sect. 5 and Sect. 6 follow this philosophy.

3.2 *Some Classes of Surfaces Preserved by the Sister Correspondence*

Daniel correspondence behaves really well with respect to many geometric conditions, which enables the translation of some problems for H-surfaces in some $\mathbb{E}(\kappa, \tau)$ to the same problem in another space. The equivariance (see Prop. 1) and the stability (see (12)) are good examples of this. Note that any condition that is purely intrinsic to the H-surface is also preserved because the correspondence is isometric, e.g., the total curvature or the area growth of intrinsic balls.

3.2.1 Cylinders and Multigraphs

Vertical H-cylinders and vertical H-multigraphs are characterized by the conditions $\nu \equiv 0$ and $\nu \neq 0$, respectively. Since ν is preserved by the sister correspondence, then so are these families of H-surfaces.

Complete multigraphs in $\mathbb{E}(\kappa, \tau)$ are indeed graphs [69, Thm. 1], so in particular complete graphs also form a preserved class. However, it is not true in general that a sister surface of a graph is a graph, as we will discuss later in Sect. 6.1. It is worth highlighting that any sister surface of an entire graph with critical mean curvature is again an entire graph [18, Cor. 4.6.4], though this is not true in general in the subcritical case, as the following example shows:

Example 1 Let $0 \leq H < \frac{1}{2}$ and consider the half-space model for $\mathbb{H}^2 \times \mathbb{R}$. The surface P_0 given by $x^2 + y^2 = 1$ is a vertical plane that can be parametrized isometrically as $(r, h) \mapsto (\tanh(r), \operatorname{sech}(r), h)$, where r is the hyperbolic distance to $(1, 0)$ in $\mathbb{H}^2$ and h is the projection onto the factor $\mathbb{R}$. Consider the surface Σ parametrized by $(t, r) \mapsto \big(e^t \tanh(r), e^t \operatorname{sech}(r), h(r)\big)$, where

$$h(r) = \frac{1}{\sqrt{1-4H^2}} \operatorname{arcsinh}\left(\frac{\sqrt{1-4H^2} - 2H\sinh(r)}{2H + \sqrt{1-4H^2}\sinh(r)}\right) + \frac{2Hr}{\sqrt{1-4H^2}}$$

for $t \in \mathbb{R}$ and $r > -2\operatorname{arctanh}(2H)$. This surface is invariant by the hyperbolic translations $(x, y, z) \mapsto (e^t x, e^t y, z)$ and has constant mean curvature H. If $H > 0$, then $h(r)$ tends to $+\infty$ as $r \to -2\operatorname{arctanh}(2H)$ or $r \to +\infty$, so Σ is a complete graph over the subset of $\mathbb{H}^2$ of points on the concave side of a complete curve α of constant curvature $2H$. If $H = 0$, then Σ is the graph over a half-plane discovered by Sa Earp [98, Eqn. (32)], in which case $h(r) \to 0$ as $r \to +\infty$. Either way, Σ is mirror symmetric about the vertical planes P_t of equation $x^2 + y^2 = e^{2t}$ for all $t \in \mathbb{R}$, whence Σ is foliated by the congruent curves $\gamma_t = P_t \cap \Sigma$.

The conjugate minimal surface $\widetilde{\Sigma}$ in $\mathbb{E}(4H^2 - 1, H)$ is therefore foliated by the horizontal geodesics $\widetilde{\gamma}_t$. Note that two curves γ_t and γ_s lie at bounded distance from each other and this distance is realized asymptotically at their endpoints in $\alpha \times \{+\infty\}$. Thus, the one-parameter group of hyperbolic translations that leave Σ invariant corresponds to a one-parameter group of parabolic screw motions that leave $\widetilde{\Sigma}$ invariant. In particular, we deduce that $\widetilde{\Sigma}$ is the entire minimal graph associated with the function $u(x, y) = lx$ for some $l \in \mathbb{R}$, also called a *tilted plane* (see [9, §3]).

We remark that if $H > 0$, then changing H to $-H$ also gives a noncongruent H-graph on the convex side of the equidistant curve α. Up to ambient isometries, there are only three tilted planes according to $l \in \{-1, 0, 1\}$. The complete H-graphs that project onto the concave and convex sides of α should correspond to the cases $l = \pm 1$, because the tilted plane with $l = 0$ is the parabolic helicoid $P_{0,4H^2+\kappa,H}$ so its conjugate H-surface in $\mathbb{H}^2 \times \mathbb{R}$ is the entire graph $P_{H,-1,0}$. These last examples will be discussed in Sect. 3.2.2.

Given an immersion $\phi : \Sigma \to \mathbb{E}(\kappa, \tau)$, the Jacobian determinant of the projection $\pi|_\Sigma : \Sigma \to \mathbb{M}^2(\kappa)$ satisfies $|\mathrm{Jac}(\pi|_\Sigma)| = |\nu|$. By a simple change of variable, if $\nu \in L^1(\Sigma)$, this implies that $\int_\Sigma |\nu|$ is the area of $\pi(\Sigma)$ taking into account its possible multiplicity. In particular, this area is preserved by the correspondence.

Example 2 ([67, §4.3]) Assume that $\kappa + 4H^2 < 0$. An ideal Scherk graph in $\mathbb{E}(\kappa, \tau)$ is an H-graph defined over an ideal polygon $\Omega \subset \mathbb{H}^2(\kappa)$ whose boundary consists of $2n$ curves (with common ideal endpoints) and constant curvature alternatively equals to $2H$ and $-2H$ with respect to the inward-pointing normal to $\partial\Omega$. Some additional Jenkins-Serrin conditions on Ω are needed to ensure that such a graph exists, but this is not our point here. Ideal Scherk graphs can be characterized as the only complete H-multigraphs whose projection has finite area [67, Prop. 3], in which case we have $\text{Area}(\Omega) = \frac{2(n-1)\pi}{-\kappa-4H^2}$. Therefore, the class of ideal Scherk graphs is preserved by the sister correspondence, and so is the number $2n$ of boundary components.

3.2.2 Surfaces with Zero Abresch-Rosenberg Differential

If Σ is topologically a sphere, then it is simply connected, and the correspondence applies globally [14, Ex. 5.16]. This means that H-spheres in $\mathbb{E}(\kappa, \tau)$, all of which are equivariant [1], form a preserved class of surfaces. More generally, a surface Σ in $\mathbb{E}(\kappa, \tau)$ has zero Abresch-Rosenberg differential if and only if the following function $q \in C^\infty(\Sigma)$ identically vanishes (see [23, Lem. 2.2]):

$$\frac{q}{\kappa - 4\tau^2} = \frac{1}{4}\left(\frac{4H^2+\kappa}{\kappa-4\tau^2} - \nu^2\right)\left(4H^2+\kappa+3(\kappa-4\tau^2)\nu^2 - 4K\right) - \|\nabla\nu\|^2, \tag{14}$$

whence q is trivially preserved by the correspondence. In [1, 23, 59], it is shown that surfaces with $q \equiv 0$ are equivariant, and in [19, Prop. 2.4], the sister correspondence was employed to prove that, except for vertical cylinders with critical mean curvature, they belong to one of the following three families:

- The rotationally invariant surfaces $S_{H,\kappa,\tau}$, locally given in $M(\kappa, \tau)$ by

$$X(u, v) = \left(v\cos(u), v\sin(u), \int_0^v \frac{-4Hs\sqrt{1+\tau^2s^2}\,\mathrm{d}s}{(4+\kappa s^2)\sqrt{1-H^2s^2}}\right). \tag{15}$$

  This is half of an H-sphere if $4H^2 + \kappa > 0$ or an entire H-graph otherwise.
- The screw-motion invariant surfaces $C_{H,\kappa,\tau}$ in the case $4H^2 + \kappa < 0$. They are complete H-surfaces globally parametrized in the model $M(\kappa, \tau)$ by

$$X(u, v) = \left(v\cos(u), v\sin(u), \frac{4\tau}{\kappa}u \pm \int_{\frac{4H}{|\kappa|}}^v \frac{16H\sqrt{16\tau^2+\kappa^2s^2}\,\mathrm{d}s}{\kappa s(4+\kappa s^2)\sqrt{\kappa^2s^2-16H^2}}\right), \tag{16}$$

  The family $C_{H,\kappa,\tau}$ contains helicoids ($H = 0$) and rotational catenoids ($\tau = 0$).

- The parabolic helicoids $P_{H,\kappa,\tau}$ in the case $4H^2 + \kappa < 0$. They are the entire H-graphs given in the halfspace model of $\mathbb{E}(\kappa, \tau)$ by the global parametrization

$$X(u, v) = (u, v, a\log(v)), \qquad \text{with } a = \frac{2H\sqrt{-\kappa + 4\tau^2}}{-\kappa\sqrt{-4H^2 - \kappa}}. \tag{17}$$

Each of the families $S_{H,\kappa,\tau}$, $C_{H,\kappa,\tau}$, and $P_{H,\kappa,\tau}$ is preserved by the correspondence (see [19, Rmk. 2.5]). On the other hand, the parabolic helicoids $P_{H,\kappa,\tau}$ along with vertical planes and horizontal slices are the only elements of the following classes:

- Isoparametric H-surfaces
- H-surfaces with constant principal curvatures
- H-surfaces which are homogeneous by ambient isometries
- H-surfaces with constant angle function (see [23, Thm.2.2])

Therefore, each of these geometric conditions is preserved by the correspondence. We remark that the family of H-surfaces with constant Gauß curvature is also preserved (the Gauß curvature is intrinsic), but it contains a few inhomogeneous examples (see [15, Thm. 3.3 and 4.6]).

3.2.3 Ruled Minimal Surfaces

Given a minimal immersion $\widetilde{\phi} : \Sigma \to \mathbb{E}(4H^2 + \kappa, H)$ ruled by horizontal geodesics, Lemma 2 implies that the conjugate H-immersion $\phi : \Sigma \to \mathbb{M}^2(\kappa) \times \mathbb{R}$ is foliated by lines of mirror symmetry lying in vertical planes. It easily follows that ϕ and $\widetilde{\phi}$ are equivariant, which establishes an isometric conjugation between minimal surfaces ruled by horizontal geodesics and H-surfaces in product spaces invariant by one-parameter groups of isometries acting in the horizontal direction (i.e., such that they fix the factor $\mathbb{R}$). Here, we will focus on a few examples that will play the role of barriers in our constructions and that illustrate how conjugation works. Ruled minimal surfaces in any $\mathbb{E}(\kappa, \tau)$-space have been classified by Kim et al. in [57].

Example 3 (The Umbrella $\mathcal{U}_p$) Given $p \in \mathbb{E}(\widetilde{\kappa}, \widetilde{\tau})$, the umbrella $\mathcal{U}_p$ is the union of all horizontal geodesics through p. It follows that $\mathcal{U}_p$ coincides with the rotationally invariant surface $S_{0,\widetilde{\kappa},\widetilde{\tau}}$ defined in Sect. 3.2.2, whence it is an entire minimal graph if $\widetilde{\kappa} \leq 0$ and a minimal sphere if $\widetilde{\kappa} > 0$. Note that $\mathcal{U}_p$ is everywhere horizontal if $\widetilde{\tau} = 0$. If $\widetilde{\tau} \neq 0$, it is horizontal only at its center p if $\widetilde{\kappa} \leq 0$ and at p and its *antipodal* point if $\widetilde{\kappa} > 0$. In the model $M(\widetilde{\kappa}, \widetilde{\tau})$, the umbrella centered at $p = (0, 0, 0)$ contains the graph of the function $u(x, y) = 0$, though this is not the complete umbrella if $\widetilde{\kappa} > 0$.

From Sect. 3.2.2, we deduce that the conjugate H-surface of an umbrella in $\mathbb{E}(4H^2 + \kappa, H)$ is the rotationally invariant surface $S_{H,\kappa,0}$ in $\mathbb{M}^2(\kappa) \times \mathbb{R}$, which is an H-sphere if $4H^2 + \kappa > 0$ or an entire H-graph otherwise.

Example 4 (The Invariant Surface $I_{\widetilde{\gamma}}$) Given a horizontal geodesic $\widetilde{\gamma}$ in $\mathbb{E}(\widetilde{\kappa},\widetilde{\tau})$, let $I_{\widetilde{\gamma}}$ be the surface formed by all horizontal geodesics orthogonal to $\widetilde{\gamma}$. It follows that $I_{\widetilde{\gamma}}$ is an entire minimal graph if $\widetilde{\kappa}\leq 0$ or $\widetilde{\tau}=0$; otherwise, it is a minimal embedded torus that will be described as a spherical helicoid in Example 5. The surface $I_{\widetilde{\gamma}}$ is invariant by translations along $\widetilde{\gamma}$, as in Lemma 1. If $\widetilde{\gamma}$ is the x-axis in the model $M(\widetilde{\kappa},\widetilde{\tau})$ and $\widetilde{\kappa}\leq 0$, then $I_{\widetilde{\gamma}}$ is the entire graph of the function

$$u(x,y)=\begin{cases}\widetilde{\tau}xy & \text{if } \widetilde{\kappa}=0,\\ \frac{2\widetilde{\tau}}{\widetilde{\kappa}}\arctan\frac{2xy}{\frac{4}{\widetilde{\kappa}}+x^2-y^2} & \text{if } \widetilde{\kappa}<0.\end{cases}$$

If we consider the surface $I_{\widetilde{\gamma}}$ in $\mathbb{E}(4H^2+\kappa,H)$, then the horizontal geodesic $\widetilde{\gamma}$ becomes a curve γ in a vertical plane of symmetry of the conjugate H-surface in $\mathbb{M}^2(\kappa)\times\mathbb{R}$. Since the angle function is 1 along $\widetilde{\gamma}$, it follows that γ is also a horizontal geodesic, whence the H-surface is invariant by translations along γ. It is an H-cylinder or H-torus if $4H^2+\kappa>0$ (explicit parametrizations can be found in [62, §2]) or an entire H-graph otherwise.

Example 5 (Spherical Helicoids [70, §4]) Lawson [58, Prop. 7.2] defined the spherical helicoid of pitch $c\in\mathbb{R}$ in the round 3-sphere $\mathbb{S}^3$ as the immersion

$$\begin{aligned}\widetilde{\phi}_c:\mathbb{R}^2&\to\mathbb{S}^3\subset\mathbb{C}^2\\ (u,v)&\mapsto\big(\cos(u)e^{icv},\sin(u)e^{iv}\big).\end{aligned}$$

These immersions are characterized by being minimal if we substitute $\mathbb{S}^3$ with $\mathbb{S}^3_{\mathrm{B}}(\widetilde{\kappa},\widetilde{\tau})$ for any $\widetilde{\kappa}>0$ and $\widetilde{\tau}\neq 0$ (see [105]). They are also equivariant by screw motions and ruled by horizontal geodesics in all Berger metrics. The surface parametrized by $\widetilde{\phi}_c$ is topologically a torus or a Klein bottle if $c\in\mathbb{Q}-\{0\}$ (these are Lawson's examples $\tau_{m,n}$; see [58, §7]), a sphere if $c=0$, and a dense cylinder otherwise; however, $\widetilde{\phi}_c$ is embedded if and only if $c\in\{-1,0,1\}$. Note that $\widetilde{\phi}_1$ and $\widetilde{\phi}_{-1}$ are congruent Clifford tori in the round metric $\mathbb{S}^3$ but they are not quite alike in the Berger case: the torus $\widetilde{\phi}_1$ is flat and everywhere vertical (i.e., the preimage by π of a great circle of $\mathbb{S}^2(\widetilde{\kappa})$), yet $\widetilde{\phi}_{-1}$ is not flat, and it is vertical just along a couple of horizontal geodesics. As a matter of fact, $\widetilde{\phi}_{-1}$ is congruent to the invariant surface $I_{\widetilde{\gamma}}$ we discussed in Example 4 in the case of Berger spheres, where $\widetilde{\gamma}(t)=\frac{1}{\sqrt{2}}(e^{-it},e^{it})$. We also remark that $\widetilde{\phi}_c$ and $\widetilde{\phi}_{1/c}$ are congruent surfaces for all $c\neq 0$, so we can assume that $c\in[-1,1]$ not losing any generality.

The spherical helicoids $\widetilde{\phi}_c$ in $\mathbb{S}^3_{\mathrm{B}}(4H^2+\kappa,H)$, where $4H^2+\kappa>0$, have the rotational Delaunay H-surfaces as conjugate H-immersions ϕ_c in $\mathbb{M}^2(\kappa)\times\mathbb{R}$. This was proved in [28, Thm. 2.1] for $\kappa=0$ (Lawson correspondence) and in [70, §4] in the general case. By comparing the rotation of the normal along the rulings of $\widetilde{\phi}_c$ with the curvature of the curve that generates the Delaunay surfaces by means of Lemma 2, we obtain the following list of conjugate surfaces:

(a) ϕ_1 is the vertical H-cylinder over a curve of $\mathbb{M}^2(\kappa)$ of curvature $2H$.
(b) ϕ_c is a rotationally invariant H-unduloid if $0 < c < 1$.
(c) ϕ_0 is the rotationally invariant H-sphere.
(d) ϕ_c is a rotationally invariant H-nodoid if $-1 < c < 0$.
(e) ϕ_{-1} is an H-torus ($\kappa > 0$) or an H-cylinder ($\kappa \leq 0$) invariant by translations along a horizontal geodesic.

We refer to [86, Lem 1.3] and [62, §2] for a more specific description of Delaunay H-surfaces. It is worth emphasizing that different directions of rotation of the normal along the axis of $\widetilde{\phi}_c$ output very different surfaces, as already discussed in Lemma 3. This will be further explored in Sect. 6.1.

4 Dirichlet Problems for H-Surfaces in $\mathbb{E}(\kappa, \tau)$

The initial minimal surfaces in our conjugate constructions will be solutions of a Plateau or a Jenkins-Serrin problem in $\mathbb{E}(4H^2+\kappa, H)$, both of which will be thought of as Dirichlet problems for the minimal surface equation. Needless to say that not all geodesic polygons span graphical surfaces; if that is not the case, additional work will be required to find the desired minimal surface.

4.1 The Plateau Problem

We will solve a Plateau problem in the more general scenario of arbitrary Killing submersions in the sense of [60]. This is motivated by the fact that some of our minimal surfaces will be graphs in $\mathbb{E}(4H^2 + \kappa, H)$ with respect to other (non-unitary) Killing directions. In the sequel, we will assume that $\pi : \mathbb{E} \to M$ is a Riemannian submersion whose fibers have infinite length and coincide with the integral curves of a nowhere vanishing Killing field ξ. We refer to [20] for a more detailed discussion.

Definition 1 A Nitsche contour in $\mathbb{E}$ is a pair (Ω, Γ), where $\Omega \subset M$ is a precompact open domain and $\Gamma \subset \mathbb{E}$ is a Jordan curve admitting a piecewise-regular parametrization $\gamma : [a, b] \to \Gamma$ satisfying the following conditions:

(a) There is a partition $a = t_1 < s_1 \leq t_2 < \ldots \leq t_r < s_r \leq t_{r+1} = b$ such that $\gamma(a) = \gamma(b)$, and for any $j \in \{1, \ldots, r\}$, the component $\gamma|_{[t_j, s_j]}$ is a nowhere vertical curve and $\gamma|_{[s_j, t_{j+1}]}$ a vertical segment.
(b) The projection $\pi \circ \gamma$ parametrizes $\partial\Omega$ injectively except at vertical segments.

This means that $\partial\Omega$ is a regular curve except possibly at the points $\pi(t_i)$, which will be called the vertexes of Ω.

Graphical surfaces with prescribed mean curvature and boundary a Nitsche contour (Ω, Γ) satisfy a maximum principle which has been proved in the case of unitary Killing submersions [63, Prop. 3.8] but also holds in the general case. This relies on the fact that the mean curvature of Killing graphs can be written in divergence form similar to (10) (see [60, §3]). We remark that in this case, the Killing graph of a function $u \in C^\infty(\Omega)$ is defined in the usual way with respect to a global section $F_0 : M \to \mathbb{E}$ as $F_u(p) = \phi_{u(p)}(F_0(p))$. Note that a global section exists by the assumption on the infinite length of the fibers [103, Thm. 12.2].

Proposition 6 *Let (Ω, Γ) and (Ω, Γ') be Nitsche contours in $\mathbb{E}$ over the same domain Ω with the same set of vertexes $V \subset \partial\Omega$. Assume that $u, v \in C^\infty(\Omega)$ verify the following:*

(a) The graphs F_u and F_v have the same mean curvature over Ω.
(b) u and v extend continuously to $\overline{\Omega} - V$ giving rise to surfaces with boundaries Γ and Γ', respectively.

If $u \leq v$ in $\partial\Omega - V$, then $u \leq v$ in Ω.

In particular, given a Nitsche contour (Ω, Γ), there is at most one minimal graph over Ω with boundary Γ. Although uniqueness holds in general, we will consider some additional conditions that ensure that $\pi^{-1}(\overline{\Omega}) \subset \mathbb{E}$ is a three-dimensional mean-convex body in the sense of Meeks and Yau [80], so there exists at least one minimal surface with boundary Γ. It follows from a standard application of the maximum principle and from uniqueness that the solution is a graph, using the same argument as in [63, Thm. 3.11] (see also [70, Prop. 2]). All in all, we have the following result:

Proposition 7 *Let (Ω, Γ) be a Nitsche contour such that Ω is simply connected, $\pi^{-1}(\pi(\alpha))$ is a minimal surface for each non-vertical component $\alpha \subset \Gamma$, and the interior angle at each vertex of Ω is at most π. There is a unique minimal surface $\Sigma \subset \pi^{-1}(\overline{\Omega})$ with boundary Γ, and the interior of Σ is a Killing graph over Ω.*

Observe that the solution Σ must be invariant by any ambient isometry that preserves both the Killing submersion and the Nitsche contour (Ω, Γ) because of the uniqueness. We also observe that Meeks and Yau's solution to the Plateau problem is of class C^1 up to the boundary (see [99, Rmk. 3.4]), which enables analytic continuation of the surface across vertical and horizontal geodesics in Γ by axial symmetry in the sense of Proposition 3.

4.2 *The Jenkins-Serrin Problem*

The nonparametric Plateau problem we have discussed in Proposition 7 amounts to obtaining a function u whose graph is minimal over Ω and extends continuously to $\overline{\Omega} - V$, being V the set of vertexes of Ω with the prescribed values on $\partial\Omega - V$ induced by Γ. The Jenkins-Serrin problem also allows $\pm\infty$ values along the components of

$\partial\Omega$ and has been studied in $\mathbb{E}(\kappa,\tau)$-spaces with $\kappa < 0$ respect to the usual Killing direction by also allowing the domain $\Omega \subset \mathbb{H}^2(\kappa)$ to be unbounded.

We will first deal with the case $\kappa < 0$ and $\tau = 0$, in which the solution has been proved under very general assumptions by Mazet, Rodríguez, and Rosenberg [76]. We will suppose that $\partial\Omega$ is piecewise regular and contains finitely many components of three distinct types: $A_1, \ldots, A_{n_1} \subset \mathbb{H}^2(\kappa)$, where the value $+\infty$ is prescribed; $B_1, \ldots, B_{n_2} \subset \mathbb{H}^2(\kappa)$, where the value $-\infty$ is prescribed; and $C_1, \ldots, C_{n_3} \subset \mathbb{H}^2(\kappa)$, where continuous finite boundary values are prescribed. We will also assume that $\partial_\infty\Omega$ consists of finitely many ideal segments $D_1, \ldots, D_{n_4} \subset \partial_\infty\mathbb{H}^2(\kappa)$ where continuous finite boundary values are also prescribed. The endpoints of the segments A_i, B_i, C_i, and D_i will be called the vertexes of Ω. There are two necessary conditions for the existence of solution: on the one hand, each component A_i or B_i must be a geodesic segment; on the other hand, there cannot be two consecutive components of type A_i meeting at an interior angle less than π and likewise for those of type B_i. Additionally, it is assumed that the arcs C_i are convex with respect to the inner conormal to Ω along C_i. Under these conditions, we will say that Ω is a *general Jenkins-Serrin domain*.

A polygonal domain $\mathcal{P}$ in $\mathbb{H}^2(\kappa)$ is a domain whose boundary $\partial\mathcal{P}$ consists of finitely many geodesic segments. We will say that it is inscribed in a general Jenkins-Serrin domain Ω if $\mathcal{P} \subset \overline{\Omega}$ and all vertexes of $\mathcal{P}$ are vertexes of Ω. Assume that the ideal vertexes of Ω are $p_1, \ldots, p_m \in \partial_\infty\mathbb{H}^2(\kappa)$, and for each $i \in \{1, \ldots, m\}$, let $H_i \subset \mathbb{H}^2(\kappa)$ be a domain with boundary a horocycle asymptotic to p_i, small enough so that it only intersects the components of $\partial\Omega$ with endpoint at p_i and $H_i \cap H_j = \emptyset$ for all $j \neq i$. Under these assumptions, the following finite lengths characterize the existence of solution:

$$\alpha(\mathcal{P}) = \sum_{i=1}^{n_1} \operatorname{Length}(\cup_{i=1}^{n_1}(A_i \cap \partial\mathcal{P}) - \cup_{j=1}^{m} H_j),$$

$$\beta(\mathcal{P}) = \sum_{i=1}^{n_2} \operatorname{Length}(\cup_{i=1}^{n_2}(B_i \cap \partial\mathcal{P}) - \cup_{j=1}^{m} H_j),$$

$$\gamma(\mathcal{P}) = \operatorname{Length}(\partial\mathcal{P} - \cup_{j=1}^{m} H_j).$$

Theorem 6 ([76, Thm. 4.9 and 4.12]) *Let $\Omega \subset \mathbb{H}^2(\kappa)$ be a general Jenkins-Serrin domain on which we consider the above Dirichlet problem:*

(a) If $n_3 = n_4 = 0$, then the problem has a solution if and only if $\alpha(\Omega) = \beta(\Omega)$ and $\max\{\alpha(\mathcal{P}), \beta(\mathcal{P})\} < \frac{1}{2}\gamma(\mathcal{P})$ for all polygonal domains $\mathcal{P} \neq \overline{\Omega}$ inscribed in Ω.

(b) Otherwise, the problem has a solution if and only if $\max\{\alpha(\mathcal{P}), \beta(\mathcal{P})\} < \frac{1}{2}\gamma(\mathcal{P})$ for all polygonal domains $\mathcal{P}$ inscribed in Ω.

Nelli and Rosenberg [84] proved this result for convex relatively compact domains, and Collin and Rosenberg [13] extended it for unbounded convex domains all of whose vertexes are ideal and with $n_4 = 0$. It is worth mentioning that convexity at the vertexes is a condition that ensures that the surface is of class C^1 up to the boundary and enables the completion of the surface given by Proposition 3, as proved by Sa Earp and Toubiana [100]. This also implies that solutions to Jenkins-Serrin problems over convex domains have finite radial limits on the corners (see [10, Lem. 4]). In the non-convex case, the solution given by Theorem 6 is a smooth graph over the interior of Ω but might fail to be of class C^1 at a vertical geodesic projecting to a non-convex vertex. This was noticed by Finn [26, Thm. 3] and Eclat and Lancaster [22] for minimal surfaces in $\mathbb{R}^3$.

A very general maximum principle for solutions of the Jenkins-Serrin problem in $\mathbb{H}^2(\kappa)\times\mathbb{R}$ can be found in [76, Thm. 4.16]. However, we will consider the following earlier version of Collin and Rosenberg, which is enough when the prescribed boundary data comes from a geodesic polygon with vertical and horizontal (possibly ideal) components. It clearly implies that the solution given by Theorem 6 in that case is unique if there are nonideal horizontal geodesics (which are components of type C_i). On the other hand, the solution can be shown to be unique up to vertical translations if there are no such horizontal geodesics [76, Thm. 4.12]. Note that there are no components of type D_i if the boundary is a geodesic polygon.

Proposition 8 ([13]) *Let $\Omega \subset \mathbb{H}^2(\kappa)$ be a general Jenkins-Serrin domain with $n_3 \neq 0$ and $n_4 = 0$ and such that any two components of $\partial\Omega$ meeting at a common ideal vertex are asymptotic to each other at that vertex (in hyperbolic distance).*

Assume that $u, v \in C^\infty(\Omega)$ span minimal graphs over Ω and extend continuously to $\overline{\Omega} - V$ (where V is the set of vertexes of Ω) with possible infinite values along some of the A_i or B_i. If $u \leq v$ on $\partial\Omega - V$, then $u \leq v$ on Ω.

Theorem 6 and Proposition 8 are believed to hold true also in the case $\tau \neq 0$ (observe that both statements translate literally to this more general case and all the needed tools seem to be available in $\widetilde{\mathrm{SL}}_2(\mathbb{R})$). However, they have not hitherto been proved in this generality. The two results in this direction so far have been given by Younes [108], who has proved existence and uniqueness in the case of a bounded convex domain, and by Melo [81], who proves only existence over convex unbounded domains having only ideal vertexes and no arcs of type D_i. Note also that Younes and Melo's surfaces can be used as barriers to give ad hoc solutions to more general Jenkins-Serrin problems as in [11, Lem. 3.2].

As a final remark, it is worth noticing that the solution Σ to any of the aforesaid Jenkins-Serrin problems can be obtained as the limit of a double sequence of minimal surfaces Σ_n^k. The surface Σ_n^k can be taken as the solution of a Plateau problem over a Nitsche contour (Ω_n, Γ_n^k), where the domains $\Omega_n \subset \mathbb{H}^2(\kappa)$ are bounded and converge to Ω and the curves Γ_n^k replace the target boundary components with prescribed values $\pm\infty$ with a sequence of horizontal geodesics that diverges in the desired direction. The surface Σ_n^k is a solution to a Plateau problem in the sense of Proposition 7 and converges to a minimal graph Σ^k over Ω as $n \to \infty$. The surfaces Σ^k in turn converge to the solution Σ as $k \to \infty$. Since the convergence is of class C^m on compact subsets for all m, we can use the description given by Lemmas 2 and 3 plus the continuity of the conjugation in Proposition 2 to analyze the ideal geodesics as limits of nonideal geodesics. In particular, we obtain the following result:

Proposition 9 ([11, Cor. 2.4]) *Assume that* $4H^2 + \kappa < 0$*, and let* $\widetilde{\Sigma} \subset \mathbb{E}(4H^2 + \kappa, H)$ *be the solution of a Jenkins-Serrin problem with boundary a polygon consisting of vertical and horizontal (possibly ideal) geodesics. Let* $\Sigma \subset \mathbb{H}^2(\kappa) \times \mathbb{R}$ *be the conjugate (possibly non-embedded)* H*-multigraph:*

(a) *Ideal vertical geodesics in* $\partial_\infty \widetilde{\Sigma}$*, if any, become ideal horizontal curves in* $\partial_\infty \Sigma$ *of constant curvature* $\pm 2H$ *at height* $\pm\infty$*.*
(b) *Ideal horizontal geodesics in* $\partial_\infty \widetilde{\Sigma}$ *become ideal vertical geodesics of* $\partial_\infty \Sigma$*.*

5 Compact H-Surfaces in $\mathbb{S}^2(\kappa) \times \mathbb{R}$

Now, we will focus on constructions where the ideas we have discussed in the previous sections apply. The first group of examples we will deal with concern the construction of compact or periodic H-surfaces such that the fundamental piece is compact and comes from the solution of a Plateau problem. As pointed out in the introduction, we are interested in providing embedded examples with different topologies in $\mathbb{S}^2(\kappa) \times \mathbb{R}$ or in a quotient by a vertical translation.

5.1 Horizontal Delaunay H-Surfaces

This section is devoted to the construction of H-surfaces in $\mathbb{S}^2(\kappa)\times\mathbb{R}$ and $\mathbb{H}^2(\kappa)\times\mathbb{R}$ provided that $4H^2+\kappa > 0$ and $H > 0$ by conjugating minimal surfaces in $M(4H^2+\kappa, H)$. The name *horizontal Delaunay* is motivated by the fact that these H-surfaces resemble classical Delaunay surfaces in Euclidean space. Although they are not equivariant if $\kappa \neq 0$ (in contrast with the vertical case discussed in Example 5), they are invariant under a discrete group of translations along a horizontal geodesic called the axis of the surface. Horizontal Delaunay surfaces comprise a deformation of the H-spheres into the H-cylinders by means of unduloid-type surfaces, and they also contain non-embedded surfaces of nodoid-type. The next result describes the moduli space of horizontal Delaunay H-surfaces:

Theorem 7 ([70, 72]) *Fix* $\kappa \in \mathbb{R}$ *and a horizontal geodesic* $\Lambda \subset \mathbb{M}^2(\kappa) \times \{0\}$*. There exists a family* $\Sigma^*_{\lambda,H}$*, parametrized by* $\lambda \geq 0$ *and* $H > 0$ *such that* $4H^2 + \kappa > 0$*, of complete* H*-surfaces in* $\mathbb{M}^2(\kappa) \times \mathbb{R}$*, invariant under a discrete group of translations along* Λ *with respect to which they are cylindrically bounded. They are also symmetric about the totally geodesic surfaces* $\mathbb{M}^2(\kappa) \times \{0\}$ *and* $\Lambda \times \mathbb{R}$*. Moreover, as sketched in Fig. 4,*

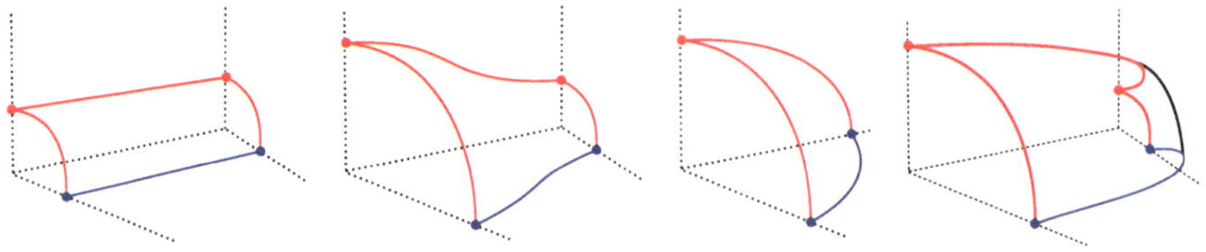

Fig. 4 Schematic representation of the fundamental piece of horizontal Delaunay H-surfaces in $\mathbb{M}^2(\kappa) \times \mathbb{R}$ (see Theorem 7). From left to right: cylinder (torus if $\kappa > 0$), unduloid, sphere, and nodoid

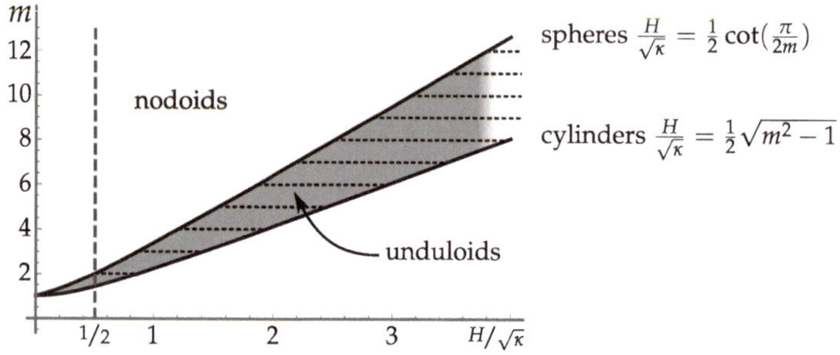

Fig. 5 The darker region is the moduli space of $\Sigma^*_{\lambda,H} \subset \mathbb{S}^2(\kappa) \times \mathbb{R}$, $\lambda \in [0, \frac{\pi}{2}]$, in terms of $\frac{H}{\sqrt{\kappa}}$ and m. Dotted horizontal segments represent the families $\mathcal{T}_m$ of embedded tori. The vertical dashed line indicates that no such tori exist if $H > \frac{\sqrt{\kappa}}{2}$

(i) $\Sigma^*_{0,H}$ *is the H-cylinder (H-torus if $\kappa > 0$) invariant under the continuous one-parameter group of translations along Λ.*
(ii) $\Sigma^*_{\lambda,H}$ *is an unduloid-type surface if* $0 < \lambda < \frac{\pi}{2}$.
(iii) $\Sigma^*_{\frac{\pi}{2},H}$ *is a stack of tangent rotationally invariant H-spheres centered on Λ.*
(iv) $\Sigma^*_{\lambda,H}$ *is a nodoid-type surface if* $\lambda > \frac{\pi}{2}$.

In the case $\kappa > 0$, we find (among the horizontal unduloids) many families of embedded H-tori which continuously deform stacks of tangent H-spheres evenly distributed along a horizontal geodesic Λ into an equivariant H-torus. The embeddedness is shown by spotting a function in the kernel of the common stability operator of the conjugate surfaces that is induced simultaneously by two one-parameter groups of isometric deformations: one in the initial space $M(4H^2+\kappa, H)$ and the other one in the target space $\mathbb{M}^2(\kappa) \times \mathbb{R}$. We will see later that this function carries insightful geometric information that also implies that horizontal unduloids are properly embedded if $\kappa \leq 0$ (see [72, Prop. 4.4]).

Theorem 8 ([72, Thm. 1.2]) *Fix $\kappa > 0$. For each integer $m \geq 2$, there is a family $\mathcal{T}_m$ of embedded H-tori in $\mathbb{S}^2(\kappa) \times \mathbb{R}$ (see Fig. 5) parametrized as*

$$\mathcal{T}_m = \left\{ \Sigma^*_{\lambda_m(H),H} : \cot(\tfrac{\pi}{2m}) < \tfrac{2H}{\sqrt{\kappa}} \leq \sqrt{m^2-1} \right\},$$

where $H \mapsto \lambda_m(H)$ *strictly decreases continuously from* $\frac{\pi}{2}$ *to* 0*:*

1. *The family* $\mathcal{T}_m$ *is a continuous deformation (in which* H *varies) from a stack of* m *tangent spheres evenly distributed along* Λ *to an equivariant torus.*
2. *The surfaces* $\Sigma^*_{\lambda_m(H),H}$*, along with* H*-spheres* $\Sigma^*_{\pi/2,H}$ *and* H*-cylinders* $\Sigma^*_{0,H}$*, are the only compact embedded* H*-surfaces among all* $\Sigma^*_{\lambda,H}$ *(for all* $\kappa \in \mathbb{R}$*).*

5.1.1 Construction of the Minimal Surface in $M(4H^2 + \kappa, H)$

Assume that $H, \kappa \in \mathbb{R}$ are fixed such that $H > 0$ and $4H^2 + \kappa > 0$ (we will omit the dependence on H in the sequel unless otherwise stated). To obtain an H-surface in $\mathbb{M}^2(\kappa) \times \mathbb{R}$ by means of the correspondence, we will begin by constructing a compact minimal surface $\widetilde{\Sigma}_\lambda$, $\lambda \in \mathbb{R}$, with boundary a polygon $\widetilde{\Gamma}_\lambda \subset M(4H^2 + \kappa, H)$ consisting of three horizontal geodesics $\widetilde{h}_0$, $\widetilde{h}_1$, and $\widetilde{h}_2$ and one vertical geodesic $\widetilde{v}$, whose vertexes will be labeled as $\widetilde{1}$, $\widetilde{2}$, $\widetilde{3}$, and $\widetilde{4}$, as shown in Fig. 6. More precisely,

- $\widetilde{h}_0$ is a quarter of a horizontal geodesic, which can be parametrized, up to an ambient isometry, by $\widetilde{h}_0(s) = \frac{2}{\sqrt{4H^2+\kappa}}\left(0, \frac{\cos(2s)}{1+\sin(2s)}, 0\right)$, for $s \in \left[0, \frac{\pi}{2}\right]$.

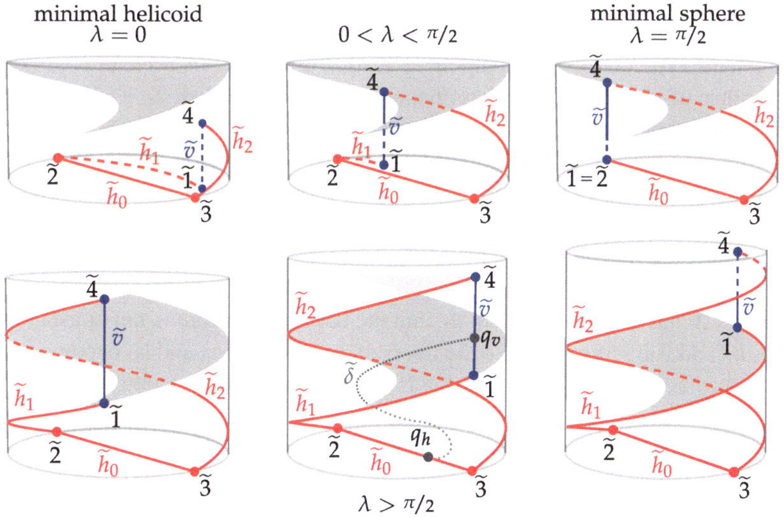

Fig. 6 A faithful representation of the polygon $\widetilde{\Gamma}_\lambda$ for different values of λ. The barriers T (vertical cylinder) and S (helicoid, in gray) demarcate the mean-convex solid Ω. Top row: polygons $\widetilde{\Gamma}_\lambda$ for $\lambda \leq \frac{\pi}{2}$ whose conjugate surfaces are unduloids and limit cases $\lambda = 0$ (left), which corresponds to the H-cylinder, and $\lambda = \frac{\pi}{2}$ (right), which corresponds to the H-sphere. Bottom row: polygon $\widetilde{\Gamma}_\lambda$ for $\lambda > \frac{\pi}{2}$ whose conjugate H-surfaces are nodoids. The dotted line in the central figure represents the curve $\widetilde{\delta}$ of zeros of the angle function (see Proposition 10)

- $\widetilde{h}_1$ and $\widetilde{h}_2$ are horizontal geodesics starting orthogonally at the endpoints of $\widetilde{h}_0$ with signed lengths $\frac{1}{2\sqrt{4H^2+\kappa}}(\lambda - \frac{\pi}{2})$ and $\frac{1}{2\sqrt{4H^2+\kappa}}(\lambda + \frac{\pi}{2})$ in the directions $\widetilde{h}_0'(0) \times \xi$ and $-\widetilde{h}_0'(\frac{\pi}{2}) \times \xi$, respectively. In particular, $\widetilde{h}_1$ goes in opposite directions according to the sign of this length.
- $\widetilde{v}$ is the vertical geodesic joining the endpoints of $\widetilde{h}_1$ and $\widetilde{h}_2$.

Observe that $\widetilde{\Gamma}_\lambda$ and $\widetilde{\Gamma}_{-\lambda}$ are congruent for all $\lambda > 0$ by the isometry $(x, y, z) \mapsto (-x, y, -z)$, so we can assume that $\lambda \geq 0$ without loss of generality.

Remark 7 The polygon $\Theta(\widetilde{\Gamma}_\lambda)$, where $\Theta : M(4H^2 + \kappa, H) \to \mathbb{S}^3_{\mathrm{B}}(4H^2 + \kappa, H)$ is the local isometry given by (5), has self-intersections if $\lambda \geq \frac{7\pi}{2}$, so the resulting Plateau problem is ill-posed in $\mathbb{S}^3_{\mathrm{B}}(4H^2 + \kappa, H)$. This is the reason why the locally isometric Cartan model $M(4H^2 + \kappa, H)$ is used throughout this section.

The polygon $\widetilde{\Gamma}_\lambda$ was first consider in [70] for $0 \leq \lambda \leq \frac{\pi}{2}$ and then extended for $\lambda \geq \frac{\pi}{2}$ in [72]. We remark that the mean-convex body used to solve the Plateau problem in [70] is no longer valid for $\lambda > \frac{\pi}{2}$ and a different approach was developed to show the existence of $\widetilde{\Sigma}_\lambda$, which we present here.

Consider the following two minimal surfaces in $M(4H^2 + \kappa, H)$: the vertical cylinder T of equation $x^2 + y^2 = \frac{4}{4H^2+\kappa}$ and the surface $S = I_{\widetilde{h}_0}$ (see Example 4) of equation $x\cos(\frac{4H^2+\kappa}{2H}z) + y\sin(\frac{4H^2+\kappa}{2H}z) = 0$. Observe that $\Theta(T)$ is congruent to the Clifford torus ϕ_1 in Example 5 and $\Theta(S)$ is congruent to $\widetilde{\phi}_{-1}$ in Example 5 (see the shaded surface in Fig. 5), where Θ is the local isometry defined in (5). The surface S divides the interior domain of the cylinder T in two connected components. The connected component W that contains $\widetilde{v}$ is a mean-convex solid in the sense of [80], so there exists an embedded minimal disk $\widetilde{\Sigma}_\lambda \subset W$ solution to the Plateau problem with boundary $\partial\widetilde{\Sigma}_\lambda = \widetilde{\Gamma}_\lambda \subset W$.

If $0 \leq \lambda \leq \frac{\pi}{2}$, then $\widetilde{\Gamma}_\lambda$ is a Nitsche contour with respect to the usual Killing submersion, but this is not true in general. To overcome this issue, consider the Killing vector field $\widetilde{X}$ associated with the group of screw motions that leave the surface S invariant. Then $\widetilde{X}$ has no zeros and gives rise to a Killing submersion $\pi_0 : M(\kappa, \tau) \to (\mathbb{R}^2, \mathrm{d}s^2)$ in the sense of [60] (note that the metric $\mathrm{d}s^2$ has not constant curvature, $\widetilde{X}$ has not constant length, and the bundle curvature is not constant). The curves $\widetilde{h}_0$ and $\widetilde{v}$ are transversal to the fibers of π_0, whereas $\widetilde{h}_1$ and $\widetilde{h}_2$ become vertical for π_0 (they are integral curves of $\widetilde{X}$). Hence, $\widetilde{\Gamma}_\lambda$ is a Nitsche contour with respect to π_0 in the sense of Definition 1 for all $\lambda \geq 0$. Proposition 7 implies that $\widetilde{\Sigma}_\lambda$ is a unique surface in W with boundary $\widetilde{\Gamma}_\lambda$, and it is everywhere transversal to $\widetilde{X}$. Uniqueness in turn implies that $\widetilde{\Sigma}_\lambda$ depends continuously on $\lambda \geq 0$ (and so does its sister surface Σ_λ thanks to Proposition 2).

We highlight the following special cases, depicted in Fig. 6 (see left and right drawings in the top row):

- If $\lambda = 0$, then $\widetilde{\Sigma}_0$ is part of a spherical helicoid with axis $\widetilde{v}$, that is, $\Theta(\widetilde{\Sigma}_0)$ is congruent to part of the spherical helicoid $\widetilde{\phi}_{-1}$ of equation $\mathrm{Im}(z^2 + w^2) = 0$ (see Fig. 6 top left).

- If $\lambda = \frac{\pi}{2}$, then $\widetilde{\Sigma}_{\frac{\pi}{2}}$ is an open piece of a minimal sphere. More precisely, $\Theta(\widetilde{\Sigma}_{\frac{\pi}{2}})$ is part of the minimal sphere $\mathrm{Im}(z - w) = 0$ (see Fig. 6 top right).

Remark 8 If $\kappa = 0$, then $M(4H^2 + \kappa, H)$ is locally isometric to the 3-sphere $\mathbb{S}^3(H^2)$ and the completion $\widetilde{\Sigma}_\lambda^*$ is a spherical helicoid (see Example 5) for all $\lambda \geq 0$ (observe that $\widetilde{v}$ and $\widetilde{h}_0$ have the same length and $\widetilde{\Sigma}_\lambda$ is invariant under the composition of translations and suitable rotations about $\widetilde{h}_2$). This is a consequence of the isotropy of the 3-sphere $\mathbb{S}^3(H^2)$ that allows rotations around any geodesic. This argument obviously fails if $\kappa \neq 0$ (the segments $\widetilde{v}$ and $\widetilde{h}_0$ no longer have the same length and there is no screw-motion group with axis $\widetilde{h}_2$).

5.1.2 Analysis of the Angle Function

As pointed out in Sect. 3, the angle function $\nu_\lambda : \widetilde{\Sigma}_\lambda \to [-1, 1]$ gives precious information about the shape of the boundary curves of the conjugate contour Γ_λ. We are mainly interested in the points where ν_λ has values ± 1 or 0, if any. We will also establish the monotonicity properties of ν_λ as a function of λ along the horizontal geodesic arcs of $\widetilde{\Gamma}_\lambda$ in the common boundary of intersection $\widetilde{\Gamma}_{\lambda_1} \cap \widetilde{\Gamma}_{\lambda_2}$, $\lambda_1 \neq \lambda_2$. In the sequel, we will choose the normal $\widetilde{N}$ to $\widetilde{\Sigma}_\lambda$ such that $\nu_\lambda(\widetilde{3}) = -1$.

Proposition 10 ([70, Lem. 3] and [72, Prop. 3.3]) *Let ν_λ be the angle function of the compact minimal disk $\widetilde{\Sigma}_\lambda$ spanning $\widetilde{\Gamma}_\lambda$ such that $\nu_\lambda(\widetilde{3}) = -1$:*

(a) The only points in which ν_λ takes the values ± 1 are $\widetilde{2}$ and $\widetilde{3}$. More precisely, if $0 < \lambda < \frac{\pi}{2}$, then $\nu_\lambda(\widetilde{2}) = -1$; if $\lambda > \frac{\pi}{2}$, then $\nu_\lambda(\widetilde{2}) = 1$.
(b) The set of points in which ν_λ vanishes consists of $\widetilde{v}$ and, in the case $\lambda > \frac{\pi}{2}$, also of an interior regular curve $\widetilde{\delta} \subset \widetilde{\Sigma}_\lambda$ with endpoints in $\widetilde{v}$ and $\widetilde{h}_0$ (see Fig. 6).
(c) Given p in the horizontal boundary of $\widetilde{\Gamma}_\lambda$, the function $\lambda \mapsto \nu_\lambda(p)$ is continuous in the interval where it is defined:

- *It is strictly increasing (possibly changing sign) if $p \in \widetilde{h}_0$ for all $\lambda > 0$.*
- *It is positive and strictly increasing if $p \in \widetilde{h}_1$ and $\lambda > \frac{\pi}{2}$. It is negative and strictly increasing if $p \in \widetilde{h}_1$ and $0 < \lambda < \frac{\pi}{2}$.*
- *It is negative and strictly increasing if $p \in \widetilde{h}_2$ for all $\lambda > 0$.*

Although we will not provide a full proof of Proposition 10, it essentially relies on strategies 1, 2, and 3 discussed in Sect. 3.1.2. We will illustrate this by sketching the analysis of the interior zeros of ν. To this end, we intersect $\widetilde{\Sigma}_\lambda$ and the tangent vertical cylinder T_p at some interior point p with $\nu(p) = 0$.

If $0 < \lambda < \frac{\pi}{2}$, then the (at least) four rays in $T_p \cap \widetilde{\Gamma}_\lambda$ emanating from p end up either in $\widetilde{v}$ or in $\widetilde{h}_0 \cup \widetilde{h}_1 \cup \widetilde{h}_2$. If two of the rays reach $\widetilde{v}$, they enclose a region of $\widetilde{\Sigma}_\lambda$ (along with a segment of $\widetilde{v}$); otherwise, there are two of the rays that reach the same point of $\widetilde{h}_0 \cup \widetilde{h}_1 \cup \widetilde{h}_2$, and they also enclose a region of $\widetilde{\Sigma}_\lambda$. Either way, such a region has boundary on the vertical cylinder T_p, and we easily find a contradiction

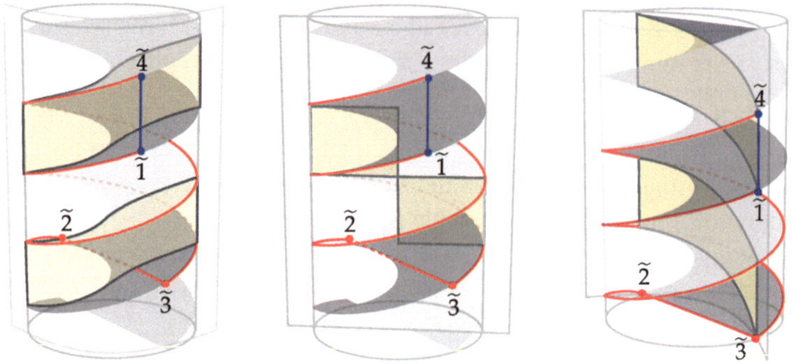

Fig. 7 Each figure indicates two connected components of $(T_p \cap W) - S$ for a Clifford torus T_p inside the mean-convex solid W. From left to right: a general case, a case in which T_p contains the z-axis, and a case in which T_p contains $\widetilde{v}$

with the maximum principle with respect to other vertical cylinders [72, Lem. 1.6]. This means that there are no interior points with $\nu = 0$ if $0 < \lambda < \frac{\pi}{2}$.

Assume now that $\lambda > \frac{\pi}{2}$ and apply the same reasoning. However, since $\Sigma_\lambda \subset W$, the rays emanating from p lie in the connected component of $(T_p \cap W) - S$ containing p, which is a vertical quadrilateral with boundary in $S \cup T_p$: three of its sides lie in S if T'_p contains the z-axis (see Fig. 7 center); otherwise, only two of the sides lie in S (see Fig. 7 left and right). It is not difficult to realize that if more than four rays arise from p, then there will be enclosed regions in contradiction with the maximum principle as discussed above. This implies that there are no points with $\nu(p) = \nabla\nu(p) = 0$, whence the curves of $\nu = 0$ do not bifurcate. The argument can be further refined to prove that there is exactly one regular curve $\widetilde{\delta}$ where $\nu = 0$, but we will not include the details here.

As for item (c), we will also say some words about the application of the maximum principle in the boundary. Assume first that $0 < \lambda_1 < \lambda_2 < \frac{\pi}{2}$. This case is easier since $\widetilde{\Sigma}_{\lambda_1}$ is a barrier from above for $\widetilde{\Sigma}_{\lambda_2}$ along the common horizontal boundary, and this enables a direct comparison of the normals. It is important to mention that the surface $\mathcal{I}_{\widetilde{h}_0}$, whose angle function is equal to -1 along $\widetilde{h}_0$ (see Example 4), is nothing but $\widetilde{\Sigma}_0$ and acts as a barrier to $\widetilde{\Sigma}_{\lambda_2}$ from below. This sandwich between $\widetilde{\Sigma}_0$ and $\widetilde{\Sigma}_{\lambda_1}$ yields the sign and monotonicity stated in Proposition 10. In the case $\frac{\pi}{2} < \lambda_1 < \lambda_2$, the discussion is similar, but we have to observe that $\widetilde{\Sigma}_{\lambda_1}$ acts as a barrier from below for $\widetilde{\Sigma}_{\lambda_2}$ as graphs in the direction of the helicoidal Killing vector field $\widetilde{X}$ along $\widetilde{h}_1$ and $\widetilde{h}_2$.

5.1.3 The Conjugate H-Immersion

Let $\Sigma_\lambda \subset \mathbb{M}^2(\kappa) \times \mathbb{R}$ be the conjugate H-surface of $\widetilde{\Sigma}_\lambda$ constructed in the previous section. By Lemmas 2 and 3, Σ_λ is a compact H-surface whose boundary Γ_λ consists of three curves h_0, h_1, and h_2, contained in vertical planes P_{23}, P_{12}, and

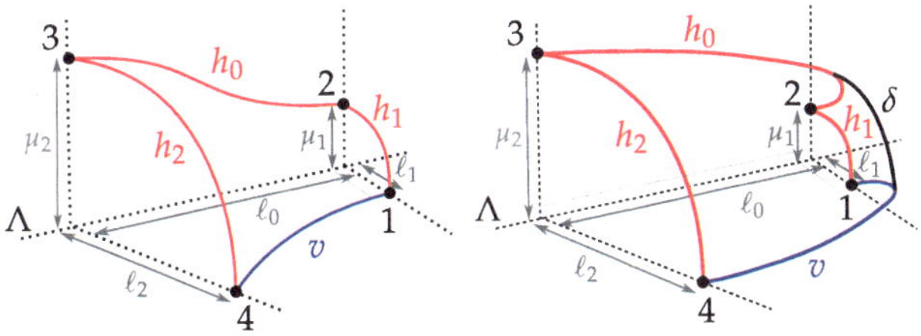

Fig. 8 Conjugate contour Γ_λ for $0 < \lambda < \frac{\pi}{2}$ (left) and $\lambda > \frac{\pi}{2}$ (right)

P_{34}, respectively, and a curve v lying in a slice P_{14}, which will be assumed to be $\mathbb{M}^2(\kappa) \times \{0\}$ after a vertical translation. The vertexes of Σ_λ will be denoted by 1–4 in correspondence with $\widetilde{1}$–$\widetilde{4}$ (see Fig. 8). All interior angles of Γ_λ are equal to $\frac{\pi}{2}$, so Proposition 3 gives a complete H-surface Σ_λ^* after successive mirror symmetries about the boundary components.

Our analysis of the angle function ν_λ indeed implies that Fig. 8 is a faithful depiction of Γ_λ, at least concerning the horizontal components. By decomposing $h_i = (\beta_i, z_i) \in \mathbb{M}^2(\kappa) \times \mathbb{R}$ component-wise for $i \in \{0, 1, 2\}$, Lemma 2 and Proposition 10 reveal that β_1, β_2, z_0, z_1, and z_2 are one-to-one. If $0 \leq \lambda \leq \frac{\pi}{2}$, then β_0 is also one-to-one, while it consists of two one-to-one subcurves if $\lambda > \frac{\pi}{2}$ (by splitting at the point where the angle function ν_λ changes sign).

To understand the dependence of Σ_λ on the parameter λ, we shall also consider the following quantities that represent the (algebraic) lengths of β_i and z_i, respectively, as shown in Fig. 8:

$$\ell_i(\lambda) = -\int_{\widetilde{h}_i} \nu_\lambda, \qquad \mu_i(\lambda) = \int_{\widetilde{h}_i} \sqrt{1 - \nu_\lambda^2} \tag{18}$$

Proposition 10 also reveals that the functions $\lambda \mapsto \ell_i(\lambda)$ satisfy the following monotonicity properties:

(a) $\lambda \mapsto \ell_0(\lambda)$ is strictly decreasing and positive on $[0, +\infty)$.
(b) $\lambda \mapsto \ell_1(\lambda)$ is strictly decreasing on $[0, +\infty)$ with $\ell_1(\frac{\pi}{2}) = 0$.
(c) $\lambda \mapsto \ell_2(\lambda)$ is strictly increasing and positive on $[0, +\infty)$.

Remark 9 If $\kappa = 0$, then Remark 8 ensures that $\widetilde{\Sigma}_\lambda$ is a spherical helicoid and hence Σ_λ^* is a Delaunay surface (see Example 5). Due to the above geometric description, Σ_λ^* stays at bounded distance from the straight line $\Lambda = P_{23} \cap (\mathbb{R}^2 \times \{0\})$, and it is symmetric with respect to P_{23} and $\mathbb{R}^2 \times \{0\}$, so in particular Σ_λ^* is rotationally invariant about Λ.

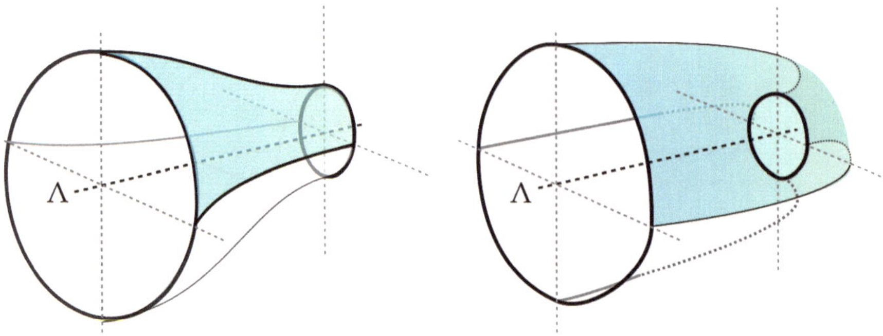

Fig. 9 Fundamental annulus A_λ constructed by extending the fundamental piece Σ_λ by means of mirror symmetries across P_{23} and $\mathbb{M}^2(\kappa) \times \{0\}$. The horizontal geodesic $\Lambda = P_{23} \cap (\mathbb{M}^2(\kappa) \times \{0\})$ is called the *axis* of Σ_λ^*

5.1.4 Compactness

Since the vertical planes P_{12} and P_{34} are orthogonal to P_{23}, it follows that Σ_λ^* is invariant under horizontal translations of length $2\ell_0(\lambda)$ along the horizontal geodesic $\Lambda = P_{23} \cap \mathbb{M}^2(\kappa) \times \{0\}$ (see Figs. 8 and 9).

Assuming that $\kappa > 0$, the surface Σ_λ^* is compact if and only if $\ell_0(\lambda)$ is a rational multiple of $\frac{2\pi}{\sqrt{\kappa}}$, the length of a great circle of $\mathbb{S}^2(\kappa)$. Since $\lambda \mapsto \ell_0(\lambda)$ is a positive continuous strictly decreasing function (see [72, Cor. 3.7] for the details), it follows that there are many compact examples in the family Σ_λ^* for $\lambda \geq 0$. Observe that the monotonicity of $\lambda \mapsto \ell_0(\lambda)$ evidences that P_{12} and P_{34} never coincide. Hence, if $\kappa \leq 0$, Σ_λ^* is a proper non-compact H-surface for all $\lambda \geq 0$.

Remark 10 The *maximum height* over the slice $\mathbb{M}^2(\kappa) \times \{0\}$ of horizontal unduloids Σ_λ^*, $0 \leq \lambda \leq \frac{\pi}{2}$, is given by $\mu_2(\lambda)$ and varies continuously from the height of the H-tori Σ_0^* to the height of the sphere $\Sigma_{\frac{\pi}{2}}^*$ (see [72, Prop. 4.5]). Notice that Σ_λ^* is singly periodic in a horizontal direction, and the monotonicity properties of Proposition 10 show that the maximum height occurs at the vertex 3 (see Fig. 8). Now, given $\lambda_1 < \lambda_2$, a comparison between Σ_{λ_2} and a $(\lambda_2 - \lambda_1)$-translated copy of Σ_{λ_1} using the flow of the helicoidal Killing vector field $\widetilde{X}$ (see Sect. 5.1.1) enables a comparison of the angle functions of both surfaces along their common boundary. Then, formula (18) ensures that $\lambda \mapsto \mu_2(\lambda)$ is strictly increasing.

Aledo, Espinar, and Gálvez [2] proved that a H-graph, with $4H^2 + \kappa > 0$, over a compact open domain whose boundary lies in the slice $\mathbb{M}^2(\kappa) \times \{0\}$ can reach at most the height of the H-sphere and equality holds if and only if the surface is a rotationally invariant hemisphere. Horizontal unduloids provide the first H-graphs with height in between those of the cylinder and the sphere.

5.1.5 Embeddedness

We will finish our sketch of the proof of Theorems 7 and 8 by saying which horizontal Delaunay surfaces are embedded. Observe first that our previous description of Γ_λ implies that horizontal nodoids ($\lambda > \frac{\pi}{2}$) are not even Alexandrov-embedded, so we can focus on unduloids with $0 < \lambda < \frac{\pi}{2}$. To this end, consider the *fundamental annulus* A_λ defined as the H-annulus in $\mathbb{M}^2(\kappa) \times \mathbb{R}$ that consists of four copies of Σ_λ obtained by mirror symmetries about P_{23} and $\mathbb{M}^2(\kappa) \times \{0\}$ (see Fig. 9).

To capture some global properties of A_λ, throughout this section, we will leave the model $M(4H^2 + \kappa, H)$ aside and assume that $\widetilde{\Sigma}_\lambda$ is immersed in $\mathbb{S}^3_{\mathrm{B}}(4H^2 + \kappa, H) \subset \mathbb{C}^2$ via the local isometry Θ given by (5). Under this assumption, the interior of $\widetilde{\Sigma}_\lambda \subset \mathbb{S}^3_{\mathrm{B}}(4H^2 + \kappa, H)$ is transversal to the Killing field $\widetilde{X}_{(z,w)} = \frac{i}{2}(-z, w)$ that defines a global Killing submersion π_0 we have used in the solution of the Plateau problem (in the Berger model, this submersion is topologically the Hopf fibration in another non-vertical direction). Therefore, the smooth function $\widetilde{u} = \langle \widetilde{X}, \widetilde{N} \rangle$ lies in the kernel of the stability operator L of $\widetilde{\Sigma}^*_\lambda$.

The function $\widetilde{u}$ is positive in the interior of $\widetilde{\Sigma}_\lambda$ and vanishes along $\widetilde{h}_1$ and $\widetilde{h}_2$ (recall that $\widetilde{\Sigma}_\lambda$ is a Killing graph in the direction of $\widetilde{X}$ and both $\widetilde{h}_1$ and $\widetilde{h}_2$ are integral curves of $\widetilde{X}$). Furthermore, $\widetilde{u}$ is preserved by the axial symmetries about $\widetilde{v}$ and $\widetilde{h}_0$ and changes sign by those about $\widetilde{h}_1$ and $\widetilde{h}_2$. This follows from a careful analysis of the relation between the axial symmetries about the boundary curves of $\widetilde{\Gamma}_\lambda$ and the screw-motion group associated with $\widetilde{X}$ (see [72, Prop. 1.6]). Since these axial symmetries correspond to mirror symmetries of A_λ in $\mathbb{M}^2(\kappa) \times \mathbb{R}$, it follows that $\widetilde{u} > 0$ in the interior of A_λ and vanishes identically along ∂A_λ. Hence, by classical elliptic theory, the first eigenvalue of the stability operator is $\lambda_1(A_\lambda) = 0$ and $\lambda_1(D) < 0$ for any open domain $D \subset \Sigma^*_\lambda$ containing A_λ. In other words, the annulus A_λ is a maximal stable domain of Σ^*_λ (for all $\lambda > 0$).

On the other hand, we can consider the function $u = \langle X, N \rangle$, where X is now the Killing field in $\mathbb{M}^2(\kappa) \times \mathbb{R}$ associated with the group of translations along the axis $\Lambda = P_{23} \cap (\mathbb{M}^2(\kappa) \times \{0\})$. Since $Lu = 0$ and u vanishes at ∂A_λ, we infer that u belongs to the eigenspace associated with $\lambda_1(A_\lambda) = 0$. This eigenspace is one-dimensional, so there exists $a_\lambda \in \mathbb{R}$ such that $u = a_\lambda \widetilde{u}$. Notice that u is only identically zero if $\lambda = 0$ because it is the only case where Σ^*_λ is equivariant. Hence, for $\lambda > 0$, we have that u does not vanish on the interior of A_λ, that is, the interior of A_λ is transversal to X, in particular h_0 and v are also transversal to X since they lie in the interior of A_λ:

1. The fundamental annulus A_λ is an H-graph in the direction of X (i.e., it intersects each integral curve of X at most once), and hence embedded, provided that $\kappa \leq 0$ or $\kappa > 0$ and $H \geq \frac{\sqrt{\kappa}}{2}$ (see [72, §4.3]).
   Notice that embeddedness finds an essential obstruction whenever Σ_λ runs over any of the poles defined by the great circle Λ (see Fig. 10). This situation is prevented by assuming that $2H > \sqrt{\kappa}$ thanks to the monotonicity properties in Sect. 5.1.4 because $\Sigma_{\pi/2}$ is the H-sphere whose *radius* $\ell_2(\frac{\pi}{2})$ is at most a quarter of the length of a great circle of $\mathbb{S}^2(\kappa)$ if $2H \geq \sqrt{\kappa}$.

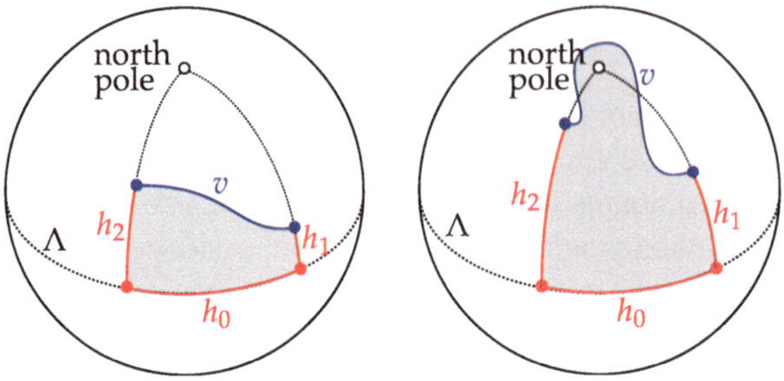

Fig. 10 Different possible projections of Σ_λ to $\mathbb{S}^2(\kappa) \times \{0\}$. In the first case, the completion Σ_λ^*, $0 \leq \lambda \leq \frac{\pi}{2}$, is embedded. The second case is not possible if $H > \frac{\sqrt{\kappa}}{2}$

2. It remains to analyze the possible type II self-intersections (after completing the fundamental annulus by further reflections); observe that they actually happen for nodoids with $\lambda > \frac{\pi}{2}$. If $0 \leq \lambda \leq \frac{\pi}{2}$, the geodesic curvature κ_g^P of v (computed as a curve of the slice $P = \mathbb{M}^2(\kappa) \times \{0\}$ with respect to its normal vector field N pointing inside the domain of the multigraph) admits the upper bound $\kappa_g^P \leq (4H^2 - \kappa)/4H$ (i.e., it is bounded by the geodesic curvature of the equator of the H-sphere; see [62, Thm. 3.3]). This implies that A_λ lies in the wedge bounded by P_{12} and P_{34}: if v escaped this region, then its length should be larger than it actually is to be able to meet P_{12} and P_{34} orthogonally from inside the wedge (see [70, p. 714]).

The argument sketched here yields the embeddedness of the unduloids if $\kappa \leq 0$ (see [72, Prop. 4.4]). However, if $\kappa > 0$, then we need to guarantee that $\ell_0(\lambda)$ is not only a rational multiple of $\frac{2\pi}{\sqrt{\kappa}}$ (so Σ_λ^* is compact; see Sect. 5.1.4) but also that $\ell_0(\lambda) = \frac{\pi}{m\sqrt{\kappa}}$ for some $m \in \mathbb{N}$, i.e., Σ_λ^* consists of $2m$ copies of A_λ and closes its period in one turn around Λ. The monotonicity properties in Sect. 5.1.4 and the well-known behavior of the cases $\lambda = 0$ and $\lambda = \frac{\pi}{2}$ (see Sect. 5.1.1) give the estimate

$$\frac{2}{\sqrt{\kappa}} \arctan \frac{\sqrt{\kappa}}{2H} = \ell_0(\tfrac{\pi}{2}) < \ell_0(\lambda) < \ell_0(0) = \frac{\pi}{\sqrt{4H^2+\kappa}}. \tag{19}$$

This easily implies that $H > \frac{\sqrt{\kappa}}{2}$ (see Fig. 5). For a fixed integer $m \geq 2$, the inequality (19) holds true if and only if $\frac{2H}{\sqrt{\kappa}} \in (\cot(\frac{\pi}{2m}), \sqrt{m^2-1})$, in which case there is a unique $\lambda = \lambda_m(H)$ such that $\ell_0(\lambda_m(H)) = \frac{\pi}{m\sqrt{\kappa}}$ because $\lambda \mapsto \ell_0(\lambda)$ is continuous and strictly decreasing. This gives rise to the family $\mathcal{T}_m$ in the statement of Theorem 8, with the following limit cases:

- If $\frac{2H}{\sqrt{\kappa}} = \cot(\frac{\pi}{2m})$, then $m = \frac{\pi}{2\arctan(\frac{\sqrt{\kappa}}{2H})}$, and hence $\ell_0(\frac{\pi}{2}) = \ell_0(\lambda)$. This means that $\lambda = \frac{\pi}{2}$ and the surface reduces to a stack of m tangent H-spheres.
- Likewise, if $\frac{2H}{\sqrt{\kappa}} = \sqrt{m^2-1}$, then $\lambda = 0$, and the surface is an H-torus.

5.2 Compact H-Surfaces of Arbitrary Genus in $\mathbb{S}^2(\kappa) \times \mathbb{R}$

The second conjugate construction we present in this survey concerns compact embedded H-surfaces with arbitrary genus $g \geq 0$ in $\mathbb{S}^2(\kappa) \times \mathbb{R}$ and mean curvature $H < \frac{\sqrt{\kappa}}{2}$. The idea is to produce a compact fundamental piece $\Sigma_{H,g}$ whose projection fits in a fundamental triangle of a regular tessellation of $\mathbb{S}^2(\kappa)$ by regular polygons (see Sect. 5.2.1). The complete H-surface $\Sigma^*_{H,g}$ after reflection about its symmetry planes inherits all the symmetries of the tessellation of $\mathbb{S}^2(\kappa)$, whence it is compact. In the case of genus 0 or 1, we already have the rotationally invariant H-spheres and H-tori, so our result is relevant for $g \geq 2$. Observe that the horizontal Delaunay tori given by Theorem (8) satisfy the opposite inequality $H > \frac{\sqrt{\kappa}}{2}$.

Therefore, it is worth saying something about the assumption $H < \frac{\sqrt{\kappa}}{2}$, which initially showed up in the continuity argument used to adjust $\Sigma_{H,g}$ to the shape of a fundamental triangle but then proved to be a natural constraint. The value $H = \frac{\sqrt{\kappa}}{2}$ is geometrically relevant in $\mathbb{S}^2(\kappa) \times \mathbb{R}$ because it is the value of H for which H-spheres are bigraphs over a hemisphere, and hence two of them are tangent along a whole equator. As a matter of fact, we can think of the map $H \mapsto \Sigma^*_{H,g}$, for a fixed $g \geq 2$, as a desingularization of two such tangent spheres as $H \to \frac{1}{2}\sqrt{\kappa}$. This number is also natural in the proof of embeddedness, which uses the convexity of the boundary components in [62, Cor. 3.5]. Recall that in the proof of embeddedness in Theorem (8), the opposite condition $H > \frac{\sqrt{\kappa}}{2}$ has also appeared as a natural constraint that prevents the surfaces to surpass the north pole and ensures that the H-tori close one of their periods.

The main result of this section is the following theorem:

Theorem 9 *Let* $0 < H < \frac{\sqrt{\kappa}}{2}$ *and an integer* $g \geq 0$. *There exists a compact embedded* H*-surface* $\Sigma^*_{H,g}$ *of genus* g *in* $\mathbb{S}^2(\kappa) \times \mathbb{R}$ *that is a bigraph over a slice and inherits all the symmetries of a* $(2, g+1)$*-tessellation (see Sect. 5.2.1). Furthermore, if* $g \geq 2$,

- *The limit of* $\Sigma^*_{H,g}$ *as* $H \to \frac{\sqrt{\kappa}}{2}$ *is a pair of* $\frac{\sqrt{\kappa}}{2}$*-spheres tangent along an equator.*
- *The limit of* $\Sigma^*_{H,g}$ *as* $H \to 0$ *is a double cover of* $\mathbb{S}^2(\kappa) \times \{0\}$ *with singularities at* $g + 1$ *points evenly distributed along an equator.*

Remark 11 It is possible to obtain a similar result for the rest of tessellations of the sphere associated with regular polyhedra. However, just a few genera can be recovered in this way, plus different restrictions for the mean curvature appear depending on the tessellation (see Table 4 and [71, Thm. 1.1]).

The same construction can be carried out for H-surfaces in $\mathbb{M}^2(\kappa) \times \mathbb{R}$ with $\kappa \leq 0$ and an arbitrary regular tessellation of $\mathbb{M}^2(\kappa) \times \mathbb{R}$ by regular polygons [71, Thm. 1.1]. In the case $\kappa = 0$, our construction reduces to Lawson's doubly periodic 1-surfaces in Euclidean space $\mathbb{R}^3$ [58, Thm. 9] (see also [28, §3]). In the case $\kappa < 0$, we obtain new properly immersed surfaces of subcritical, critical, and supercritical constant mean curvature in a slab of $\mathbb{H}^2(\kappa) \times \mathbb{R}$, though we have not been able to analyze their embeddedness.

5.2.1 Regular Tessellations

Given integers $m, k \geq 2$, a regular (m, k)-tessellation is a tiling of $\mathbb{M}^2(\kappa)$ by regular m-gons such that k of them meet at each vertex (see Fig. 11 for an example of a $(2, 3)$-tessellation of the sphere). The centers and vertexes of the m-gons will be called the centers and vertexes of the tessellation. A straightforward application of the Gauß-Bonnet formula to one of the m-gons reveals that the sign of $\frac{1}{k} + \frac{1}{m} - \frac{1}{2}$ agrees with the sign of κ, in which case the (m, k)-tessellation of $\mathbb{M}^2(\kappa)$ actually exists.

In the case of $\mathbb{S}^2(\kappa)$, the inequality $\frac{1}{k} + \frac{1}{m} > \frac{1}{2}$ is quite restrictive, for it only allows the tessellations associated with the Platonic solids and also two infinite families: the *beach ball* tessellations (with $m = 2$ and arbitrary k) and the *degenerated* case for $k = 2$ and arbitrary m). The possible configurations are shown in Table 4. Notice that the (m, k) and the (k, m)-tessellation are dual by swapping centers and vertexes and hence they have the same isometry group.

In a (m, k)-tessellation of $\mathbb{S}^2(\kappa)$, each m-gon can be decomposed into $2m$ congruent triangles by joining the center with the vertexes and the midpoints of the sides. Each of these triangles has angles $\frac{\pi}{k}$, $\frac{\pi}{m}$, and $\frac{\pi}{2}$ and will be called a *fundamental triangle* because the whole tessellation can be recovered by symmetries about its sides. The idea is to construct an H-bigraph in $\mathbb{S}^2(\kappa) \times \mathbb{R}$ symmetric with respect to $\mathbb{S}^2(\kappa) \times \{0\}$ and with a curve at height zero around each of the centers (see Fig. 11). This gives a surface of genus the number of polygons of the tessellation minus one (see the last column of Table 4), so we will focus on the $(2, g + 1)$-tessellation of $\mathbb{S}^2(\kappa)$ for $g \geq 2$, which leads to the proof of Theorem 9 that will be sketched throughout this section.

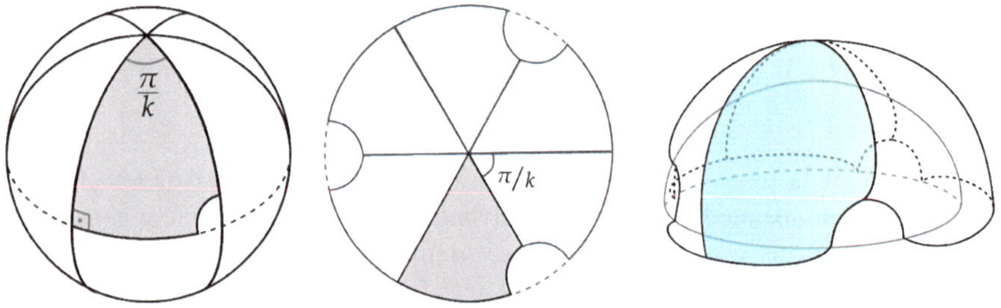

Fig. 11 From left to right for $k = 3$: $(2, k)$-tessellation of the sphere (see Sect. 5.2.1), view of the $(2, k)$-tessellation from the north pole and a sketch of $\Sigma_{H,g}$ $(g = 2)$ in the conformal model $\mathbb{R}^3 \setminus \{0\}$ of $\mathbb{S}^2(\kappa) \times \mathbb{R}$ (right). The shaded area in the first two figures is the projection of $\Sigma_{H,g}$ over its slice of symmetry $\mathbb{S}^2(\kappa) \times \{0\}$

Table 4 For each (m, k)-tessellation, [71, Thm. 1.1] gives compact orientable H-surfaces in $\mathbb{S}^2(\kappa) \times \mathbb{R}$ provided that $0 < \frac{4H^2}{\kappa} < \alpha(m, k)$ (see also Remark 11). Theorem 9 corresponds to the case of a *beach ball* tessellation

Initial tessellation	(m, k)	$\alpha(m, k)$	Genus of Σ
Beach ball	$(2, g+1)$	1	g
Degenerated	$(g+1, 2)$	$\cot^2(\frac{\pi}{2+2g})$	1
Tetrahedron	$(3, 3)$	$2+\sqrt{3}$	3
Hexahedron	$(4, 3)$	$5+2\sqrt{6}$	5
Octahedron	$(3, 4)$	$3+2\sqrt{2}$	7
Dodecahedron	$(5, 3)$	$8+4\sqrt{3}+3\sqrt{5}+2\sqrt{15}$	11
Icosahedron	$(3, 5)$	$4+\sqrt{5}+2\sqrt{5+2\sqrt{5}}$	19

5.2.2 Construction of the Minimal Surface in $\mathbb{S}^3_{\mathrm{B}}(4H^2+\kappa, H)$

We will fix positive real numbers H and κ, and the target genus $g \geq 2$ in what follows so we will omit the dependence on these data. Instead, we will take a parameter $0 < \rho \leq \frac{\pi}{\sqrt{4H^2+\kappa}}$ that will provide one degree of freedom to be used later in a continuity argument. The dependence on ρ will be written in functional notation to make it clear that it represents an auxiliary parameter.

Consider a convex spherical triangle $\widetilde{\Delta}(\rho) \subset \mathbb{S}^2(4H^2+\kappa)$ with two angles $\frac{\pi}{g+1}$ and $\frac{\pi}{2}$ adjacent to a side of length ρ. This triangle defines a geodesic quadrilateral $\widetilde{\Gamma}(\rho) \subset \mathbb{S}^3_{\mathrm{B}}(4H^2+\kappa, H)$ with three horizontal sides $\widetilde{h}_1$, $\widetilde{h}_2$, and $\widetilde{h}_3$ projecting to the sides of $\widetilde{\Delta}(\rho)$ (being ρ the length of $\widetilde{h}_2$) and a vertical segment $\widetilde{v}$ joining the ends of $\widetilde{h}_1$, and $\widetilde{h}_3$. The vertexes of $\widetilde{\Gamma}(\rho)$ will be denoted by $\widetilde{1}$, $\widetilde{2}$, $\widetilde{3}$, and $\widetilde{4}$, as shown in Fig. 12 (top left).

The pair $(\widetilde{\Delta}(\rho), \widetilde{\Gamma}(\rho))$ is a Nitsche graph (see Definition 1) such that $W(\rho) = \pi^{-1}(\widetilde{\Delta}(\rho))$ is a mean-convex set, so Proposition 7 ensures the existence and uniqueness of a minimal disk $\widetilde{\Sigma}(\rho) \subset W(\rho)$ with boundary $\widetilde{\Gamma}(\rho)$ for any $0 < \rho \leq \frac{\pi}{\sqrt{4H^2+\kappa}}$, which is also a graph over the interior of $\widetilde{\Delta}(\rho)$. We will assume without loss of generality that the angle function ν_ρ of $\widetilde{\Delta}(\rho)$ is negative over the interior of $\widetilde{\Delta}(\rho)$. We can analyze ν_ρ by means of the boundary maximum principle with respect to ∂W and by Strategy 2, discussed in Sect. 3.1.2.

Proposition 11 ([71, §3.1]) *Let $\nu_\rho \leq 0$ be the angle function of the compact minimal disk $\widetilde{\Sigma}(\rho)$ spanning $\widetilde{\Gamma}(\rho)$:*

(a) The only points at which ν_ρ vanishes are those in the curve $\widetilde{v}$.
(b) The only points at which ν_ρ takes the value -1 are $\widetilde{2}$ and $\widetilde{3}$.

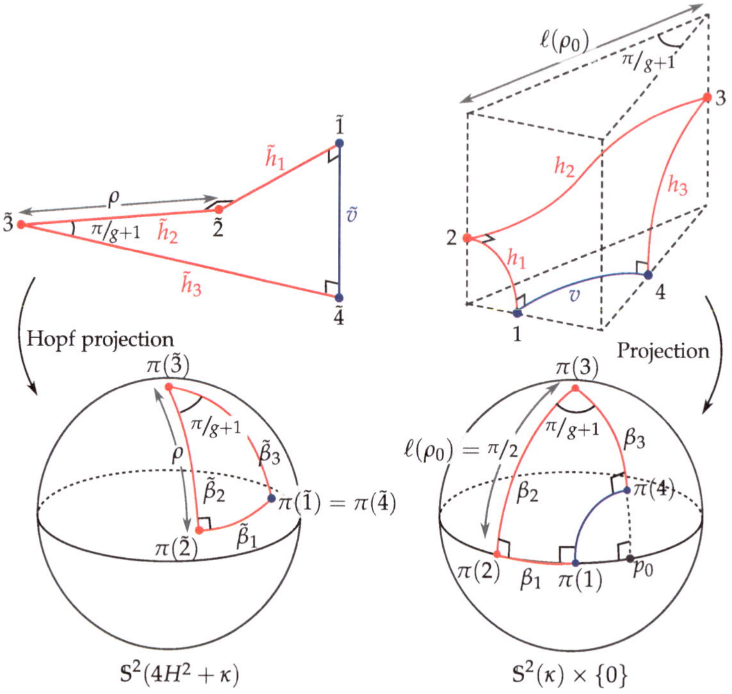

Fig. 12 Geodesic polygon $\widetilde{\Gamma}(\rho)$ (top left), its sister contour $\Gamma(\rho)$ (top right) in the target case $\ell(\rho_0) = \frac{\pi}{2\sqrt{\kappa}}$ (see Sect. 5.2.3), and their projections to $\mathbb{S}^2(4H^2+\kappa)$ via the Hopf fibration (bottom left) and to $\mathbb{S}^2(\kappa)$ (bottom right)

5.2.3 The Conjugate H-Immersion

Thanks to Lemmas 2 and 3, the conjugate H-surface $\Sigma(\rho)$ in $\mathbb{S}^2(\kappa) \times \mathbb{R}$ is bounded by a contour $\Gamma(\rho)$ formed by the curves h_1, h_2, and h_3 (corresponding to $\widetilde{h}_1$, $\widetilde{h}_2$, $\widetilde{h}_3$) contained in vertical planes and v (corresponding to $\widetilde{v}$) contained in a horizontal slice, which will be assumed to be $\mathbb{S}^2(\kappa) \times \{0\}$ after a vertical translation (see Fig. 12 top right). Their endpoints will be denoted by 1–4, in correspondence with $\widetilde{1}$–$\widetilde{4}$.

Write $h_i = (\beta_i, z_i) \in \mathbb{S}^2(\kappa) \times \mathbb{R}$ for $i \in \{1, 2, 3\}$ as in Lemma 2. From the properties of ν_ρ in Proposition 11, it follows that the curves β_i are one-to-one, and the height components z_i are strictly monotonic. On the one hand, it is easy to show that β_1, β_2, and β_3 are part of the geodesics containing the sides of a spherical triangle $\Delta(\rho) \subset \mathbb{S}^2(\kappa)$ with two angles equal to $\frac{\pi}{g+1}$ and $\frac{\pi}{2}$, which coincide with the angles made by h_1 and h_2 and by h_2 and h_3, respectively (see Fig. 12 bottom right). On the other hand, v is strictly convex as a curve of $\mathbb{S}^2(\kappa) \times \{0\}$ with respect to $-N$ as a conormal along v by Lemma 3. As a consequence, the curve $\pi \circ v$ is embedded and contained in $\Delta(\rho)$. Were it not the case, $\pi \circ v$ would intersect itself or other points of some β_i producing a convex loop in $\pi(\Gamma(\rho))$. However, $\pi(\Sigma(\rho))$ must Alexandrov-embedded as a domain of $\mathbb{S}^2(\kappa)$ since π restricted to $\Sigma(\rho)$ is an immersion (its Jacobian is the angle function ν_ρ, which does not vanish in the interior). In Sect. 5.2.4, we will prove that $\Sigma(\rho)$ is a graph and confirm that Fig. 12 is a faithful picture of $\Gamma(\rho)$.

Proposition 3 allows us to produce a complete H-surface $\Sigma^*(\rho)$ by successive mirror symmetries about the totally geodesic surfaces containing h_1, h_2, h_3, and v. The symmetry group of $\Sigma^*(\rho)$ depends on the shape of the triangle $\Delta(\rho)$, which is determined by the distance $\ell(\rho)$ from $\pi(2)$ to $\pi(3)$ given by

$$\ell(\rho) = -\int_{h_2} \nu_\rho = -\int_{\widetilde{h}_2} \nu_\rho, \tag{20}$$

since the angle function ν_ρ has been chosen negative. Observe that $\widetilde{\Gamma}(\rho)$ depends continuously on ρ, so the uniqueness in the Plateau problem (solved by $\widetilde{\Sigma}(\rho)$) and Proposition 2 yields that $\rho \mapsto \ell(\rho)$ is a continuous function. This enables a continuity argument to prove that there always exists a value of ρ_0 such that $\ell(\rho_0) = \frac{\pi}{2\sqrt{\kappa}}$ (see Fig. 12 bottom right), in which case $\Sigma^*(\rho_0)$ is compact since it has the symmetries of the $(2, g+1)$-tessellation. Let us analyze the limit cases:

- As $\rho \to 0$, the length of $\widetilde{h}_2$ converges to zero, so $\ell(\rho)$ gets arbitrarily close to zero by Equation (20).
- As $\rho \to \frac{\pi}{\sqrt{4H^2+\kappa}}$, the surface $\widetilde{\Sigma}(\rho)$ converges to $\frac{1}{4(g+1)}$ of the horizontal umbrella centered at $\widetilde{3}$, whence $\Sigma(\rho)$ is a sector of angle $\frac{\pi}{g+1}$ of the upper half of an H-sphere $S_{H,\kappa,0}$ in $\mathbb{S}^2(\kappa) \times \mathbb{R}$ (see Example 3). This means that $\ell(\frac{\pi}{\sqrt{4H^2+\kappa}})$ is the radius of the spherical circle of $\mathbb{S}^2(\kappa)$ over which $S_{H,\kappa,0}$ is a bigraph, that is, $\ell(\frac{\pi}{\sqrt{4H^2+\kappa}}) = \frac{2}{\sqrt{\kappa}} \arctan \frac{\sqrt{\kappa}}{2H}$, cp. Equation (19).

By the intermediate value theorem, $\ell(\rho)$ takes all values in $(0, \frac{2}{\sqrt{\kappa}} \arctan \frac{\sqrt{\kappa}}{2H})$, though it might take each value more than once. We finish the argument by realizing that this interval contains the target value $\frac{\pi}{2\sqrt{\kappa}}$ if and only if $H < \frac{\sqrt{\kappa}}{2}$. Observe that we can discuss the topology of the complete surface $\Sigma^*(\rho_0)$ analytically by means of Gauß-Bonnet formula if $\ell(\rho_0) = \frac{\pi}{2\sqrt{\kappa}}$: since $8(g+1)$ copies of $\Sigma(\rho_0)$ are needed to get a compact surface and the total curvature of each piece is $\int_{\Sigma(\rho_0)} K = \frac{\pi}{g+1} - \frac{\pi}{2}$, we easily deduce that the genus of $\Sigma^*(\rho_0)$ is g.

If $H \to \frac{\sqrt{\kappa}}{2}$, then $\frac{2}{\sqrt{\kappa}} \arctan \frac{\sqrt{\kappa}}{2H} \to \frac{\pi}{2\sqrt{\kappa}}$, which forces $\rho_0 \to \frac{\pi}{\sqrt{4H^2+\kappa}}$, and the constructed surface $\Sigma(\rho_0)$ converges to a subset of an $\frac{\sqrt{\kappa}}{2}$-sphere, so $\Sigma^*(\rho_0)$ becomes a pair of tangent $\frac{\sqrt{\kappa}}{2}$-spheres. If $H \to 0$, then we will immerse $\widetilde{\Sigma}_{\rho_0}$ immersed in the local model $M(4H^2 + \kappa, H)$ via the isometry in (5). As the bundle curvature of $M(4H^2 + \kappa, H)$ tends to zero, so does Length$(\widetilde{v})$ $=$ $2H$Area$(\widetilde{\Delta}(\rho))$ (see the discussion about the geometric meaning of the bundle curvature in Sect. 2.3). The maximum principle and the stability of the piece $\widetilde{\Sigma}(\rho_0)$ imply that its angle function ν_{ρ_0} converges uniformly to -1, so the conjugate piece $\Sigma(\rho_0)$ becomes in the limit a slice (with singularities at the vertexes of the tessellation). We remark that we have used the model $M(4H^2+\kappa, H)$ instead of the Berger sphere $\mathbb{S}^3_{\mathrm{B}}(4H^2+\kappa, H)$ to study the limit due to the fact that, as $H \to 0$, the Berger sphere collapses onto $\mathbb{S}^2(\kappa)$ while the Cartan model smoothly converges to $\mathbb{S}^2(\kappa)\times\mathbb{R}$.

5.2.4 Embeddedness

Finally, we will show that the constructed fundamental piece $\Sigma(\rho_0)$ is actually a graph over some domain of $\mathbb{S}^2(\kappa)$ (so there are no self-intersections of type I (see Sect. 3.1.3)) and lies in the prism $\Delta(\rho_0)\times\mathbb{R}$ (so there are no self-intersections of type II (see Sect. 3.1.3)), where $\Delta(\rho_0)$ is the desired triangle with angles $\frac{\pi}{g+1}$, $\frac{\pi}{2}$ and $\frac{\pi}{2}$. This clearly implies that the completion $\Sigma^*(\rho_0)$ is embedded (see Fig. 12 right). To this end, a standard application of the maximum principle reveals that it suffices to show that $\Gamma(\rho_0)$ is a graph and lies in $\Delta(\rho_0)\times\mathbb{R}$. After our discussion in Sect. 5.2.2, it will be enough to prove that $\pi(1)$ and $\pi(4)$ lie in $\Delta(\rho_0)$, i.e., the geodesics β_1 and β_3 do not reach the point p_0 shown in Fig. 12.

In the case of $\pi(1)$, this is a consequence of the fact that the distance from $\pi(2)$ to p_0 is $\frac{\pi}{(g+1)\sqrt{\kappa}}$, whereas the length of β_1 can be estimated as

$$\mathrm{Length}(\beta_1) = -\int_{\widetilde{h}_1} \nu \leq \mathrm{Length}(\widetilde{h}_1) \leq \frac{\pi}{(g+1)\sqrt{4H^2+\kappa}} < \frac{\pi}{(g+1)\sqrt{\kappa}}. \tag{21}$$

The last inequality in (21) follows from the fact that $\widetilde{h}_1$ has maximum length when $\widetilde{h}_2$ is a quarter of a great circle of $\mathbb{S}^2(4H^2+\kappa)$. As for $\pi(4)$, since varying ρ in the construction produces a foliation of a region of the Berger sphere $\mathbb{S}^3_{\mathrm{B}}(4H^2+\kappa, 4H)$, this implies that ν_ρ along β_3 depends monotonically on ρ. The maximum value of $\mathrm{Length}(\beta_3) = -\int_{\widetilde{h}_3}\nu$ is thus attained when $\rho = \frac{\pi}{\sqrt{4H^2+\kappa}}$, since for this value we integrate the largest function on the largest interval. In particular, $\mathrm{Length}(\beta_3)$ is less than $\frac{\pi}{2\sqrt{\kappa}}$, the radius of the disk over which the $\frac{\sqrt{\kappa}}{2}$-sphere is a bigraph. This means that $\pi(4)$ also lies in the boundary of $\Delta(\rho_0)$.

5.3 *Compact Minimal Surfaces in* $\mathbb{S}^2(\kappa)\times\mathbb{S}^1(\eta)$

This construction concerns periodic minimal surfaces in $\mathbb{S}^2(\kappa)\times\mathbb{R}$ that are compact in the quotient by a vertical translation of a certain length $2h$, i.e., in the homogeneous three-manifold $\mathbb{S}^2(\kappa)\times\mathbb{S}^1(\eta)$, where $\mathbb{S}^1(\eta)$ is a circle of curvature $\eta = \frac{\pi}{h}$.

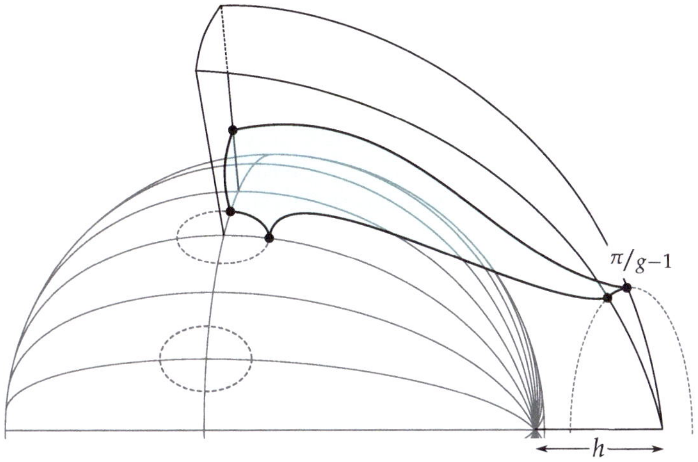

Fig. 13 A sketch of the fundamental piece of the minimal surface $\Sigma_{g,\eta}$ of Theorem 10 represented in the conformal model $\mathbb{R}^3 - \{(0,0,0)\}$ of $\mathbb{S}^2(\kappa) \times \mathbb{R}$. See also the numerical approximations in Fig. 25

Hoffman, Traizet, and White [42, Thm. 1] obtained a class of properly embedded minimal surfaces in $\mathbb{S}^2(\kappa) \times \mathbb{R}$ called *periodic genus g helicoids* as well as infinitely many noncongruent compact orientable embedded minimal surfaces in $\mathbb{S}^2(\kappa) \times \mathbb{S}^1(\eta)$ with arbitrary genus $g \geq 2$ and arbitrary $\eta > 0$ [42, Thm. 2]. Note that as we are dealing with minimal surfaces, non-orientable examples might also exist. Rosenberg [95, §4] constructed non-orientable compact minimal surfaces in $\mathbb{S}^2(\kappa) \times \mathbb{S}^1(\eta)$, for all $\eta > 0$, with even Euler characteristic, and in [68] it is proved that there cannot be examples with odd Euler characteristic.

We will obtain other compact minimal surfaces of arbitrary genus $g \geq 3$ that can be thought of as Schwarz P-surfaces in $\mathbb{S}^2(\kappa) \times \mathbb{S}^1(\eta)$ (see Fig. 13). As in the constructions in Sect. 5.1 and Sect. 5.2, we will produce a fundamental piece fitting a tile of a tessellation of $\mathbb{S}^2(\kappa) \times \mathbb{R}$ giving the desired genus (in the quotient) after extending it by symmetries about its boundary components. The conjugate technique starts with another minimal surface in $\mathbb{S}^2(\kappa) \times \mathbb{R}$ and involves a continuity argument that only allows us to get the result for η large enough (see Sect. 5.3.3). Moreover, in principle, this family might fail to be continuous as in Sect. 5.1.

The main result is the following theorem (see also Fig. 13):

Theorem 10 ([68, Prop. 3]) *For any integer $g \geq 3$ and $\eta > 2\sqrt{\kappa}$ big enough depending on g and κ, there exists a compact embedded orientable minimal surface $\Sigma_{g,\eta}$ with genus g in $\mathbb{S}^2(\kappa) \times \mathbb{S}^1(\eta)$ that is a bigraph over a slice and inherits all the symmetries of a $(2, g-1)$-tessellation of $\mathbb{S}^2(\kappa)$ (see Sect. 5.2.1).*

Remark 12 In the cases $g = 0$ and $g = 1$, there is no such a Schwarz P-surface, but in these cases we can easily produce a compact minimal surface of genus

g by considering the $\mathbb{S}^2(\kappa) \times \{0\}$ or a vertical torus. The case $g = 2$ is more subtle, since the arguments in the proof of Theorem 10 fail, and we believe that no such Schwarz P-surface with these prescribed symmetries will exist. Recall that Hofmann, Traizet, and White have provided compact embedded examples of all genera. We also presented a similar but more symmetric construction in [68, §3.2] to produce compact minimal surfaces in $\mathbb{S}^2(\kappa) \times \mathbb{S}^1(\eta)$ with arbitrary odd genus $2k - 1$ for $k \geq 2$ and η large enough, which inherit the symmetries of a $(2, k)$-tessellation, but they are not congruent those in Theorem 10. Besides, a genus 7 compact minimal example in $\mathbb{S}^2(\kappa) \times \mathbb{S}^1(\eta)$ with the symmetries of the (4, 3)-tessellation (i.e., the hexahedron; see Table 4) is also constructed.

On the other hand, the strange condition $\eta > 2\sqrt{\kappa}$ shows up in the spherical geometry when one tries to produce the initial geodesic polygon, so there cannot be a Schwarz P-surface in $\mathbb{S}^2(\kappa) \times \mathbb{S}^1(\eta)$ with $\eta \leq 2\sqrt{\kappa}$ enjoying the symmetries we have prescribed.

5.3.1 Construction of the Minimal Surface in $\mathbb{S}^2(\kappa) \times \mathbb{R}$

We will fix $\eta > 0$ and the genus $g \geq 3$ in the sequel, not writing the dependence on these variables. Consider three real parameters $0 \leq \widetilde{a}, \widetilde{b} \leq \frac{\pi}{2\sqrt{\kappa}}$ and $\rho > 0$, and define the geodesic triangle $\widetilde{\Delta}(\widetilde{a}, \widetilde{b}) \subset \mathbb{S}^2(\kappa) \times \{0\}$ with two sides of lengths $\widetilde{a}$ and $\widetilde{b}$ meeting at an angle of $\frac{\pi}{2}$. The opposite angles will be denoted by $\widetilde{\alpha}$ and $\widetilde{\beta}$. We can produce a geodesic polygon $\widetilde{\Gamma}(\widetilde{a}, \widetilde{b}, \rho)$ by adding two vertical geodesic segments of length ρ at the vertexes $\widetilde{\alpha}$ and $\widetilde{\beta}$. Therefore, $\widetilde{\Gamma}(\widetilde{a}, \widetilde{b}, \rho)$ is a closed curve in $\mathbb{S}^2(\kappa) \times \mathbb{R}$ whose vertexes will be denoted by $\widetilde{1}, \widetilde{2}, \widetilde{3}, \widetilde{4}$, and $\widetilde{5}$ as shown in Fig. 14

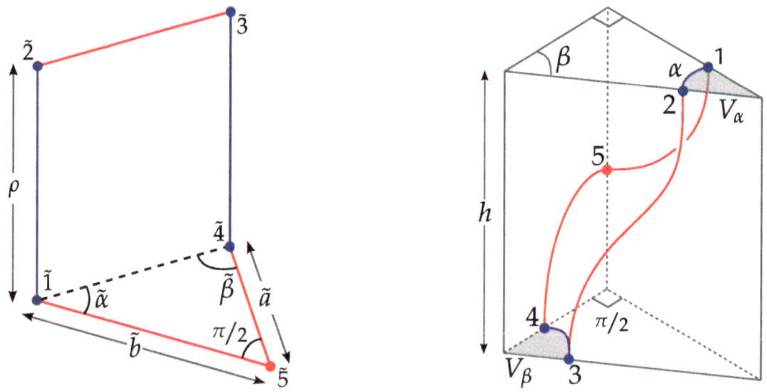

Fig. 14 The polygon $\widetilde{\Gamma}(\widetilde{a}, \widetilde{b}, \rho)$ (left) and its conjugate $\Gamma(\widetilde{a}, \widetilde{b}, \rho)$ (right) that lies inside the prism demarcated by the symmetry planes. Note that by construction $h \leq \ell(23) = \ell(\widetilde{23}) \leq \frac{\pi}{2\sqrt{\kappa}}$ so we get the restriction $\eta = \frac{\pi}{h} \geq 2\sqrt{\kappa}$ in Theorem 10. The restriction $\rho \leq \frac{\pi}{2\sqrt{\kappa}}$ is unrelated to this and will be used to prove embeddedness. Moreover, $\alpha = \text{area}(V_\alpha) + \widetilde{\alpha}$ and $\beta = \text{area}(V_\beta) + \widetilde{\beta}$

left. Notice that $\widetilde{\Gamma}(\widetilde{a}, \widetilde{b}, \rho)$ is still well defined for $0 \leq \widetilde{a}, \widetilde{b} \leq \frac{\pi}{\sqrt{\kappa}}$ not both equal to $\frac{\pi}{\sqrt{\kappa}}$ (a half of the length of a great circle of $\mathbb{S}^2(\kappa)$), though imposing the restriction $0 \leq \widetilde{a}, \widetilde{b} \leq \frac{\pi}{2\sqrt{\kappa}}$ is essential for some of the arguments in our construction (observe that this restriction on $\widetilde{a}$ and $\widetilde{b}$ implies that the angles $\widetilde{\alpha}$ and $\widetilde{\beta}$ are at most $\frac{\pi}{2}$).

Remark 13 One can also consider the boundary curve $\widetilde{\Gamma}(\widetilde{a}, \widetilde{b}, \rho)$ constructed likewise using horizontal and vertical geodesics in $\mathbb{R}^3$. This leads, by conjugation, to the classical Schwarz P-surface, whose name is motivated by the fact that it is invariant under a primitive cubic lattice C, when choosing $\widetilde{a} = \widetilde{b} = \rho$ and $\widetilde{\alpha} = \widetilde{\beta} = \frac{\pi}{4}$ (e.g., see [51, §1.6.2]). The quotient of the conjugate surface under the lattice C has genus 3 and consists of 16 copies of the fundamental piece with $\alpha = \beta = \frac{\pi}{4}$ and $a = b$ (cp. Fig. 14 right). Here, the actual lengths of $\widetilde{a}$ and a are not relevant, since we can change them by homotheties of $\mathbb{R}^3$. Note also that this construction in $\mathbb{R}^3$ can be seen as a limit of our construction as $\kappa \to 0$.

By construction, the triangle $\widetilde{\Delta}(\widetilde{a}, \widetilde{b})$ and the polygon $\widetilde{\Gamma}(\widetilde{a}, \widetilde{b}, \rho)$ form a Nitsche contour, so Proposition 7 ensures the existence and uniqueness of a minimal disk $\widetilde{\Sigma}(\widetilde{a}, \widetilde{b}, \rho) \subset \widetilde{\Delta}(\widetilde{a}, \widetilde{b}) \times \mathbb{R}$ with boundary $\widetilde{\Gamma}(\widetilde{a}, \widetilde{b}, \rho)$, whose interior is a graph over $\widetilde{\Gamma}(\widetilde{a}, \widetilde{b}, \rho)$. We will assume that its angle function $\nu_{\widetilde{a},\widetilde{b},\rho}$ is positive in the interior without loss of generality. By the Strategy 2 in Sect. 3.1.2 and the boundary maximum principle, we deduce that the angle function only vanishes along the vertical segments $\widetilde{12}$ and $\widetilde{34}$, whereas it only takes the value 1 at the vertex $\widetilde{5}$.

5.3.2 The Conjugate Minimal Surface

Let $\Sigma(\widetilde{a}, \widetilde{b}, \rho)$ be the conjugate minimal surface in $\mathbb{S}^2(\kappa) \times \mathbb{R}$ whose boundary $\Gamma(\widetilde{a}, \widetilde{b}, \rho)$ consists of curves contained in vertical or horizontal totally geodesic surfaces that Σ meets orthogonally (see Lemmas 2 and 3). We will denote by 1, 2, 3, 4, and 5 the corresponding vertexes of $\Gamma(\widetilde{a}, \widetilde{b}, \rho)$ and by P_{ij} the vertical plane or horizontal slice containing i and j. Up to a vertical translation, we will assume that $P_{34} = \mathbb{S}^2(\kappa) \times \{0\}$.

We will sketch the proof that $\Sigma(\widetilde{a}, \widetilde{b}, \rho)$ is embedded and contained in a triangular prism of $\mathbb{S}^2(\kappa) \times \mathbb{R}$ (see [68, Lem. 3] for further details). To prove this, we will additionally assume that $\rho \leq \frac{\pi}{2\sqrt{\kappa}}$.

Observe that P_{45} and P_{51} are different since they form an angle of $\frac{\pi}{2}$. Moreover, P_{12} and P_{34} do not coincide (so they are at a certain distance $h > 0$), and the height of the point 5 is between 0 and h; otherwise, we would find a contradiction with the maximum principle when comparing with slices $\mathbb{S}^2(\kappa) \times \{t\}$. Due to the behavior of the angle function, the curves 23, 45, and 51 project one-to-one to $\mathbb{S}^2(\kappa) \times \{0\}$. The normal to $\widetilde{\Sigma}(\widetilde{a}, \widetilde{b}, \rho)$ has different directions of rotation along $\widetilde{12}$ and $\widetilde{34}$, so the curves 12 and 34 are convex arcs that $\Sigma(\widetilde{a}, \widetilde{b}, \widetilde{h})$ meets from above and below, respectively, by Lemma 3. This easily leads to the fact that P_{45}, P_{23}, and P_{15} demarcate a spherical triangle $\Delta(\widetilde{a}, \widetilde{b}, \rho)$ in the projection to $\mathbb{S}^2(\kappa)$, such that

$\Sigma(\widetilde{a}, \widetilde{b}, \rho) \subset \Delta(\widetilde{a}, \widetilde{b}, \rho) \times \mathbb{R}$. We will call α and β the angles of this triangle, as indicated in Fig. 14.

It remains to prove that 12 and 34 are embedded and P_{23}, P_{15}, and P_{45} are pairwise distinct. We will explain the argument for 12, being the case 34 analogous. The total curvature of 12 is $\int_{12} \kappa_g = -\widetilde{\alpha}$, i.e., minus the total rotation of the normal along $\widetilde{12}$ (note that κ_g is computed with respect to the normal N such that $\nu > 0$ in the interior). If 12 were not embedded, then, by convexity, it would contain a loop enclosing a domain D, where Gauß-Bonnet formula yields $\kappa \operatorname{Area}(D) \geq \pi - \widetilde{\alpha} \geq \frac{\pi}{2}$. On the other hand, $\operatorname{Length}(\partial D) \leq \rho \leq \frac{\pi}{2\sqrt{\kappa}}$ by assumption, which contradicts the isoperimetric inequality in $\mathbb{S}^2(\kappa)$ that reads $4\pi \operatorname{Area}(D) - \kappa \operatorname{Area}(D)^2 \leq \operatorname{Length}(\partial D)^2$.

Once we know that both curves 12 and 34 are embedded, they must enclose some domains V_α and V_β as in Fig. 14, demarcated by the horizontal geodesics orthogonal at their endpoints. Gauß-Bonnet formula applied to these domains V_α and V_β gives the relations $\alpha = \operatorname{area}(V_\alpha) + \widetilde{\alpha}$ and $\beta = \operatorname{area}(V_\beta) + \widetilde{\beta}$. The same argument as above shows that the angles α and β do not exceed π whenever $\rho \leq \frac{\pi}{2\sqrt{\kappa}}$, so the planes P_{23}, P_{45}, and P_{51} are pairwise different.

5.3.3 Compactness

As we have mentioned in the introduction, our plan to prove Theorem 10 is to fit $\Sigma(\widetilde{a}, \widetilde{b}, \rho)$ in the prism with base half of a 2-gon in the $(2, g-1)$-tessellation of the sphere $\mathbb{S}^2(\kappa) \times \{0\}$ (see Fig. 13). This amounts to saying that $\alpha = \frac{\pi}{g-1}$ and $\beta = \frac{\pi}{2}$ (see Fig. 14 right). To prove that these conditions are satisfied, we will use a degree argument inspired by the work of Karcher, Pinkall, and Sterling [55].

We will assume that $0 < \rho < \frac{\pi}{2\sqrt{\kappa}}$ is fixed, so the conjugate minimal surface $\Sigma(\widetilde{a}, \widetilde{b}, \rho)$ continuously depends on $\widetilde{a}$ and $\widetilde{b}$ (see Proposition 2), whence there exists a continuous function $f_\rho : (0, \frac{\pi}{2\sqrt{\kappa}}] \times (0, \frac{\pi}{2\sqrt{\kappa}}] \to \mathbb{R}^2$ such that $f_\rho(\widetilde{a}, \widetilde{b}) = (\alpha, \beta)$. Let c_1, c_2, c_3, and c_4 be four straight segments parametrized by

$$c_1(t) = \tfrac{1}{\sqrt{\kappa}}(t, \tfrac{\pi}{2}), \quad t \in [\tfrac{1}{2(g-1)}, \tfrac{\pi}{g-1}], \quad c_2(t) = \tfrac{1}{\sqrt{\kappa}}(\tfrac{\pi}{g-1}, t), \quad t \in [\tfrac{1}{2}, \tfrac{\pi}{2}],$$

$$c_3(t) = \tfrac{1}{\sqrt{\kappa}}(t, \tfrac{1}{2}), \quad t \in [\tfrac{1}{2(g-1)}, \tfrac{\pi}{g-1}], \quad c_4(t) = \tfrac{1}{\sqrt{\kappa}}(\tfrac{1}{2(g-1)}, t), \quad t \in [\tfrac{1}{2}, \tfrac{\pi}{2}],$$

which form a closed rectangle $R \subset (0, \frac{\pi}{2\sqrt{\kappa}}] \times (0, \frac{\pi}{2\sqrt{\kappa}}]$. We claim that there exists $\rho > 0$ such that the image of $f_\rho(R)$ is a closed curve around $(\frac{\pi}{g-1}, \frac{\pi}{2})$.

By spherical trigonometry, it is easy to obtain that $\widetilde{\beta} = \frac{\pi}{2}$ along c_1, $\widetilde{\alpha} > \frac{\pi}{g-1}$ along c_2, $\widetilde{\beta} < \frac{\pi}{2}$ along c_3, and $\widetilde{\alpha} < \frac{\pi}{g-1}$ along c_4. Also, we know that $\alpha > \widetilde{\alpha}$ and $\beta > \widetilde{\beta}$ (see Sect. 5.3.2), so the image of $f_\rho \circ c_1$ is above the horizontal line at height $\frac{\pi}{2}$ and the image of $f_\rho \circ c_2$ is to the right of the vertical line at $\frac{\pi}{g-1}$. Finally, since $\alpha \to \widetilde{\alpha}$ and $\beta \to \widetilde{\beta}$ for $\rho \to 0$, we deduce that there exists ρ_0 small enough so that the image of $f_{\rho_0} \circ c_3$ is below the horizontal line at height $\frac{\pi}{2}$ and the image of $f_{\rho_0} \circ c_4$ is to the left of the vertical line at $\frac{\pi}{k}$, and the claim is verified. Observe that this argument holds true for any $\rho \leq \rho_0$.

By continuity, for any $\rho \leq \rho_0$, there exists $\widetilde{a}$ and $\widetilde{b}$ (that depend on ρ) such that $f_\rho(\widetilde{a}, \widetilde{b}) = (\frac{\pi}{g-1}, \frac{\pi}{2})$ so the polygon $\Gamma(\widetilde{a}, \widetilde{b}, \rho)$ fits in the $(2, g-1)$-tessellation. Then, the complete surface $\Sigma^*(\widetilde{a}, \widetilde{b}, \rho)$ obtained by successive reflections of Σ inherits the desired symmetries; in particular, it is compact in the quotient. Finally, a similar computation using the Gauß-Bonnet theorem as in Sect. 5.2.3 ensures that quotient minimal surface in $\mathbb{S}^2(\kappa) \times \mathbb{S}^1(\frac{h}{\pi})$ has genus precisely g, $g \geq 3$. By the continuous dependence of h on the parameters and taking into account that both $\widetilde{\Sigma}(\widetilde{a}, \widetilde{b}, \rho)$ and $\Sigma(\widetilde{a}, \widetilde{b}, \rho)$ converge to a horizontal slice as $\rho \to 0$, the desired compact surfaces exist for all h small enough, i.e., for η large enough.

6 Complete H-Surfaces in $\mathbb{H}^2(\kappa) \times \mathbb{R}$

In this section, we will obtain minimal and constant mean curvature k-noids and k-nodoids in $\mathbb{H}^2 \times \mathbb{R}$ by conjugating the solution of a Jenkins-Serrin problem. The main difference with respect to the above constructions in Sect. 5 is that additionally we have to deal with the asymptotic behavior of the surface via the analysis of ideal horizontal and vertical geodesics in the initial minimal surface (see Proposition 9).

6.1 Genus Zero (H, k)-noids and (H, k)-nodoids in $\mathbb{H}^2(\kappa) \times \mathbb{R}$

The first construction concerns H-surfaces in $\mathbb{H}^2(\kappa) \times \mathbb{R}$ of genus 0 with an arbitrary number of ends $k \geq 3$ asymptotic to vertical H-cylinders and $4H^2 + \kappa \leq 0$. Before stating the main theorem, it is worth saying that these surfaces, in the minimal case, reduce to the minimal k-noids and saddle towers constructed by Morabito and Rodríguez [83] and also by Pyo [92] independently, whose names are inspired by their counterparts in $\mathbb{R}^3$ given by Jorge and Meeks [45] and Karcher [50]. The (H, k)-noids in Theorem 11 were constructed first by Plehnert in [90] and by Daniel and Hauswirth [17] for $k = 2$. However, the main contribution of our theorem is the new family of (H, k)-nodoids, with a similar behavior at infinity but approaching the asymptotic H-cylinders from the convex side. Some of them have self-intersections and give rise to counterexamples to the Krust property for subcritical H-surfaces.

We will use the term H-catenodoid to refer to $(H, 2)$-nodoids, inspired by the fact that the H-nodoids we have constructed in Sect. 5.1 seem to converge to a H-catenodoid of critical mean curvature as $4H^2 + \kappa \to 0$. In this limit, we choose $\widetilde{2}$ (see Fig. 6) as accumulation point, and the parameter $\lambda > \frac{\pi}{2}$ should also be chosen appropriately along the sequence. Likewise, the limit of the H-unduloids as $4H^2 + \kappa \to 0$ seems to be a catenoid of critical mean curvature. The situation is similar if we look at rotationally invariant H-surfaces with subcritical and supercritical mean curvature (see [82, Fig. A and B], noticing that Montaldo and Onnis choose $\kappa = -2$ and the mean curvature as the sum of the principal curvatures, not their average).

Theorem 11 ([11, Thm. 1.2]) *Assume that $4H^2 + \kappa \leq 0$. For each integer $k \geq 2$, there is a continuous two-parameter family of proper Alexandrov-embedded H-surfaces $\Sigma^*_{a,b} \subset \mathbb{H}^2(\kappa) \times \mathbb{R}$, $a, b \in (0, \infty]$ not both equal to ∞. These surfaces are invariant by mirror symmetries about a horizontal plane and k equiangular vertical planes:*

(a) *If $a, b < \infty$, then $\Sigma^*_{a,b}$ are called* saddle towers. *They are singly periodic, having genus 0 and $2k$ ends in the quotient of $\mathbb{H}^2(\kappa) \times \mathbb{R}$ by a vertical translation, and each end is asymptotic in the quotient to a half of a H-cylinder.*
(b) *If $a = \infty$ and $b < \infty$, then $\Sigma^*_{\infty,b}$ are called* (H, k)-noids *(or H-catenoids if $k = 2$). They have genus 0 and k ends. If $4H^2 + \kappa < 0$, then each end is embedded and contained in the* concave *side of an H-cylinder to which it is asymptotic. If $4H^2+\kappa = 0$, then each end is tangent to the asymptotic boundary $\partial_\infty \mathbb{H}^2(\kappa) \times \mathbb{R}$ along a vertical ideal geodesic.*
(c) *If $a < \infty$ and $b = \infty$, then $\Sigma^*_{a,\infty}$ are called* (H, k)-nodoids *(or H-catenodoids if $k = 2$). They have genus 0 and k ends. If $4H^2 + \kappa < 0$, then each end is embedded and contained in the* convex *side of a H-cylinder to which it is asymptotic. If $4H^2 + \kappa = 0$, then such H-cylinders disappear at infinity.*

In the case $H = 0$, the (H, k)-nodoids are congruent to the (H, k)-nodoids, but the subtle difference between these two families in the case $H > 0$ will be transparent via conjugation. It relies on the key role of the orientation in $\mathbb{E}(4H^2 + \kappa, H)$, whose bundle curvature is nonzero.

6.1.1 The Construction of the Minimal Surface in $\mathbb{E}(4H^2 + \kappa, H)$

We will fix κ and H such that $4H^2+\kappa \leq 0$, as well as the integer $k \geq 2$ in the sequel, so we will omit the dependence on these parameters as in previous constructions. Also, we will work on the global Cartan model $M(4H^2 + \kappa, H)$ given in Sect. 2.3.

Given $a, b > 0$, let $\widetilde{T}_{a,b} \subset \mathbb{M}^2(4H^2 + \kappa)$ be a geodesic triangle of vertexes $\widetilde{p}_0 = (0, 0)$, $\widetilde{p}_1$ and $\widetilde{p}_2$ labeled counterclockwise such that the angle at p_0 is equal to $\frac{\pi}{k}$ and the geodesic segments $\widetilde{p}_0\widetilde{p}_1$ and $\widetilde{p}_0\widetilde{p}_2$ have lengths a and b, respectively. After an orientation-preserving isometry, we will assume that $\widetilde{p}_1$ lies on the x-axis. We call ℓ the length of the side $\widetilde{p}_1\widetilde{p}_2$. We can extend this triangle to the case $a = \infty$ or $b = \infty$ by setting $\widetilde{T}_{\infty,b} = \mathrm{cl}(\cup_{a>0}\widetilde{T}_{a,b})$ and $\widetilde{T}_{a,\infty} = \mathrm{cl}(\cup_{b>0}\widetilde{T}_{a,b})$, respectively, where $\mathrm{cl}(G)$ denotes the topological closure of some subset $G \subset \mathbb{M}^2(4H^2 + \kappa)$. Note that $\widetilde{p}_1$ or $\widetilde{p}_2$ becomes ideal if $4H^2 + \kappa < 0$ or just disappears if $4H^2 + \kappa = 0$ since Nil_3 has not a notion of ideal boundary, and the triangle just becomes a truncated strip. Either way, we can lift the vertexes of $\widetilde{T}_{a,b}$ by means of the zero section as $\widetilde{q}_i = F_0(\widetilde{p}_i) = (\widetilde{p}_i, 0) \in \mathbb{E}(4H^2 + \kappa, H)$ for $i \in \{0, 1, 2\}$, and we denote by $\widetilde{q}_0\widetilde{q}_i$ the horizontal geodesic in $\mathbb{E}(4H^2 + \kappa, H)$ joining $\widetilde{q}_0$ and $\widetilde{q}_i$.

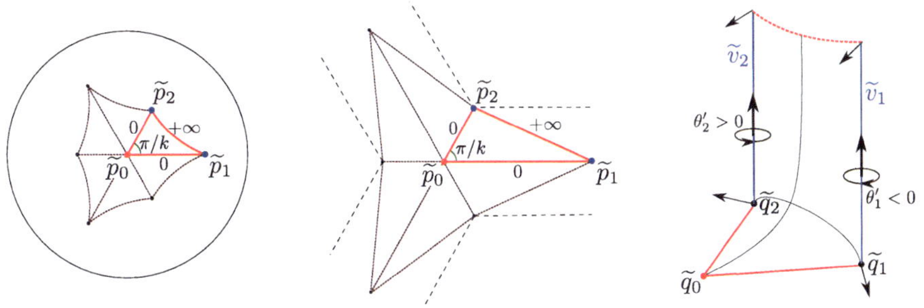

Fig. 15 The shaded triangle $\widetilde{T}_{a,b}$, $a, b < \infty$, and the Jenkins-Serrin boundary values in the subcritical case (left) and the critical case (center) for $k = 3$. Sketch of the solution $\widetilde{R}_{a,b}$ and the rotation of the normal along $\widetilde{v}_1$ and $\widetilde{v}_2$ (right)

Our minimal fundamental piece is the vertical minimal graph $\widetilde{R}_{a,b}$ that solves the Jenkins-Serrin problem with prescribed boundary values 0 over $\widetilde{p}_0\widetilde{p}_1$ and $\widetilde{p}_0\widetilde{p}_2$ and $+\infty$ over $\widetilde{p}_1\widetilde{p}_2$ (see Fig. 15). We call $\widetilde{\Sigma}_{a,b}$ the vertical minimal graph obtained after successive reflections about the horizontal geodesics $\widetilde{q}_0\widetilde{q}_1$ and $\widetilde{q}_0\widetilde{q}_2$, which are contained in $\widetilde{R}_{a,b}$. We call $\widetilde{\Omega}_{a,b}$ the domain of $\mathbb{M}^2(4H^2+\kappa)$ onto which $\widetilde{\Sigma}_{a,b}$ projects, and we denote by $\widetilde{p}_i$, for $i \in \{1, \ldots 2k\}$ the vertexes of $\widetilde{\Omega}_{a,b}$ labeled counterclockwise. Observe that the vertexes $\widetilde{p}_{2i-1}$ (resp. $\widetilde{p}_{2i}$) are ideal if $a = \infty$ (resp. $b = \infty$) when $4H^2 + \kappa < 0$ or disappear if $4H^2 + \kappa = 0$. The existence of the solution $\widetilde{\Sigma}_{a,b}$ is guaranteed by [11, Lem. 3.2 and 3.6] (see also Sect. 4.2).

Remark 14 If $4H^2 + \kappa = 0$, the uniqueness of solution for the Jenkins-Serrin problem in the cases $a = \infty$ or $b = \infty$ is not clear. In that case, we take the graphs $\widetilde{\Sigma}_{\infty,b}$ and $\widetilde{\Sigma}_{a,\infty}$ as limit of the graphs $\widetilde{\Sigma}_{a,b}$ making the family continuous. The case $k = 2$ is specially relevant, since the solutions $\widetilde{\Sigma}_{a,\infty}$ and $\widetilde{\Sigma}_{\infty,b}$ are explicit. They belong to a broader family $\mathcal{H}_\mu$ of minimal surfaces that are foliated by non-geodesic straight lines in $\mathrm{Nil}_3(\frac{\sqrt{-\kappa}}{2})$ orthogonal to a horizontal geodesic, having different directions of rotation if $\mu < \frac{-1}{2}$ or $\mu > \frac{1}{2}$ (see Fig. 16). These surfaces are described in [11, Lem. 3.3] and resemble *horizontal helicoids* with arbitrary distance between two consecutive vertical geodesics, so they can be used as barrier to solve the Jenkins-Serrin problem in a limit of a double sequence of minimal surfaces for $k > 3$. The surfaces $\mathcal{H}_\mu$ with $\mu < -\frac{1}{2}$ solve the case $a = \infty$ and have been previously found by Daniel and Hauswirth [17, §7] by other methods.

We will now analyze the angle function $\nu_{a,b}$ of $\widetilde{\Sigma}_{a,b}$. To this end, we can restrict to the fundamental piece $\widetilde{R}_{a,b}$, in whose interior $\nu_{a,b}$ will be assumed positive. We will only consider the case $a, b < \infty$ because if $a = \infty$ or $b = \infty$, the surface $\widetilde{\Sigma}_{a,b}$ can be analyzed as a limit of the finite case. However, the following lemma is also expected to hold true in the cases $a = \infty$ or $b = \infty$. It is important to remark that the finite boundary of $\widetilde{R}_{a,b}$ consists of the vertical geodesics $\widetilde{v}_i = \pi^{-1}(\widetilde{p}_i)$ for $i \in \{1, \ldots, 2k\}$, which are the only points where $\nu_{a,b}$ vanishes.

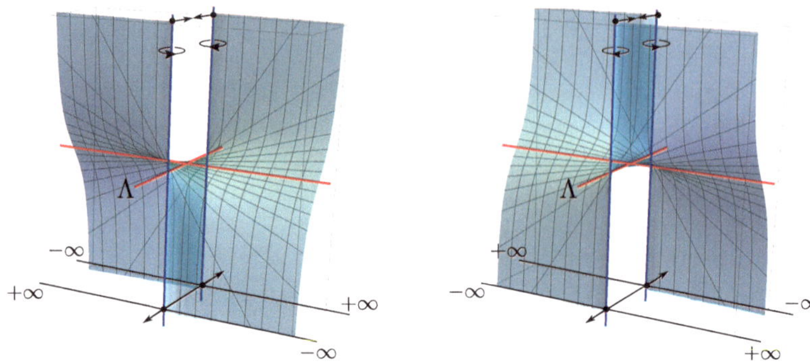

Fig. 16 The surfaces $\widetilde{\Sigma}_{\infty,b}$ (left) and $\widetilde{\Sigma}_{a,\infty}$ (right) in the case $4H^2+\kappa=0$ and $k=2$ are horizontal helicoids in $\mathrm{Nil}_3(\frac{\sqrt{-\kappa}}{2})$ foliated by straight lines orthogonal to Λ

Proposition 12 ([11, Lem. 4.4]) *Let $a, b \in (0, \infty)$ and $H > 0$, and define $\nu_{a,b}$ as the angle function of $\widetilde{R}_{a,b} \subset \mathbb{E}(4H^2+\kappa, H)$ that is positive in the interior of $\widetilde{R}_{a,b}$:*

(a) *If $k = 2$, then $\nu_{a,b}$ takes the value 1 only at $\widetilde{q}_0$.*
(b) *If $k \geq 3$, then $\nu_{a,b}$ only takes the value 1 at $\widetilde{q}_0$ and at other some point $\widehat{q}_1 \in \widetilde{q}_0\widetilde{q}_1$.*

6.1.2 The Conjugate *H*-Immersion

Let $\Sigma_{a,b}$ be the conjugate H-surface in $\mathbb{H}^2(\kappa)\times\mathbb{R}$, which is a multigraph over a (possibly non-embedded; see Fig. 19 right) domain $\Omega_{a,b} \subset \mathbb{H}^2(\kappa)$. Since $\widetilde{\Sigma}_{a,b}$ is invariant by axial symmetries about the geodesics $\widetilde{q}_i\widetilde{q}_{k+i}$ for $i \in \{1,\dots,k\}$, Lemma 2 says that $\Sigma_{a,b}$ has mirror symmetry with respect to k vertical planes meeting at a common vertical line, say the z-axis, arranged symmetrically. Moreover, Lemma 3 and Proposition 9 show that the boundary components of $\Sigma_{a,b}$ are the $2k$ complete (possibly ideal) horizontal curves $v_1,\dots,v_{2k}$ along with $2k$ ideal vertical geodesics joining the endpoints of v_i and v_{i+1} for $i \in \{1,\dots,2k\}$. We will assume hereafter that $v_2 \subset \mathbb{H}^2(\kappa)\times\{0\}$ (resp. $v_1 \subset \mathbb{H}^2(\kappa)\times\{0\}$) if $a=\infty$ (resp. $b=\infty$) (see Figs. 17 and 18):

1. If $\widetilde{p}_i$ is not ideal, then let θ_i be the angle of rotation of $\widetilde{N}$ along the vertical geodesic $\widetilde{v}_i$. It satisfies $\theta_i' < 0$ (resp. $\theta_i' > 0$) if i is odd (resp. even) (see Figs. 15, 17, and 18). Lemma 3 implies that $\kappa_g > 2H$ (resp. $\kappa_g < 2H$), being κ_g the geodesic curvature of v_i as a curve in a horizontal plane with respect to the unit normal N of $\Sigma_{a,b}$. Lemma 3 also shows that the projection of N points to the exterior (resp. interior) of $\Omega_{a,b} \subset \mathbb{H}^2(\kappa)$ if i is odd (resp. even).
2. If $\widetilde{p}_i$ is ideal (only possible in the case $4H^2+\kappa < 0$), then we can reason similarly for a sequence of graphs over finite triangles $\widetilde{T}_{a_n,b}$ or $\widetilde{T}_{a,b_n}$, with $a_n, b_n < \infty$, converging to $T_{a,b}$. Since θ_i' converges uniformly to zero as $n\to\infty$, Lemma 3

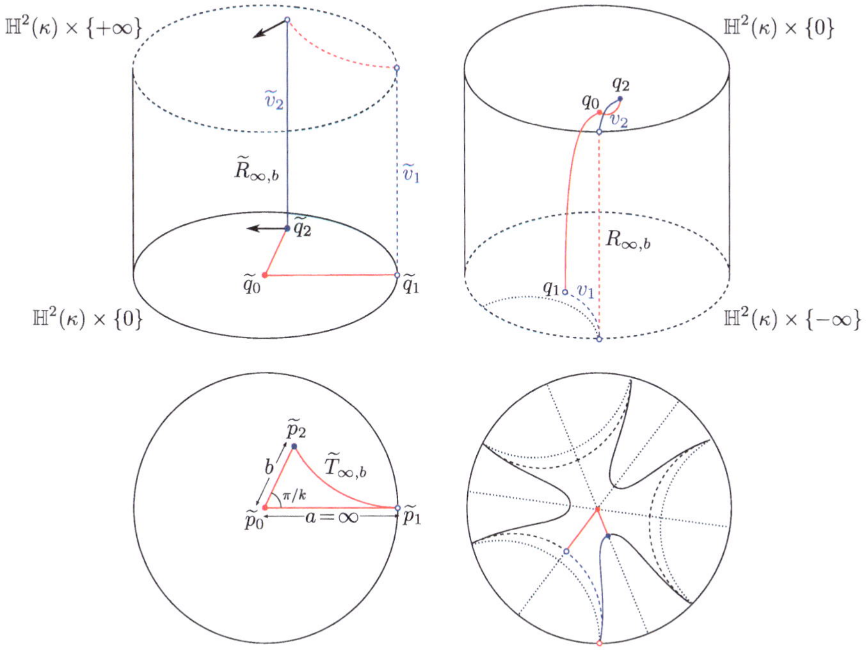

Fig. 17 The fundamental piece $\widetilde{R}_{\infty,b}$ and its conjugate $R_{\infty,b}$ of a (H,k)-noid for $k=3$ and $4H^2+\kappa<0$ and their projections to $\mathbb{H}^2(\kappa)$. Observe that the asymptotic H-cylinders are approached from the concave side. Dotted lines are geodesics

reveals that the geodesic curvature of v_i is $2H$ with respect to N. As a limit of curves in the assumption of item (1), we infer that the projection of N points to the exterior (resp. interior) of the domain $\Omega_{a,b}$ along v_i if i is odd (resp. even). Hence, the geodesic curvature of v_i with respect to an inner conormal to $\Omega_{a,b}$ is $-2H$ (resp. $2H$) if i is odd (resp. even). This confirms that (H,k)-noids (resp. (H,k)-nodoids) are asymptotic to the H-cylinders from the concave side (resp. convex side) (see Figs. 17 and 18).

The complete surfaces $\Sigma^*_{a,b}$ are obtained after reflecting $\Sigma_{a,b}$ about the horizontal plane of symmetry. From the properties sketched so far, we deduce the description of the surfaces given by Theorem 11.

Remark 15 Proposition 12 shows that $\Sigma_{a,b}$ behaves differently when $H=0$ and $H>0$. In the former case, the angle function $\nu_{a,b}$ only takes the value 1 at $\widetilde{q}_0$ for all $k\geq 2$. However, in the later case, the height function of the conjugate surface (i.e., the projection to the factor $\mathbb{R}$) has a local minimum at the conjugate of $\widetilde{q}_0$ and saddle points at the conjugate point of $\widehat{q}_1$ and its symmetric points for all $k\geq 3$. So, the shape of $\Sigma_{a,b}$ is somewhat different in the cases $H=0$ and $H>0$. For instance, this situation leads to type II self-intersections for certain values $a,b<\infty$ (i.e., in

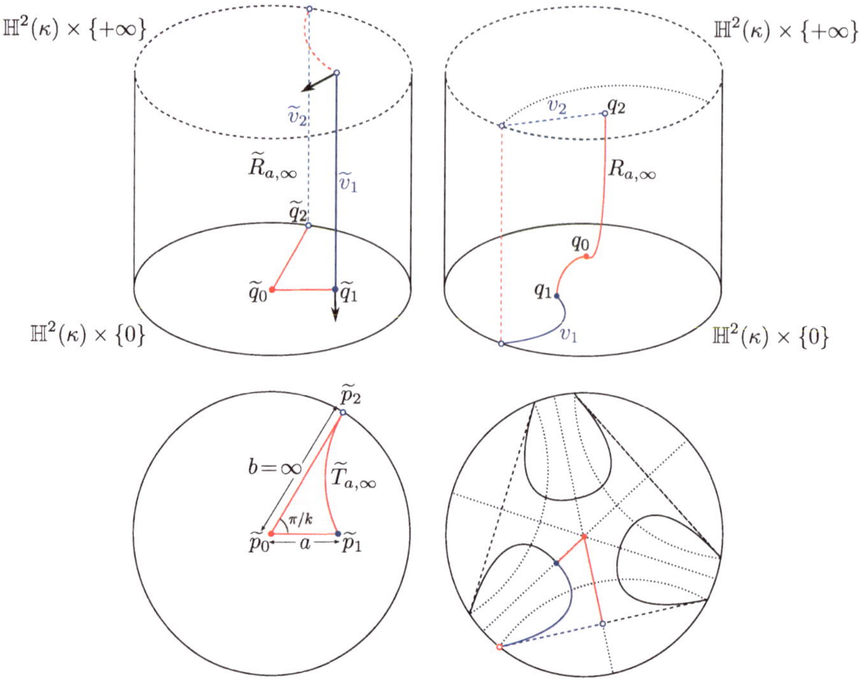

Fig. 18 The fundamental piece $\widetilde{R}_{a,\infty}$ and its conjugate $R_{a,\infty}$ of a (H, k)-nodoid for $k = 3$ and $4H^2 + \kappa < 0$ and their projections to $\mathbb{H}^2(\kappa)$. Observe that the asymptotic H-cylinders are approached from the convex side. Dotted lines are geodesics

the case of saddle towers) and $H > 0$ because the fundamental piece escapes the slab where the boundary is contained (see [11, Prop. 4.6]).

Remark 16 To see the differences between the cases $4H^2+\kappa < 0$ and $4H^2+\kappa = 0$, we define the function $\rho = \rho(a, b)$ (resp. $d = d(a, b)$) as the distance in $\mathbb{H}^2(\kappa)$ from the center $\pi(q_0)$ of $\Omega_{a,b}$, where q_0 is the conjugate point of $\widetilde{q}_0$, to the curves $\pi(v_1)$ (resp. $\pi(v_2)$) in the projection of the (possibly asymptotic) boundary of $\Sigma_{a,b}$. In [11, Lem. 4.1], it is shown that for $4H^2 + \kappa < 0$ the functions $a \mapsto d(a, \infty)$ and $b \mapsto d(\infty, b)$ are strictly increasing and range from 0 to a finite number d_∞. If $4H^2 + \kappa = 0$, then $d(a, \infty) = \rho(\infty, b) = \infty$ for all $a, b \in (0, \infty)$, so the H-cylinders asymptotic to $\Sigma_{a,\infty}$ and $\Sigma_{\infty,b}$ disappear at infinity when $4H^2 + \kappa = 0$.

We also remark that if such horocylinders did not disappear, then we would have encountered a contradiction to the halfspace theorem for surfaces of critical mean curvature given by Hauswirth, Rosenberg, and Spruck [37].

6.1.3 Embeddedness

We begin by observing that the embeddedness of $\Sigma^*_{a,b}$ in case $H = 0$ easily follows from the Krust property (see Proposition 5), so we will restrict to the case $H > 0$. We have already pointed out that type II self-intersections may occur even if $\Sigma_{a,b}$ is embedded (see Remark 15). Using Proposition 12 and the maximum principle with respect to horizontal planes, it is not difficult to see that this behavior is not possible for $k = 2$ (see [11, Prop. 4.6] for the details). Said this, we cannot discard type I self-intersections either, because the counterexamples to the Krust property will actually come from this type of self-intersections. We will analyze the cases $a = \infty$ and $b = \infty$ separately:

1. The embeddedness of (H, k)-noids: As the geodesics $\widetilde{v}_{2i-1}$ are ideal and the angle of rotation of the normal along $\widetilde{v}_{2i}$ satisfies $\theta'_{2i} > 0$ for $i \in \{1, \ldots, k\}$, Lemma 3 and the maximum principle with respect to horizontal planes give the inclusion $\Sigma_{a,b} \subset \mathbb{H}^2(\kappa) \times (-\infty, 0]$, where the symmetry curves of $\Sigma_{a,b}$ are contained in $\mathbb{H}^2(\kappa) \times \{0\}$. By Proposition 4, if $\int_{\widetilde{v}_{2i}} \theta'_{2i} \leq \pi$, then the conjugate curves v_{2i} are embedded, and by the symmetries of the surface, the same happens to $\Sigma_{a,b}$. Therefore, the embeddedness of the (H, k)-noids is proved when $\int_{\widetilde{v}_{2i}} \theta'_{2i} \leq \pi$, i.e., when $\widetilde{\Omega}_{a,b}$ is a convex domain of $\mathbb{M}^2(4H^2 + \kappa)$. This condition is always satisfied for $k = 2$, so all H-catenoids are embedded.
2. The embeddedness of (H, k)-nodoids: If $k \geq 3$, we do not have in general the inclusion $\Sigma_{a,b} \subset \mathbb{H}^2(\kappa) \times [0, +\infty)$ if we assume that v_1 lies in $\mathbb{H}^2(\kappa) \times \{0\}$. The maximum principle with respect to horizontal slices does not apply. However, for $4H^2 + \kappa < 0$ and a large enough that inclusion can be proved using the continuity of the conjugation (Proposition 2). Notice that $\Sigma_{a,\infty}$ converges to a Scherk H-graph $\Sigma_{\infty,\infty}$ as $a \to \infty$ after appropriate vertical translations (see Fig. 19) and the geodesic curvatures of v_{2i-1} in the horizontal plane of symmetry converge to $2H$ with respect to the exterior conormal. In other words, $\Sigma_{a,\infty}$ is embedded for a large enough.

   On the other hand, the hyperbolic distance from $\pi(v_2)$ to the origin $\pi(q_0)$ ranges from 0 to d_∞ as a runs from 0 to ∞ (see Remark 16). Since $\Sigma_{a,\infty}$ lies in the convex side of the cylinders over v_{2i}, there exists $a_1 > 0$, depending on k and H, such that the asymptotic equidistant curves v_2 and v_4 in $\Sigma_{a_1,\infty}$ share at least one endpoint at infinity. This implies that the endpoints of the curve v_3 coincide when $a = a_1$ and there are type I self-intersections if and only if $a < a_1$ (see Fig. 19). By symmetry, the same happens for any v_{2i-1}, so $\Sigma_{a,\infty}$ (and hence $\Sigma^*_{a,\infty}$) is not embedded in that case. If $4H^2 + \kappa = 0$, the curves $v_{2i} \subset \Sigma^*_{a,b}$ get close to horocycles at the same time that they diverge to infinity (see Remark 16). As $\Sigma^*_{a,\infty}$ lies locally in the convex side of the horocylinders, the curves v_{2i-1} cannot be embedded.

All in all, if $4H^2 + \kappa < 0$, then $\Sigma_{a,\infty}$ is not embedded if $a < a_1$; if $4H^2 + \kappa = 0$, then $\Sigma_{a,\infty}$ is never embedded. In the case $k = 2$, this non-embeddedness yields the counterexamples to the Krust property with $4H^2 + \kappa \leq 0$, since the surface $\widetilde{\Sigma}_{a,b}$ is a vertical graph over a convex quadrilateral. In the case $4H^2 + \kappa > 0$, there are

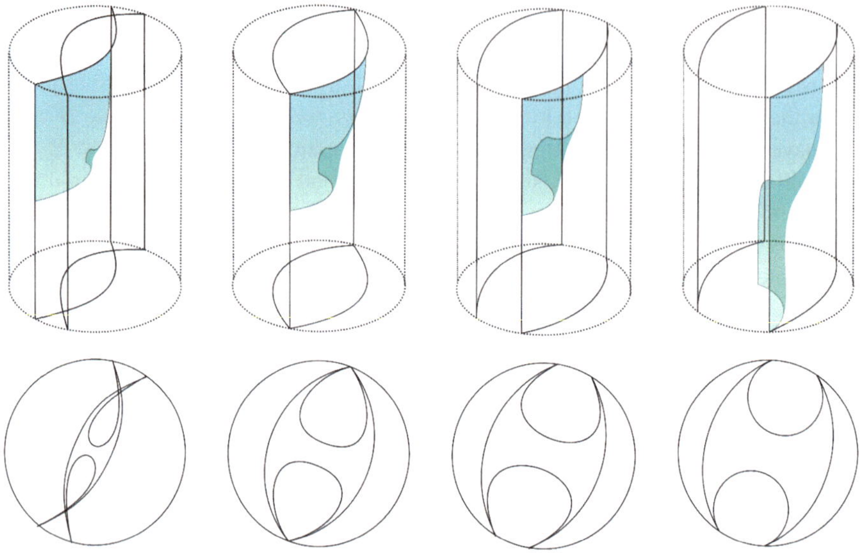

Fig. 19 The fundamental piece $R_{a,\infty}$ of an H-catenodoid and its projection to $\mathbb{H}^2(\kappa)$. From left to right: $a < a_1$ (non-embedded), $a = a_1$ (vertical planes touching at infinity), $a > a_1$ (embedded), $a = \infty$ (limit Scherk H-graph after suitable vertical translations)

also counterexamples to the Krust property constructed by similar methods (see [11, §5]).

Remark 17 It is expected that (H, k)-noids and (H, k)-nodois have finite total curvature as in the minimal case. By the properties of conjugation, this problem is equivalent that the solutions of the Jenkins-Serrin problem $\widetilde{R}_{\infty,b}$ and $\widetilde{R}_{a,\infty}$ have finite total curvature in $\widetilde{\mathrm{SL}}_2(\mathbb{R})$. However, this problem is much more involved than in $\mathbb{H}^2(\kappa) \times \mathbb{R}$ because the Gauss curvature of minimal surfaces in $\widetilde{\mathrm{SL}}_2(\mathbb{R})$ may change sign (see the Gauss equation in Sect. 2.4).

6.2 Genus One Minimal k-noids in $\mathbb{H}^2(\kappa) \times \mathbb{R}$

The last application we present in this survey is a construction of minimal surfaces in $\mathbb{H}^2(\kappa) \times \mathbb{R}$ with finite total curvature by means of conjugating a solution to a Jenkins-Serrin problem in $\mathbb{H}^2(\kappa) \times \mathbb{R}$. They are analogous to the genus 1 minimal k-noids in $\mathbb{R}^3$ obtained by Mazet [75], and their construction is inspired by the work of Plehnert [89], who got similar surfaces in $\mathbb{H}^2(\kappa) \times \mathbb{R}$ with critical mean curvature.

In the minimal case, we take advantage of the Krust property (see Proposition 5) as well as of the fact that our surfaces have finite total curvature, which enables a finer control of the asymptotic behavior. Hauswirth, Nelli, Sa Earp, and Toubiana [35] and Hauswirth, Menezes, and Rodríguez [34] proved that a complete minimal surface immersed in $\mathbb{H}^2(\kappa) \times \mathbb{R}$ has finite total curvature if and only if it is proper and has finite topology and each of its ends is asymptotic to an admissible polygon consisting of finitely many complete vertical and horizontal ideal geodesics, in which case the total curvature must be a negative multiple of 2π (see [36, Thm. 3.1]). Such a polygon is particularly well controlled when we want to obtain the surface by conjugation in view of Proposition 9. Also, [34, Thm. 6] says that we have to prescribe a symmetry with respect to a horizontal slice if we want our surfaces to have finite total curvature and embedded ends.

It is worth pointing out that the literature does not contain many examples of minimal surfaces with finite total curvature in $\mathbb{H}^2(\kappa) \times \mathbb{R}$, and the surfaces given here are the first examples with genus 1 and an arbitrary number of ends $k \geq 3$. There are two important remarks to this claim. On the one hand, our result cannot be extended to the case $k = 2$ since it would contradict the uniqueness of the horizontal catenoids given by Hauswirth, Nelli, Sa Earp, and Toubiana [35] (as usual, the condition $k \geq 3$ will appear as a natural restriction in the conjugate construction). On the other hand, Martín, Mazzeo, and Rodríguez [73], by means of gluing methods, obtained properly embedded minimal surfaces with finite total curvature in $\mathbb{H}^2(\kappa) \times \mathbb{R}$ of genus g and k ends asymptotic to vertical planes, for arbitrary genus $g \geq 0$. Nonetheless, in their result, k is not arbitrary in principle but sufficiently large depending on g and κ.

Theorem 12 ([10, Thm. 1]) *For each $k \geq 3$, there exists a one-parameter family Σ^*_φ, with $\frac{\pi}{k} \leq \varphi \leq \frac{\pi}{2}$, of properly Alexandrov-embedded minimal surfaces in $\mathbb{H}^2(\kappa) \times \mathbb{R}$ with genus 1 and k ends. They are invariant by mirror symmetries about a horizontal plane and about k equiangular vertical planes and have finite total curvature $-4k\pi$. Moreover, each of their ends is embedded and asymptotic to a vertical plane.*

This construction can be adapted to produce minimal surfaces in $\mathbb{H}^2(\kappa) \times \mathbb{R}$ invariant by an arbitrary vertical translation, with genus 1 and finite total curvature in the quotient of $\mathbb{H}^2(\kappa) \times \mathbb{R}$ by the vertical translation (see [10, Thm. 2]). These are the genus one counterparts of Morabito and Rodríguez' saddle towers [83].

6.2.1 The Minimal Surface in $\mathbb{H}^2(\kappa) \times \mathbb{R}$

Given $a > 0$ and $0 < \varphi < \frac{\pi}{2}$, consider the triangle $\widetilde{\Delta}(a, \varphi) \subset \mathbb{H}^2(\kappa)$ with one ideal vertex $\widetilde{p}_1$ and two interior vertexes $\widetilde{p}_2$ and $\widetilde{p}_3$, such that the finite edge $\widetilde{p}_2\widetilde{p}_3$ has length a and the angle in the vertex $\widetilde{p}_2 = (0, 0)$ is equal to φ (see Fig. 20 bottom left). We will work in the global Cartan model given in Sect. 2.3 assuming that $\widetilde{p}_2 = (0, 0)$ and $\widetilde{p}_1 = (\frac{2}{\sqrt{-\kappa}}, 0)$.

Our initial minimal piece is the unique minimal vertical graph $\widetilde{\Sigma}(a, \varphi, b)$ in $\mathbb{H}^2(\kappa) \times \mathbb{R}$ that solves the Jenkins-Serrin problem over $\widetilde{\Delta}(a, \varphi)$ with boundary data b over $\widetilde{p}_2\widetilde{p}_3$, $+\infty$ over $\widetilde{p}_1\widetilde{p}_3$ and 0 over $\widetilde{p}_1\widetilde{p}_2$. The existence and uniqueness of solution are guaranteed by Theorem 6 and Proposition 9. The finite boundary

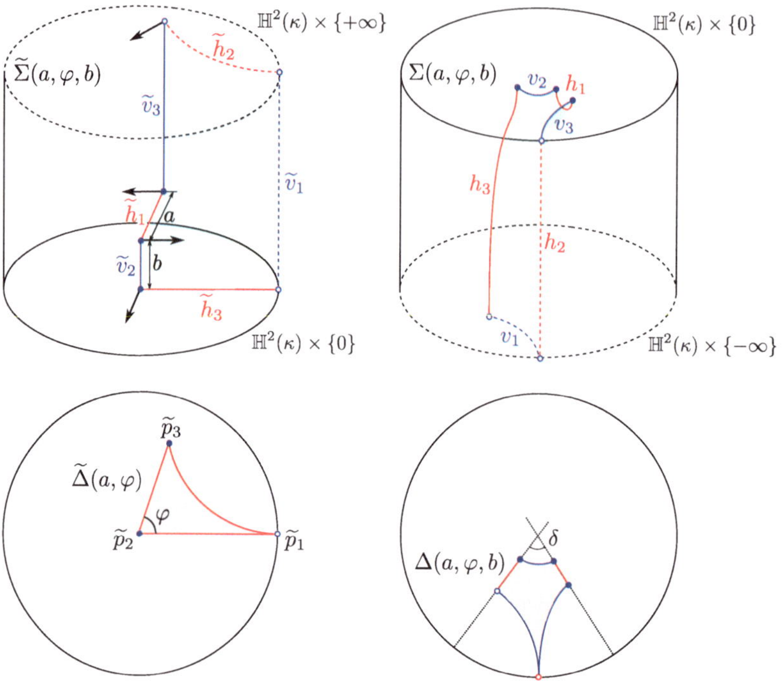

Fig. 20 The conjugate surfaces $\widetilde{\Sigma}(a,\varphi,b)$ and $\Sigma(a,\varphi,b)$ and their projections to $\mathbb{H}^2(\kappa)$. The arrows represent the normal $\widetilde{N}$ at the endpoints of $\widetilde{v}_2$ and $\widetilde{v}_3$. The period $\mathcal{P}_1$ is zero when v_2 and v_3 are at the same height; the period $\mathcal{P}_2$ is the cosine of δ, if such an angle exists

of $\widetilde{\Sigma}(a,\varphi,b)$ consists of the vertical segment $\widetilde{v}_2$ of length b projecting to $\widetilde{p}_2$, the vertical half-line $\widetilde{v}_3$ projecting to $\widetilde{p}_3$, and the horizontal geodesics $\widetilde{h}_1$ and $\widetilde{h}_3$ contained in $\mathbb{H}^2(\kappa)\times\{b\}$ and $\mathbb{H}^2(\kappa)\times\{0\}$ that project to $\widetilde{p}_2\widetilde{p}_3$ and $\widetilde{p}_1\widetilde{p}_2$, respectively. Moreover, the asymptotic boundary of $\widetilde{\Sigma}(a,\varphi,b)$ consists of a vertical half-line $\widetilde{v}_1$ projecting to $\widetilde{p}_1$ and the horizontal geodesic $\widetilde{h}_2$ contained in $\mathbb{H}^2(\kappa)\times\{+\infty\}$ that projects to $\widetilde{p}_1\widetilde{p}_3$ (see Fig. 20). The interior of $\widetilde{\Sigma}(a,\varphi,b)$ is a graph, where we will assume that the angle function is positive. Next, we analyze the horizontal and vertical points as in the previous constructions (again, this analysis follows from the boundary maximum principle together with Strategy 2 in Sect. 3.1.2).

Proposition 13 ([10, Lem. 2]) *Let $\nu_{a,\varphi,b}$ be the angle function of the minimal surface $\widetilde{\Sigma}(a,\varphi,b)$, which will be assumed positive in the interior:*

(a) *The points with $\nu_{a,\varphi,b}=0$ are precisely those at $\widetilde{v}_2\cup\widetilde{v}_3$.*
(b) *There is exactly one point with $\nu_{a,\varphi,b}=1$ and it belongs to $\widetilde{h}_1$.*

We also remark here that if $b>0$, the angles of rotation of θ_2 and θ_3 of the normal $\widetilde{N}$ along $\widetilde{v}_2$ and $\widetilde{v}_3$ satisfy $\theta_2'>0$ and $\theta_3'>0$, respectively (see Fig. 20).

6.2.2 The Conjugate Minimal Surface

Since $\widetilde{\Delta}(a, \varphi)$ is convex, the Krust property (Proposition 5) implies that the conjugate surface $\Sigma(a, \varphi, b)$ is also a vertical minimal graph in $\mathbb{H}^2(\kappa)\times\mathbb{R}$ over some domain $\Delta(a, \varphi, b) \subset \mathbb{H}^2(\kappa)$. By Lemma 3, the curves v_2 and v_3 in $\partial\Sigma(a, \varphi, b)$ are contained in horizontal slices, being their projections convex with respect to the inner-pointing conormal to $\Delta(a, \varphi, b)$ along its boundary. Moreover, $\Sigma(a, \varphi, b)$ lies locally above the horizontal planes containing v_2 and v_3. By Lemma 2, the conjugate curves h_1 and h_3 lie in vertical planes, and it can be shown that the component of h_1 in the factor $\mathbb{R}$ has a minimum at the unique point where $\nu = 1$. The asymptotic boundary of $\Sigma(a, \varphi, b)$ is composed of the half horizontal geodesics v_1 in $\mathbb{H}^2(\kappa) \times \{-\infty\}$ and of the ideal vertical half-line h_2 (see Fig. 20).

We aim at showing that appropriate parameters give rise to a complete minimal surface $\Sigma^*(a, \varphi, b)$ of genus 1 in $\mathbb{H}^2(\kappa) \times \mathbb{R}$ after extending $\Sigma(a, \varphi, b)$ by mirror symmetries over the horizontal and vertical planes. We will sketch the proof that for each $\frac{\pi}{k} < \varphi < \frac{\pi}{2}$, there exist a_φ and b_φ such that $\Sigma^*_\varphi = \Sigma^*(a_\varphi, \varphi, b_\varphi)$ has the desired properties. This will be accomplished if the following two periods are closed, inspired by [89, §6.3]:

1. **First period problem.** We call $\mathcal{P}_1(a, \varphi, b)$ the difference of heights of the endpoints of h_1, which must be zero so each end of $\Sigma(a, \varphi, b)$ is an annulus. Parametrizing $h_1 : [0, a] \to \mathbb{H}^2(\kappa) \times \mathbb{R}$ with endpoints $h_1(0) \in v_2$, $h_1(a) \in v_3$ and unit speed, the properties of the conjugation yield

$$\mathcal{P}_1(a, \varphi, b) = \int_{h_1} \langle h_1', \xi\rangle = \int_{\widetilde{h}_1} \langle \eta, \xi\rangle, \tag{22}$$

   where $\eta = -J\widetilde{h}_1'$ is the unit inward conormal vector to $\widetilde{\Sigma}(a, \varphi, b)$ along $\widetilde{h}_1$.
2. **Second period problem.** Assume that the vertical planes containing the symmetry curves h_1 and h_3 intersect each other at a non-oriented angle δ, and call $\mathcal{P}_2$ the cosine of the angle δ. To give an analytic expression for $\mathcal{P}_2$, we will consider the half-space model (see Sect. 2.3.2).
   Parametrize $v_2(t) = (x(t), y(t), 0)$ for $t \in [0, b]$, and assume, up to an ambient isometry, that h_3 and v_2 lie in the vertical plane $\{x = 0\}$ and the horizontal plane $\{z = 0\}$, respectively, and also $x(0) = 0$, $y(0) = 1$, and $x(t) < 0$ when t is close to 0 (see Fig. 21). Let $\psi \in C^\infty[0, b]$ be the angle of rotation of v_2 with respect to the horocycle foliation in the sense of Remark 6, where we choose the initial angle $\psi(0) = \pi$. The second period is given by

$$\mathcal{P}_2(a, \varphi, b) = \cos(\delta) = \frac{x(b)\sin(\psi(b))}{y(b)} - \cos(\psi(b)). \tag{23}$$

   Incidentally, the right-hand side of (23) is well defined even when the vertical planes containing the symmetry curves h_1 and h_3 do not intersect. However, provided that the first period is solved, it can be shown that the vertical planes containing the symmetry curves h_1 and h_3 intersect each other with an angle δ if and only if $\mathcal{P}_2(a, \varphi, b) = \cos(\delta)$ (see [10, Lem. 6]).

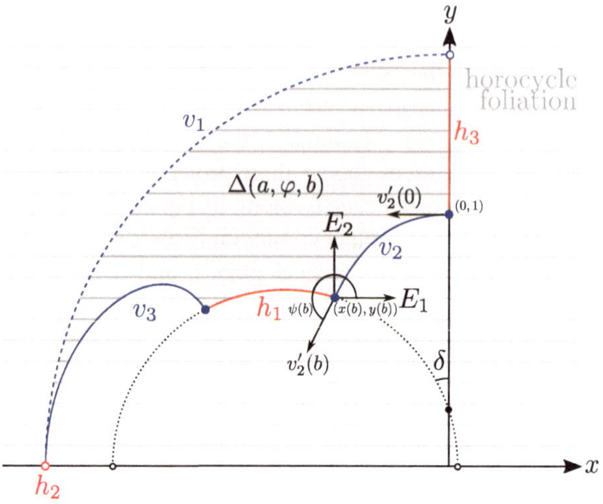

Fig. 21 The angle $\psi(b)$ of rotation of v_2 with respect to the horocycle foliation at $v_2(b)$, where we identify $\mathbb{H}^2 \times \{0\} \equiv \mathbb{H}^2$. The surface $\Sigma(a, \varphi, b)$ projects onto the shaded region $\Delta(a, \varphi, b)$ with boundary. The complete geodesic containing the projection of h_1 appears in dotted line in the case it intersects the y-axis with (non-oriented) angle δ

6.2.3 Solving the Period Problems

We will start by discussing the first period. We will assume that b is any nonnegative real number and we have the restrictions $0 < \varphi < \frac{\pi}{2}$ and $0 < a < a_{\max}(\varphi) = 2\,\text{arctanh}\,(\cos(\varphi))$. By hyperbolic trigonometry, this means that the angle of $\widetilde{\Delta}(a, \varphi)$ at $\widetilde{p}_3$ is strictly less than φ. This assumption is natural since a simple comparison argument shows that the first period problem cannot be solved if $a > a_{\max}(\varphi)$ (see [10, Rmk. 1]). The following result guarantees that both period problems can be solved simultaneously:

Lemma 4 ([10, Lem. 5 and 6]) *Let* $\Omega = \left\{(a, \varphi) \in \mathbb{R}^2 : 0 < \varphi < \frac{\pi}{2}, 0 < a < a_{\max}(\varphi)\right\}$*:*

(a) *There exists a unique function* $f : \Omega \to \mathbb{R}_+$ *such that* $\mathcal{P}_1(a, \varphi, f(a, \varphi)) = 0$ *for all* $(a, \varphi) \in \Omega$*, which is continuous. For a fixed* $\varphi_0 \in (0, \frac{\pi}{2})$*, it has limits*

$$\lim_{a \to a_{\max}(\varphi_0)} f(a, \varphi_0) = +\infty, \qquad \lim_{(a,\varphi) \to (0,\varphi_0)} f(a, \varphi) = 0.$$

(b) *If* $\varphi_0 \in (0, \frac{\pi}{2})$ *and* $b = f(a, \varphi_0)$*, then the inequalities* $x(t) < 0$ *and* $\pi < \psi(t) < 2\pi$ *hold true for all* $t \in (0, b]$ *(i.e., along the curve* v_2*; see Sect. 6.2.2). We have the limits*

$$\lim_{a \to 0} \mathcal{P}_2(a, \varphi_0, f(a, \varphi_0)) = \cos(\varphi_0), \qquad \lim_{a \to a_{\max}(\varphi_0)} \mathcal{P}_2(a, \varphi_0, f(a, \varphi_0)) = +\infty.$$

We will highlight some ideas in the proof but we refer to [10] for further details. In the proof of item (a), the existence of the function f is based on a comparison along the boundary using Strategy 3 in Sect. 3.1.2. We show that the first period function $\mathcal{P}_1$ is strictly decreasing with respect to the third argument since we can compare $\widetilde{\Sigma}(a,\varphi,b_1)$ and $\widetilde{\Sigma}(a,\varphi,b_2)$ for $0 < b_1 < b_2$ and their conormals along the horizontal geodesic $\widetilde{h}_1$ by the boundary maximum principle. Moreover, we show that $\mathcal{P}_1(a,\varphi,0) > 0$ and $\mathcal{P}_1(a,\varphi,b) < 0$ for $b > 0$ large enough. The monotonicity of $\mathcal{P}_1$ implies that there is a unique $b_0 > 0$ such that $\mathcal{P}_1(a,\varphi,b_0) = 0$, and this uniqueness implies in turn the uniqueness and continuity of f.

The computation of the limits in (a) is also based on the same comparison idea with the limit surfaces. For instance, in the limit as $a \to 0$, we use Proposition 2 and Remark 4 to rescale the space while keeping a constant. The limit surface lies in Euclidean space $\mathbb{R}^3$ and is a minimal graph over a truncated strip to which we can also apply a similar comparison argument.

As for item (b), the result is based on a careful use of the formula $\theta' = -\kappa_g^P = \psi' + \sqrt{-\kappa}\cos(\psi)$ given by Lemma 3 and Remark 6, where θ is the angle of rotation of the normal $\widetilde{N}$ along $\widetilde{v}_2$ and ψ is the angle of rotation of v_2 with respect to the horocycle foliation. On the one hand, we can integrate $\theta' = -\kappa_g^P$ to get $\int_0^b \kappa_g(t)dt = -\int_0^b \theta'(t)dt = -\varphi$. This gives estimates for the total geodesic curvature of subsets of v_2, which can be used to prove the inequalities $x(t) < 0$ and $\pi < \psi(t) < 2\pi$ via the Gauß-Bonnet formula (applied to appropriate domains; see [10, Fig. 5]). On the other hand, we can integrate $\theta' = \psi' + \sqrt{-\kappa}\cos(\psi)$ to obtain $\varphi = \psi(b) - \pi + \sqrt{-\kappa}\int_0^b \cos(\psi(t))dt$. In particular, $\psi(b) \to \varphi + \pi$ and $(x(b), y(b)) \to (0,1)$ as $b \to 0$. Using this and the limits in item (a), the first limit in item (b) can be easily deduced. We will skip the proof of the second one, which is more technical and involves the limit of surfaces.

In view of Lemma 4, it is not difficult to see how to finish the proof of Theorem 12. Given $k \geq 3$, for each $\frac{\pi}{k} < \varphi < \frac{\pi}{2}$, we choose $b = f(a,\varphi)$ to solve the first period problem. Observe that $\mathcal{P}_2(a,\varphi,f(a,\varphi))$ tends to $\cos(\varphi)$ when $a \to 0$ and tends to $+\infty$ when $a \to a_{\max}(\varphi)$; since $\cos(\varphi) < \cos(\frac{\pi}{k})$ and $\mathcal{P}_2$ is continuous, there exists some $a_\varphi \in (0, a_{\max}(\varphi))$ such that $\mathcal{P}_2(a,\varphi,f(a_\varphi,\varphi)) = \cos(\frac{\pi}{k})$, though it might not be unique. Therefore, we choose $b_\varphi = f(a_\varphi,\varphi)$ so that $\Sigma_\varphi^* = \Sigma^*(a_\varphi,\varphi,b_\varphi)$ solves both period problems. By a similar argument to that of Collin and Rosenberg in [13, Rmk. 7] using Fatou's Lemma, it follows that Σ_φ^* has finite total curvature. This is also a consequence of the characterization of minimal surfaces with finite total curvature of Hauswirth, Menezes, and Rodríguez in [34], since Σ_φ^* is proper and has finite topology and each of its end is asymptotic to a vertical plane; in particular, Σ_φ^* is asymptotic to an admissible polygon at infinity. Either way, the fact that Σ_φ^* has finite total curvature enables a better understanding of the asymptotic behavior. For instance, it follows from [34] that if an end of Σ_φ^* is embedded, then it is a horizontal graph in the sense of [34, Def. 7 and 8].

6.2.4 Embeddedness

The analysis of the second period function does not allow us to prove the uniqueness of a_φ since we have not been able to control the dependence of the second problem with respect to the parameter a. Nevertheless, we expect that there do exist values of φ for which the complete surface Σ^*_φ will be embedded. Observe that the fundamental piece $\Sigma(a_\varphi, \varphi, b_\varphi)$ is a vertical graph contained in the half-space $\mathbb{H}^2(\kappa) \times (-\infty, 0]$, but we can find self-intersections of type II after reflecting it about the vertical planes of symmetry, even if all periods are closed. This actually happens if $\varphi \to \frac{\pi}{k}$ because it implies that $a_\varphi \to 0$, and the surfaces Σ^*_φ converge, after rescaling, to a genus 1 minimal k-noid in $\mathbb{R}^3$, which is not globally embedded.

We can ensure that Σ^*_φ is embedded if the value a_φ that solves both period problem is bigger than the quantity $a_{\text{emb}}(\varphi) = \operatorname{arcsinh}(\cot(\varphi))$. By a simple application of hyperbolic trigonometry, the inequality $a_\varphi \geq a_{\text{emb}}(\varphi)$ amounts to saying that the angle of $\widetilde{\Delta}(a_\varphi, \varphi)$ at $\widetilde{p}_3$ is at most $\frac{\pi}{2}$, so that the fundamental piece is still a vertical graph over a convex domain after extending it by axial symmetry about $\widetilde{h}_1$. Although we are not able to prove that there are values of a_φ that satisfy this inequality, we expect that the surface Σ^*_φ is embedded if φ is close to $\frac{\pi}{2}$.

The fact that the ends of Σ^*_φ are embedded is a consequence of the fact that each of them is contained in four copies of the fundamental piece that form a symmetric embedded bigraph. This claim follows from the fact that two of these four pieces come from the fundamental piece extended by axial symmetric about $\widetilde{h}_2$ and the extended surface projects to a convex quadrilateral of $\mathbb{H}^2(\kappa)$. The Krust property guarantees that the conjugate surface is graph, and the other two copies needed to produce the aforesaid bigraph are their symmetric ones with respect to the slice $\mathbb{H}^2(\kappa) \times \{0\}$ containing $\widetilde{v}_2$ and $\widetilde{v}_3$.

Remark 18 If $\mathcal{P}_2(a, \varphi, f(a, \varphi)) \geq 1$, then the completion $\Sigma^*(a, \varphi, f(a, \varphi))$ is a surface invariant by a discrete group of parabolic or hyperbolic translations (depending on whether $\mathcal{P}_2 = 1$ or $\mathcal{P}_2 > 1$, respectively), instead of a discrete group of rotations. We call these examples *parabolic* and *hyperbolic ∞-noids*, respectively (see Fig. 22). The former are obtained when the vertical planes of symmetry of

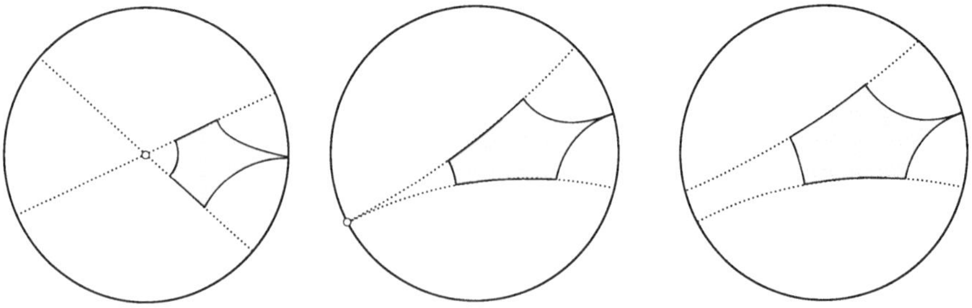

Fig. 22 The fundamental domains of a three-noid (left), a parabolic ∞-noid (center), and a hyperbolic ∞-noid (right)

$\Sigma(a, \varphi_0, f(a, \varphi_0))$ are asymptotic, whereas in the latter these planes lie at positive distance. Parabolic and hyperbolic ∞-noids have genus 0 and infinitely many ends, each of them asymptotic to a vertical plane and having finite total curvature. We also remark that these surfaces induce surfaces with finite total curvature in some quotients of $\mathbb{H}^2(\kappa) \times \mathbb{R}$ in which minimal surfaces of finite total curvature have been described by Hauswirth and Menezes [33]. In the case of hyperbolic ∞-noids, we can always find values of the parameters such that $a > a_{\text{emb}}(\varphi)$, so that family always contains embedded examples.

7 Numerical Examples of Minimal Surfaces in Product Spaces

The aim of this last section is to present some numerical experiments that help us to visualize the minimal surfaces constructed in Sect. 5.3. To this end, we will use Kenneth Brakke's *Surface Evolver* [7] (version 2.70), which is publicly available in https://facstaff.susqu.edu/brakke/evolver/evolver.html. This software has been successfully used to approximate both minimal and nonzero constant mean curvature surfaces in the Euclidean space (see, for example, [29, 44] and Brakke's gallery of triply periodic minimal surfaces [6]). Surface Evolver is also able to perform the computation of the adjoint[1] of a minimal surface in $\mathbb{R}^3$ as well as the conjugate[2] discrete constant mean curvature surface in $\mathbb{R}^3$ of a minimal surface in $\mathbb{S}^3$. The scripts are implemented following the algorithm developed by Pinkall and Polthier [87] and Oberknapp and Polthier [85]. These procedures are based on the computation of the discrete conjugate harmonic map, which is possible in space forms thanks to the close relation between harmonic maps and minimal surfaces. However, this approach is not available in $\mathbb{E}(\kappa, \tau)$-spaces.

Surface Evolver is an interactive software to minimize energies of triangulated surfaces subject to constraints and boundary conditions. The default energy is the surface tension or the area functional, but Surface Evolver is able to operate with many other quantities like gravitation or even user-defined ones. A *surface* is implemented as a triangulation, initially defined by the user in an input *datafile* by prescribing the vertexes and the incidence relations between edges and faces of the triangulation. The program *evolves* the initial surface by minimizing the energy towards a possible local minimum close to the initial configuration by a gradient descent method. However, the numerical algorithm can also find critical saddle points.

To avoid problems with the triangulation in the evolution process, Surface Evolver provides commands for vertex averaging (`V`) and equitriangulation (`u`) as well as commands to modify the triangulation by eliminating elongated triangles

[1] See http://facstaff.susqu.edu/brakke/evolver/html/scripts.htm#adjoint.cmd.

[2] See http://facstaff.susqu.edu/brakke/evolver/html/scripts.htm#cmccousin.cmd.

(K), small edges (l and t), or faces (w). The usual evolution consists in iterating by gradient descent with the g command, refining the triangulation when necessary with the r command, and using the previous commands to keep the triangulation in good shape throughout the process. In general, after a reasonable amount of iterations, the energy stalls and the resulting surface are usually near a critical point of the energy. However, some subtleties might be in place (see, for example, the evolution of the unstable catenoid in the Surface Evolver manual).

For our purposes, a key feature of Surface Evolver is its ability to operate with any Riemannian metric in three coordinates. However, according to its manual[3] *"the metric is used solely to calculate lengths and areas."* For instance, it is not used for computing the enclosed volume so in order to get a volume constraint the user needs to define his own *named quantity* (see Sect. 7.3 for further details).

We consider in the punctured space $\mathbb{R}^3_* = \mathbb{R}^3 - \{(0,0,0)\}$ the conformal metric $g = \frac{1}{x^2+y^2+z^2} g_0$, where g_0 denotes the usual flat metric and (x, y, z) are the standard coordinates. It follows that $\mathbb{S}^2 \times \mathbb{R}$ is isometric to $(\mathbb{R}^3_*, g)$ via the map $F(p, t) = e^t p$, i.e., $\mathbb{S}^2 \times \mathbb{R}$ is conformally flat. We use this identification from now on:

- Horizontal slices $\mathbb{S}^2 \times \{t_0\}$, $t_0 \in \mathbb{R}$, are in correspondence with spheres S of radius e^{t_0} centered at the origin. Moreover, the reflection about the slice $\mathbb{S}^2 \times \{t_0\}$ corresponds to an inversion in $\mathbb{R}^3$ with respect to the sphere S.
- Vertical cylinders $\gamma \times \mathbb{R}$, where γ is a geodesic of $\mathbb{S}^2$, correspond to affine planes $P \subset \mathbb{R}^3$ through the origin. Moreover, the reflection about $\gamma \times \mathbb{R}$ corresponds to the Euclidean reflection about P.
- Vertical geodesics $\{p\} \times \mathbb{R} \subset \mathbb{S}^2 \times \mathbb{R}$ correspond to straight lines through the origin, and rotations about them correspond to rotations in the Euclidean space.

We present in the following sections two numerical experiments: The first one, concerning the minimal sphere $\mathbb{S}^2 \times \{t_0\} \subset \mathbb{S}^2 \times \mathbb{R}$, is a toy example that helps us to understand better how Surface Evolver operates with the new metric and to know its limitations. The aim of the second one is to get an approximation of a singly periodic minimal surface that produces the compact genus $g \geq 3$ minimal surface $\Sigma_{g,\eta}$ in the quotient $\mathbb{S}^2 \times \mathbb{S}^1(\eta)$ for certain η (see Sect. 5.2 and Theorem 10).

The Surface Evolver datafiles used in both experiments are publicly available at https://arxiv.org/src/2203.13162/anc.

7.1 Evolution to the Minimal Sphere

Our first goal is to evolve an initial parallelepiped to a sphere centered at the origin that corresponds, via the isometry F, with a slice $\mathbb{S}^2 \times \{t_0\}$. As we will see, the choice of initial parallelepiped will determine the slice (i.e., the radius of the sphere) in the

[3] See http://facstaff.susqu.edu/brakke/evolver/html/model.htm.

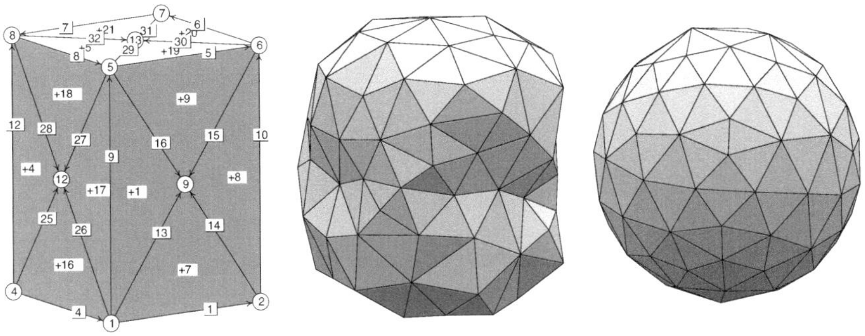

Fig. 23 Initial triangulation (left) where vertexes are marked in circled numbers while edges in squared ones (faces have a plus or minus sign depending on its orientation). Triangulation after refining (center) using `r`, `u`, and `V` commands. Final state of the evolution (right) after 100 steps with the `g` command

final evolution. The slices $\mathbb{S}^2 \times \{t_0\}$, $t_0 \in \mathbb{R}$, are stable minimal surfaces [106] so Surface Evolver is expected to approximate properly such a surface.

We start loading a datafile into Surface Evolver with a parallelepiped inscribed in the sphere of radius 1 centered at the origin, where we specify the command `conformal_metric 1/(x^2 + y^2 + z^2)` at the beginning of the file to set the aforesaid conformal metric. We first refine the rough initial triangulation (see Fig. 23 left) three times and then use the `V` and `u` commands (see Fig. 23 center). Finally, we evolve the surface 100 times (see Fig. 23) using the `g` command. After that, each step in the gradient descent method only decreases the area by approximately 10^{-4}, giving a value about 12.6557. This approximates the expected value $4\pi \approx 12.5664$ within an error of order 10^{-2}.

Surface Evolver can show the Euclidean volume enclosed by the surface. In the final step of evolution, such enclosed volume is approximately 1.225 so the surface approaches a sphere of radius 0.66. If we start with the same parallelepiped but inscribed in a sphere of radius 2, the same evolution yields a sphere of approximately the same area 12.6557 (which is the expected behavior) but of Euclidean volume approximately 9.796, i.e., the radius of the sphere is approximately 1.327. Finally, Surface Evolver is able to compute the Euclidean discrete mean curvature of the triangulated surface giving an average of 1.517 and 0.758 in the first and second cases which approximately agrees with the computed radii.

7.2 *Singly Periodic Minimal Surfaces in* $\mathbb{S}^2 \times \mathbb{R}$

Now, we are interested in getting an approximation of the compact minimal surfaces in $\mathbb{S}^2 \times \mathbb{S}^1(\eta)$ with arbitrary genus $g \geq 3$ obtained in Sect. 5. We recall that $\eta \geq 2\sqrt{\kappa}$ has to be large enough to guarantee the existence of the surface (see

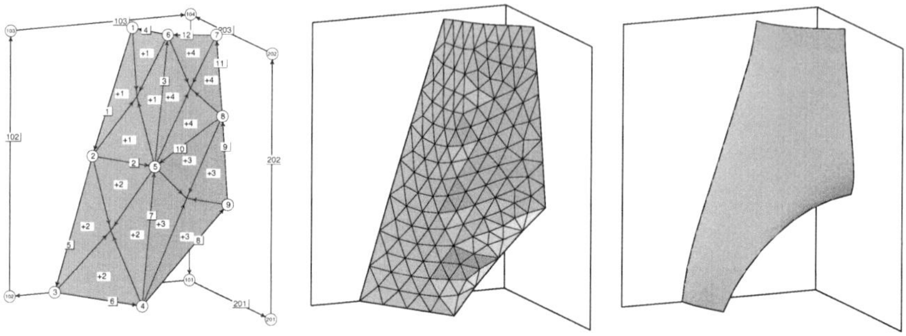

Fig. 24 Initial triangulation (left) for genus $g = 3$: the edges $\overline{49}$ and $\overline{17}$ are, respectively, constrained to the slice $\mathbb{S}^2 \times \{0\}$ (sphere of radius 1) and the slice $\mathbb{S}^2 \times \{h\}$ (sphere of radius e^h, $h \approx 0.7$ in the figure); the edges $\overline{13}$, $\overline{34}$, and $\overline{79}$ are constrained to the vertical planes of symmetry (planes $y = 0$, $z = 0$, and $-x\sin(\frac{\pi}{g-1}) + y\cos(\frac{\pi}{g-1}) = 0$, respectively). Triangulation after refining using `r`, `u`, and `V` commands (center). Final state of evolution after 150 steps using the `g` commands and taking care of the triangulation each 50 steps (right). The symmetry planes are drawn to help visualization

Theorem 10). Those surfaces are obtained as the quotient of singly periodic (by a vertical translation) minimal surfaces in $\mathbb{S}^2 \times \mathbb{R}$. We will use Surface Evolver to approximate the latter (see Figs. 24 and 25).

The triangulation defined in the datafile used to generate Figs. 24 and 25 depends on several parameters that provide an easy way to test the evolution process for different initial configurations. It is possible to change the genus, which actually controls the angle between the symmetry planes drawn in Fig. 24, as well as the *height* h of the fundamental piece (i.e., half the length of the vertical translation that leaves the surface invariant; see Figs. 13 and 14 right). After loading the datafile in Surface Evolver, we first improve the initial triangulation (with `u` and `V` commands), and then we evolve the surface 150 steps (with `g` command) taking care of the triangulation each 50 steps. At this stage, we observe that the area only decreases by 0.01 each step. Further iteration with the `g` commands just decreases the area slightly, which insinuates that the surface might be in a critical saddle point. We also notice that the *scale factor* (a real number that controls the size of the motion at each step of the iteration) becomes small due to the fact that the area of some faces approaches zero as the triangulation accumulates around the boundary $\overline{49}$ (see Fig. 24), which is constrained to the sphere of radius one. This suggests that there is an obstacle to the evolution.[4] However, trying to overcome this issue either by activating the *conjugate gradient method* (with `U` command) or by removing the small faces (with `w` command), as suggested in Surface Evolver manual, and then evolving the surface (with `g` command) produce a collapse near the boundary $\overline{49}$.

[4] The same behavior is observed in the free boundary example `free_bdry.fe` (see http://facstaff.susqu.edu/brakke/evolver/workshop/html/day2.htm) where the scale factor converges to zero as we iterate with the `g` command.

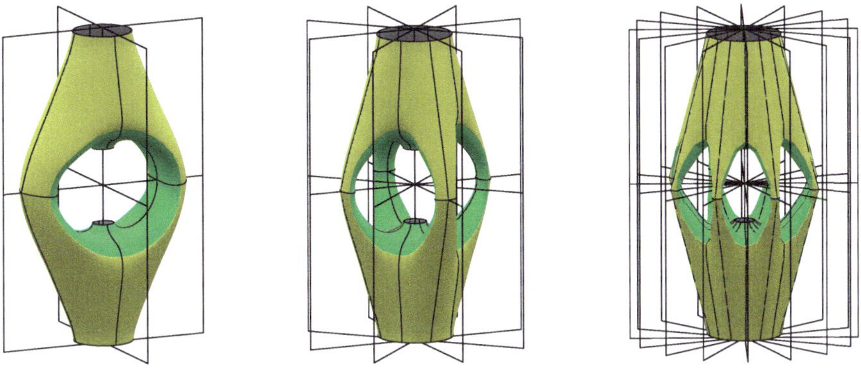

3

Fig. 25 From left to right: visual approximation in $(\mathbb{R}^3_*, g)$ of three singly periodic minimal surfaces of $\mathbb{S}^2 \times \mathbb{R}$. These surfaces, from left to right, produce compact minimal examples of genus 3, 5, and 9 in the quotient $\mathbb{S}^2 \times \mathbb{S}^1(\frac{\pi}{h})$ ($h \approx 0.7$ in the figure). The green and yellow parts are congruent in $\mathbb{S}^2 \times \mathbb{R}$, i.e., they are congruent in the figure by an inversion of the sphere of radius 1

7.3 *Final Remarks and Future Work*

The experiments of the previous section have produced surfaces that resemble the theoretical ones (see the detailed description in Sect. 5.3.2, and compare Figs. 14 and 24). However, due to the change of metric, the program exhibits some limitations in order to check the precision of the approximation. On the one hand, even after prescribing the metric, Surface Evolver computes the discrete mean curvature at each vertex of the triangulation with respect to the Euclidean metric, as shown in Sect. 7.1. As a consequence, we cannot check the precision of the approximation by computing the deviation of the discrete mean curvature from zero. On the other hand, the change of metric does not affect other quantities like the volume (see next paragraph) or the implementation of the Willmore functional,[5] both of which computed in the Euclidean metric. The latter could have been used to easily check if the evolution is near a minimal surface. Other problem we have found (see Sect. 7.2) is that the scale factor approaches zero as we iterate using the command `g` so the conducted experiments invariably lead to a collapse near one of the necks of the surface. This could indicate some instability in the discrete free boundary problem.

However, Surface Evolver is flexible enough to allow the user to define his own *quantities*. In this sense and as a future work, it will be interesting to extend Surface Evolver's functionality to overcome the mentioned issues. For instance, an implementation of the enclosed volume with respect to the new metric will be

[5] `star_perp_sq_mean_curvature`, see http://facstaff.susqu.edu/brakke/evolver/workshop/doc/quants.htm#star_perp_sq_mean_curvature.

extremely useful to get approximations of the H-surfaces constructed in Sect. 5.2 decreasing the area functional with a constrain on the volume. Another quite interesting solution is to look for an algorithm able to compute the discrete conjugate of a minimal surface in $\mathbb{E}(\kappa, \tau)$. We also propose a final approach, motivated by the fact that the initial minimal surface seems to be more tractable (Surface Evolver finds satisfactorily the unique solution to a Plateau problem given by Proposition 7). Therefore, it should be possible to get good approximations of the angle function (resp. rotation of the normal) along the horizontal (resp. vertical) geodesics of the boundary, so the different period problems that we have encountered in our constructions can be potentially solved numerically a priori. This would provide us with precious information to be plugged into the initial configuration of the conjugate surface.

Acknowledgments The authors are supported by project PID2019.111531GA.I00; the first author is also partially supported by project PID2020.117868GB.I00, both funded by MCIN/AEI/10.13039/501100011033. The second author is also supported by the Ramón y Cajal program of MCIN/AEI and by a FEDER-UJA project (ref. 1380860); the third author is also supported by the Programa Operativo FEDER Andalucía 2014–2020, Grant No. E-FQM-309-UGR18.

References

1. Abresch, U., Rosenberg, H.: A Hopf differential for constant mean curvature surfaces in $\mathbb{S}^2 \times \mathbb{R}$ and $\mathbb{H}^2 \times \mathbb{R}$. Acta Math. **193**(2), 141–174 (2004)
2. Aledo, J.A., Espinar, J.M., Gálvez, J.A.: Height estimates for surfaces with positive constant mean curvature in $\mathbb{M}^2 \times \mathbb{R}$. Ill. J. Math. **52**(1), 203–211 (2008)
3. Andrews, B., Li, H.: Embedded constant mean curvature tori in the three-sphere. J. Differ. Geom. **99**(2), 169–189 (2015)
4. Bers, L.: Survey of local properties of solutions of elliptic partial differential equations. Commun. Pure Appl. Math. **9**, 339–350 (1956). https://doi.org/10.1002/cpa.3160090306
5. Bobenko, A.I., Heller, S., Schmitt, N.: Constant mean curvature surfaces based on fundamental quadrilaterals. Math. Phys. Anal. Geom. **24**(4), 37 (2021). https://doi.org/10.1007/s11040-021-09397-z
6. Brakke, K.A.: Triply periodic minimal surfaces. http://facstaff.susqu.edu/brakke/evolver/examples/periodic/periodic.html
7. Brakke, K.A.: The surface evolver. Exp. Math. **1**(2), 141–165 (1992)
8. Cartan, É.: Leçons Sur La Géométrie Des Espaces de Riemann. Gauthier-Villars, Paris (1946)
9. Castro-Infantes, J.: On the asymptotic Plateau problem in $SL_2(R)$. J. Math. Anal. Appl. **507**(2), 125831 (2022). https://doi.org/10.1016/j.jmaa.2021.125831
10. Castro-Infantes, J., Manzano, J.M.: Genus one minimal k-noids and saddle towers in $\mathbb{H}^2 \times \mathbb{R}$. J. Inst. Math. Jussieu. **22**(5), 2155–2175 (2023). https://doi.org/10.1017/S1474748021000591
11. Castro-Infantes, J., Manzano, J.M., Rodríguez, M.: A construction of constant mean curvature surfaces in $\mathbb{H}^2 \times \mathbb{R}$ and the Krust property. Int. Math. Res. Not. **2022**(19), 14605–14638 (2022). https://doi.org/10.1093/imrn/rnab353
12. Choi, H.I., Schoen, R.: The space of minimal embeddings of a surface into a three-dimensional manifold of positive Ricci curvature. Invent. Math. **81**(3), 387–394 (1985). https://doi.org/10.1007/BF01388577
13. Collin, P., Rosenberg, H.: Construction of harmonic diffeomorphisms and minimal graphs. Ann. Math. (2) **172**(3), 1879–1906 (2010). https://doi.org/10.4007/annals.2010.172.1879

14. Daniel, B.: Isometric immersions into 3-dimensional homogeneous manifolds. Comment. Math. Helv. **82**(1), 87–131 (2007)
15. Daniel, B., Domingos, I., Vitório, F.: Constant mean curvature Isometric Immersions into $\mathbb{S}^2 \times \mathbb{R}$ and $\mathbb{H}^2 \times \mathbb{R}$ and related results. Ann. Inst. Fourier, Grenoble **73**(1), 203–249 (2023). https://doi.org/10.5802/aif.3521
16. Daniel, B., Fernández, I., Mira, P.: The Gauss map of surfaces in $\widetilde{\mathrm{PSL}}_2(\mathbb{R})$. Calc. Var. Partial Differ. Equ. **52**(3–4), 507–528 (2015). https://doi.org/10.1007/s00526-014-0721-1
17. Daniel, B., Hauswirth, L.: Half-space theorem, embedded minimal annuli and minimal graphs in the Heisenberg group. Proc. Lond. Math. Soc. **98**(2), 445–470 (2008). https://doi.org/10.1112/plms/pdn038
18. Daniel, B., Hauswirth, L., Mira, P.: Lectures Notes on Homogeneous 3-Manifolds. Korea Institute for Advanced Study, Seoul, Korea (2009)
19. Domínguez-Vázquez, M., Manzano, J.M.: Isoparametric surfaces in $\mathbb{E}(\kappa, \tau)$-spaces. Ann. Sc. Norm. Super. Pisa Cl. Sci. (5) **22**(1), 269–285 (2021)
20. Del Prete, A., Manzano, J.M., Nelli, B.: The Jenkins–Serrin problem in 3-manifolds with a Killing vector field. arXiv:2306.12195
21. Dorfmeister, J., Pedit, F., Wu, H.: Weierstrass type representation of harmonic maps into symmetric spaces. Commun. Anal. Geom. **6**(4), 633–668 (1998). https://doi.org/10.4310/CAG.1998.v6.n4.a1
22. Elcrat, A.R., Lancaster, K.E.: On the behavior of a nonparametric minimal surface in a nonconvex quadrilateral. Arch. Ration. Mech. Anal. **94**(3), 209–226 (1986). DOI 10.1007/BF00279863
23. Espinar, J., Rosenberg, H.: Complete constant mean curvature surfaces in homogeneous spaces. Comment. Math. Helv. **86**(3), 659–674 (2011). https://doi.org/10.4171/CMH/237
24. Fernández, I., Mira, P.: A characterization of constant mean curvature surfaces in homogeneous 3-manifolds. Differ. Geom. Appl. **25**(3), 281–289 (2007)
25. Fernández, I., Mira, P.: Holomorphic quadratic differentials and the Bernstein problem in Heisenberg space. Trans. Am. Math. Soc. **361**(11), 5737–5752 (2009). https://doi.org/10.1090/S0002-9947-09-04645-5
26. Finn, R.: Remarks relevant to minimal surfaces, and to surfaces of prescribed mean curvature. J. Anal. Math. **14**, 139–160 (1965). https://doi.org/10.1007/BF02806384
27. Gálvez, J.A., Martínez, A., Mira, P.: The Bonnet problem for surfaces in homogeneous 3-manifolds. Commun. Anal. Geom. **16**(5), 907–935 (2008)
28. Große-Brauckmann, K.: New surfaces of constant mean curvature. Math. Z. **214**(4), 527–565 (1993)
29. Große-Brauckmann, K.: Gyroids of constant mean curvature. Exp. Math. **6**(1), 33–50 (1997)
30. Große-Brauckmann, K., Kusner, R.B., Sullivan, J.M.: Triunduloids: Embedded constant mean curvature surfaces with three ends and genus zero. J. Reine Angew. Math. **564**, 35–61 (2003)
31. Große-Brauckmann, K., Polthier, K.: Constant mean curvature surfaces derived from Delaunay's and Wente's examples. In: Visualization and Mathematics (Berlin-Dahlem, 1995), pp. 119–134. Springer, Berlin (1997)
32. Große-Brauckmann, K., Wohlgemuth, M.: The gyroid is embedded and has constant mean curvature companions. Calc. Var. Partial Differ. Equ. **523**, 499–523 (1996)
33. Hauswirth, L., Menezes, A.: On doubly periodic minimal surfaces in $\mathrm{H}^2 \times \mathrm{R}$ with finite total curvature in the quotient space. Ann. Mat. Pura Appl. (4) **195**(5), 1491–1512 (2016). https://doi.org/10.1007/s10231-015-0524-9
34. Hauswirth, L., Menezes, A., Rodríguez, M.: On the characterization of minimal surfaces with finite total curvature in $\mathrm{H}^2 \times \mathrm{R}$ and PSL_2 (R). Calc. Var. Partial Differ. Equ. **58**(2), 80, 24 (2019). https://doi.org/10.1007/s00526-019-1505-4
35. Hauswirth, L., Nelli, B., Sa Earp, R., Toubiana, E.: A Schoen theorem for minimal surfaces in $\mathbb{H}^2 \times \mathbb{R}$. Adv. Math. **274**, 199–240 (2015). https://doi.org/10.1016/j.aim.2014.12.030
36. Hauswirth, L., Rosenberg, H.: Minimal surfaces of finite total curvature in H×R. In: Matemática Contemporânea, vol. 31, pp. 65–80 (2006)

37. Hauswirth, L., Rosenberg, H., Spruck, J.: On complete mean curvature $\frac{1}{2}$ surfaces in $\mathbb{H}^2 \times \mathbb{R}$. Commun. Anal. Geom. **16**(5), 989–1005 (2008). https://doi.org/10.4310/CAG.2008.v16.n5.a4
38. Hauswirth, L., Sa Earp, R., Toubiana, E.: Associate and conjugate minimal immersions in $\boldsymbol{M} \times \boldsymbol{R}$. Tohoku Math. J. **60**(2), 267–286 (2008). https://doi.org/10.2748/tmj/1215442875
39. Heller, L., Heller, S., Traizet, M.: Complete families of embedded high genus CMC surfaces in the 3-sphere (2021). arXiv:2108.10214 [math]
40. Heller, S., Schmitt, N.: Deformations of symmetric CMC surfaces in the 3-sphere. Exp. Math. **24**(1), 65–75 (2015). https://doi.org/10.1080/10586458.2014.954294
41. Hitchin, N.J.: Harmonic maps from a 2-torus to the 3-sphere. J. Differ. Geom. **31**(3), 627–710 (1990)
42. Hoffman, D., Traizet, M., White, B.: Helicoidal minimal surfaces of prescribed genus. Acta Math. **216**(2), 217–323 (2016). https://doi.org/10.1007/s11511-016-0139-z
43. Hsiang, W.y., Hsiang, W.T.: On the uniqueness of isoperimetric solutions and imbedded soap bubbles in noncompact symmetric spaces, I. Invent. Math. **98**, 39–58 (1989)
44. Hsu, L., Kusner, R., Sullivan, J.: Minimizing the squared mean curvature integral for surfaces in space forms. Exp. Math. **1**(3), 191–207 (1992)
45. Jorge, L.P., Meeks III, W.H.: The topology of complete minimal surfaces of finite total Gaussian curvature. Topology **22**(2), 203–221 (1983). https://doi.org/10.1016/0040-9383(83)90032-0
46. Kapouleas, N.: Compact constant mean curvature surfaces in Euclidean three-space. J. Differ. Geom. **33**(3), 683–715 (1991)
47. Kapouleas, N.: Constant mean curvature surfaces constructed by fusing Wente tori. Invent. Math. **119**(3), 443–518 (1995). https://doi.org/10.1007/BF01245190
48. Kapouleas, N.: Minimal surfaces in the round three-sphere by doubling the equatorial two-sphere, I. J. Differ. Geom. **106**(3), 393–449 (2017). https://doi.org/10.4310/jdg/1500084022
49. Kapouleas, N., Yang, S.D.: Minimal surfaces in the three-sphere by doubling the Clifford torus. Am. J. Math. **132**(2), 257–295 (2010). https://doi.org/10.1353/ajm.0.0104
50. Karcher, H.: Embedded minimal surfaces derived from Scherk's examples. Manuscr. Math. **62**(1), 83–114 (1988). https://doi.org/10.1007/BF01258269
51. Karcher, H.: The triply periodic minimal surfaces of Alan Schoen and their constant mean curvature companions. Manuscr. Math. **64**(3), 291–357 (1989). https://doi.org/10.1007/BF01165824
52. Karcher, H.: Hyperbolic surfaces of constant mean curvature one with compact fundamental domains. In: Global Theory of Minimal Surfaces, Clay Math. Proc., vol. 2, pp. 311–323. American Mathematical Society, Providence (2005)
53. Karcher, H.: Introduction to conjugate Plateau constructions. In: Global Theory of Minimal Surfaces, vol. 2, pp. 137–161. American Mathematical Society, Providence (2005)
54. Karcher, H.: Construction of minimal surfaces. Surveys in Geometry, pp. 1–96. University of Tokyo, 1989, and Lecture Notes No. 12, SFB256, Bonn, 1989
55. Karcher, H., Pinkall, U., Sterling, I.: New minimal surfaces in S^3. J. Differ. Geom. **28**(2), 169–185 (1988)
56. Karcher, H., Polthier, K.: Construction of triply periodic minimal surfaces. Philos. Trans. Roy. Soc. Lond. Ser. A **354**(1715), 2077–2104 (1996). https://doi.org/10.1098/rsta.1996.0093
57. Kim, Y.W., Koh, S.E., Lee, H.Y., Shin, H., Yang, S.D.: Helicoidal Killing fields, helicoids and ruled minimal surfaces in homogeneous three-manifolds. J. Korean Math. Soc. **55**(5), 1235–1255 (2018). https://doi.org/10.4134/JKMS.j170671
58. Lawson Jr., H.B.: Complete minimal surfaces in S^3. Ann. Math. **92**, 335–374 (1970)
59. Leite, M.L.: An elementary proof of the Abresch-Rosenberg theorem on constant mean curvature immersed surfaces in $\mathbb{S}^2 \times \mathbb{R}$ and $\mathbb{H}^2 \times \mathbb{R}$. Q. J. Math. **58**(4), 479–487 (2007)
60. Lerma, A.M., Manzano, J.M.: Compact stable surfaces with constant mean curvature in Killing submersions. Ann. Mat. Pura Appl. (4) **196**(4), 1345–1364 (2017). https://doi.org/10.1007/s10231-016-0619-y

61. Leung, D.S.P.: The reflection principle for minimal submanifolds of Riemannian symmetric spaces. J. Differ. Geom. **8**, 153–160 (1973)
62. Manzano, J.: Estimates for constant mean curvature graphs in $M \times \mathbb{R}$. Rev. Mat. Iberoam. **29**(4), 1263–1281 (2013). https://doi.org/10.4171/RMI/756
63. Manzano, J.M.: Superficies de curvature media constante en espacios homogéneos. Ph.D. Thesis, Universidad de Granada (2012)
64. Manzano, J.M.: On the classification of Killing submersions and their isometries. Pac. J. Math. **270**(2), 367–392 (2014). https://doi.org/10.2140/pjm.2014.270.367
65. Manzano, J.M.: Dual quadratic differentials and entire minimal graphs in Heisenberg space. Ann. Global Anal. Geom. **55**(2), 197–213 (2019). https://doi.org/10.1007/s10455-018-9623-3
66. Manzano, J.M.: Invariant constant mean curvature tubes around a horizontal geodesic in $\mathbb{E}(\kappa, \tau)$-spaces. arXiv:2305.09014
67. Manzano, J.M., Nelli, B.: Height and area estimates for constant mean curvature graphs in $\mathbb{E}(\kappa, \tau)$-spaces. J. Geom. Anal. **27**(4), 3441–3473 (2017). https://doi.org/10.1007/s12220-017-9810-7
68. Manzano, J.M., Plehnert, J., Torralbo, F.: Compact embedded minimal surfaces in $\mathbb{S}^2 \times \mathbb{S}^1$. Commun. Anal. Geom. **24**(2), 409–429 (2016)
69. Manzano, J.M., Rodríguez, M.M.: On complete constant mean curvature vertical multigraphs in $E(\kappa, \tau)$. J. Geom. Anal. **25**(1), 336–346 (2015)
70. Manzano, J.M., Torralbo, F.: New examples of constant mean curvature surfaces in $\mathbb{S}^2 \times \mathbb{R}$ and $\mathbb{H}^2 \times \mathbb{R}$. Mich. Math. J. **63**(4), 701–723 (2014)
71. Manzano, J.M., Torralbo, F.: Compact embedded surfaces with constant mean curvature in $\mathbb{S}^2 \times \mathbb{R}$. Am. J. Math. **142**(6), 1981–1994 (2020). https://doi.org/10.1353/ajm.2020.0050
72. Manzano, J.M., Torralbo, F.: Horizontal Delaunay surfaces with constant mean curvature in $\mathbb{S}^2 \times \mathbb{R}$ and $\mathbb{H}^2 \times \mathbb{R}$. Camb. J. Math. **10**(3), 657–688 (2022). https://doi.org/10.4310/CJM.2022.v10.n3.a2
73. Martín, F., Mazzeo, R., Rodríguez, M.M.: Minimal surfaces with positive genus and finite total curvature in $\mathbb{H}^2 \times \mathbb{R}$. Geom. Topol. **18**(1), 141–177 (2014). https://doi.org/10.2140/gt.2014.18.141
74. Martín, F., Rodríguez, M.M.: Non-simply connected minimal planar domains in $\mathbb{H}^2 \times \mathbb{R}$. Trans. Am. Math. Soc. **365**(12), 6167–6183 (2013). https://doi.org/10.1090/S0002-9947-2013-05794-7
75. Mazet, L.: The Plateau problem at infinity for horizontal ends and genus 1. Indiana Univ. Math. J. **55**(1), 15–64 (2006). https://doi.org/10.1512/iumj.2006.55.2583
76. Mazet, L., Rodríguez, M.M., Rosenberg, H.: The Dirichlet problem for the minimal surface equation, with possible infinite boundary data, over domains in a Riemannian surface. Proc. Lond. Math. Soc. (3) **102**(6), 985–1023 (2011). https://doi.org/10.1112/plms/pdq032
77. Mazet, L., Rodríguez, M.M., Rosenberg, H.: Periodic constant mean curvature surfaces in $\mathbb{H}^2 \times \mathbb{R}$. Asian J. Math. **18**(5), 829–858 (2014). https://doi.org/10.4310/AJM.2014.v18.n5.a4
78. Meeks III, W.H., Pérez, J.: Constant mean curvature surfaces in metric Lie groups. In: Geometric Analysis: Partial Differential Equations and Surfaces, vol. 570, pp. 25–110. American Mathematical Society, Providence (2012). https://doi.org/10.1090/conm/570/11304
79. Meeks III, W.H., Yau, S.T.: The classical Plateau problem and the topology of three-dimensional manifolds. The embedding of the solution given by Douglas-Morrey and an analytic proof of Dehn's lemma. Topology **21**(4), 409–442 (1982). https://doi.org/10.1016/0040-9383(82)90021-0
80. Meeks III, W.H., Yau, S.T.: The existence of embedded minimal surfaces and the problem of uniqueness. Math. Z. **179**(2), 151–168 (1982). https://doi.org/10.1007/BF01214308
81. Melo, S.: Minimal graphs in $\widetilde{\mathrm{PSL}}_2(\mathbb{R})$ over unbounded domains. Bull. Braz. Math. Soc. (N.S.) **45**(1), 91–116 (2014). https://doi.org/10.1007/s00574-014-0042-1
82. Montaldo, S., Onnis, I.I.: Invariant CMC surfaces in $\mathbb{H}^2 \times \mathbb{R}$. Glasg. Math. J. **46**(02), 311–321 (2004)

83. Morabito, F., Rodríguez, M.M.: Saddle towers and minimal k-noids in $\mathbb{H}^2 \times \mathbb{R}$. J. Inst. Math. Jussieu **11**(2), 333–349 (2012). https://doi.org/10.1017/S1474748011000107
84. Nelli, B., Rosenberg, H.: Minimal surfaces in $\mathbb{H}^2 \times \mathbb{R}$. Bull. Braz. Math. Soc. (N.S.) **33**(2), 263–292 (2002). https://doi.org/10.1007/s005740200013
85. Oberknapp, B., Polthier, K.: An algorithm for discrete constant mean curvature surfaces. In: Visualization and Mathematics (Berlin-Dahlem, 1995), pp. 141–161. Springer, Berlin (1997)
86. Pedrosa, R.H.L., Ritoré, M.: Isoperimetric domains in the Riemannian product of a circle with a simply connected space form and applications to free boundary problems. Indiana Univ. Math. J. **48**(4), 1357–1394 (1999)
87. Pinkall, U., Polthier, K.: Computing discrete minimal surfaces and their conjugates. Exp. Math. **2**(1), 15–36 (1993)
88. Pinkall, U., Sterling, I.: On the classification of constant mean curvature tori. Ann. Math. (2) **130**(2), 407–451 (1989). https://doi.org/10.2307/1971425
89. Plehnert, J.: Surfaces with constant mean curvature 1/2 and genus one in $\mathbb{H}^2 \times \mathbb{R}$. Preprint (2013)
90. Plehnert, J.: Constant mean curvature k-noids in homogeneous manifolds. Ill. J. Math. **58**(1), 233–249 (2014). https://doi.org/10.1215/ijm/1427897176
91. Polthier, K.: Sonderforschungsbereich 256. Grape (2013)
92. Pyo, J.: New complete embedded minimal surfaces in $\mathbb{H}^2 \times \mathbb{R}$. Ann. Global Anal. Geom. **40**(2), 167–176 (2011). https://doi.org/10.1007/s10455-011-9251-7
93. Rakotoniaina, C.: Cut locus of the B-spheres. Ann. Global Anal. Geom. **3**(3), 313–327 (1985)
94. Rodríguez, M.: Minimal surfaces with limit ends in $\mathbb{H}^2 \times \mathbb{R}$. J. Reine Angew. Math. **685**, 123–141 (2013). https://doi.org/10.1515/crelle-2012-0010
95. Rosenberg, H.: Minimal surfaces in $\mathbb{M}^2 \times \mathbb{R}$. Ill. J. Math. **46**(4), 1177–1195 (2002)
96. Rosenberg, H., Souam, R., Toubiana, E.: General curvature estimates for stable H-surfaces in 3-manifolds and applications. J. Differ. Geom. **84**(3), 623–648 (2010)
97. Rossman, W.: Mean curvature one surfaces in hyperbolic space, and their relationship to minimal surfaces in Euclidean space. J. Geom. Anal. **11**(4), 669–692 (2001). https://doi.org/10.1007/BF02930762
98. Sa Earp, R.: Parabolic and Hyperbolic Screw Motion Surfaces in $\mathbb{H}^2 \times \mathbb{R}$. J. Aust. Math. Soc. **85**(1), 113–143 (2008). https://doi.org/10.1017/S1446788708000013
99. Sa Earp, R., Toubiana, E.: A reflection principle for minimal surfaces in smooth three manifolds. Preprint (2019)
100. Sa Earp, R., Toubiana, E.: Classical Schwarz reflection principle for Jenkins-Serrin type minimal surfaces. Ann. Global Anal. Geom. **57**(2), 365–379 (2020). https://doi.org/10.1007/s10455-020-09704-x
101. Schoen, A.: Infinite periodic minimal surfaces without self-intersections. NASA Technical Note No. TN D-5 (1970)
102. Souam, R., Toubiana, E.: Totally umbilic surfaces in homogeneous 3-manifolds. Comment. Math. Helv. **84**, 673–704 (2009)
103. Steenrod, N.: The Topology of Fibre Bundles. Princeton Mathematical Series, vol. 14. Princeton University Press, Princeton (1951)
104. Torralbo, F.: Rotationally invariant constant mean curvature surfaces in homogeneous 3-manifolds. Differ. Geom. Appl. **28**(5), 593–607 (2010). https://doi.org/10.1016/j.difgeo.2010.04.007
105. Torralbo, F.: Compact minimal surfaces in the Berger spheres. Ann. Global Anal. Geom. **41**(4), 391–405 (2012). https://doi.org/10.1007/s10455-011-9288-7
106. Torralbo, F., Urbano, F.: On stable compact minimal submanifolds. Proc. Am. Math. Soc. **142**(2), 651–658 (2014). https://doi.org/10.1090/S0002-9939-2013-11810-1
107. Weber, M.: The Minimal Surface Archive. https://minimal.sitehost.iu.edu/archive/index.html.
108. Younes, R.: Minimal surfaces in $\widetilde{PSL}_2(\mathbb{R})$. Ill. J. Math. **54**(2), 671–712 (2010)

Integral Geometry of Pairs of Lines and Planes

Julià Cufí, Eduardo Gallego, and Agustí Reventós

Abstract In this paper, we present some results obtained previously in Cufí et al. (J. Math. Anal. Appl. 458(1):436–451 (2018); Mathematika 65(4):874–896 (2019); Rend. Circ. Mat. Palermo (2) 69(3):1115–1130 (2020); Arch. Math. (Basel) 117(5):579–591 (2021)) related to convex sets in the plane and in the space.

In the plane, we deal with Hurwitz's inequality, which provides an upper bound of the isoperimetric deficit of a convex set K in terms of the area of the evolute of the boundary of K. We improve this inequality finding strictly positive lower bounds for the Hurwitz's deficit, these bounds involving the visual angle of the boundary of K. In a different look, we provide a unified approach that encompasses some integral formulas for functions of the visual angle of a compact convex set due to Crofton, Hurwitz, and Masotti. Also, we interpret these formulas from the point of view of Integral Geometry of pairs of lines.

In the space, we deal with integrals of invariant measures of pairs of planes, expressing some of these integrals in terms of functions of the visual dihedral angle of the convex set. As a consequence of our results, we evaluate the deficit in a Crofton-type inequality due to Blaschke.

Keywords Convex set · Visual angle · Invariant measure · Constant width · Dihedral visual angle

MSC Classification: 52A10, 52A15, 53C65

1 Introduction

Here, we present some results obtained previously in [3–6] related to convex sets in the plane and in the space. For the case of the plane, our contribution is strongly related to the celebrated work by Hurwitz [10] in which he introduced the use of

J. Cufí · E. Gallego (✉) · A. Reventós
Departament de Matemàtiques, Universitat Autònoma de Barcelona, Barcelona, Catalonia
e-mail: Julia.Cufi@uab.cat; Eduardo.Gallego@uab.cat; Agusti.Reventos@uab.cat

A. Alarcón et al. (eds.), *New Trends in Geometric Analysis*,
RSME Springer Series 10, https://doi.org/10.1007/978-3-031-39916-9_4

Fourier series to deal with geometrical problems, some of them related to the visual angle of a convex set. In the case of the space, Fourier series must be substituted by spherical harmonics and the visual angle by the dihedral visual angle. In both cases, we adopt the point of view of Integral Geometry according to Santaló [12].

The results in Sect. 3 are related to the classical isoperimetric inequality

$$\Delta = L^2 - 4\pi F \geq 0,$$

where L is the length of a simple closed plane curve Γ enclosing a region of area F. In the case that Γ bounds a convex set K, Hurwitz [10] established a kind of reverse isoperimetric inequality, namely,

$$L^2 - 4\pi F \leq \pi |F_e|, \tag{1}$$

where F_e is the algebraic area ($F_e \leq 0$) enclosed by the evolute of Γ. Moreover, equality holds in (1) if and only if Γ is a circle or a curve parallel to an astroid.

We improve Hurwitz's inequality (1) finding strictly positive lower bounds for *Hurwitz's deficit* $\pi|F_e| - \Delta$. These bounds involve the *visual angle* of K from a point P, which is the angle between the two tangents from P to the boundary of K (see Theorem 1). In the constant width case, we prove, in Theorem 2, the inequality $L^2 - 4\pi F \leq \frac{4}{9}\pi|F_e|$.

Before addressing what we do in Sect. 4, let us remember that in 1868 Crofton showed [2] the well-known formula

$$\int_{P \notin K} (\omega - \sin\omega)\, dP = \frac{L^2}{2} - \pi F, \tag{2}$$

where $\omega = \omega(P)$ is the visual angle of K from the point P.

Later on, Hurwitz in 1902 [10] considered again the integral of some functions of the visual angle. In particular, he gave a new proof of Crofton's formula using the Fourier series of the radius of curvature ρ of the boundary ∂K. He also established the equality

$$\int_{P \notin K} \sin^3\omega\, dP = \frac{3}{4}L^2 + \frac{1}{4}\pi^2\gamma_2^2 \tag{3}$$

with $\gamma_k^2 = \alpha_k^2 + \beta_k^2$, where α_k and β_k are the Fourier coefficients of ρ.

In 1955, Masotti [11] considered a Crofton-type formula computing

$$\int_{P \notin K} (\omega^2 - \sin^2\omega)\, dP$$

in terms of the area of K, the length of ∂K, and the Fourier coefficients of the radius of curvature of ∂K. Santaló in 1976 [12, I.4.5] gave lower and upper bounds for the above integral.

In Sect. 4.1, we provide a unified approach that encompasses the previous results and allows us to obtain new integral formulas for functions of the visual angle. The basic tool is the integral formula given in Theorem 3. Using this theorem, we obtain integral formulas for any power of $\sin\omega$ (Theorem 5) and for the function $\omega^m - \sin^m \omega$ (see Eq. 19).

In Sect. 4.2, we deal with a general type of integral formulas of the visual angle including those we have just commented above, from the point of view of Integral Geometry according to Crofton and Santaló [12]. The purpose is twofold: to provide an interpretation of these formulas in terms of integrals of functions with respect to the canonical measure in the space of pairs of lines and to give new simpler proofs of them (see Propositions 7, 8, and 9).

The main goal of Sect. 5 is to study integrals of invariant measures with respect to Euclidean motions in the Euclidean space $\mathbb{E}^3$, extended to the set of pairs of planes meeting a compact convex set. To carry out this objective, we express these integrals in terms of functions of the dihedral visual angle of the convex set from a line and integrate them with respect to an invariant measure in the space of lines. The main tool we use is spherical harmonics. In this sense, Theorem 8 plays an important role. Then, we assign to any invariant measure on the space of pairs of planes an appropriate function of the dihedral visual angle of a given convex set. The integral of this function with respect to the measure on the space of lines gives the integral of the above measure extended to those planes meeting the convex set (see Theorem 9). In Sect. 5.2, we relate this last result to Blaschke's work [1, p. 75] in Theorem 10.

We thank the people in charge of the REAG network (Red Española de Análisis Geométrico) for the opportunity to present our recent work here.

2 Preliminaries

2.1 *Support Function*

A set $K \subset \mathbb{R}^n$ is *convex* if it contains the complete segment joining every two points in the set. We shall consider nonempty compact convex sets. The *support function* of K is defined as

$$p_K(u) := \sup\{\langle x, u\rangle \ : \ x \in K\} \qquad \text{for} \quad u \in \mathbb{R}^n.$$

For a vector u in the unit sphere S^{n-1}, the number $p_K(u)$ is the signed distance of the support hyperplane to K with outer normal vector u from the origin. The distance is negative if and only if u points into the open half-space containing the origin (cf. [13]).

In the case of the plane, we shall denote by $p(\varphi)$ the 2π-periodic function obtained by evaluating $p_K(u)$ on $u = (\cos\varphi, \sin\varphi)$. Note that ∂K is the envelope of the one parametric family of lines given by $x\cos\varphi + y\sin\varphi = p(\varphi)$. When p is a C^2 function the radius of curvature $\rho(\varphi)$ of ∂K is given by $p(\varphi) + p''(\varphi)$. Then, convexity is equivalent to $p(\varphi) + p''(\varphi) \geq 0$. We say that a C^2 support function p defines a *strictly convex* set if $p(\varphi) + p''(\varphi) > 0$ for every value of φ.

It can be seen that the length L of ∂K and the area F of K are given by

$$L = \int_0^{2\pi} p\,d\varphi, \quad F = \frac{1}{2}\int_0^{2\pi} p(p + p'')d\varphi.$$

In general, a one parameter family of lines $x\cos t + y\sin t = f(t)$, where f is a differentiable function, defines a curve in the plane. In this setting, the curve is not necessarily closed nor convex. When a curve $\gamma(t)$, $a \leq t \leq b$, is defined as the envelope of a family of lines of this type, for a function f of class C^2, we say that $f(t)$ is the *generalized support function* of the curve. The *area with multiplicities* swept by the radius vector of the curve is given by

$$F = \frac{1}{2}\int_a^b f(f + f'')dt, \tag{4}$$

as a simple computation shows.

Let $p(\varphi)$ be the support function of a strictly convex set K. Then, $p_r(\varphi) = p(\varphi) + r$ defines for each real r a *parallel curve* to ∂K. If the origin is in the interior of K, then p is a strictly positive function. If $r > 0$, the function p_r corresponds to the outer parallel set at distance r. When $r < 0$, the curve given by p_r is not necessarily convex (this is the case when $|r| > \min(\rho)$, ρ being the radius of curvature of ∂K).

A special type of convex sets are those of *constant width*, which is those convex sets whose orthogonal projection on any direction has the same length w. In terms of the support function p of K, constant width means that $p(\varphi) + p(\varphi + \pi) = w$. Expanding p in Fourier series

$$p(\varphi) = a_0 + \sum_{n=1}^{\infty} a_n \cos(n\varphi) + b_n \sin(n\varphi), \tag{5}$$

one obtains that K has constant width if and only if $a_n = b_n = 0$ for all even $n > 0$.

The *Steiner point* of K is defined by the vector-valued integral

$$s(K) = \frac{1}{\pi}\int_0^{2\pi} p(\varphi)N(\varphi)\,d\varphi,$$

where $N(\varphi) = (\cos\varphi, \sin\varphi)$ denotes the normal vector to ∂K. The Steiner point is rigid motion equivariant; this means that $s(gK) = gs(K)$ for every rigid motion g. It is known that $s(K)$ lies in the interior of K (cf. [8, p. 56]). In terms of the Fourier coefficients of $p(\varphi)$ given in (5), the Steiner point is $s(K) = (a_1, b_1)$. Hence, taking the Steiner point as a new origin, one has

$$p(\varphi) = a_0 + \sum_{n\geq 2} (a_n \cos n\varphi + b_n \sin n\varphi)\,.$$

2.2 Measure of Lines in the Plane

We denote by $\mathcal{A}_{2,1}$ the set of straight lines in the plane. For each straight line $G \in \mathcal{A}_{2,1}$ that does not pass through the origin, let P be the point of G at a minimum distance from the origin. We take as coordinates for G the polar coordinates (p, φ) of the point P, with $p > 0$ and $0 \leq \varphi < 2\pi$. Notice that p and φ can also be seen as functions in this space of lines, and we shall write $p(G)$, $\varphi(G)$ for the corresponding coordinates of the straight line G.

The invariant measure in the set of lines of the plane not containing the origin is given by a constant multiple of

$$dG = dp \wedge d\varphi. \tag{6}$$

In fact this measure is, except for a constant factor, the only one invariant under Euclidean motions (see [12], Section I.3.1). In the space of ordered pairs of lines $\mathcal{A}_{2,1} \times \mathcal{A}_{2,1}$, we consider the canonical measure $dG_1 \wedge dG_2$. For every function $\tilde{f}(G_1, G_2)$, we can consider the measure $\tilde{f}(G_1, G_2)\, dG_1 \wedge dG_2$. We prove in Proposition 5 that this measure is invariant under Euclidean motions if and only if $\tilde{f}(G_1, G_2) = f(\varphi(G_2) - \varphi(G_1))$ with f an even π-periodic function on $\mathbb{R}$.

2.3 Spherical Harmonics

Let us recall that a *spherical harmonic* of degree n on the unit sphere S^2 is the restriction to S^2 of an harmonic homogeneous polynomial of degree n. It is known that every continuous function on S^2 can be uniformly approximated by finite sums of spherical harmonics (see, for instance, [8]).

More precisely, the function $p(u)$ can be written in terms of spherical harmonics as

$$p(u) = \sum_{n=0}^{\infty} \pi_n(p)(u), \tag{7}$$

where $\pi_n(p)$ is the projection of the support function p on the vector space of spherical harmonics of degree n. An orthogonal basis of this space is given in terms of the longitude θ and the colatitude φ in S^2 by

$$\{\cos(j\theta)(\sin\varphi)^j\, P_n^{(j)}(\cos\varphi), \quad \sin(j\theta)(\sin\varphi)^j\, P_n^{(j)}(\cos\varphi): \quad 0 \le j \le n\}$$

where $P_n^{(j)}$ denotes the jth derivative of the nth Legendre polynomial P_n (cf. [8]).

It can be seen that $\pi_0(p) = \mathcal{W}/2 = M/(4\pi)$ where $\mathcal{W} = 1/(4\pi)\int_{S^2} w(u)\,du$ is the *mean width* of K and M is the *mean curvature* of K. Moreover, $\pi_1(p) = \langle s(K), \cdot\rangle$ where $s(K)$ denotes the Steiner point of K (cf. [8, p. 182]). It is clear that $\pi_0(p)$ is invariant under Euclidean motions and that $\pi_1(p)$ is not. It is known that $\pi_n(p)$ is invariant under translations for every $n \neq 1$ (cf. [13, p. 5]). One can easily check that K has constant width if and only if $\pi_n(p) = 0$ for $n \neq 0$ even.

3 Lower Bounds for the Hurwitz's Deficit

In order to study the Hurwitz's deficit $\pi|F_e| - \Delta$ of a convex set K, we introduce *Wirtinger's deficit* W_q of a C^1 function $q(\varphi)$ of period 2π,

$$W_q = \int_0^{2\pi} (q'^2 - q^2)\,d\varphi.$$

Note that by (4), $W_q = -2F$ where F is the area with multiplicities enclosed by the curve defined by the generalized support function q.

Recall that Wirtinger's inequality (see [8]) states that if $\int_0^{2\pi} q(\varphi)d\varphi = 0$, then $W_q \ge 0$. In particular, we always have $W_{q'} \ge 0$. In [3], we give a relationship between Wirtinger and Hurwitz's deficits:

Proposition 1 *Let K be a compact strictly convex set of area F bounded by a curve $\Gamma = \partial K$ of class C^2 and length L. Let p be the support function of K and let F_e be the area with multiplicities enclosed by the evolute of Γ. Then*

$$\pi|F_e| - \Delta = \frac{\pi}{2}(W_{q'} - 4W_q)$$

where $q(\varphi) = p(\varphi) - \frac{L}{2\pi}$ and $\Delta = L^2 - 4\pi F$.

Remark 1 Let F be the area enclosed by the curve with generalized support function the 2π-periodic function q, and let F_e be the area enclosed by the evolute of this curve, both areas counted with multiplicities. The equalities $W_{q'} = -2F_e$ and $W_q = -2F$ give

$$\frac{1}{2}(W_{q'} - W_q) = F - F_e.$$

Thus, for closed curves with positive curvature, we have

$$F - F_e = \frac{1}{2}\int_0^{2\pi} (q + q'')^2\, d\varphi = \frac{1}{2}\int_0^L \rho\, ds \tag{8}$$

where $\rho = q + q''$ is the radius of curvature and L the length of the curve. We have used the relation $ds = \rho\, d\varphi$. Equality (8) for the case of simple closed curves that bound a strictly convex domain was proved in [10].

The next lemma compares Wirtinger's deficit of a given function with that of its derivative. The proof follows the standard pattern of the proof of Wirtinger's inequality using Fourier series.

Lemma 1 *Let $q = q(\varphi)$ a 2π-periodic C^2 function. Then,*

$$W_{q'} \geq 4W_q + \frac{2}{\pi}\left(\int_0^{2\pi} q\, d\varphi\right)^2 \geq 0.$$

Moreover, the first inequality is an equality if and only if

$$q(\varphi) = a_0 + a_1 \cos\varphi + b_1 \sin\varphi + a_2 \cos 2\varphi + b_2 \sin 2\varphi,$$

for some constants $a_0, a_1, b_1, a_2, b_2 \in \mathbb{R}$.

Remark that the first inequality in Lemma 1 improves Wirtinger's inequality for the derivative of 2π-periodic functions.

As a consequence of Proposition 1, one obtains the well-known Hurwitz's inequality (1).

We proceed now to find a lower bound for the Hurwitz's deficit $\pi|F_e| - \Delta$ so improving (1). If

$$p(\varphi) = a_0 + \sum_{n\geq 1} a_n \cos n\varphi + b_n \sin n\varphi$$

is the Fourier series of the support function of K, it is known that the quantities $c_n^2 = a_n^2 + b_n^2$, for $n \geq 2$, are invariant under the group of plane motions.

Consider ω the *visual angle* of ∂K from P and let dP be the area measure. Writing

$$I_n = \int_{P\notin K} \left(-2\sin(\omega) + \frac{n+1}{n-1}\sin(n-1)\omega - \frac{n-1}{n+1}\sin(n+1)\omega\right) dP,$$

it is proved in [10][1] that

$$I_n = L^2 + (-1)^n \pi^2 (n^2 - 1) c_n^2, \; n \geq 2 \tag{9}$$

L being the length of the boundary of K. For instance, if $n = 2$ one gets

$$\frac{4}{3} \int_{P \notin K} \sin^3 \omega \, dP = L^2 + 3\pi^2 c_2^2. \tag{10}$$

Moreover, this visual angle satisfies the Crofton's formula (see [10])

$$\int_{P \notin K} (\omega - \sin \omega) dP = \frac{L^2}{2} - \pi F. \tag{11}$$

We can prove now the following result involving (9) and (11):

Theorem 1 ([3]) *Let K be a compact strictly convex set of area F bounded by a curve $\Gamma = \partial K$ of class C^2 and length L. Let F_e be the area with multiplicities enclosed by the evolute of Γ and let Δ be the isoperimetric deficit. Then*

$$\pi |F_e| - \Delta \geq \frac{5}{4} L^2 + 5 \int_{P \notin K} (\omega - \sin \omega - \frac{2}{3} \sin^3 \omega) dP. \tag{12}$$

The right-hand side of this inequality is a strictly positive quantity except when $\pi |F_e| - \Delta = 0$ in which case it also vanishes.

Proof It can be seen that

$$\pi |F_e| - \Delta = \frac{\pi}{2} (W_{q'} - 4 W_q) = \frac{\pi}{2} \left(4 \int_0^{2\pi} q^2 d\varphi - 5 \int_0^{2\pi} q'^2 d\varphi + \int_0^{2\pi} q''^2 d\varphi \right),$$

where $q(\varphi) = p(\varphi) - L/2\pi$ and $p(\varphi)$ is the support function of K with respect to the Steiner point.

In terms of the Fourier coefficients of p,

$$\pi |F_e| - \Delta = \frac{\pi^2}{2} \sum_{n \geq 3} (n^4 - 5n^2 + 4) c_n^2.$$

[1] There is a misprint with the sign in Hurwitz's paper. Moreover, the c_n coefficients appearing in (9) are different from those in Hurwitz's paper because the latter correspond to the Fourier series of the curvature radius function.

Observe now that for $n \geq 3$, we have $n^4 - 5n^2 + 4 \geq 5(n^2 - 1)$, with equality only for $n = 3$. Therefore,

$$\begin{aligned}\pi|F_e| - \Delta &\geq \frac{5\pi^2}{2}\sum_{n\geq 3}(n^2-1)c_n^2 = \frac{5\pi^2}{2}\left(\sum_{n\geq 2}(n^2-1)c_n^2 - 3c_2^2\right)\\ &= \frac{5}{4}L^2 - 5\pi F - \frac{15\pi^2}{2}c_2^2 = \frac{15}{4}L^2 - 5\pi F - \frac{10}{3}\int_{P\notin K}\sin^3\omega\, dP.\end{aligned} \tag{13}$$

Using Crofton's formula (11), the last expression can be written as

$$\frac{5}{4}L^2 + 5\int_{P\notin K}(\omega - \sin\omega - \frac{2}{3}\sin^3\omega)dP$$

and the inequality in the theorem is proved. Moreover, the sum $\sum_{n\geq 3}(n^2-1)c_n^2$ in (13) vanishes if and only if $c_n = 0$ for $n \geq 3$ as well as $\pi|F_e| - \Delta$. □

The equality in Theorem 1 is considered in the following result:

Proposition 2 *Equality in* (12) *holds if and only if for the compact strictly convex set K one of the following assertions holds:*

a) K is a disk or it is bounded by a curve parallel to an astroid.
b) K is bounded by a curve parallel to a Steiner curve (deltoid).
c) K is parallel to the Minkowski sum of compact sets of the above types.

Although Hurwitz's inequality (1) cannot be improved for general convex domains, it is possible to obtain a stronger inequality for convex sets of constant width. In fact, we have

Theorem 2 ([3]) *Let K be a compact strictly convex set of constant width and area F bounded by a curve $\Gamma = \partial K$ of class C^2 and length L. Let F_e be the area with multiplicities of the evolute of Γ. Then*

$$L^2 - 4\pi F \leq \frac{4}{9}\pi|F_e|. \tag{14}$$

Equality holds if and only if Γ is a circle or a curve parallel to a Steiner curve at distance $L/2\pi$.

We also obtain an inequality better than (14) in terms of the visual angle (see [3, Theorem 5.3]).

4 Integral Formulas for the Visual Angle

As seen in the previous section, the view of a convex set from an exterior point gives interesting geometric information about this set. Now, we study some aspects of the visual angle of a convex set.

4.1 On Crofton and Hurwitz's Formulas

In this subsection, we provide a unified approach for some integral formulas for functions of the visual angle of a convex set due to Crofton, Hurwitz, and Massoti obtaining also new integral formulas for this kind of functions. The basic tool is the integral formula given in

Theorem 3 ([4]) *Let K be a compact convex set with boundary of class C^2, and let L be the length of ∂K. Let $c_k^2 = a_k^2 + b_k^2$ where a_k and b_k are the Fourier coefficients of the support function of K. Then, for every continuous function of the visual angle $f(\omega)$ on $[0, \pi]$ such that $f(\omega) = O(\omega^3)$, as ω tends to zero, one has*

$$\int_{P \notin K} f(\omega)\, dP$$
$$= \left(\int_0^\pi \frac{f(\omega)(1+\cos\omega)^2}{\sin^3\omega}\, d\omega \right) \frac{L^2}{2\pi} + \pi \sum_{k \geq 2} \left(\int_0^\pi \frac{f(\omega) h_k(\omega)}{\sin^3\omega}\, d\omega \right) c_k^2,$$

where h_k, for $k \geq 2$, are the universal functions given in (15).

Proof For each point $P \notin K$, let φ be the angle at the origin formed by the normal to one of the tangents from P to ∂K with the x axis and ω the visual angle from P; the pair (φ, ω) can be considered as a system of coordinates in $\mathbb{R}^2 \setminus K$.

We shall denote by A, A_1 the contact points of the tangents from P to ∂K and by $p = p(\varphi)$ the support function of K with respect to an origin O inside K (see Figure (1)). Let us write $T = PA$ and $T_1 = PA_1$.

The area element dP of $\mathbb{R}^2 \setminus K$ is given by

$$dP = \frac{TT_1}{\sin\omega}\, d\varphi \wedge d\omega.$$

This expression of the area element, introduced by Crofton in [2], appears also in [12, I.2.2].

Hence, the integral on $\mathbb{R}^2 \setminus K$ of a suitable function of the visual angle $f(\omega)$ is given by

$$\int_{P \notin K} f(\omega)\, dP = \int_0^\pi \int_0^{2\pi} \frac{f(\omega)}{\sin\omega} TT_1\, d\varphi\, d\omega = \int_0^\pi \frac{f(\omega)}{\sin\omega} \left(\int_0^{2\pi} TT_1\, d\varphi \right) d\omega.$$

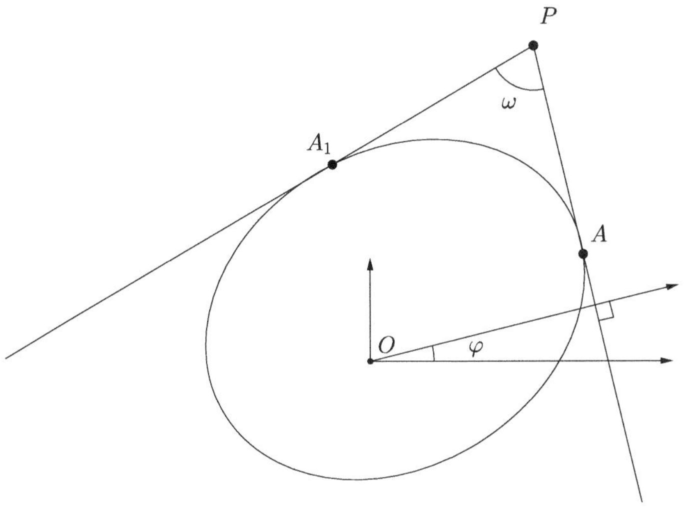

Fig. 1 The visual angle ω

Now, we will write the product TT_1 in terms of the Fourier coefficients of $p(\varphi)$ given in (5) and the Fourier coefficients of $p_1(\varphi) := p(\pi + \varphi - \omega)$ given by

$$p_1(\varphi) = a_0 + \sum_{k>0} (A_k \cos k\varphi + B_k \sin k\varphi)$$

which are related to the coefficients of $p(\varphi)$ by

$$\begin{aligned} A_k &= (-1)^{k+1}(-a_k \cos k\omega + b_k \sin k\omega), \\ B_k &= (-1)^{k+1}(-a_k \sin k\omega - b_k \cos k\omega). \end{aligned}$$

Substituting these Fourier series in the expressions of T and T_1, a straightforward but long calculation gives

$$\int_0^{2\pi} TT_1 \, d\varphi = \frac{1}{\sin^2 \omega} \left(\frac{L^2}{2\pi}(1 + \cos \omega)^2 + \pi \sum_{k>0} c_k^2 h_k(\omega) \right),$$

where $c_k^2 = a_k^2 + b_k^2$ and

$$\begin{aligned} h_k(\omega) = 2 \cos \omega + (-1)^{k+1} \big(&- \cos k\omega (1 + \cos^2 \omega) \\ &- 2k \sin k\omega \sin \omega \cos \omega + k^2 \cos k\omega \sin^2 \omega \big). \end{aligned} \tag{15}$$

This ends the proof. □

If we consider the area $F(\omega)$ enclosed by the locus C_ω of the points from which the convex set K is viewed under the same angle ω and introduced the functions

$$g_k(\omega) = 1 + \frac{(-1)^k}{2}\left((k+1)\cos(k-1)\omega - (k-1)\cos(k+1)\omega\right),$$

the previous Theorem can be stated as

Proposition 3 *With the same hypothesis of Theorem 3, we have*

$$\int_{P\notin K} f(\omega)\,dP = -\left[f(\omega)F(\omega)\right]_{0_+}^{\pi_-} + \frac{L^2}{2\pi}M(f) + \pi\sum_{k\geq 2}\beta_k(f)c_k^2,$$

where

$$M(f) = \int_0^\pi \frac{f'(\omega)}{1-\cos\omega}\,d\omega \quad \textit{and} \quad \beta_k(f) = \int_0^\pi \frac{f'(\omega)g_k(\omega)}{\sin^2\omega}\,d\omega. \tag{16}$$

As an application of Proposition 3, we can easily prove Crofton's formula

$$\int_{P\notin K}(\omega - \sin\omega)\,dP = -\pi F + \frac{L^2}{2}. \tag{17}$$

Indeed, $M(\omega - \sin\omega) = \pi$ and $\beta_k(\omega - \sin\omega) = \int_0^\pi g_k(x)/(1+\cos(x))\,dx = 0$, as can be easily seen integrating by parts and using elementary trigonometric identities. Since

$$-\lim_{\omega\to\pi} f(\omega)F(\omega) + \lim_{\omega\to 0} f(\omega)F(\omega) = -\pi F,$$

the formula follows.

Also, from Proposition 3, we obtain a new proof (see [4]) of the classical equality of Hurwitz given in (9).

In [11], Masotti gives without proof a Crofton-type formula evaluating $\int_{P\notin K}(\omega^2 - \sin^2\omega)\,dP$. We derive here Masotti's formula from Proposition 3. To this end, consider the function $f(\omega) = \omega^2 - \sin^2\omega$ which clearly satisfies the hypothesis of Theorem 3, and let us compute $[f(\omega)F(\omega)]_0^\pi$, $M(f)$ and the integrals $\int_0^\pi f'(\omega)\cos(j\omega)\,d\omega$, for j integer. We have

$$\lim_{\omega\to\pi} f(\omega)F(\omega) = \pi^2 F$$

and

$$\lim_{\omega\to 0} f(\omega)F(\omega) = \lim_{\omega\to 0}\frac{\omega^2 - \sin^2\omega}{\sin^2\omega}\left(\frac{L^2}{2\pi}(1+\cos\omega) + \pi\sum_{k\geq 2}c_k^2 g_k(\omega)\right) = 0,$$

since the term inside the parentheses is bounded. Hence, $[f(\omega)F(\omega)]_0^\pi = \pi^2 F$.

On the other hand,

$$M(f) = \int_0^\pi \frac{f'(\omega)}{1-\cos\omega}\,d\omega = \int_0^\pi \frac{2\omega - \sin(2\omega)}{1-\cos\omega}\,d\omega$$
$$= \Big[\sin^2(\omega/2) - 3\cos^2(\omega/2) - 2\omega\cot(\omega/2)\Big]_0^\pi = 8.$$

Moreover, for $j \neq 2$,

$$\int_0^\pi f'(\omega)\cos(j\omega)\,d\omega = \int_0^\pi (2\omega - \sin(2\omega))\cos(j\omega)\,d\omega$$
$$= \left[\frac{2}{j^2}(\cos(j\omega) + j\omega\sin(j\omega)) - \frac{\cos((j-2)\omega)}{2(j-2)} + \frac{\cos((j+2)\omega)}{2(j+2)}\right]_0^\pi$$
$$= \frac{8(1-(-1)^j)}{j^2(j^2-4)},$$

and

$$\int_0^\pi (2\omega - \sin(2\omega))\cos(2\omega)\,d\omega = 0.$$

It follows that

$$\sum_{j=1,\,\mathrm{odd}}^{k-1} \int_0^\pi f'(\omega)\cos(j\omega)\,d\omega = \sum_{j=1,\,\mathrm{odd}} \frac{16}{j(j^2-4)} = \frac{4k^2}{1-k^2}.$$

Summing up, we obtain

Theorem 4 (Masotti, [11]) *Let K be a compact convex set of area F with boundary of class C^2 and length L. Let $c_k^2 = a_k^2 + b_k^2$ where a_k and b_k are the Fourier coefficients of the support function of K. Then*

$$\int_{P\notin K} (\omega^2 - \sin^2\omega)\,dP = -\pi^2 F + \frac{4L^2}{\pi} + 8\pi \sum_{k\geq 2,\,\mathrm{even}} \left(\frac{1}{1-k^2}\right) c_k^2. \tag{18}$$

Moreover, the equality

$$\int_{P\notin K} (\omega^2 - \sin^2\omega)\,dP = -\pi^2 F + \frac{4L^2}{\pi}$$

holds if and only if the compact convex set K has constant width.

Now, we compute the integral of $\sin^m(\omega)$ for integer values of m greater than 3. The case $m = 3$, due to Hurwitz, was given in (10). We have the following theorem:

Theorem 5 ([4]) *Let K be a compact convex set with boundary of class C^2 and length L. Write $c_k^2 = a_k^2 + b_k^2$ where a_k and b_k are the Fourier coefficients of the support function of K. Then, for $m \geq 3$,*

$$\int_{P\notin K} \sin^m \omega \, dP = M(\sin^m \omega)\frac{L^2}{2\pi} + \frac{m!\pi^2}{2^{m-1}(m-2)} \sum_{k\geq 2,\, even} \frac{(-1)^{\frac{k}{2}+1}(k^2-1)}{\Gamma(\frac{m+1+k}{2})\Gamma(\frac{m+1-k}{2})} c_k^2,$$

where $M(\sin^m(\omega))$ comes from (16). For m odd, the index k in the sum runs only from 2 to $m - 1$.

In the special case of convex sets of constant width, we get

Proposition 4 *Let K be a compact convex of constant width with boundary of class C^2 and length L. Then, for $m \geq 3$,*

$$\int_{P\notin K} \sin^m \omega \, dP = \frac{\pi \, m!}{2^{m-1}(m-2)\Gamma(\frac{m+1}{2})^2} \frac{L^2}{2\pi}.$$

We also consider the integral

$$\int_{P\notin K} (\omega^m - \sin^m \omega) \, dP.$$

For $m = 1$ and $m = 2$, these are the integrals appearing in Crofton's formula (17) and in the Masotti's integral formula (18), respectively. For the general case, we obtain from Proposition 3

$$\int_{P\notin K} (\omega^m - \sin^m \omega) \, dP = -\pi^m F + M_m \frac{L^2}{2\pi} + \pi \sum_{k\geq 2} \beta_k c_k^2, \tag{19}$$

where $M_m = M(\omega^m - \sin^m \omega)$ and $\beta_k = \beta_k(\omega^m - \sin^m \omega)$ are given in (16).

We are able to prove that $\beta_k \leq 0$ for $k \geq 2$. So we get the following inequality:

Theorem 6 ([4]) *Let K be a compact convex set with boundary of class C^2, area F, and length of the boundary L, and let $\omega = \omega(P)$ be the visual angle from the point P. Then*

$$\int_{P\notin K} (\omega^m - \sin^m \omega) \, dP \leq -\pi^m F + M_m \frac{L^2}{2\pi}, \qquad m \geq 1,$$

where $M_m = \int_0^\pi \frac{(\omega^m - \sin^m \omega)'}{1 - \cos \omega} d\omega$. Equality holds only for circles.

4.2 Measure of Pairs of Lines in the Plane

The classical proof of Crofton's formula (2) comes from the study of the measure of pairs of lines intersecting a convex set. In order to obtain new formulas involving the visual angle, we consider the measures $\tilde{f}(G_1, G_2)dG_1 \wedge dG_2$ for every function $\tilde{f}(G_1, G_2)$ defined on the space of ordered pair of lines $\mathcal{A}_{2,1} \times \mathcal{A}_{2,1}$. We want now to characterize when these measures are invariant under Euclidean motions. Using the polar coordinate $\varphi(G)$ introduced in (6), we have

Proposition 5 *The measure $\tilde{f}(G_1, G_2)dG_1 \wedge dG_2$ is invariant under the group of Euclidean motions if and only if $\tilde{f}(G_1, G_2) = f(\varphi(G_2) - \varphi(G_1))$ with f an even π-periodic function on $\mathbb{R}$.*

Proof Let (p_i, φ_i) be the coordinates of G_i, and define the function g by $g(p_1, \varphi_1, p_2, \varphi_2) = \tilde{f}(G_1, G_2)$. The invariance of the measure is equivalent to the equality $g(p_1, \varphi_1, p_2, \varphi_2) = g(p_1', \varphi_1', p_2', \varphi_2')$ for each Euclidean motion sending the pair of lines with coordinates $(p_1, \varphi_1, p_2, \varphi_2)$ to the pair of lines with coordinates $(p_1', \varphi_1', p_2', \varphi_2')$. First of all, let us show that g does not depend on p_1, p_2. In fact, for every straight line $G = G(p, \varphi)$ and an arbitrary $a > 0$, there is a parallel line to G with coordinates (a, φ). Given two straight lines $G_1 = G(p_1, \varphi_1), G_2 = G(p_2, \varphi_2)$ and two numbers $a_1, a_2 > 0$, let G_1' and G_2' be the corresponding parallel lines with coordinates $(a_1, \varphi_1), (a_2, \varphi_2)$. Performing the translation that sends the point $G_1 \cap G_2$ to the point $G_1' \cap G_2'$, we have that $g(p_1, \varphi_1, p_2, \varphi_2) = g(a_1, \varphi_1, a_2, \varphi_2)$, and so g does not depend on p_1 and p_2.

Given now the line $G(p, \varphi)$ if we perform, for instance, the translation given by the vector $-(p + \epsilon)(\cos\varphi, \sin\varphi)$, $\epsilon > 0$, the translated line has coordinates $(\epsilon, \varphi + \pi)$. Therefore, the function g must be π-periodic with respect to the arguments φ_1, φ_2. Due to the invariance under rotations, it follows that $g(p_1, \varphi_1, p_2, \varphi_2) = g(p_1, 0, p_2, \varphi_2 - \varphi_1)$ and so $g(p_1, \varphi_1, p_2, \varphi_2) = f(\varphi_2 - \varphi_1) = f(\varphi(G_2) - \varphi(G_1))$ with f a π-periodic function. Finally, the invariance under symmetries implies that f is an even function.

Conversely, it is clear that if f is an even π-periodic function, then the measure $f(\varphi(G_2) - \varphi(G_1))dG_1 \wedge dG_2$ is invariant under Euclidean motions. □

Our goal is now to integrate measures, not necessarily invariant under Euclidean motions, over the set of pairs of lines meeting a compact convex set K. We shall consider measures of the form $\tilde{f}(G_1, G_2)dG_1 \wedge dG_2 = f(\varphi(G_2) - \varphi(G_1))dG_1 \wedge dG_2$ with f a continuous 2π-periodic function. Notice that $\varphi(G_2) - \varphi(G_1)$ gives one of the two angles between lines G_1 and G_2. In Theorem 7, we give a formula to compute the integral of the above measures in terms of both the Fourier coefficients of f and of the support function of K.

Theorem 7 ([5]) *Let K be a compact convex set with C^1 boundary of length L. Let f be a 2π-periodic continuous function on $\mathbb{R}$ with Fourier expansion*

$$f(\varphi) = \sum_{n \geq 0} A_n \cos(n\varphi) + B_n \sin(n\varphi).$$

Then

$$\int_{G_i \cap K \neq \emptyset} f(\varphi(G_2) - \varphi(G_1))\, dG_1 dG_2 = A_0 L^2 + \pi^2 \sum_{n \geq 1} A_n c_n^2,$$

with $c_n^2 = a_n^2 + b_n^2$ *where* $a_n and b_n$ *are the Fourier coefficients of the support function* p *of* K.

The original proof of Crofton's formula, via Integral Geometry, involves a measure on the space of pairs of lines. The aim now is to interpret the formulas of Masotti, powers of sine, and Hurwitz in [4] in terms of integrals of measures in the space of pairs of lines.

The classical proof of Crofton's formula is based on the change of variables in the space of ordered pairs of lines given by

$$(p_1, \varphi_1, p_2, \varphi_2) \longrightarrow (P, \alpha_1, \alpha_2)$$

where P is the intersection point of the two straight lines and $\alpha_i \in [0, \pi]$ are the angles which determine the directions of the lines. With these new coordinates, proceeding as in [12, I.4.3], one has

$$dG_1 \wedge dG_2 = |\sin(\alpha_2 - \alpha_1)|\, d\alpha_1 \wedge d\alpha_2 \wedge dP.$$

Using this change of variables, we have

Proposition 6 *Let* f *be an even* π*-periodic continuous function on* $\mathbb{R}$, *and let* H *be the* C^2 *function on* $[0, \pi]$ *satisfying the conditions* $H''(x) = f(x) \cdot \sin(x)$, $x \in [0, \pi]$ *and* $H(0) = H'(0) = 0$. *Then, one has*

$$\int_{G_i \cap K \neq \emptyset} f(\varphi(G_2) - \varphi(G_1))\, dG_1 dG_2 = 2H(\pi)F + 2 \int_{P \notin K} H(\omega)\, dP. \tag{20}$$

Using this Proposition, we give the announced interpretation of the following formulas:

4.2.1 Crofton's Formula

Taking $H(x) = x - \sin(x)$, it follows that $f = 1$ in Proposition 6 and since $H(\pi) = \pi$ using (20), we get

Proposition 7 *The following equality holds:*

$$\int_{G_i \cap K \neq \emptyset} dG_1 dG_2 = 2\pi F + 2 \int_{P \notin K} (\omega - \sin \omega)\, dP.$$

4.2.2 Masotti's Formula

Taking $H(x) = x^2 - \sin^2(x)$, one gets $H''(x)/\sin(x) = 4\sin(x)$. So the function $f(x) = 4|\sin(x)|$, $x \in \mathbb{R}$, satisfies the hypothesis of Proposition 6, and Eq. (20) gives

Proposition 8 *The following equality holds*

$$2\int_{G_i \cap K \neq \emptyset} |\sin(\varphi(G_2) - \varphi(G_1))| dG_1 dG_2 = \pi^2 F + \int_{P \notin K} (\omega^2 - \sin^2 \omega)\, dP.$$

4.2.3 Powers of Sine Formula

Finally, in an analogous way, we can interpret the integral of any power of the sine of the visual angle. Effectively for $H(x) = \sin^m(x)$, it follows that

$$H''(x)/\sin(x) = m(m-1)\sin^{m-3}(x) - m^2 \sin^{m-1}(x).$$

So taking $f(x) = m(m-1)|\sin^{n-3}(x)| - m^2|\sin^{m-1}(x)|$, the hypotheses of Proposition 6 are satisfied, and by (20), we have

Proposition 9 *The following equality holds:*

$$2\int_{P \notin K} \sin^m(\omega)\, dP =$$

$$= \int_{G_i \cap K \neq \emptyset} \Big(m(m-1)|\sin^{m-3}(\varphi(G_2) - \varphi(G_1))| - m^2|\sin^{m-1}(\varphi(G_2) - \varphi(G_1))|\Big)$$

$$dG_1 dG_2.$$

An interesting consequence of our results is that when f is a π-periodic function, according to Proposition 6, the integral $\int_{G_i \cap K \neq \emptyset} f(\varphi(G_2) - \varphi(G_1))\, dG_1 dG_2$ is a linear combination of integrals extended outside K of the functions of the visual angle $H_k(\omega)$, where

$$H_k(x) = \frac{1}{2(k^2-1)} \left(f_k(x) + 2(\sin x - x)\right), \quad k \geq 2,$$

and $H_1(x) = (1/8)(2x - \sin(2x))$, $f_k(\omega)$ being the functions of Hurwitz given by

$$f_k(\omega) = -2\sin\omega + \frac{k+1}{k-1}\sin((k-1)\omega) - \frac{k-1}{k+1}\sin((k+1)\omega), \quad k \geq 2.$$

Summarizing, it appears that the functions of Crofton and Hurwitz are some kind of *basis* for the integral of any π-periodic function with respect to the measure $dG_1 \wedge dG_2$ over the set of pairs of lines meeting a given compact convex set.

5 Integral Formulas for the Dihedral Angle

So far, we have been working with pairs of lines in the plane. From now on, we will consider pairs of planes in the space. The aim of this section is to write the integral of an isometry-invariant measure over the pairs of planes meeting a convex set K as an integral of an appropriate function of the *dihedral visual angle*.

5.1 Invariant Measures in the Set of Ordered Pairs of Planes in the Space

We consider measures in the space $\mathcal{A}_{3,2} \times \mathcal{A}_{3,2}$ of pairs of planes in $\mathbb{E}^3$ of the form $m_{\tilde{f}} := \tilde{f}(E_1, E_2) dE_1 \wedge dE_2$ with dE_i the normalized isometry-invariant measures in $\mathcal{A}_{3,2}$ as considered in [12]. We want to study which functions $\tilde{f}$ give an isometry-invariant measure, that is, a measure $m_{\tilde{f}}$ satisfying $m_{\tilde{f}}(B) = m_{\tilde{f}}(gB)$ for every Euclidean motion g. For instance, it is known that for a given compact convex set K, one has $\int_{E \cap K \neq \emptyset} dE = M$. So when $\tilde{f}(E_1, E_2) = 1$, we have

$$\int_{K \cap E_i \neq \emptyset} dE_1 dE_2 = M^2 = 4\pi^2 \mathcal{W}^2, \tag{21}$$

where M and $\mathcal{W}$ are the mean curvature and the mean width of K, respectively. A first result in this direction is

Proposition 10 *The measure given by $\tilde{f}(E_1, E_2) dE_1 \wedge dE_2$ in $\mathcal{A}_{3,2} \times \mathcal{A}_{3,2}$ is invariant under isometries of $\mathbb{E}^3$ if and only if $\tilde{f}(E_1, E_2) = f(\langle u_1, u_2 \rangle)$ where $\pi(E_i)^\perp = \operatorname{span}\{u_i\}$, $i = 1, 2$ and $f : [-1, 1] \to \mathbb{R}$ is an even measurable function.*

Let K be a compact convex set in the Euclidean space $\mathbb{E}^3$. According to equality (21), it is a natural question to evaluate

$$\int_{E_i \cap K \neq \emptyset} \tilde{f}(E_1, E_2) dE_1 dE_2,$$

where $\tilde{f}(E_1, E_2) dE_1 \wedge dE_2$ is an isometry-invariant measure on $\mathcal{A}_{3,2} \times \mathcal{A}_{3,2}$. This can be done in terms of the coefficients of the expansion of the support function of K in spherical harmonics and the coefficients of the Legendre series of the measurable even function $f : [-1, 1] \to \mathbb{R}$ such that $\tilde{f}(E_1, E_2) = f(\langle u_1, u_2 \rangle)$.

The following result is a special case, with a different notation, of Theorem 5 in [9], whose proof is based on the Funk-Hecke Theorem [8, p. 98].

Theorem 8 *Let K be a compact convex set with support function p given in terms of spherical harmonics by (7). Let $\tilde{f}(E_1, E_2)dE_1\,dE_2$ be an isometry-invariant measure on $\mathcal{A}_{3,2} \times \mathcal{A}_{3,2}$ and $f : [-1, 1] \to \mathbb{R}$ an even measurable function such that $\tilde{f}(E_1, E_2) = f(\langle u_1, u_2\rangle)$ where $\pi(E_i)^{\perp} = \mathrm{span}\{u_i\}$, $i = 1, 2$. Then*

$$\int_{E_i \cap K \neq \emptyset} \tilde{f}(E_1, E_2)dE_1\,dE_2 = \frac{\lambda_0}{4\pi}M^2 + \sum_{n=1}^{\infty} \lambda_{2n} \|\pi_{2n}(p)\|^2, \tag{22}$$

where $\lambda_{2n} = 2\pi \int_{-1}^{1} f(t)P_{2n}(t)\,dt$ with P_{2n} the Legendre polynomial of degree $2n$.

Recall that when K is a convex set of constant width $\mathcal{W}$, one has $\pi_n(p) = 0$ for $n \neq 0$ even. Therefore, in this case,

$$\int_{E_i \cap K \neq \emptyset} f(\langle u_1, u_2\rangle)\,dE_1\,dE_2 = \frac{\lambda_0}{4\pi}M^2 = \lambda_0 \pi \mathcal{W}^2.$$

The measure $dE_1\,dE_2$ in the space of pairs of planes in $\mathbb{E}^3$ can be written according to Santaló (cf. [12], section II.12.6) as

$$dE_1 \wedge dE_2 = \sin^2(\alpha_2 - \alpha_1)\,d\alpha_1\,d\alpha_2\,dG, \tag{23}$$

where α_i are the angles of E_i about G.

Introducing the *visual dihedral angle* of a convex set K from a line G not meeting K as the angle $\omega = \omega(G)$, $0 \leq \omega \leq \pi$, between the half-planes through G tangent to K and using (23), we can prove the following:

Theorem 9 ([6]) *Let K be a compact convex set, and let $f : [-1, 1] \longrightarrow \mathbb{R}$ be an even continuous function. Let H be the C^2 function on $[-\pi, \pi]$ satisfying*

$$H''(x) = f(\cos(x))\sin^2(x), \quad -\pi \leq x \leq \pi, \quad H(0) = H'(0) = 0.$$

Then,

$$\int_{E_i \cap K \neq \emptyset} f(\langle u_1, u_2\rangle)\,dE_1\,dE_2 = \pi H(\pi)F + 2\int_{G \cap K = \emptyset} H(\omega)\,dG, \tag{24}$$

where u_i are unit normal vectors to the planes E_i, $\omega = \omega(G)$ is the dihedral visual angle from the line G, and F is the area of the boundary of K.

5.2 Crofton's Formula in the Space

In Blaschke's work [1, p. 75], the following Crofton-Herglotz formula is given

$$\int_{G\cap K=\emptyset} (\omega^2 - \sin^2\omega)\, dG = 2M^2 - \frac{\pi^3 F}{2}. \tag{25}$$

We can easily recover (25) from Theorem 9. In fact, considering $f(t) = 1$, one gets $H(x) = (x^2 - \sin^2 x)/4$, and equality (24) gives

$$M^2 = \int_{E_i\cap K\neq\emptyset} dE_1\, dE_2 = \frac{1}{4}\pi^3 F + \frac{1}{2}\int_{G\cap K=\emptyset} (\omega^2 - \sin^2\omega)\, dG.$$

Formula (25) reveals the significance of the function of the dihedral visual angle $\omega^2 - \sin^2\omega$. One can ask what role the function $\omega - \sin\omega$ does it play; this function, interpreting ω as the visual angle in the plane, is significant thanks to Crofton's formula (2).

In [1, p. 85], Blaschke shows that

$$\int_{G\cap K=\emptyset} (\omega - \sin\omega) dG = \frac{1}{4}\int_{u\in S^2} L_u^2 du - \frac{\pi^2}{2} F, \tag{26}$$

where L_u is the length of the boundary of the projection of K on $\operatorname{span}\{u\}^{\perp}$.

It can be easily seen that $\int_{u\in S^2} L_u du = 2\pi M$, and from this equality, applying Schwarz's inequality, one gets

$$\int_{u\in S^2} L_u^2 du \geq \pi M^2. \tag{27}$$

Introducing (27) into (26), one obtains

$$\int_{G\cap K=\emptyset} (\omega - \sin\omega) dG \geq \frac{\pi}{4}(M^2 - 2\pi F), \tag{28}$$

an inequality given by Blaschke.

As a consequence of Theorem 9, we can now evaluate the deficit in both inequalities (27) and (28).

Theorem 10 ([6]) *Let K be a compact convex set with support function p, area of its boundary F, and mean curvature M. Let L_u be the length of the boundary of the projection of K on* $\operatorname{span}\{u\}^{\perp}$*, and let $\omega = \omega(G)$ be the dihedral visual angle of K from the line G. Then,*

1. $\displaystyle\int_{u\in S^2} L_u^2 du = \pi M^2 + 4\pi \sum_{n=1}^{\infty} \frac{\Gamma(n+1/2)^2}{\Gamma(n+1)^2} \|\pi_{2n}(p)\|^2,$

2. $\displaystyle\int_{G\cap K=\emptyset}(\omega-\sin\omega)dG=\frac{\pi}{4}(M^2-2\pi F)+\pi\sum_{n=1}^{\infty}\frac{\Gamma(n+1/2)^2}{\Gamma(n+1)^2}\|\pi_{2n}(p)\|^2,$

with $\pi_{2n}(p)$ the projection of the support function p of K on the vector space of spherical harmonics of degree $2n$.

Moreover, equality holds in both (27) *and* (28) *if and only if K is of constant width.*

Proof We consider $f(t)=1/\sqrt{1-t^2}$. For this function, the corresponding H in Theorem 9 is $H(x)=|x|-|\sin x|$. Applying equality (24) and Theorem 8 with the corresponding λ_{2n}'s given by

$$\lambda_{2n}=2\pi\int_{-1}^{1}f(t)P_{2n}(t)dt=2\pi\frac{\Gamma(n+1/2)^2}{\Gamma(n+1)^2}$$

(cf. [7], 7.226), item *2* follows. Equality in item *1* is a consequence of item *2* and (26).

The statement about equality in (27) and (28) is a consequence of the fact that K is of constant width if and only if $\pi_{2n}(p)=0$ for $n\neq 0$. □

In [6, Proposition 6.2], we decompose the integral of an invariant measure of pairs of planes as

$$\int_{E_i\cap K\neq\emptyset}f(\langle u_1,u_2\rangle)\,dE_1\,dE_2=a_0M^2+\sum_{m=2}^{\infty}a_{2m}\int_{G\cap K=\emptyset}\sin^{2m}\omega\,dG,$$

where u_i are normal vectors to the planes E_i, ω the dihedral visual angle from the line G, and F denotes the area of the boundary of K and certain coefficients a_{2n} depending on f. From this (cf. [6, Proposition 6.3]), one sees that

$$\int_{E_i\cap K\neq\emptyset}f(\langle u_1,u_2\rangle)\,dE_1\,dE_2=a_0M^2+\sum_{m=2}^{\infty}a_{2m}\int_{E_i\cap K\neq\emptyset}h_m(\langle u_1,u_2\rangle)\,dE_1\,dE_2,$$

where

$$h_m(t)=m(2mt^2-1)(1-t^2)^{m-2},\ m>1.$$

So we have exhibited a simple family of polynomial functions that are in some sense a basis for the integrals in Theorem 8. In fact, every invariant integral can be written as an infinite linear combination of the integrals of the invariant measures given by the polynomials h_m.

Funding The authors were partially supported by grants 2021SGR01015 (Generalitat de Catalunya) and PID2021-125625NB-I00 (Ministerio de Ciencia e Innovación).

References

1. Blaschke, W.: Vorlesungen uber Integralgeometrie, 3rd edn. VEB Deutscher Verlag der Wissenschaften, Berlin (1955)
2. Crofton, M.W.: On the theory of local probability. Phil. Trans. R. Soc. Lond. **158**, 181–199 (1868)
3. Cufí, J., Gallego, E., Reventós, A.: A note on Hurwitz's inequality. J. Math. Anal. Appl. **458**(1), 436–451 (2018)
4. Cufí, J., Gallego, E., Reventós, A.: On the integral formulas of Crofton and Hurwitz relative to the visual angle of a convex set. Mathematika **65**(4), 874–896 (2019)
5. Cufí, J., Gallego, E., Reventós, A.: Integral geometry about the visual angle of a convex set. Rend. Circ. Mat. Palermo (2) **69**(3), 1115–1130 (2020)
6. Cufí, J., Gallego, E., Reventós, A.: Integral geometry of pairs of planes. Arch. Math. (Basel) **117**(5), 579–591 (2021)
7. Gradshteyn, I.S., Ryzhik, I.M.: Table of Integrals, Series, and Products, 7th edn. Elsevier/Academic Press, Amsterdam (2007)
8. Groemer, H.: Geometric applications of Fourier series and spherical harmonics. Encyclopedia of Mathematics and Its Applications, vol. 61. Cambridge University Press, Cambridge (1996)
9. Hug, D., Schneider, R.: Integral geometry of pairs of hyperplanes or lines. Arch. Math. (Basel) **115**(3), 339–351 (2020)
10. Hurwitz, A.: Sur quelques applications géométriques des séries de Fourier. Annales scientifiques de l'É.N.S. **19**(3e série), 357–408 (1902)
11. Masotti Biggiogero, G.: La Geometria Integrale. Rend. Sem. Mat. Fis. Milano **25**, 164–231 (1955) (1953/54)
12. Santaló, L.A.: Integral Geometry and Geometric Probability, 2nd edn. Cambridge University Press, Cambridge (2004)
13. Schneider, R.: Convex bodies: the Brunn-Minkowski theory. Encyclopedia of Mathematics and Its Applications, vol. 151, expanded edn. Cambridge University Press, Cambridge (2014)

Homogeneous Hypersurfaces in Symmetric Spaces

José Carlos Díaz-Ramos, Miguel Domínguez-Vázquez, and Tomás Otero

Abstract A hypersurface of a Riemannian manifold is called homogeneous if it is an orbit of an isometric action on the ambient manifold. Homogeneous hypersurfaces have remarkable geometric properties, providing the simplest examples of hypersurfaces with constant mean curvature. Thus, they are crucial for the investigation of more general types of submanifolds in ambient spaces with large isometry groups.

In this survey article, we present an introduction to some of the basic geometric, topological, and algebraic features of homogeneous hypersurfaces, describing what is known about their classification problem in symmetric spaces and explaining the main tools needed for their study in the context of symmetric spaces of noncompact type.

Keywords Symmetric space · Noncompact type · Homogeneous submanifold · Isometric action · Cohomogeneity one action · Isoparametric hypersurface · Minimal submanifold · Constant principal curvatures · Projective space · Hyperbolic space · Parabolic subgroup

1 Introduction

Minimal, and more generally, constant mean curvature hypersurfaces play a fundamental role in Riemannian submanifold geometry. As solutions to variational problems involving areas and volumes, they arise naturally in various contexts such as physics, biology, or optimal design. Their mathematical investigation has a long history and constitutes one of the most important trends in current research in geometric analysis. Some fundamental techniques in their study, such as the use

J. C. Díaz-Ramos · M. Domínguez-Vázquez (✉) · T. Otero
CITMAga, Santiago de Compostela, Spain

Department of Mathematics, Universidade de Santiago de Compostela, Santiago, Spain
e-mail: josecarlos.diaz@usc.es; miguel.dominguez@usc.es; tomas.otero.casal@usc.es

A. Alarcón et al. (eds.), *New Trends in Geometric Analysis*,
RSME Springer Series 10, https://doi.org/10.1007/978-3-031-39916-9_5

of barriers, the maximum principle, or the bifurcation results, are based on the good knowledge that we have of certain examples of hypersurfaces with constant mean curvature and with a high degree of symmetry (hyperplanes, spheres, cylinders, catenoids, horospheres, etc.). Thus, one of the first natural steps in the investigation of submanifolds of a given Riemannian space is to determine some classes of constant mean curvature hypersurfaces that are invariant under a large group of isometries of the ambient space. When such group is large enough to act transitively on the hypersurface, the latter is called a homogeneous hypersurface, and the isometric action on the ambient space is said to be of cohomogeneity one. This is the case of hyperplanes, spheres, and cylinders in the Euclidean space or of horospheres in the hyperbolic space, among other examples.

By definition, the investigation of homogeneous hypersurfaces only makes sense in spaces with a large isometry group. Therefore, homogeneous spaces constitute the natural context for this investigation. Indeed, apart from spaces of constant curvature, the theory of surfaces with constant mean curvature in homogeneous 3-manifolds has undergone important advances (see, for example, [71]).

For arbitrary dimensions, a particularly elegant class of homogeneous manifolds is that of symmetric spaces. Locally, symmetric spaces are characterized by the property that curvature is invariant under parallel transport. Globally, the defining property of symmetric spaces is the existence of isometric central inversions around any point, which readily imply the existence of a transitive group of isometries. Symmetric spaces were classified by Élie Cartan [21] into several infinite families, some exceptional examples, and their products. Of course, this includes the space forms but also the isotropic (or two-point homogeneous) Riemannian manifolds, various (compact and noncompact) Grassmannians, compact Lie groups with bi-invariant metrics, and different moduli spaces of algebraic structures: real structures of a complex vector space, complex structures on a real vector space, positive definite symmetric matrices, etc. Symmetric spaces constitute a distinguished class in Berger's classification of holonomy groups [17] but also an appropriate setting for several problems of geometric analysis [54]. Their study also arises naturally in other areas such as number theory and algebraic geometry [58, 86, 101].

In view of the crucial role played by homogeneous hypersurfaces in the classical submanifold theory of space forms, we believe that the investigation of homogeneous hypersurfaces in symmetric spaces constitutes one of the first steps in the long-term program of developing a *submanifold theory of symmetric spaces*. The centrality of these spaces in Mathematics, along with their fascinating geometric, algebraic, and analytic properties, gives us a glimpse of a field yet to be explored.

The aim of this text is to provide a survey on homogeneous hypersurfaces, their generalizations, and their classification problem in symmetric spaces, with a focus on the noncompact setting. Thus, we will start by discussing the definition, general properties, and important topological and geometric properties of homogeneous hypersurfaces in Sect. 2. In Sect. 3, we will review the notion and fundamental geometric and algebraic aspects of symmetric spaces (Sects. 3.1–3.2), mainly of those of noncompact type (Sect. 3.3), and the algebraic theory of parabolic subalgebras (Sect. 3.4). Section 4 will be devoted to report on the classification problem of homogeneous hypersurfaces in symmetric spaces of compact type. Here, we will provide an introductory discussion to the problem in spheres through various interesting examples (Sect. 4.1), and then we will describe the classification on the other compact symmetric spaces, focusing on the rank one case (Sect. 4.2). In Sect. 5, we will review the classification problem in symmetric spaces of noncompact type of rank one (the hyperbolic spaces over the normed division algebras), whereas in Sect. 6 we will present what is known in the higher rank case. Finally, in Sect. 7, we provide a list of open problems.

2 Homogeneous Hypersurfaces

Let M be a Riemannian manifold and $\mathrm{Isom}(M)$ its isometry group, which is known to be a Lie group. A connected, injectively immersed submanifold P of M is called *(extrinsically) homogeneous* if for any $p, q \in P$ there exists an isometry φ of M such that $\varphi(p) = q$ and $\varphi(P) = P$. Note that if $P = M$, we recover the standard notion of (intrinsic) homogeneity of a Riemannian manifold. By considering the subgroup of isometries of M that leave the submanifold P invariant, one easily sees that P is homogeneous if and only if P is an orbit of an isometric action on M, that is, there exists a subgroup H of $\mathrm{Isom}(M)$ such that $P = H \cdot p$, for some (and hence, for any) $p \in P$. Hereafter, by $H \cdot p$, we denote the orbit of an action $H \times M \to M$ of a group H through a point p of M, and by $H_p = \{h \in H : h \cdot p = 0\}$, we denote the isotropy group (or stabilizer) at p. Of course, $H \cdot p \cong H/H_p$ is a bijection, which is indeed a diffeomorphism if the H-action on M is smooth, when endowing the set of left cosets $H/H_p = \{hH_p : h \in H\}$ with a natural differentiable structure. Moreover, P is properly embedded (equivalently, closed and embedded) in M if and only if $H = \{\varphi \in \mathrm{Isom}(M) : \varphi(P) = P\}$ is closed in $\mathrm{Isom}(M)$ (in particular, an embedded Lie subgroup of $\mathrm{Isom}(M)$); this in turn means that P is an orbit of a proper isometric smooth action on M. See [65, Chapter 21], [69], and [72, §6] for further information on smooth, proper, and isometric actions.

The family of orbits of a smooth isometric action of a connected Lie group H on a Riemannian manifold M determines what is called a *singular Riemannian foliation* of M. This is a decomposition of M into connected, injectively immersed submanifolds (leaves) that are locally equidistant to each other and such that there is a collection of smooth vector fields on M spanning all tangent spaces to all leaves. In the case of a smooth isometric H-action on M, the collection $\{X^* : X \in \mathfrak{h}\}$ of Killing fundamental vector fields on M, induced by elements X in the Lie algebra $\mathfrak{h}$ of H, span all tangent spaces to all orbits. Here, X^* is given by $X^*_p = \frac{d}{dt}|_{t=0} \operatorname{Exp}(tX) \cdot p$, where $\operatorname{Exp}\colon \mathfrak{h} \to H$ denotes the Lie group exponential map. Moreover, if γ is a geodesic in M that is orthogonal at $\gamma(0)$ to one orbit, $H \cdot \gamma(0)$, then for any fundamental vector field X^*, we have $\frac{d}{dt}\langle \dot\gamma, X^* \rangle = \langle \dot\gamma, \nabla_{\dot\gamma} X^* \rangle = 0$, since ∇X^* is skew-adjoint as X^* is Killing. Hence, $\langle \dot\gamma, X^* \rangle = 0$, and thus, any geodesic orthogonal to one orbit remains orthogonal to any other orbit that it meets. This means that the orbits are locally equidistant to each other. Orbit foliations, that is, singular Riemannian foliations induced by isometric actions, are also called homogeneous foliations. See [1] for more information on these concepts and properties.

Although Lie group theory plays a fundamental role in the analysis of homogeneous submanifolds and isometric actions, from a geometric perspective, we are ultimately interested in the orbit foliations of smooth isometric actions, and not so much in the (possibly multiple) groups that give rise to the same orbit foliation. Thus, when discussing isometric actions, we will usually consider actions to be equivalent if they have the same orbits. More precisely, we will say that two isometric actions of groups H_1 and H_2 on M are *orbit equivalent* if there is an isometry φ of M such that $\varphi(H_1 \cdot p) = H_2 \cdot \varphi(p)$ for all $p \in M$, that is, φ maps the H_1-orbits to the H_2-orbits. Thus, two smooth isometric actions are orbit equivalent if and only if their orbit foliations are congruent in M.

From now on, unless otherwise stated, *isometric actions will be assumed to be smooth and proper*, and *homogeneous submanifolds will be closed and embedded.*

The *cohomogeneity* of an isometric action is the lowest codimension of its orbits. Thus, a (proper) action has cohomogeneity zero precisely when it is transitive. An orbit of an isometric action is called *regular* if its codimension agrees with the cohomogeneity and is called *singular* otherwise.

Remark 1 Among regular orbits, we can distinguish two types: principal orbits and exceptional orbits. Given an isometric H-action on M, the H-orbit through p is *principal* if the isotropy group at p, $H_p = \{h \in H : h \cdot p = p\}$, is contained in any other isotropy group H_q, $q \in M$, up to conjugation in H. A nonprincipal regular orbit is called *exceptional*. The union of all principal orbits constitutes an open dense subset of M. If M is simply connected and complete and H is connected, then there are no exceptional orbits. See [1, Chapter 3 and Corollary 5.35] for further details.

2.1 Homogeneous Hypersurfaces and Cohomogeneity One Actions

A *homogeneous hypersurface* of M is a regular orbit of a cohomogeneity one action on M. It is known that a cohomogeneity one action on a complete connected Riemannian manifold has exactly zero, one, or two singular orbits. Indeed, the space of orbits $M/H = \{H \cdot p : p \in M\}$ of a cohomogeneity one H-action is homeomorphic to $\mathbb{R}$, $\mathbb{S}^1$, $[0, \infty)$ or $[0, 1]$, and nonprincipal orbits correspond to the boundary of such spaces [3]. Depending on the geometry and topology of the ambient manifold M, some of these possibilities may be excluded (see Remark 2).

Example 1 The following items provide very simple examples of cohomogeneity one actions with orbit spaces homeomorphic to $\mathbb{R}$, $\mathbb{S}^1$, $[0, \infty)$, and $[0, 1]$, respectively:

(a) The action of $(\mathbb{R}^{n-1},+)$ on $\mathbb{R}^n$ by translations: $h \cdot p = p + (h, 0)$, where $h \in \mathbb{R}^{n-1}$, $p \in \mathbb{R}^n$. All orbits are regular (parallel hyperplanes).
(b) The action of SO_2 on a torus of revolution around the z-axis in $\mathbb{R}^3$, given by $A \cdot p = (A(p_1, p_2), p_3)$, where $A \in \mathsf{SO}_2$ and $p = (p_1, p_2, p_3) \in \mathbb{R}^3$. All orbits are regular (circles).
(c) The standard action of SO_n on $\mathbb{R}^n$ by rotations around the origin. The origin is precisely the only singular orbit, whereas the concentric spheres about it are the regular orbits.
(d) The action of SO_n on the unit sphere $\mathbb{S}^n$ of $\mathbb{R}^{n+1}$: $A \cdot p = (A(p_1, \dots, p_n), p_{n+1})$, where $A \in \mathsf{SO}_n$ and $p = (p_1, \dots, p_{n+1})$. The north and south poles of $\mathbb{S}^n$ are the two singular orbits, and the parallels are the regular orbits.

None of these actions has exceptional orbits (and hence the boundary points of their orbit spaces correspond to singular orbits). Here, we have three actions with exceptional orbits:

(e) The action (d) above descends to a cohomogeneity one action of SO_n on the real projective space $\mathbb{R}\mathsf{P}^n$. This action has the same orbit space, namely, $[0, 1]$, but only one singular orbit (the image of the poles under the projection map $\pi : \mathbb{S}^n \to \mathbb{R}\mathsf{P}^n$) and one exceptional orbit (the projection of the equator).
(f) The action of U_1 on the infinite Möbius band $\mathbb{R}^2/\{(x, y) \sim (-x, y + 2\pi)\}$, given by $e^{i\theta} \cdot [(x, y)] = [(x, y + 2\theta)]$, has orbit space $[0, \infty)$ and exceptional orbit $\mathsf{U}_1 \cdot [(0, 0)]$.
(g) The action in (f) descends to a U_1-action on the Klein bottle $\mathbb{R}^2/\{(x, y) \sim (-x, y + 2\pi) \sim (x + 2\pi, y)\}$ with orbit space $[0, 1]$, two exceptional orbits $\mathsf{U}_1 \cdot [(0, 0)]$ and $\mathsf{U}_1 \cdot [(\pi, 0)]$, and no singular orbit.

As exemplified by the previous actions, the singular Riemannian foliation induced by a cohomogeneity one action on a complete connected manifold M is of a very particular type. It is a decomposition of M into mutually equidistant, properly embedded leaves, all of them of codimension one (the regular orbits/leaves), except at most two (the singular orbits/leaves). Each regular orbit is a *tube* around any of the singular orbits. Here, by tube of radius r around a submanifold P of M, we mean the subset of M given by

$$P^r = \{\exp(r\xi) : \xi \in \nu P, \ |\xi| = 1\},$$

where exp is the Riemannian exponential map of M and νP is the normal bundle of P. If P is a hypersurface, each connected component of P^r is called a *parallel* or *equidistant hypersurface* to P. Locally and for small enough r, a tube P^r of radius r around P is a hypersurface. If Q is a hypersurface of M, ξ is a smooth unit normal vector field along Q, and $Q^r_\xi = \{\exp(r\xi_p) : p \in Q\}$ is a submanifold of codimension higher than one in M, then Q^r_ξ is said to be a *focal submanifold* of Q. Thus, if P is a submanifold of codimension greater than one, then P is a focal submanifold of any of its codimension one, immersed tubes P^r. Observe that a codimension one, immersed tube around a submanifold P of codimension k in M is diffeomorphic to $P \times \mathbb{S}^{k-1}$.

Remark 2 The existence of a cohomogeneity one H-action on a complete connected Riemannian manifold M imposes some topological restrictions on M; we refer to [1, §6.3], [20, Chapter IV, Theorems 8.1–8.2], and [6, §2.9.3] for more information. If M/H is $\mathbb{R}$ or $\mathbb{S}^1$, then all orbits are mutually diffeomorphic and principal, and M is a fiber bundle over M/H (which is trivial if $M/H \cong \mathbb{R}$) and with fiber a principal orbit. In particular, if M is simply connected, M/H cannot be $\mathbb{S}^1$; indeed, if M is a Hadamard manifold, the only possibilities for M/H are $\mathbb{R}$ and $[0, \infty)$ (see [5, p. 212]). If $M/H \cong [0, \infty)$, then M is diffeomorphic to a tubular neighborhood of the only nonprincipal H-orbit, say $H \cdot p$, and hence $M \cong (H \cdot p) \times_{H_p} V$ is a Euclidean space bundle over such nonprincipal orbit $H \cdot p$. If $M/H \cong [0, 1]$, then there are two nonprincipal orbits, say $H \cdot p_+$ and $H \cdot p_-$, and M admits a decomposition as a union of two disk bundles

$$M \cong (H \times_{H_{p_+}} \mathbb{D}_-) \cup_{H/K} (H \times_{H_{p_-}} \mathbb{D}_+),$$

where $H_{p_\pm}$ are the isotropy groups at $p_\pm$; K is the isotropy at a point of a principal orbit H/K; $\mathbb{D}_\pm$ are two disks centered at the origin of the normal spaces to $H \cdot p_\pm$ at $p_\pm$, respectively; and the union of the disk bundles is made along the principal orbit H/K. This decomposition into two disk bundles is fundamental for various constructions and classifications (see, for instance, [47, 50, 80]).

2.2 Geometric Properties of Homogeneous Hypersurfaces

Homogeneous hypersurfaces and, in general, orbits of cohomogeneity one actions have some nice geometric properties. Since the shape operators (at different points) of a homogeneous hypersurface P of M are conjugate by the differentials of isometries of M, their eigenvalues are independent of the point, that is, P has *constant principal curvatures*. As the orbits of an isometric action are locally equidistant and nearby orbits to a regular one are regular, the nearby (locally defined) equidistant hypersurfaces to a homogeneous hypersurface are also (open subsets of) homogeneous hypersurfaces and therefore also with constant principal curvatures. This implies that a homogeneous hypersurface P is *isoparametric*: the locally defined, nearby parallel hypersurfaces to P have constant mean curvature. Isoparametric hypersurfaces have a long history arising from a problem in geometric optics, with contributions by Levi-Civita, Segre, and Cartan in the 1930s, and with many beautiful results obtained over the last five decades. We refer to [26, 27, 34, 90, 98, 99] and the references therein for more information on this topic.

Actually, the classification of homogeneous hypersurfaces in Euclidean and real hyperbolic spaces follows from the respective Segre's [87] and Cartan's [22] classifications of isoparametric hypersurfaces in such spaces. For Euclidean spaces

$\mathbb{R}^n$, this classification states that isoparametric hypersurfaces are open subsets of affine hyperplanes $\mathbb{R}^{n-1}$, spheres $\mathbb{S}^{n-1}$, or cylinders $\mathbb{S}^k \times \mathbb{R}^{n-k-1}$, $k = 1, \dots, n-2$. Since homogeneous hypersurfaces are always isoparametric and the previous complete hypersurfaces of $\mathbb{R}^n$ are homogeneous, they also exhaust all homogeneous hypersurfaces in Euclidean spaces. Thus, up to orbit equivalence, the cohomogeneity one actions on a Euclidean space $\mathbb{R}^n$ are the standard actions of $\mathbb{R}^{n-1}$ (Example 1 (a)), of SO_n (Example 1 (c)), and of $\mathsf{SO}_{k+1} \times \mathbb{R}^{n-k-1}$. Cartan's classification for hyperbolic spaces will be reviewed in Sect. 5.

We remark that whereas in spaces of constant curvature a hypersurface is isoparametric if and only if it has constant principal curvatures, this is not true in general. Examples of isoparametric hypersurfaces with nonconstant principal curvatures (and hence inhomogeneous hypersurfaces) have been found in several symmetric spaces, such as complex and quaternionic projective spaces [36, 39], and many symmetric spaces of noncompact type [29, 30, 35, 41]; see also Remarks 4, 5, 6, 7, and 8. Conversely, we do not know of any nonisoparametric hypersurface with constant principal curvatures in symmetric spaces, although there do exist examples for some particular conformally flat metrics [81]. There are also important spaces where isoparametric hypersurfaces are known to be homogeneous, such as the homogeneous 3-manifolds with 4-dimensional isometry group or the product of two round 2-spheres [40, 100], besides Euclidean and real hyperbolic spaces.

There is, however, an important characterization of isoparametric hypersurfaces that holds in any Riemannian manifold. Specifically, a hypersurface P of M is isoparametric if and only if P is (maybe only locally) a regular level set of an *isoparametric function* on (an open subset of) M. Here, a smooth function $f : M \to \mathbb{R}$ is called isoparametric if f is not constant on any open subset of M and it satisfies the system of partial differential equations

$$|\nabla f|^2 = a \circ f, \qquad \Delta f = b \circ f, \tag{1}$$

for some real functions a, b of real variable, with a smooth and b continuous. In other words, the norm of the gradient and the Laplacian of f are constant along the level sets of f. The collection of level sets of an isoparametric function is called an isoparametric family of hypersurfaces. We refer the reader to [1, §5.5], [6, §2.9.2], and [48] for more information on isoparametric functions.

We would like to emphasize that as homogeneous hypersurfaces are isoparametric, they are also given as level sets of isoparametric functions. This result, which would be only local in principle, is indeed global. More precisely, given a cohomogeneity one action on a complete and simply connected Riemannian manifold M, its orbit foliation is recovered as the collection of level sets of an isoparametric function on M, as follows from [1, Theorem 5.68]. Of course, the converse is not true due to the existence of inhomogeneous isoparametric families of hypersurfaces. If M is compact, any isoparametric family of hypersurfaces on M has at least a minimal hypersurface in the family, which is unique if M has positive Ricci curvature [49]. In particular, any cohomogeneity one action on a compact Riemannian manifold M has a minimal regular orbit.

The fact that homogeneous hypersurfaces (or, more generally, isoparametric hypersurfaces) arise as regular level sets of solutions to the equations (1) makes that these geometric objects appear naturally in relation to certain overdetermined problems of partial differential equations; see [59] for a survey. These include parabolic equations, as in the study of the heat flow [83–85] or of stationary isothermic surfaces [70, 82], and elliptic equations, as in some problems in mathematical physics [77] and in various overdetermined boundary value problems (including generalizations of the outstanding Schiffer conjecture [89]). Indeed, the homogeneity (respectively, isoparametricity) of geodesic spheres plays a crucial role in a partial symmetry result proved in [37] for overdetermined boundary value problems for semilinear elliptic equations on small domains of two-point homogeneous spaces (resp. harmonic spaces). Here, by symmetry result, we mean a Serrin-type theorem [88] showing that bounded solution domains to certain overdetermined problems must be balls (in the case of [37], such domains are assumed to be small perturbations of small geodesic balls).

Finally, we mention that not only homogeneous hypersurfaces have interesting geometric properties but also their focal submanifolds (i.e., the singular orbits of the corresponding cohomogeneity one actions). It was stated by Wang [102] and proved by Ge and Tang [49] that the focal submanifolds of an isoparametric family of hypersurfaces are minimal. However, if the hypersurfaces of such an isoparametric family have, in addition, constant principal curvatures, each one of their focal submanifolds has a stronger geometric property: their shape operators for all unit normal vectors are isospectral, i.e., they have the same principal curvatures and corresponding multiplicities [49]. This geometric property was called *CPC* (which stands for "constant principal curvatures") in [11]. In particular, *focal submanifolds of homogeneous hypersurfaces are CPC*. Notice that any CPC submanifold is *austere*, that is, their principal curvatures counted with multiplicities are invariant under change of sign. The notion of austere submanifold was introduced by Harvey and Lawson [52, Definition 3.15]. Clearly, austere submanifolds are minimal. In spaces of constant sectional curvature, CPC submanifolds of codimension higher than one are precisely the focal submanifolds of isoparametric hypersurfaces (equivalently, of hypersurfaces with constant principal curvatures). This is not true in general as, in many spaces (e.g., nonflat complex space forms), tubes around certain totally geodesic (and hence CPC) submanifolds are not isoparametric and have nonconstant principal curvatures (cf. Sect. 5.3 and Remark 5). Recently, it was proved that focal submanifolds of isoparametric hypersurfaces need not be austere [41].

3 Symmetric Spaces

In this section, we provide a short introduction to Riemannian symmetric spaces, with special focus on those of noncompact type. There are several references for the reader interested in obtaining further information on this topic. Two classical references are Helgason's book [53] and Loos' books [67, 68]. As introductory texts, we refer to Eschenburg's survey [45] and Ziller's notes [105]. Some nice chapters on symmetric spaces can be found in the books by Berndt, Console, and Olmos [6], Besse [17], O'Neill [76], and Wolf [103].

3.1 Definition and Fundamental Properties

Let M be a connected Riemannian manifold. Given a point $p \in M$, we can consider the geodesic ball $B_p(r)$ centered at p of radius r, for $r > 0$ small enough. On such ball, we can define the smooth map $\sigma_p\colon B_p(r) \to B_p(r)$ that sends $q = \exp_p(v)$ to $\sigma_p(q) = \exp_p(-v)$, where $v \in T_pM$, $|v| < r$. This map σ_p is nothing but a geodesic reflection about p. It is clearly an involution, i.e., $\sigma_p^2 = \mathrm{id}$.

A *Riemannian symmetric space* is a connected Riemannian manifold M such that, for any $p \in M$, the geodesic reflection σ_p at p is defined globally on M and is an isometry of M. Thus, symmetric spaces are characterized by the existence of central symmetries around any point. From this definition, one can easily see that symmetric spaces are complete (geodesics can be extended by using geodesic reflections) and homogeneous (given $p_1, p_2 \in M$, by completeness, there is a geodesic segment joining them, and if q is its midpoint, then $\sigma_q(p_1) = p_2$).

We fix from now on an arbitrary point $o \in M$, which is sometimes called the origin or the base point of M. The homogeneity and the connectedness of M imply that the Lie group $G = \mathrm{Isom}(M)^0$, the connected component of the identity of the isometry group $\mathrm{Isom}(M)$ of M, acts transitively on M. Let $K = \{g \in G : g(o) = o\}$ be the isotropy group at the origin o, which can be shown to be a compact Lie group. Hence, M is diffeomorphic to the set of left cosets $G/K = \{gK : g \in G\}$ endowed with some natural differentiable structure. Note that under the diffeomorphism $M \cong G/K$, the origin o corresponds to the coset eK, where e is the identity of G. See [65, Chapter 21] for more information on homogeneous spaces.

The map $s\colon G \to G$, $s(g) = \sigma_o g \sigma_o$, is a well-defined involutive Lie group automorphism of G. It satisfies $G_s^0 \subset K \subset G_s$, where $G_s = \{g \in G : s(g) = g\}$ and G_s^0 denotes its connected component of the identity. Its differential $\theta = s_*\colon \mathfrak{g} \to \mathfrak{g}$ is an involutive Lie algebra automorphism, the so-called Cartan involution associated with the symmetric space (of course, θ depends on the choice of o). The Lie algebra $\mathfrak{k}$ of the isotropy group K is precisely the $(+1)$-eigenspace of θ. If we denote by $\mathfrak{p}$ the (-1)-eigenspace of θ, then $\mathfrak{g} = \mathfrak{k} \oplus \mathfrak{p}$ is the eigenspace decomposition of θ, called the *Cartan decomposition* of $\mathfrak{g}$. Since θ is an automorphism, it is easy to check that $[\mathfrak{k}, \mathfrak{k}] \subset \mathfrak{k}$, $[\mathfrak{k}, \mathfrak{p}] \subset \mathfrak{p}$, and $[\mathfrak{p}, \mathfrak{p}] \subset \mathfrak{k}$.

Consider the smooth map $\phi\colon G \to M$, $\phi(g) = g(o)$. Its differential ϕ_{*e} at the identity element e induces a vector space isomorphism $\mathfrak{p} \cong T_oM$. Moreover, the linearization of the isotropy action $K \times M \to M$, $k \cdot p = k(p)$, at o turns out to be an isometric linear action $K \times T_oM \to T_oM$, $k \cdot v = k_{*o}v$. This is called the *isotropy representation* of $M \cong G/K$ at o. The isotropy representation turns out to be equivalent to the adjoint representation of K on $\mathfrak{p}$, namely, the action $K \times \mathfrak{p} \to \mathfrak{p}$ given by $k \cdot X = \mathrm{Ad}(k)X$. We will also call this action the isotropy representation of M.

The curvature tensor R of a symmetric space M at the base point o admits a simple description as

$$R(X, Y)Z = -[[X, Y], Z], \qquad X, Y, Z \in \mathfrak{p} \cong T_oM. \tag{2}$$

It turns out that the curvature tensor of a symmetric space is parallel with respect to the Levi-Civita connection, $\nabla R = 0$. Riemannian manifolds with this property are called *locally symmetric*, and the complete ones turn out to be quotients of symmetric spaces by discrete group actions.

Formula (2) leads to a very simple characterization of the totally geodesic submanifolds of symmetric spaces: they are (up to congruence in M) of the form $S = \exp_o \mathfrak{s}$, where $\mathfrak{s}$ is a subspace of $\mathfrak{p} \cong T_oM$ such that $[[\mathfrak{s}, \mathfrak{s}], \mathfrak{s}] \subset \mathfrak{s}$. Such a subspace $\mathfrak{s}$ of $\mathfrak{p}$ is called a *Lie triple system*. However, determining such Lie triple systems is a very difficult problem, and indeed the classification of totally geodesic submanifolds is still an outstanding problem; see [9, 10], and [64] for important recent contributions. In the particular case when $\mathfrak{s}$ is abelian, then the associated totally geodesic submanifold is flat by (2). Among all the abelian subspaces of $\mathfrak{p}$, the maximal ones have the same dimension. The associated totally geodesic submanifolds are called maximal flats of the symmetric space. The common dimension of such maximal flats is called the *rank* of the symmetric space.

Remark 3 It is rather common to express symmetric spaces as quotient manifolds G/K where G is not necessarily exactly $\mathrm{Isom}(M)^0$. For instance, the complex hyperbolic space $\mathbb{C}H^n$ is usually presented as $\mathsf{SU}_{1,n}/\mathsf{S}(\mathsf{U}_1\mathsf{U}_n)$ instead of $(\mathsf{SU}_{1,n}/\mathbb{Z}_{n+1})/(\mathsf{S}(\mathsf{U}_1\mathsf{U}_n)/\mathbb{Z}_{n+1})$. The common practice is to present a symmetric space in terms of a so-called symmetric pair (G, K), where K is compact, there is an involutive automorphism s of G such that $G_s^0 \subset K \subset G_s$, and G acts almost effectively on $M = G/K$ (i.e., there is at most a discrete subgroup of elements of G that act trivially on G/K). Of course, if M is a symmetric space, then $(\mathrm{Isom}(M)^0, \mathrm{Isom}(M)_o^0)$ is a symmetric pair. These subtleties will not play an important role in this article.

3.2 *Types of Symmetric Spaces*

A symmetric space $M \cong G/K$ is called *(isotropy) irreducible* if the restriction of its isotropy representation to the connected component of the identity of K is irreducible. This is equivalent to the property that the Riemannian universal cover $\widetilde{M}$ of M (which is again a symmetric space) is not a nontrivial product of symmetric spaces, unless $\widetilde{M} = \mathbb{R}^n$ is a Euclidean space.

Recall that the Killing form of a Lie algebra $\mathfrak{g}$ is the symmetric bilinear form $\mathcal{B}\colon \mathfrak{g} \times \mathfrak{g} \to \mathbb{R}$ given by $\mathcal{B}(X, Y) = \operatorname{tr}(\operatorname{ad}(X) \circ \operatorname{ad}(Y))$, where $\operatorname{ad}(X) = [X, \cdot]$. It is not difficult to check that $\mathfrak{k}$ and $\mathfrak{p}$ are orthogonal subspaces with respect to $\mathcal{B}$. A symmetric space $M \cong G/K$ is said to be of *compact type*, of *noncompact type*, or of *Euclidean type* if $\mathcal{B}|_{\mathfrak{p}\times\mathfrak{p}}$ is negative definite, positive definite, or identically zero, respectively. If M is irreducible, Schur's lemma implies that $\mathcal{B}|_{\mathfrak{p}\times\mathfrak{p}}$ is proportional to the inner product on $\mathfrak{p} \cong T_o M$ induced by the symmetric metric of M. According to the sign of the proportionality constant, M falls into one of the three possible types. If M is not irreducible, there is no guarantee that it is of one of the three types.

If M is of compact type, then G is a compact semisimple Lie group, and M is compact and has nonnegative sectional curvature. If M is of noncompact type, then it turns out that G is a noncompact real semisimple Lie group (with no compact factors), and M is a Hadamard manifold (it is diffeomorphic to a Euclidean space and has nonpositive sectional curvature). If M is of Euclidean type, its Riemannian universal cover is a Euclidean space $\mathbb{R}^n$. In general, the Riemannian universal cover of a symmetric space M splits as a Riemannian product of symmetric spaces $\widetilde{M} = M_+ \times M_- \times M_0$, where M_+ is of compact type, M_- is of noncompact type, and M_0 is a Euclidean space.

There is a notion of duality between the classes of symmetric spaces of compact type and of noncompact type. Specifically, there is a one-to-one correspondence between symmetric spaces of noncompact type and simply connected symmetric spaces of compact type. This duality can be made explicit in terms of the Lie algebras and groups involved, although here we will not enter into details. Dual symmetric spaces have the same rank and equivalent isotropy representations, and hence, duality preserves irreducibility. However, it is important to remark that both types of symmetric spaces have very different topological and geometrical properties.

Example 2 We illustrate the notion of duality through some examples:

(a) The real hyperbolic space $\mathbb{R}H^n \cong \mathsf{SO}^0_{1,n}/\mathsf{SO}_n$ is of noncompact type and has two dual symmetric spaces of compact type: the sphere $\mathbb{S}^n \cong \mathsf{SO}_{n+1}/\mathsf{SO}_n$ and the real projective space $\mathbb{R}\mathsf{P}^n \cong \mathsf{SO}_{n+1}/\mathsf{O}_n$. These spaces have rank one.
(b) The other rank one (nonflat) symmetric spaces are the projective and the hyperbolic spaces over the division algebras of the complex numbers $\mathbb{C}$, the quaternions $\mathbb{H}$, and the octonions $\mathbb{O}$. Thus, the complex spaces $\mathbb{C}\mathsf{P}^n = \mathsf{SU}_{n+1}/\mathsf{S}(\mathsf{U}_1\mathsf{U}_n)$ and $\mathbb{C}\mathsf{H}^n = \mathsf{SU}_{1,n}/\mathsf{S}(\mathsf{U}_1\mathsf{U}_n)$, the quaternionic spaces $\mathbb{H}\mathsf{P}^n = \mathsf{Sp}_{n+1}/\mathsf{Sp}_1\mathsf{Sp}_n$ and $\mathbb{H}\mathsf{H}^n = \mathsf{Sp}_{1,n}/\mathsf{Sp}_1\mathsf{Sp}_n$, and the Cayley planes $\mathbb{O}\mathsf{P}^2 = \mathsf{F}_4/\mathsf{Spin}_9$ and $\mathbb{O}\mathsf{H}^2 = \mathsf{F}_4^{-20}/\mathsf{Spin}_9$ constitute three pairs of dual symmetric spaces of rank one.
(c) Any compact Lie group K endowed with a bi-invariant metric is a symmetric space of compact type. An associated symmetric pair is $(K \times K, \Delta K)$, where $\Delta K = \{(k, k) : k \in K\}$. Its dual symmetric space of noncompact type is of the form $K^{\mathbb{C}}/K$, where $K^{\mathbb{C}}$ denotes the complexification of K. For example, SO_n

(and also its universal cover Spin_n) and $\mathsf{SO}_n(\mathbb{C})/\mathsf{SO}_n$ are dual to each other, $n \geq 3$.

(d) The compact space $\mathsf{SU}_n/\mathsf{SO}_n$ of special Lagrangian subspaces of $\mathbb{C}^n$ is dual to the noncompact space $\mathsf{SL}_n(\mathbb{R})/\mathsf{SO}_n$ of all positive definite symmetric matrices of determinant 1 and order n.

For the complete list of irreducible symmetric spaces (up to coverings), we refer to [53, pp. 516, 518]. See [103] for a discussion of locally symmetric spaces of compact type.

3.3 Symmetric Spaces of Noncompact Type: Root Space and Iwasawa Decompositions

Symmetric spaces of noncompact type constitute a rich family of Hadamard manifolds that generalize the hyperbolic spaces. We refer to [4, 34], [44, Chapter 2], [53, Chapter VI], and [60, Chapter VI, §4-5] for more information on different aspects of these spaces.

Let $M \cong G/K$ be a (not necessarily irreducible) symmetric space of noncompact type. Let $\mathfrak{g} = \mathfrak{k} \oplus \mathfrak{p}$ be the Cartan decomposition of the Lie algebra $\mathfrak{g}$ of G determined by the choice of a base point $o \in M$. The Killing form $\mathcal{B}$ of $\mathfrak{g}$ makes $\mathfrak{k}$ and $\mathfrak{p}$ orthogonal, restricts to a positive definite inner product on $\mathfrak{p}$ by definition of noncompact type, and turns out to be negative definite when restricted to $\mathfrak{k}$. Thus, by changing its sign on $\mathfrak{k}$, we get a positive definite inner product on $\mathfrak{g}$. This inner product $\mathcal{B}_\theta$ can alternatively be defined by $\mathcal{B}_\theta(X, Y) = -\mathcal{B}(\theta X, Y)$, for each X, $Y \in \mathfrak{g}$, where θ is the Cartan involution.

Let $\mathfrak{a}$ be an arbitrary maximal abelian subspace of $\mathfrak{p}$. Recall that $\dim \mathfrak{a}$ is the rank of M. The endomorphisms $\operatorname{ad}(H) = [H, \cdot]$ of $\mathfrak{g}$, where $H \in \mathfrak{a}$, turn out to be self-adjoint with respect to $\mathcal{B}_\theta$, and they commute with each other (since ad is a Lie algebra homomorphism and $\mathfrak{a}$ is abelian). Thus, such endomorphisms of $\mathfrak{g}$ diagonalize simultaneously. Their common eigenspaces are called the *restricted root spaces*, and their nonzero eigenvalues (which are linear in $H \in \mathfrak{a}$) are called the *restricted roots* of $\mathfrak{g}$. More precisely, for each linear functional $\lambda \in \mathfrak{a}^*$, consider the subspace of $\mathfrak{g}$ given by

$$\mathfrak{g}_\lambda = \{X \in \mathfrak{g} : [H, X] = \lambda(H)X \text{ for all } H \in \mathfrak{a}\}.$$

Then, any $\mathfrak{g}_\lambda \neq 0$ is a restricted root space, and any $\lambda \neq 0$ with $\mathfrak{g}_\lambda \neq 0$ is a restricted root. Note that $0 \neq \mathfrak{a} \subset \mathfrak{g}_0$. Let us denote by

$$\Sigma = \{\lambda \in \mathfrak{a}^* : \lambda \neq 0, \ \mathfrak{g}_\lambda \neq 0\}$$

the set of restricted roots of $\mathfrak{g}$. Hence, we can write the $\mathcal{B}_\theta$-orthogonal direct sum decomposition

$$\mathfrak{g} = \mathfrak{g}_0 \oplus \left(\bigoplus_{\lambda \in \Sigma} \mathfrak{g}_\lambda \right),$$

known as the *restricted root space decomposition* of $\mathfrak{g}$. The multiplicity m_λ of a restricted root λ is the dimension of its root space, $m_\lambda = \dim \mathfrak{g}_\lambda$. In what follows, we will omit the word "restricted".

Roots and root space decompositions enjoy several nice properties, such as the following:

(a) $[\mathfrak{g}_\lambda, \mathfrak{g}_\mu] \subset \mathfrak{g}_{\lambda+\mu}$, for any $\lambda, \mu \in \Sigma \cup \{0\}$.
(b) $\theta \mathfrak{g}_\lambda = \mathfrak{g}_{-\lambda}$, for any $\lambda \in \Sigma \cup \{0\}$. Hence, $\lambda \in \Sigma$ if and only if $-\lambda \in \Sigma$.
(c) $\mathfrak{g}_0 = \mathfrak{k}_0 \oplus \mathfrak{a}$, where $\mathfrak{k}_0 = \mathfrak{g}_0 \cap \mathfrak{k} = N_\mathfrak{k}(\mathfrak{a}) = Z_\mathfrak{k}(\mathfrak{a})$ is both the normalizer and the centralizer of $\mathfrak{a}$ in $\mathfrak{k}$.

Moreover, the finite subset Σ of $\mathfrak{a}^*$ formed by the roots has various symmetry properties. Firstly, we can endow $\mathfrak{a}^*$ with an inner product given by $\langle \lambda, \mu \rangle = \mathcal{B}_\theta(H_\lambda, H_\mu)$, for any $\lambda, \mu \in \mathfrak{a}^*$, and where $H_\lambda \in \mathfrak{a}$ is defined by the relation $\mathcal{B}_\theta(H_\lambda, H) = \lambda(H)$ for all $H \in \mathfrak{a}$. With this inner product, one can show that Σ is an *abstract root system* on the Euclidean space $(\mathfrak{a}^*, \langle \cdot, \cdot \rangle)$. This means (see [60, §II.5]):

(a) $\mathfrak{a}^* = \operatorname{span} \Sigma$.
(b) The number $a_{\alpha\beta} = 2\langle \alpha, \beta \rangle / \langle \alpha, \alpha \rangle$ is an integer for any $\alpha, \beta \in \Sigma$.
(c) $\beta - a_{\alpha\beta}\, \alpha \in \Sigma$, for any $\alpha, \beta \in \Sigma$.

This system is called *nonreduced* if there is $\lambda \in \Sigma$ such that $2\lambda \in \Sigma$. In this case, 2λ is called a nonreduced root. Root systems can be classified, and this is indeed the basis for the classification of real semisimple Lie algebras and of symmetric spaces.

The information of a root system can be codified in a smaller set of roots. By considering an open halfspace of $\mathfrak{a}^*$ containing exactly half of the roots in Σ (recall that Σ is invariant under the reflection about the origin), we can declare as positive those roots lying in such halfspace. If we denote by Σ^+ this set of positive roots, we then have $\Sigma = \Sigma^+ \sqcup (-\Sigma^+)$. Among the elements of Σ^+, there are some that cannot be expressed as sum of exactly two positive roots. These are called the *simple roots*, and we denote by Λ its collection. It turns out that Λ is a basis for $\mathfrak{a}^*$, and hence, its cardinality $|\Lambda|$ is precisely the rank of M. Any root λ in Σ turns out to be a linear combination of elements of Λ with integer coefficients, all of them nonnegative (when $\lambda \in \Sigma^+$) or all of them nonpositive (when $\lambda \in -\Sigma^+$).

The set Λ of simple roots allows to construct the Dynkin diagram of Σ (and, ultimately, of the symmetric space M). This is a graph consisting in as many nodes as elements in Λ. Two nodes are joined by a simple (respectively, double or triple) edge whenever the angle between the corresponding simple roots is $2\pi/3$ (respectively, $3\pi/4$ or $5\pi/6$). Finally, if the system is nonreduced, any simple root whose double is also a root is represented by a double node (two concentric circles). We refer to [6, pp. 336-340] for a list of all possible connected Dynkin diagrams, together with the multiplicities of the simple (and nonreduced) roots and with the associated irreducible symmetric spaces of noncompact type. There is also a theory of roots for symmetric spaces of compact type, and both theories behave well under duality, cf. [68, pp. 119, 146].

The sum of the root spaces associated with positive roots,

$$\mathfrak{n} = \bigoplus_{\lambda \in \Sigma^+} \mathfrak{g}_\lambda,$$

is a nilpotent Lie subalgebra of $\mathfrak{g}$, by the properties of the root space decomposition. Since $\mathfrak{a}$ normalizes $\mathfrak{n}$, we have that $\mathfrak{a} \oplus \mathfrak{n}$ is a solvable Lie subalgebra of $\mathfrak{g}$. The *Iwasawa decomposition theorem* for Lie algebras ensures that

$$\mathfrak{g} = \mathfrak{k} \oplus \mathfrak{a} \oplus \mathfrak{n}$$

is a vector space direct sum (but it is not orthogonal, and none of the addends is an ideal of $\mathfrak{g}$). Let A and N denote the connected subgroups of G with Lie algebras $\mathfrak{a}$ and $\mathfrak{n}$, respectively. Then AN is the connected subgroup of G with Lie algebra $\mathfrak{a} \oplus \mathfrak{n}$. The Iwasawa decomposition at the Lie group level states that the multiplication map

$$K \times A \times N \to G, \qquad (k, a, n) \mapsto kan,$$

is a diffeomorphism and the Lie groups A, N, and AN are diffeomorphic to Euclidean spaces.

Recall the smooth map $\phi \colon G \to M$, $\phi(g) = g(o)$, from Sect. 3.1. By the Iwasawa decomposition, its restriction to AN is a diffeomorphism $\phi|_{AN} \colon AN \to M$. Let us denote by g the symmetric Riemannian metric of M, and consider its pullback metric $(\phi|_{AN})^*\mathsf{g}$ on AN. This metric, which will be denoted by $\langle \cdot, \cdot \rangle$ in what follows, happens to be left-invariant on the Lie group AN. Therefore, we have that *any symmetric space of noncompact type M is isometric to a certain solvable Lie group AN endowed with a particular left-invariant metric*. This in particular implies, as we had already advanced, that a symmetric space of noncompact type M is diffeomorphic to a Euclidean space. By formula (2), one can actually show that such an M is nonpositively curved and hence it is a Hadamard manifold. This enables us to regard any of these spaces as an open ball endowed with certain metric, similarly as with the ball model of the real hyperbolic space.

For certain problems, it can be useful to regard a symmetric space of noncompact type M as an open dense subset of a larger compact topological space $M \sqcup M(\infty)$ that is homeomorphic to a closed ball. The *ideal boundary* $M(\infty)$ of M is defined to be the set of *points at infinity* of M, namely, the equivalence classes of complete, unit-speed geodesics of M under the relation $\gamma_1 \sim \gamma_2$ if $\{d(\gamma_1(t), \gamma_2(t)) : t \geq 0\}$ is bounded. One can endow $M \sqcup M(\infty)$ with the so-called cone topology, so that $M \sqcup M(\infty)$ becomes homeomorphic to a closed ball whose interior corresponds to M and whose boundary is $M(\infty)$. Two geodesics are called asymptotic if they converge to the same point at infinity, i.e., if they belong to the same equivalence class. If $M = G/K$, the action of G on M can be naturally extended to $M(\infty)$ by taking $g \cdot [\gamma] := [g \cdot \gamma]$.

Symmetric spaces of noncompact type, as particular instances of Hadamard manifolds, admit certain codimension one foliations by the so-called horospheres. Specifically, given any complete, unit-speed geodesic γ in M, one can consider the *Busemann function* $f_\gamma : M \to \mathbb{R}$ given by $f_\gamma(p) = \lim_{t\to+\infty}\big(d(p,\gamma(t)) - t\big)$. The *horosphere foliation* determined by γ is the regular Riemannian foliation of M given by the collection of level sets of the Busemann function f_γ. It is known that any horosphere foliation of a symmetric space of noncompact type M is homogeneous: it is the orbit foliation of the cohomogeneity one action on M given by the codimension one subgroup of AN with Lie algebra $(\mathfrak{a} \ominus \ell) \oplus \mathfrak{n}$, for some specific one-dimensional subspaces ℓ of $\mathfrak{a}$; see [42, Remark 5.4] for more information.

By making use of the solvable model of a symmetric space of noncompact type, one can provide an explicit formula for the Levi-Civita connection on $AN \cong M$ and also relate the left-invariant metric $\langle \cdot, \cdot \rangle = (\phi|_{AN})^*\mathsf{g}$ on AN to the inner product $\mathcal{B}_\theta$. These tools, along with a careful analysis of root space decompositions, are very useful in the investigation of submanifold geometry of symmetric spaces of noncompact type. We refer to [34] for further details.

3.4 *Parabolic Subgroups and Subalgebras, and Boundary Components*

The investigation of cohomogeneity one actions on symmetric spaces of noncompact type that we will review in Sect. 6 depends on a number of concepts and notation related to the theory of the so-called parabolic subalgebras of real semisimple Lie algebras. Here, we present a quick introduction to this topic. We refer to [6, §13.2], [18, §I.1], [44, §2.17], and [60, §VII.7] for more information.

Geometrically speaking, we say that a Lie subgroup Q of G is parabolic if $Q = G$ or Q is the stabilizer G_x of a point at infinity $x \in M(\infty)$. From the algebraic viewpoint, it can be proved that a Lie subalgebra $\mathfrak{q}$ of $\mathfrak{g}$ is the Lie algebra of a parabolic subgroup Q of G precisely if it contains a subalgebra of $\mathfrak{g}$ conjugate to $\mathfrak{k}_0 \oplus \mathfrak{a} \oplus \mathfrak{n}$ (recall that $\mathfrak{k}_0 = N_{\mathfrak{k}}(\mathfrak{a})$). In this case, we say that $\mathfrak{q}$ is a *parabolic subalgebra* of $\mathfrak{g}$.

Our interest in parabolic subalgebras arises from their explicit description in terms of roots and root spaces, which we explain now. Up to conjugacy in G, a parabolic subalgebra of $\mathfrak{g}$ can be constructed from the choice of a subset $\Phi \subset \Lambda$ of simple roots of $\mathfrak{g}$. Let $\Sigma_\Phi = \Sigma \cap \operatorname{span}\Phi$ be the root subsystem generated by Φ, and consider the positivity notion on Σ_Φ induced by that of Σ, that is, $\Sigma_\Phi^+ = \Sigma^+ \cap \Sigma_\Phi$. Define the following subalgebras of $\mathfrak{g}$:

$$\mathfrak{l}_\Phi = \mathfrak{g}_0 \oplus \left(\bigoplus_{\lambda \in \Sigma_\Phi} \mathfrak{g}_\lambda \right), \qquad \mathfrak{a}_\Phi = \bigcap_{\alpha \in \Phi} \ker \alpha, \qquad \mathfrak{n}_\Phi = \bigoplus_{\lambda \in \Sigma^+ \setminus \Sigma_\Phi^+} \mathfrak{g}_\lambda,$$

which are reductive (in the sense of it being invariant with respect to a Cartan involution of $\mathfrak{g}$, cf. [62]), abelian, and nilpotent, respectively. The subalgebra $\mathfrak{l}_\Phi$ is the centralizer and the normalizer of $\mathfrak{a}_\Phi$ in $\mathfrak{g}$ and normalizes $\mathfrak{n}_\Phi$. Thus, $\mathfrak{q}_\Phi = \mathfrak{l}_\Phi \oplus \mathfrak{n}_\Phi$ is a subalgebra of $\mathfrak{g}$ containing $\mathfrak{k}_0 \oplus \mathfrak{a} \oplus \mathfrak{n}$. We say that $\mathfrak{q}_\Phi$ is the parabolic subalgebra of $\mathfrak{g}$ associated with the subset $\Phi \subset \Lambda$. The decomposition $\mathfrak{q}_\Phi = \mathfrak{l}_\Phi \oplus \mathfrak{n}_\Phi$ is known as the Chevalley decomposition of $\mathfrak{q}_\Phi$. The subalgebra $\mathfrak{m}_\Phi = \mathfrak{l}_\Phi \ominus \mathfrak{a}_\Phi$ (hereafter, $\ominus$ denotes orthogonal complement with respect to the inner product $\mathcal{B}_\theta$) is a reductive subalgebra of $\mathfrak{g}$ that normalizes $\mathfrak{a}_\Phi \oplus \mathfrak{n}_\Phi$. Hence, we have a decomposition $\mathfrak{q}_\Phi = \mathfrak{m}_\Phi \oplus \mathfrak{a}_\Phi \oplus \mathfrak{n}_\Phi$, which is known as the Chevalley decomposition of $\mathfrak{q}_\Phi$. By a result of Borel and Tits [19], any parabolic subalgebra of a real semisimple Lie algebra $\mathfrak{g}$ is conjugate to one of the subalgebras $\mathfrak{q}_\Phi$, for some $\Phi \subset \Lambda$.

The orthogonal projection $\mathfrak{k}_\Phi = \pi_\mathfrak{k}(\mathfrak{m}_\Phi)$ of $\mathfrak{m}_\Phi$ onto $\mathfrak{k}$ turns out to be a maximal compact subalgebra of $\mathfrak{m}_\Phi$. It can be written as

$$\mathfrak{k}_\Phi = \mathfrak{q}_\Phi \cap \mathfrak{k} = \mathfrak{l}_\Phi \cap \mathfrak{k} = \mathfrak{m}_\Phi \cap \mathfrak{k} = \mathfrak{k}_0 \oplus \left(\bigoplus_{\lambda \in \Sigma_\Phi^+} \mathfrak{k}_\lambda \right),$$

where $\mathfrak{k}_\lambda = \pi_\mathfrak{k}(\mathfrak{m}_\lambda) = \mathfrak{k} \cap (\mathfrak{g}_\lambda \oplus \mathfrak{g}_{-\lambda})$. Similarly, the projection $\mathfrak{b}_\Phi = \pi_\mathfrak{p}(\mathfrak{m}_\Phi)$ of $\mathfrak{m}$ onto $\mathfrak{p}$ is a Lie triple system, which is also given by

$$\mathfrak{b}_\Phi = \mathfrak{m}_\Phi \cap \mathfrak{p} = \mathfrak{a}^\Phi \oplus \left(\bigoplus_{\lambda \in \Sigma_\Phi^+} \mathfrak{p}_\lambda \right),$$

where $\mathfrak{a}^\Phi = \mathfrak{a} \ominus \mathfrak{a}_\Phi = \bigoplus_{\alpha \in \Phi} \mathbb{R} H_\alpha$ and $\mathfrak{p}_\lambda = \pi_\mathfrak{p}(\mathfrak{m}_\lambda) = \mathfrak{p} \cap (\mathfrak{g}_\lambda \oplus \mathfrak{g}_{-\lambda})$. Associated with $\mathfrak{b}_\Phi$, one can consider the semisimple Lie algebra $\mathfrak{s}_\Phi = [\mathfrak{b}_\Phi, \mathfrak{b}_\Phi] \oplus \mathfrak{b}_\Phi$. The previous decomposition is a Cartan decomposition for $\mathfrak{s}_\Phi$, and $\mathfrak{a}^\Phi$ is a maximal abelian subspace of $\mathfrak{b}_\Phi$. The root subsystem Σ_Φ can be identified with a set of roots for $\mathfrak{s}_\Phi$ by restricting the roots of Σ_Φ to $\mathfrak{a}^\Phi$. The corresponding root spaces of $\mathfrak{s}_\Phi$ coincide with those of $\mathfrak{g}$. More precisely, we have the root space decomposition

$$\mathfrak{s}_\Phi = (\mathfrak{s}_\Phi \cap \mathfrak{k}_0) \oplus \mathfrak{a}^\Phi \oplus \left(\bigoplus_{\lambda \in \Sigma_\Phi} \mathfrak{g}_\lambda \right).$$

It is possible to give results analogous to the previous Lie algebra decompositions for the group G and the symmetric space M. For this, consider the connected Lie subgroups A_Φ, N_Φ, and S_Φ of G with Lie algebras $\mathfrak{a}_\Phi$, $\mathfrak{n}_\Phi$, and $\mathfrak{s}_\Phi$, respectively. If we define $L_\Phi = Z_G(\mathfrak{a}_\Phi)$ as the centralizer of $\mathfrak{a}_\Phi$ in G, then L_Φ is a Lie subgroup of G that normalizes N_Φ. The subgroup $Q_\Phi = N_G(\mathfrak{q}_\Phi) = L_\Phi N_\Phi$ is the parabolic subgroup of G associated with Φ. The decomposition $Q_\Phi = L_\Phi N_\Phi$ is known as the Chevalley decomposition of Q_Φ.

Define $K_\Phi = Z_K(\mathfrak{a}_\Phi)$, which is a Lie subgroup of G with Lie algebra $\mathfrak{k}_\Phi$. It is a maximal compact subgroup of L_Φ which normalizes S_Φ. The subgroup $M_\Phi = K_\Phi S_\Phi$ is a reductive subgroup of G with Lie algebra $\mathfrak{m}_\Phi$. Moreover, the multiplication map $A_\Phi \times N_\Phi \times M_\Phi \to Q_\Phi$ is a diffeomorphism, known as the Langlands decomposition of the parabolic subgroup Q_Φ.

Consider now the orbit B_Φ of the isometric action of M_Φ through o. Since $\mathfrak{b}_\Phi$ is a Lie triple system,

$$B_\Phi = M_\Phi \cdot o = S_\Phi \cdot o \cong M_\Phi / K_\Phi \cong S_\Phi / (K_\Phi \cap S_\Phi)$$

is a totally geodesic submanifold of M, called the *boundary component* of M associated with the subset of simple roots Φ. Intrinsically, B_Φ turns out to be a symmetric space of noncompact type and rank $|\Phi|$. In fact, since S_Φ is a connected semisimple Lie group and $K_\Phi \cap S_\Phi$ a maximal compact subgroup of S_Φ, $(S_\Phi, K_\Phi \cap S_\Phi)$ is a symmetric pair for B_Φ. The Langlands decomposition of Q_Φ induces a diffeomorphism at the manifold level, given by

$$A_\Phi \times N_\Phi \times B_\Phi \to M, \qquad (a, n, m \cdot o) \mapsto (anm) \cdot o.$$

This diffeomorphism is known as the *horospherical decomposition* of the symmetric space M corresponding to the subset $\Phi \subset \Lambda$ of simple roots.

The horospherical decomposition can be restated in terms of an isometric action on M with some interesting geometric properties. The connected solvable Lie group $A_\Phi N_\Phi$ acts freely and isometrically on M, and its orbits are mutually congruent minimally embedded submanifolds of M. Moreover, the totally geodesic submanifold B_Φ of M intersects each $A_\Phi N_\Phi$-orbit exactly once and perpendicularly. These properties are fundamental in a geometric extension procedure of submanifolds from B_Φ to M called canonical extension (see [35]). Its application to the extension of cohomogeneity one actions will be discussed in Sect. 6.

4 Homogeneous Hypersurfaces in Compact Symmetric Spaces

In this section, we give an overview of the classification problem of homogeneous hypersurfaces in compact symmetric spaces. We will mostly focus on the spherical case (Sect. 4.1), since it admits a more elementary approach and showcases very interesting geometric properties. Then, in Sect. 4.2, we will consider the case of the other compact symmetric spaces, with special focus on the rank one setting.

4.1 Homogeneous Hypersurfaces of Round Spheres

The classification of homogeneous hypersurfaces in round spheres had to wait more than 30 years after Segre's and Cartan's works on Euclidean and real hyperbolic spaces. The classification in round spheres was achieved by Hsiang and Lawson [55] and revisited by Takagi and Takahashi [95], who calculated the principal curvatures of such homogeneous hypersurfaces. Their works provide a very interesting family of examples, which surprisingly turns out to be characterized by a subclass of symmetric spaces, as we will comment on below. But before explaining their results, let us consider some examples.

Recall that a homogeneous hypersurface is isoparametric with constant principal curvatures, and hence, the multiplicities of such principal curvatures are constant. We will denote by g the number of distinct constant principal curvatures. Notice also that cohomogeneity one actions on $\mathbb{S}^n$, $n \geq 2$, must have orbit space of type $[0, 1]$, in view of Remark 2. The simplest example of cohomogeneity one action on spheres was given in Example 1 (d) as the standard action of SO_n on the unit sphere $\mathbb{S}^n$. Note that its regular orbits (the parallels) are totally umbilical ($g = 1$) and there are exactly two singular orbits (the poles). Observe that a geodesic of $\mathbb{S}^n$ normal to one orbit (and hence to all orbits) intersects the singular orbits at points separated by distance π. Let us see how this generalizes to more interesting examples.

Example 3 Consider the action of $H = \mathsf{U}_1 \times \mathsf{U}_1$ on $\mathbb{C}^2 \cong \mathbb{R}^4$ given by $(e^{i\theta_1}, e^{i\theta_2}) \cdot (z_1, z_2) = (e^{i\theta_1} z_1, e^{i\theta_2} z_2)$. Since it is an isometric action for the Euclidean metric on $\mathbb{R}^4$, it leaves the unit sphere of $\mathbb{R}^4$ invariant, so it induces an isometric action on the unit sphere $\mathbb{S}^3$. It is easy to calculate that the isotropy groups at points of the form $(z_1, 0)$ or $(0, z_2)$ are isomorphic to U_1, whereas the stabilizers at any other point are trivial. Thus, there are exactly two singular orbits, $H \cdot (1, 0)$ and $H \cdot (0, 1)$, which are totally geodesic circles in $\mathbb{S}^3$, and the remaining orbits are principal and diffeomorphic to tori $\mathbb{S}^1 \times \mathbb{S}^1$. Among these tori, exactly one turns out to be minimal, namely, $H \cdot \left(\frac{1}{\sqrt{2}}, \frac{1}{\sqrt{2}}\right)$: the Clifford torus. The principal orbits have $g = 2$ distinct principal curvatures. Any normal geodesic to the orbit foliation (e.g., $\gamma(t) = (\cos t, \sin t) \in \mathbb{C}^2$) intersects the singular orbits at four equidistributed points ($(\pm 1, 0)$ and $(0, \pm 1)$). This action of $\mathsf{U}_1 \times \mathsf{U}_1 \cong \mathsf{SO}_2 \times \mathsf{SO}_2$ on $\mathbb{S}^3$ can easily be generalized to a cohomogeneity one action of $\mathsf{SO}_{k+1} \times \mathsf{SO}_{n-k}$ on $\mathbb{S}^n$, $k = 1, \ldots, n-2$, with totally geodesic singular orbits $\mathbb{S}^k$ and $\mathbb{S}^{n-k-1}$ and principal orbits $\mathbb{S}^k \times \mathbb{S}^{n-k-1}$ with $g = 2$.

Example 4 Let $\mathrm{Herm}_3^0(\mathbb{R})$ denote the vector space of real symmetric matrices of order 3 and trace 0, endowed with the standard inner product $\langle X, Y \rangle = \operatorname{tr} XY$.

Hence, $\mathrm{Herm}_3^0(\mathbb{R})$ is a Euclidean space $\mathbb{R}^5$. The smooth action of $H = \mathsf{SO}_3$ on $\mathrm{Herm}_3^0(\mathbb{R})$ by conjugation, $A \cdot X = AXA^t$, is clearly isometric. We consider its induced isometric action on the unit sphere $\mathbb{S}^4$ of $\mathrm{Herm}_3^0(\mathbb{R}) \cong \mathbb{R}^5$. The subset $\Sigma = \{X \in \mathbb{S}^4 : X \text{ is diagonal}\}$ is the trace of a geodesic in $\mathbb{S}^4$ that intersects all H-orbits in $\mathbb{S}^4$ by the spectral theorem. One can easily compute the stabilizers at the points of Σ. These stabilizers are larger when $X \in \Sigma$ has two equal entries in the diagonal, which happens when X has diagonal entries $\frac{1}{\sqrt{6}}, \frac{1}{\sqrt{6}}, -\frac{2}{\sqrt{6}}$ or $-\frac{1}{\sqrt{6}}, -\frac{1}{\sqrt{6}}, \frac{2}{\sqrt{6}}$ (reordered in any way). Notice that these six points are equidistributed along the great circle Σ. For these six points X in Σ, we have $H_X \cong \mathsf{S}(\mathsf{O}_2 \times \mathsf{O}_1) \cong \mathsf{O}_2$. The other points X in Σ have finite isotropy group. Therefore, the regular orbits have dimension $\dim \mathsf{SO}_3 = 3$, and hence, we have a cohomogeneity one action on $\mathbb{S}^4$. There are two singular orbits (one passing through the diagonal matrices with diagonal entries $\frac{1}{\sqrt{6}}, \frac{1}{\sqrt{6}}, -\frac{2}{\sqrt{6}}$ and the other passing through the diagonal matrices with the opposite entries). These are diffeomorphic to real projective planes, $H/H_X \cong \mathsf{SO}_3/\mathsf{O}_3 \cong \mathbb{R}\mathsf{P}^2$, which are minimally embedded in $\mathbb{S}^4$. One can show that the principal orbits of this action have $g = 3$ principal curvatures.

Example 5 The previous action in Example 4 is the simplest one of a collection of four cohomogeneity one actions on the unit spheres $\mathbb{S}^4$, $\mathbb{S}^7$, $\mathbb{S}^{13}$, and $\mathbb{S}^{25}$ of the space $\mathrm{Herm}_3^0(\mathbb{F})$ of trace-free Hermitian matrices of order three with coefficients in some normed division algebra $\mathbb{F} \in \{\mathbb{R}, \mathbb{C}, \mathbb{H}, \mathbb{O}\}$, with inner product $\langle X, Y\rangle = \mathrm{Re}(\mathrm{tr}\, XY)$. The respective groups acting upon are SO_3, SU_3, Sp_3, and F_4. These actions produce homogeneous hypersurfaces with $g = 3$ principal curvatures (all of them with the same multiplicity $m \in \{1, 2, 4, 8\}$) which are tubes around certain minimal embeddings of the projective planes $\mathbb{R}\mathsf{P}^2$, $\mathbb{C}\mathsf{P}^2$, $\mathbb{H}\mathsf{P}^2$, and $\mathbb{O}\mathsf{P}^2$, respectively. Of particular interest is the octonionic case, as it provides one of the simplest models of the exceptional Lie group F_4, as well as of the Cayley projective plane $\mathbb{O}\mathsf{P}^2$. Indeed, F_4 can be defined as the automorphism group of the Jordan algebra $\mathrm{Herm}_3(\mathbb{O})$ with multiplication $X \circ Y = \frac{1}{2}(XY + YX)$. Similarly as in Example 4, the minimally embedded Cayley projective planes are obtained as the orbits through $\mathrm{diag}(\frac{1}{\sqrt{6}}, \frac{1}{\sqrt{6}}, -\frac{2}{\sqrt{6}})$ and $\mathrm{diag}(-\frac{1}{\sqrt{6}}, -\frac{1}{\sqrt{6}}, \frac{2}{\sqrt{6}}) \in \mathrm{Herm}_3^0(\mathbb{O})$ of the action of the automorphism group F_4 on the unit sphere $\mathbb{S}^{25}$ of $\mathrm{Herm}_3^0(\mathbb{O})$. For more information on these actions, we refer the reader to the discussion in [78, §3.3.3], which is based on [2, §3], [6, p. 86], and [51, pp. 289–292].

The homogeneous hypersurfaces described in the examples above were characterized by Cartan [23] as the only (complete) isoparametric hypersurfaces in round spheres with up to $g = 3$ distinct principal curvatures. Whereas the examples with $g \in \{1, 2\}$ arise in spheres $\mathbb{S}^n$ of any dimension ($n \geq 3$ if $g = 2$), examples with $g = 3$ are restricted to four possible dimensions $n \in \{4, 7, 13, 25\}$. Cartan [24] also initiated the study of isoparametric hypersurfaces with $g = 4$ and was able to produce two examples in $\mathbb{S}^5$ and $\mathbb{S}^9$. These are recovered in the following two constructions:

Example 6 Let $\mathcal{M}_{2\times k}(\mathbb{F})$ denote the vector space of $2 \times k$ matrices with entries in $\mathbb{F} \in \{\mathbb{R}, \mathbb{C}, \mathbb{H}\}$, endowed with the standard inner product $\langle X, Y\rangle = \operatorname{tr} XY^*$, where $(\cdot)^*$ denotes conjugate transpose. In order to settle ideas, assume $\mathbb{F} = \mathbb{R}$. Consider the isometric action of $H = \mathsf{SO}_2 \times \mathsf{SO}_k$ on $\mathcal{M}_{2\times k}(\mathbb{R})$ by $(A, B) \cdot X = AXB^*$ and then its restriction to the unit sphere $\mathbb{S}^{2k-1}$ of $\mathcal{M}_{2\times k}(\mathbb{R}) \cong \mathbb{R}^{2k}$. The geodesic of $\mathbb{S}^{2k-1}$ given by

$$\gamma(t) = \begin{pmatrix} \cos t & 0 & 0 \dots 0 \\ 0 & \sin t & 0 \dots 0 \end{pmatrix}$$

intersects all H-orbits in $\mathbb{S}^{2k-1}$ and always perpendicularly (again, it suffices to check this at one point, say for $t = 0$). One can compute the stabilizers of the points in this geodesic, obtaining that for any $t \notin \{\ell\frac{\pi}{4} : \ell \in \mathbb{Z}\}$, $\gamma(t)$ belongs to a principal orbit of codimension one in $\mathbb{S}^{2k-1}$. Moreover, for any $\ell \in \mathbb{Z}$, $\gamma\left(\ell\frac{\pi}{2}\right)$ belongs to a singular orbit of dimension k, and $\gamma\left(\frac{\pi}{4} + \ell\frac{\pi}{2}\right)$ belongs to a singular orbit of dimension $2k-3$. Again, the singular points along the normal geodesic γ are equidistributed. The homogeneous hypersurfaces arising from this action turn out to have $g = 4$ principal curvatures with multiplicities $1, 1, k-2$, and $k-2$. Cartan's example with $g = 4$ in $\mathbb{S}^5$ corresponds to $k = 3$. The discussion for $\mathbb{F} \in \{\mathbb{C}, \mathbb{H}\}$ is analogous by considering the actions of $\mathsf{U}_2 \times \mathsf{U}_k$ on the unit sphere $\mathbb{S}^{4k-1}$ of $\mathcal{M}_{2\times k}(\mathbb{C})$ and of $\mathsf{Sp}_2 \times \mathsf{Sp}_k$ on the unit sphere $\mathbb{S}^{8k-1}$ of $\mathcal{M}_{2\times k}(\mathbb{H})$. In these cases, the distribution of singular points along a normal geodesic is the same as in the real case, but now the $g = 4$ principal curvatures of the homogeneous hypersurfaces have multiplicities $2, 2, 2k-3, 2k-3$ for the complex case and $4, 4, 4k-5, 4k-5$ for the quaternionic case.

Example 7 Consider the action by conjugation of SO_5 on its Lie algebra $\mathfrak{so}_5$ of skew-symmetric matrices, namely, $A \cdot X = AXA^t$. When $\mathfrak{so}_5$ is endowed with the standard inner product $\langle X, Y\rangle = -\operatorname{tr} XY$, this action is isometric and hence induces an isometric action on the unit sphere $\mathbb{S}^9$ of $\mathfrak{so}_5$. The geodesic of $\mathbb{S}^9$ given by the block diagonal matrices

$$\gamma(t) = \frac{1}{\sqrt{2}} \operatorname{diag}\left(\begin{pmatrix} 0 & \cos t \\ -\cos t & 0 \end{pmatrix}, \begin{pmatrix} 0 & \sin t \\ -\sin t & 0 \end{pmatrix}, 0\right)$$

intersects all orbits and always perpendicularly. Similarly as in Example 6, γ meets the two singular orbits at $t \in \{\ell\frac{\pi}{4} : \ell \in \mathbb{Z}\}$, and these singular orbits have dimension 6. The principal orbits are homogeneous hypersurfaces of $\mathbb{S}^9$ with $g = 4$ principal curvatures, all of them with multiplicity 2.

All the actions above fit into a general construction: they are induced by isotropy representations of symmetric spaces of rank 2. This is, roughly speaking, what Hsiang and Lawson proved in [55] for arbitrary cohomogeneity one actions on round spheres.

Let us recall from Sect. 3 that if $M \cong G/K$ is a symmetric space, where K is the isotropy in G of some point $o \in M$, then K acts on the tangent space T_oM by the differential of the isometries in K. That is, we have a smooth action $K \times T_oM \to T_oM$ given by $k \cdot v = k_{*o}v$. This action is equivalent to the adjoint representation of K on $\mathfrak{p}$, that is, $K \times \mathfrak{p} \to \mathfrak{p}$, $k \cdot X = \mathrm{Ad}(k)X$. Each one of these actions is called the isotropy representation of M.

Since K is made of isometries of M, the isotropy representation is an isometric action on $T_oM \cong \mathfrak{p}$. Hence, it restricts to an isometric action on the unit sphere $\mathbb{S}^{\dim M-1}$ of $T_oM \cong \mathfrak{p}$. Any maximal abelian subspace $\mathfrak{a}$ of $\mathfrak{p}$ turns out to intersect all the orbits of the isotropy representation and always perpendicularly (see [6, §2.3.2] for a proof). Hence, $\mathfrak{a} \cap \mathbb{S}^{\dim M-1}$ is a totally geodesic submanifold of $\mathbb{S}^{\dim M-1}$ that intersects all the orbits of the restricted action to the unit sphere of $T_oM \cong \mathfrak{p}$ perpendicularly. By dimension reasons, if we want this restricted action on $\mathbb{S}^{\dim M-1}$ to be of cohomogeneity one, we just need to impose that $\mathfrak{a} \cap \mathbb{S}^{\dim M-1}$ has dimension 1 or, equivalently, that $\dim \mathfrak{a} = 2$. But $\dim \mathfrak{a}$ is, by definition, the rank of the symmetric space M. Hence, we are led to the conclusion that the restriction of the isotropy representation of a symmetric space M to the unit sphere of the tangent space is of cohomogeneity one precisely when M has rank two. Up to orbit equivalence, these actions exhaust all cohomogeneity one actions on spheres, by Hsiang and Lawson's theorem.

Theorem 1 (Homogeneous Hypersurfaces in Round Spheres) *Any homogeneous hypersurface of a round sphere is congruent to a principal orbit of the action obtained by restriction to the unit sphere of the isotropy representation of a symmetric space of rank two.*

For the whole list of symmetric spaces, see [53, pp. 516, 518], or [38, Table 2] for the list of the rank 2 symmetric spaces of compact type.

Example 8 The compact symmetric spaces whose isotropy representations induce the examples considered above in this section are the following:

(1) Example 1 (d): $M = \mathbb{S}^n \times \mathbb{S}^1 \cong (\mathsf{SO}_{n+1}/\mathsf{SO}_n) \times \mathbb{S}^1$, since in this case $K = \mathsf{SO}_n$ and $T_oM = T_v\mathbb{S}^n \times \mathbb{R}$.
(2) Example 3: $M = \mathbb{S}^{k+2} \times \mathbb{S}^{n-k+1}$.
(3) Example 4: $M = \mathsf{SU}_3/\mathsf{SO}_3$.
(4) Example 5: M is $\mathsf{SO}_6/\mathsf{SU}_3$, $\mathsf{SU}_6/\mathsf{Sp}_3$, or the exceptional space $\mathsf{E}_6/\mathsf{F}_4$.
(5) Example 6: M is $\mathsf{SO}_{2+k}/\mathsf{SO}_2\mathsf{SO}_k$, namely, the Grassmannian of oriented 2-planes of $\mathbb{R}^{2+k}$ or $\mathsf{U}_{2+k}/\mathsf{U}_2\mathsf{U}_k \cong \mathsf{SU}_{2+k}/\mathsf{S}(\mathsf{U}_2\mathsf{U}_k)$ or $\mathsf{Sp}_{2+k}/\mathsf{Sp}_2\mathsf{Sp}_k$, that is, the Grassmannians of complex or quaternionic 2-planes of $\mathbb{C}^{2+k}$ or $\mathbb{H}^{2+k}$, respectively.
(6) Example 7: M is the compact Lie group SO_5 with a bi-invariant metric.

In each of the previous cases, we have indicated a compact symmetric space, but there is also a noncompact symmetric space with equivalent isotropy representation, by duality. For instance, in item (3), M could be taken as $\mathsf{SL}_3(\mathbb{R})/\mathsf{SO}_3$. In this case, its Cartan decomposition $\mathfrak{g} = \mathfrak{k} \oplus \mathfrak{p}$ is nothing but the decomposition of $\mathfrak{g} = \mathfrak{sl}_3(\mathbb{R})$

into the sum of the subspace of skew-symmetric matrices $\mathfrak{k} = \mathfrak{so}_3$ and the subspace $\mathfrak{p} = \mathrm{Herm}_3^0(\mathbb{R})$ of trace-free symmetric matrices, and the isotropy representation $K \times \mathfrak{p} \to \mathfrak{p}$ agrees directly with the SO_3-action by conjugation on $\mathrm{Herm}_3^0(\mathbb{R})$ described in Example 4.

In the examples discussed in this section, we could see the following remarkable fact: a normal geodesic to the orbit foliation intersects the singular orbits in exactly $2g$ equidistributed points, where g agrees with the number of distinct principal curvatures of each one of the principal orbits. This is something that holds for any cohomogeneity one action on a round sphere (and more generally, for any isoparametric family, by Münzner's seminal work [74]).

The number g of principal curvatures of homogeneous hypersurfaces in spheres, their multiplicities, and their actual values were calculated by Takagi and Takahashi [95]. Their description can be done in terms of the restricted root system Σ associated with the symmetric space M of noncompact type whose isotropy representation induces the action. Let $X \in \mathfrak{a} \cap \mathbb{S}^{\dim M-1}$ be a point in a geodesic of $\mathbb{S}^{\dim M-1}$ that is orthogonal to the orbits of the isotropy representation $K \times \mathfrak{p} \to \mathfrak{p}$. Assume that X lies in a principal orbit. Let ξ be a unit normal vector to the hypersurface $K \cdot X$, i.e., ξ spans $T_X(\mathfrak{a} \cap \mathbb{S}^{\dim M-1})$. Then, the principal curvatures of $K \cdot X$ are of the form $\mu_\alpha = -\alpha(\xi)/\alpha(X)$, for each positive root $\alpha \in \Sigma^+$. Note that if $\alpha, 2\alpha \in \Sigma^+$, then both roots have the same associated principal curvature, $\mu_\alpha = \mu_{2\alpha}$. Thus, the number g of principal curvatures of a homogeneous hypersurface is precisely the cardinality of the set of reduced positive roots. It is a standard fact of root systems that those of rank 2 have exactly 2, 3, 4 or 6 reduced roots (see [60, Figure 2.2 in p. 151]). This immediately gives that $g \in \{1, 2, 3, 4, 6\}$; the case $g = 1$ arises since the symmetric space can have a flat factor (see Example 8 (1)), and then the associated root system is of rank 1. The multiplicities correspond to the multiplicities of the positive roots. Specifically, the multiplicity of the principal curvature μ_α associated with the reduced positive root α is $m_\alpha = \dim \mathfrak{g}_\alpha \oplus \mathfrak{g}_{2\alpha}$. We refer to [6, §2.3.2 and §2.7] for more details and to [74, §2, Satz 1] or [78, Teorema 3.8] for an alternative description in the general setting of isoparametric hypersurfaces.

4.2 *Homogeneous Hypersurfaces in the Other Compact Symmetric Spaces*

In this subsection, we will review the classification problem of homogeneous hypersurfaces in compact symmetric spaces of nonconstant curvature. We will mainly focus on the rank one setting.

The simply connected Riemannian symmetric spaces of compact type and rank one are the sphere $\mathbb{S}^n$ and the projective spaces $\mathbb{C}\mathsf{P}^n$, $\mathbb{H}\mathsf{P}^n$, and $\mathbb{O}\mathsf{P}^2$ ($n \geq 2$). They can be described by a symmetric pair (G, K) as specified in Table 1.

Table 1 Data for the sphere and the projective spaces

	$\mathbb{S}^n$	$\mathbb{C}\mathsf{P}^n$	$\mathbb{H}\mathsf{P}^n$	$\mathbb{O}\mathsf{P}^2$
G	SO_{n+1}	SU_{n+1}	Sp_{n+1}	F_4
K	SO_n	$\mathsf{S}(\mathsf{U}_1\mathsf{U}_n)$	$\mathsf{Sp}_1\mathsf{Sp}_n$	Spin_9

The classification problem for the complex projective space was solved by Takagi [96].

Theorem 2 (Homogeneous Hypersurfaces in Complex Projective Spaces) *A homogeneous hypersurface in a complex projective space $\mathbb{C}\mathsf{P}^n$ is congruent to one of the following:*

(1) *A geodesic sphere*
(2) *A tube around a totally geodesic $\mathbb{C}\mathsf{P}^k$ in $\mathbb{C}\mathsf{P}^n$, $k \in \{1, \dots, n-1\}$*
(3) *A tube around a totally geodesic $\mathbb{R}\mathsf{P}^n$ in $\mathbb{C}\mathsf{P}^n$*
(4) *A tube around the Segre embedding of $\mathbb{C}\mathsf{P}^1 \times \mathbb{C}\mathsf{P}^k$ into $\mathbb{C}\mathsf{P}^n$ with $n = 2k+1$ odd, $k \geq 1$*
(5) *A tube around the Plücker embedding of the complex Grassmannian $\mathsf{G}_2(\mathbb{C}^5)$ into $\mathbb{C}\mathsf{P}^9$*
(6) *A tube around the half spin embedding of $\mathsf{SO}_{10}/\mathsf{U}_5$ into $\mathbb{C}\mathsf{P}^{15}$*

A remarkable observation, similar to the discussion above in spheres and which follows from the work of Takagi [96] and later by Podestà and Thorbergsson [79], is that a homogeneous hypersurface in the complex projective space $\mathbb{C}\mathsf{P}^n$ is congruent to the quotient of a principal orbit of the isotropy representation of a Hermitian symmetric space of rank two. We will develop this idea a bit further before commenting on the different items of the classification given by Theorem 2.

Let $M \cong G/K$ be a Hermitian symmetric space of rank two. Being Hermitian means that M has a complex structure that is invariant under each geodesic symmetry. Then, M has even dimension, and we write $\dim M = 2n+2$. Consider the base point $o \cong eK$. The isotropy representation of M is the action $K \times T_oM \to T_oM$, $(k, v) \mapsto k_{*e}v$. Since $T_oM \cong \mathbb{R}^{2n+2} \cong \mathbb{C}^{n+1}$ and the elements of K act as linear holomorphic isometries of $\mathbb{C}^{n+1}$, this action can be restricted to an action on the unit sphere $\mathbb{S}^{2n+1} \subset \mathbb{C}^{n+1}$. As discussed in the previous subsection, this action on the unit sphere is of cohomogeneity one. Moreover, the action on $\mathbb{C}^{n+1}$ is polar and with totally real section, that is, there exists a totally real plane in $\mathbb{C}^{n+1}$ that intersects all the orbits of the isotropy representation, and, at the points of intersection, the plane and the orbits are orthogonal. Since this action maps complex lines of $\mathbb{C}^{n+1}$ to complex lines of $\mathbb{C}^{n+1}$, it descends to a cohomogeneity one action on the projectivization $\mathsf{P}(\mathbb{C}^{n+1}) \cong \mathbb{C}\mathsf{P}^n$.

In order to obtain the classification in Theorem 2, it is therefore enough to consider the classification of (possibly reducible) Hermitian symmetric spaces of rank two and calculate their induced isotropy representations on the corresponding projectivization of the tangent space at the point where the isotropy is considered. See [53, X.6], taking into account the possible coincidences between different classes.

Tubes around totally geodesic $\mathbb{C}\mathsf{P}^k$, $k \in \{0, \ldots, n-1\}$, are principal orbits of the action of $\mathsf{U}_{k+1} \times \mathsf{U}_{n-k}$. This action comes from the isotropy representation of the reducible symmetric space $\mathbb{C}\mathsf{P}^{k+1} \times \mathbb{C}\mathsf{P}^{n-k} = (\mathsf{SU}_{k+2} \times \mathsf{SU}_{n-k+1})/(\mathsf{S}(\mathsf{U}_1 \times \mathsf{U}_{k+1}) \times \mathsf{S}(\mathsf{U}_1 \times \mathsf{U}_{n-k}))$. If $k = 0$, we recover geodesic spheres.

The real oriented two-plane Grassmannian $\mathsf{G}_2^+(\mathbb{R}^{n+3}) = \mathsf{SO}_{n+3}/\mathsf{SO}_2 \times \mathsf{SO}_{n+1}$ induces an action of SO_{n+1} on $\mathbb{C}\mathsf{P}^n$ with two singular orbits: a totally geodesic real projective space $\mathbb{R}\mathsf{P}^n$ and the complex quadric $\mathsf{Q}^{n-1} = \{[z] \in \mathbb{C}\mathsf{P}^n : z_0^2 + \cdots + z_n^2 = 0\}$.

Similarly, the complex two-plane Grassmannian $\mathsf{G}_2(\mathbb{C}^{k+3}) = \mathsf{SU}_{k+3}/\mathsf{S}(\mathsf{U}_2\mathsf{U}_{k+1})$ induces an action on $\mathbb{C}\mathsf{P}^{2k+1}$, one of whose singular orbits is the Segre embedding of $\mathbb{C}\mathsf{P}^1 \times \mathbb{C}\mathsf{P}^k$ in $\mathbb{C}\mathsf{P}^{2k+1}$. This is an embedding of a product of projective spaces onto another projective space of suitable dimension, where homogeneous coordinates are multiplied out. In our case, this embedding is given by the map $\mathbb{C}\mathsf{P}^1 \times \mathbb{C}\mathsf{P}^k \to \mathbb{C}\mathsf{P}^{2k+1}$, $([z_0 : z_1], [w_0 : \cdots : w_k]) \mapsto [z_0 w_0 : \cdots : z_0 w_k : z_1 w_0 : \cdots : z_1 w_k]$.

The Plücker embedding is another classical embedding into a complex projective space. In this case, we embed a Grassmannian of k-planes into the projectivization of the space of k-forms. For $\mathsf{G}_2(\mathbb{C}^5)$ into $\mathsf{P}(\Lambda^2\mathbb{C}^5) \cong \mathbb{C}\mathsf{P}^9$, this embedding is defined by $\operatorname{span}\{v_1, v_2\} \mapsto [v_1 \wedge v_2]$. Tubes around this submanifold are homogeneous and correspond to the principal orbits of the cohomogeneity one action induced by the isotropy representation of the Hermitian symmetric space $\mathsf{SO}_{10}/\mathsf{U}_5$.

Finally, the Hermitian symmetric space $\mathsf{E}_6/\mathsf{U}_1\mathsf{Spin}_{10}$ induces a cohomogeneity one action on $\mathbb{C}\mathsf{P}^{15}$. One of the singular orbits of this action is the half spin embedding of the symmetric space $\mathsf{SO}_{10}/\mathsf{U}_5$. We refer to [25, §7.5] for further details on this embedding.

The classification problem in quaternionic projective space is attributed to D'Atri [28] and Iwata [56].

Theorem 3 (Homogeneous Hypersurfaces in Quaternionic Projective Spaces) *A homogeneous hypersurface in a quaternionic projective space $\mathbb{H}\mathsf{P}^n$ is congruent to one of the following:*

(1) *A geodesic sphere*
(2) *A tube around a totally geodesic $\mathbb{H}\mathsf{P}^k$ in $\mathbb{H}\mathsf{P}^n$, $k \in \{1, \ldots, n-1\}$*
(3) *A tube around a totally geodesic $\mathbb{C}\mathsf{P}^n$ in $\mathbb{H}\mathsf{P}^n$*

The action of $\mathsf{Sp}_{k+1} \times \mathsf{Sp}_{n-k}$ on $\mathbb{H}\mathsf{H}^k$ is of cohomogeneity one, and its principal orbits are tubes around totally geodesic quaternionic projective spaces $\mathbb{H}\mathsf{P}^k$, $k \in \{0, \ldots, n-1\}$. If $k = 0$, we retrieve geodesic spheres. The principal orbits of the action of U_{n+1} on $\mathbb{H}\mathsf{P}^n$ are tubes around a totally geodesic $\mathbb{C}\mathsf{P}^n$.

It can be shown [79] that a cohomogeneity one action on a quaternionic projective space is induced by the isotropy representation of a product of two quaternionic Kähler symmetric spaces of rank one or of an irreducible quaternionic Kähler symmetric space of rank two. Thus, an alternative way of getting the list of Theorem 3 is to look at the corresponding list of these spaces, which turns out to be $\mathbb{H}\mathsf{P}^{k+1} \times \mathbb{H}\mathsf{P}^{n-k}$ and $\mathsf{SU}_{n+3}/\mathsf{S}(\mathsf{U}_2 \times \mathsf{U}_{n+1})$.

We finish our review of homogeneous hypersurfaces in rank one symmetric spaces of compact type recalling the classification result for the Cayley projective plane given by Iwata [57].

Theorem 4 (Homogeneous Hypersurfaces in the Cayley Projective Plane) *A homogeneous hypersurface in the Cayley projective plane $\mathbb{O}\mathsf{P}^2$ is congruent to one of the following:*

(1) *A geodesic sphere*
(2) *A tube around a totally geodesic $\mathbb{H}\mathsf{P}^2$ in $\mathbb{O}\mathsf{P}^2$*

A geodesic sphere can be seen as a principal orbit of the isotropy action of Spin_9 on $\mathbb{O}\mathsf{P}^2$. This action has two singular orbits: a fixed point and a totally geodesic $\mathbb{S}^8 = \mathbb{O}\mathsf{P}^1$. The second example in this classification is congruent to a principal orbit of the action of $\mathsf{Sp}_3\mathsf{Sp}_1$, which has two singular orbits: a totally geodesic $\mathbb{H}\mathsf{P}^2$ and a minimal $\mathbb{S}^{11} = \mathsf{Sp}_3\mathsf{Sp}_1/\mathsf{Sp}_2\mathsf{Sp}_1$. As pointed out by Iwata, there are two more groups, up to conjugation, with the same orbits as $\mathsf{Sp}_3\mathsf{Sp}_1$. These are $\mathsf{Sp}_3\mathsf{U}_1$ and Sp_3. Unlike the results presented here, Iwata's classification was obtained up to conjugation by an element of F_4, not up to orbit equivalence.

Remark 4 We would like to point out that there is a classification of isoparametric families of hypersurfaces in complex projective spaces $\mathbb{C}\mathsf{P}^n$, $n \neq 15$ [36], and in quaternionic projective spaces $\mathbb{H}\mathsf{P}^n$, $n \neq 7$ [39]. It follows from these classifications that there are inhomogeneous examples of isoparametric hypersurfaces in complex and quaternionic projective spaces. However, the classification problem of isoparametric hypersurfaces in the Cayley projective plane is still open [99].

All these results were generalized by Kollross [61], who classified cohomogeneity one actions on irreducible symmetric spaces of compact type up to orbit equivalence. Thus, homogeneous hypersurfaces in an irreducible symmetric space of compact type can be obtained via a case-by-case study of all these actions in each corresponding space.

Theorem 5 (Cohomogeneity One Actions on Irreducible Symmetric Spaces of Compact Type) *A cohomogeneity one action on an irreducible symmetric space of compact type is locally orbit equivalent to one of the following:*

(1) *A Hermann action of cohomogeneity one*
(2) *The action of $\{(g, \bar{g}) : g \in \mathsf{SU}_3\}$ on SU_3*
(3) *An action induced by the isotropy representation of a symmetric space of rank two*
(4) *One of the seven exceptions corresponding to the action of $H \times K$ on G or of the action of H on G/K, where (H, K, G) is a triple of Table 2*

Let H and K be compact Lie subgroups of G. In Theorem 5 and in the discussion below, the isometric action of a product group $H \times K$ on a compact Lie group G with bi-invariant metric is given by

$$(h, k) \cdot g = hgk^{-1}, \qquad h \in H,\ k \in K,\ g \in G.$$

Table 2 Seven exceptional cohomogeneity one actions on symmetric spaces of compact type

H	G_2	G_2	U_3	Spin_9	$\mathsf{Sp}_1\mathsf{Sp}_n$	SU_3	SU_3
K	$\mathsf{SO}_3 \times \mathsf{SO}_4$	G_2	G_2	$\mathsf{SO}_2 \times \mathsf{SO}_{14}$	$\mathsf{SO}_2 \times \mathsf{SO}_{4n-2}$	SO_4	SU_3
G	SO_7	SO_7	SO_7	SO_{16}	SO_{4n}	G_2	G_2

The action of a subgroup H of G on a compact symmetric space G/K is given by $h \cdot gK = hgK$.

Let G be a compact semisimple Lie group. A subgroup K of G is called a symmetric subgroup of G if its Lie algebra is a fixed point set of an involutive automorphism of the Lie algebra of G. Then (G, K) is a symmetric pair and G/K a symmetric space of compact type if equipped with a suitable metric.

A *Hermann action* is the action of $H \times K$ on G defined above, where H and K are symmetric subgroups of G. The natural action of H on G/K is also called a Hermann action, and it turns out that the action of $H \times K$ on G is of cohomogeneity one if and only if so is the action of H on G/K (or the action of K on G/H). Thus, classifying cohomogeneity one Hermann H-actions on G/K and classifying cohomogeneity one Hermann $H \times K$-actions on G are equivalent problems. Indeed, there is a correspondence between Hermann actions on symmetric spaces of type II (or group type), that is, compact simple Lie groups, and Hermann actions on symmetric spaces of type III, that is, compact symmetric spaces with simple isometry group, and this correspondence preserves the cohomogeneity.

Not any Hermann action is of cohomogeneity one, but it is possible to determine explicitly which ones are by looking at the classification of symmetric spaces of compact type. Obvious examples that fall into this category are isotropy actions of symmetric spaces G/K of rank one and the corresponding $K \times K$ actions on G. However, there are a few more examples as shown in [61, Theorem B].

In Theorem 5 (2), the action of $\{(g, \bar{g}) : g \in \mathsf{SU}_3\}$ on SU_3 is given by $(g, \bar{g}) \cdot g' = gg'\bar{g}^{-1}$. Here, $\bar{g}$ denotes the complex conjugation of a matrix g, which induces an outer Lie group automorphism of SU_3.

Finally, we describe the actions in Theorem 5 (3). Let $\widehat{G}/\widehat{K}$ be a simply connected symmetric space of rank 2. Then, the isotropy representation of $\widehat{G}/\widehat{K}$ can be regarded as a Lie group homomorphism $\rho = \mathrm{Ad}|_{\widehat{K}} \colon \widehat{K} \to \mathsf{SO}(\widehat{\mathfrak{p}}) \cong \mathsf{SO}_n$, where $\widehat{\mathfrak{p}} \cong T_{\hat{o}}\widehat{G}/\widehat{K}$ and $n = \dim \widehat{G}/\widehat{K}$. If $\widehat{G}/\widehat{K}$ is Hermitian, then $\widehat{K} \cong K_h \cdot \mathsf{U}_1$, for some compact Lie group K_h, and we can regard the restriction of ρ to K_h as a homomorphism $\rho|_{K_h} \colon K_h \to \mathsf{SU}_n$, where $n = \dim_{\mathbb{C}} \widehat{G}/\widehat{K}$. If $\widehat{G}/\widehat{K}$ is quaternionic Kähler, then $\widehat{K} = K_q \cdot \mathsf{Sp}_1$, for some compact Lie group K_q, and we can regard the restriction of ρ to K_q as a homomorphism $\rho|_{K_q} \colon K_q \to \mathsf{Sp}_n$, where $n = \dim_{\mathbb{H}} \widehat{G}/\widehat{K}$. Then, the actions in item (3) of Theorem 5 correspond to the action of $H \times K$ on G and to the action of H on G/K, where (H, K, G) is given in Table 3 and $\widehat{G}/\widehat{K}$ is a rank two symmetric space.

Table 3 Actions induced by isotropy representations of symmetric spaces of rank two

	$\widehat{G}/\widehat{K}$ arbitrary	$\widehat{G}/\widehat{K}$ Hermitian	$\widehat{G}/\widehat{K}$ quaternionic Kähler
H	$\rho(\widehat{K})$	$\rho(K_h)$	$\rho(K_q)$
K	SO_{n-1}	$\mathsf{S}(\mathsf{U}_1 \times \mathsf{U}_{n-1})$	$\mathsf{Sp}_1 \times \mathsf{Sp}_{n-1}$
G	SO_n	SU_n	Sp_n

5 Homogeneous Hypersurfaces in Hyperbolic Spaces

In this section, we review the classification results of homogeneous hypersurfaces in rank one symmetric spaces of noncompact type. These are precisely the hyperbolic spaces over the normed real division algebras, namely, $\mathbb{R}\mathsf{H}^n$, $\mathbb{C}\mathsf{H}^n$, $\mathbb{H}\mathsf{H}^n$, and $\mathbb{O}\mathsf{H}^2$ $(n \geq 2)$.

5.1 Homogeneous Hypersurfaces in Real Hyperbolic Spaces

The classification of homogeneous hypersurfaces in real hyperbolic spaces was solved in a classical paper by Cartan [22]. Actually, Cartan's aim was to classify isoparametric hypersurfaces in Riemannian manifolds of constant curvature. He succeeded to get such classification in $\mathbb{R}\mathsf{H}^n$, but not in spheres, where the problem remained open for nearly a century. It follows from this classification that an isoparametric hypersurface in $\mathbb{R}\mathsf{H}^n$ is an open part of a homogeneous hypersurface. This implies the classification of homogeneous hypersurfaces in $\mathbb{R}\mathsf{H}^n$:

Theorem 6 (Homogeneous Hypersurfaces in Real Hyperbolic Spaces) *A homogeneous hypersurface in $\mathbb{R}\mathsf{H}^n$ is congruent to one of the following:*

(1) *A geodesic sphere*
(2) *A tube around a totally geodesic $\mathbb{R}\mathsf{H}^k$, $k \in \{1, \dots, n-2\}$, in $\mathbb{R}\mathsf{H}^n$*
(3) *A totally geodesic $\mathbb{R}\mathsf{H}^{n-1}$ or one of its equidistant hypersurfaces*
(4) *A horosphere*

Recall that the connected component of the identity of the isometry group of the real hyperbolic space $\mathbb{R}\mathsf{H}^n$ is $\mathsf{SO}^0_{1,n}$. A geodesic sphere is congruent to a principal orbit of the action of SO_n on $\mathbb{R}\mathsf{H}^n$. Similarly, a tube around a totally geodesic $\mathbb{R}\mathsf{H}^k$ in $\mathbb{R}\mathsf{H}^n$ is congruent to a principal orbit of the action of $\mathsf{SO}^0_{1,k} \times \mathsf{SO}_{n-k}$, $k \in \{1, \dots, n-2\}$. If $k = 0$, we recover the geodesic spheres, and if $k = n-1$, then $\mathsf{SO}^0_{1,n-1}$ acts with cohomogeneity one, but in this case, all orbits are principal; in particular, a totally geodesic $\mathbb{R}\mathsf{H}^{n-1}$ is also a homogeneous hypersurface. Finally, the horospheres are the orbits of the nilpotent part N of the Iwasawa decomposition of $\mathsf{SO}^0_{1,n}$ (see Sect. 3.3). It is remarkable that the horospheres are Euclidean spaces

$\mathbb{R}^{n-1}$ embedded in $\mathbb{R}\mathsf{H}^n$ in a totally umbilical way [93, p. 14]; all horospheres of $\mathbb{R}\mathsf{H}^n$ are congruent to each other.

5.2 General Approach to Homogeneous Hypersurfaces in Hyperbolic Spaces

In the rest of this section, we address the classification problem for the remaining symmetric spaces of noncompact type and rank one. In this subsection, we review the algebraic structure theory of these spaces and explain the general approach for the classification of homogeneous hypersurfaces in this setting. In the subsequent subsections, we will describe the classification results separately for each family of spaces. We will use the notation introduced in Sect. 3.3.

Let (G, K) be a symmetric pair representing the symmetric space $\mathbb{F}\mathsf{H}^n$, $\mathbb{F} \in \{\mathbb{R}, \mathbb{C}, \mathbb{H}, \mathbb{O}\}$ ($n = 2$ if $\mathbb{F} = \mathbb{O}$). Then, the root space decomposition of $\mathfrak{g}$, the Lie algebra of G, reads

$$\mathfrak{g} = \mathfrak{g}_{-2\alpha} \oplus \mathfrak{g}_{-\alpha} \oplus \mathfrak{g}_0 \oplus \mathfrak{g}_\alpha \oplus \mathfrak{g}_{2\alpha},$$

where $\mathfrak{g}_{2\alpha} = \mathfrak{g}_{-2\alpha} = 0$ in the case of the real hyperbolic space $\mathbb{R}\mathsf{H}^n$; the associated root system is otherwise nonreduced. Recall that $\mathfrak{g}_0 = \mathfrak{k}_0 \oplus \mathfrak{a}$, where $\mathfrak{a}$ is one-dimensional. We denote by K_0 the connected subgroup of K whose Lie algebra is $\mathfrak{k}_0$. Then, the possibilities for G, K, and K_0 are summarized in Table 4.

In this case, the nilpotent part of the Iwasawa decomposition of $\mathfrak{g}$ is simply $\mathfrak{n} = \mathfrak{g}_\alpha \oplus \mathfrak{g}_{2\alpha}$. If $\mathfrak{g}_\alpha = 0$, then $\mathfrak{n}$ is abelian. Otherwise, if $\mathfrak{g}_\alpha \neq 0$, it turns out that $\mathfrak{g}_{2\alpha}$ is the center of $\mathfrak{n}$ and the derived subalgebra of the nilpotent Lie algebra $\mathfrak{n}$, that is, $[\mathfrak{n}, \mathfrak{n}] = \mathfrak{g}_{2\alpha}$. We have $\dim \mathfrak{g}_{2\alpha} = \dim_{\mathbb{R}} \mathbb{F} - 1$. In fact, $\mathfrak{g}_{2\alpha}$ can be interpreted as the imaginary part of $\mathbb{F}$; following this idea, there is a Clifford algebra representation $J\colon \mathrm{Cl}(\mathfrak{g}_{2\alpha}) \to \mathrm{End}(\mathfrak{g}_\alpha)$ which turns $\mathfrak{g}_\alpha$ into a Clifford module. The restriction of J to $\mathfrak{g}_{2\alpha}$ gives rise to endomorphisms J_Z of $\mathfrak{g}_\alpha$ that are defined by the relation

$$\langle [U, V], Z\rangle = \langle J_Z U, V\rangle, \qquad U, V \in \mathfrak{g}_\alpha, Z \in \mathfrak{g}_{2\alpha}.$$

Table 4 Data for each hyperbolic space

	$\mathbb{R}\mathsf{H}^n$	$\mathbb{C}\mathsf{H}^n$	$\mathbb{H}\mathsf{H}^n$	$\mathbb{O}\mathsf{H}^2$
G	$\mathsf{SO}^0_{1,n}$	$\mathsf{SU}_{1,n}$	$\mathsf{Sp}_{1,n}$	F_4^{-20}
K	SO_n	$\mathsf{S}(\mathsf{U}_1\mathsf{U}_n)$	$\mathsf{Sp}_1\mathsf{Sp}_n$	Spin_9
K_0	SO_{n-1}	$\mathsf{S}(\mathsf{U}_1\mathsf{U}_{n-1})$	$\mathsf{Sp}_1\mathsf{Sp}_{n-1}$	Spin_7
$\mathfrak{g}_\alpha$	$\mathbb{R}^{n-1}$	$\mathbb{C}^{n-1}$	$\mathbb{H}^{n-1}$	$\mathbb{O}$
$\mathfrak{g}_{2\alpha}$	0	$\mathbb{R}$	$\mathbb{R}^3$	$\mathbb{R}^7$

See [16] for further details. Moreover, $\mathfrak{g}_\alpha \cong \mathbb{F}^{n-1}$, and the action of K_0 on $\mathfrak{g}_\alpha$ is equivalent to the standard action.

We will now describe the possible types of cohomogeneity one actions that may arise on a rank one symmetric space of noncompact type and nonconstant curvature. The fact that the following types exhaust all actions follows from the various works of Berndt, Brück, and Tamaru [5, 12, 14]. In Sect. 6, we will describe a more general approach that holds for arbitrary rank.

(1) If a cohomogeneity one action on $\mathbb{F}\mathsf{H}^n$ induces a regular foliation, then there are two options up to orbit equivalence [12]:

    (a) The *horosphere foliation*, whose leaves are the orbits of the action of the nilpotent part of the Iwasawa decomposition of G, namely, the connected subgroup N with Lie algebra $\mathfrak{n}$
    (b) The *solvable foliation*, whose leaves are the orbits of the subgroup S whose Lie algebra is $\mathfrak{s} = \mathfrak{a} \oplus \mathfrak{w} \oplus \mathfrak{g}_{2\alpha}$, where $\mathfrak{w}$ is a real hyperplane of $\mathfrak{g}_\alpha$

(2) In order to determine the cohomogeneity one actions on $\mathbb{F}\mathsf{H}^n$ that have a totally geodesic singular orbit, it is enough to determine which totally geodesic submanifolds of $\mathbb{F}\mathsf{H}^n$ have homogeneous tubes. Totally geodesic submanifolds of hyperbolic spaces have been classified. By calculating the stabilizer of each one of these submanifolds, as well as its slice representation (i.e., the linearized action on the normal space to the totally geodesic submanifold), one can conclude which ones give rise to cohomogeneity one actions [5].
(3) Finally, it remains to study cohomogeneity one actions on $\mathbb{F}\mathsf{H}^n$ with a non-totally geodesic singular orbit. Berndt and Tamaru devised in [14] a procedure to address this case. In symmetric spaces of higher rank, this method is called the nilpotent construction, cf. Sect. 6.4. In brief, the classification of cohomogeneity one actions on $\mathbb{F}\mathsf{H}^n$ with a non-totally geodesic singular orbit reduces to the classification of the subspaces $\mathfrak{w}$ of $\mathfrak{g}_\alpha$ such that $N_{K_0}(\mathfrak{w})$, the normalizer of $\mathfrak{w}$ in K_0, acts transitively on the unit sphere of $\mathfrak{w}^\perp = \mathfrak{g}_\alpha \ominus \mathfrak{w}$, the orthogonal complement of $\mathfrak{w}$ in $\mathfrak{g}_\alpha$, up to congruence by an element of K_0. In this case, the connected subgroup of $K_0AN \subset G$ whose Lie algebra is $N_{K_0}(\mathfrak{w}) \oplus \mathfrak{a} \oplus \mathfrak{w} \oplus \mathfrak{g}_{2\alpha}$ acts on $\mathbb{F}\mathsf{H}^n$ with cohomogeneity one. The subspaces $\mathfrak{w} \subset \mathfrak{g}_\alpha$ satisfying this condition have been classified in [14] for $\mathbb{F} \in \{\mathbb{C}, \mathbb{O}\}$ and in [32] for $\mathbb{F} = \mathbb{H}$.

5.3 Homogeneous Hypersurfaces in Complex Hyperbolic Spaces

The classification of homogeneous hypersurfaces in the complex case was obtained by Berndt and Tamaru in [14]. It can be stated as follows:

Theorem 7 (Homogeneous Hypersurfaces in Complex Hyperbolic Spaces) *A homogeneous hypersurface in $\mathbb{C}\mathsf{H}^n$ is congruent to one of the following:*

(1) *A geodesic sphere*
(2) *A tube around a totally geodesic $\mathbb{C}\mathsf{H}^k$ in $\mathbb{C}\mathsf{H}^n$, $k \in \{1, \ldots, n-1\}$*
(3) *A tube around a totally geodesic $\mathbb{R}\mathsf{H}^n$ in $\mathbb{C}\mathsf{H}^n$*
(4) *A horosphere*
(5) *A ruled homogeneous minimal Lohnherr hypersurface $W^{2n-1}_{\pi/2}$ or one of its equidistant hypersurfaces*
(6) *A tube around a ruled homogeneous minimal Berndt–Brück submanifold W^{2n-k}_{φ}, for $k \in \{2, \ldots, n-1\}$, $\varphi \in (0, \pi/2]$, where k is even if $\varphi \neq \pi/2$*

Tubes around totally geodesic complex hyperbolic spaces $\mathbb{C}\mathsf{H}^k$, $k \in \{0, \ldots, n-1\}$, are congruent to the principal orbits of the action of $\mathsf{S}(\mathsf{U}_{1,k} \times \mathsf{U}_{n-k})$. The particular case of $k = 0$ corresponds to geodesic spheres. The principal orbits of the group $\mathsf{SO}^0_{1,n}$ produce tubes around a totally geodesic real hyperbolic space $\mathbb{R}\mathsf{H}^n$. Note that tubes around a totally geodesic $\mathbb{R}\mathsf{H}^k$, $k \in \{1, \ldots, n-1\}$, are not homogeneous because the normal space of $\mathbb{R}\mathsf{H}^k$ is a direct sum of a nontrivial totally real and a nontrivial complex subspace of a complex vector space and isometries of $\mathbb{C}\mathsf{H}^n$ are holomorphic. The group N gives rise to a horosphere foliation, whose orbits are isometric to generalized Heisenberg groups. All of the orbits of this action are principal and congruent to each other.

Item (5) in Theorem 7 corresponds to the solvable foliation, whereas example (6) corresponds to a nilpotent construction. We review them in more detail here. Let $\mathfrak{w}$ be a real subspace of $\mathfrak{g}_\alpha \cong \mathbb{C}^{n-1}$. We denote by J the complex structure of $\mathfrak{g}_\alpha \cong \mathbb{C}^{n-1}$. The Kähler angle of a nonzero $v \in \mathfrak{w}^\perp$ is the angle between Jv and $\mathfrak{w}^\perp$. We say that $\mathfrak{w}^\perp$ has constant Kähler angle $\varphi \in [0, \pi/2]$ if the Kähler angle of any nonzero vector of $\mathfrak{w}^\perp$ is φ. Examples of subspaces with constant Kähler angle are totally real subspaces, that is, $\langle J\mathfrak{w}^\perp, \mathfrak{w}^\perp \rangle = 0$, whose Kähler angle is $\pi/2$, and complex subspaces, that is, $J\mathfrak{w}^\perp = \mathfrak{w}^\perp$, whose Kähler angle is 0. Any angle $\varphi \in (0, \pi/2)$ can be achieved, and in this case, $\dim \mathfrak{w}^\perp = k$ is an even number. Two subspaces of $\mathfrak{g}_\alpha$ with the same dimension and Kähler angle are congruent by an isometry of K_0, and a basis of such a subspace can be written as

$$\{e_1, \cos(\varphi)Je_1 + \sin(\varphi)Je_2, \ldots, e_{2k-1}, \cos(\varphi)Je_{2k-1} + \sin(\varphi)Je_{2k}\},$$

where $\{e_1, \ldots, e_{2k}\}$ is a $\mathbb{C}$-orthonormal subset in $\mathfrak{g}_\alpha \cong \mathbb{C}^{n-1}$.

It turns out that if $\mathfrak{w}^\perp$ has constant Kähler angle φ, then $N_{K_0}(\mathfrak{w})$ acts transitively on the unit sphere of $\mathfrak{w}^\perp$. Berndt and Tamaru [14] showed that the connected subgroup of $\mathsf{SU}_{1,n}$ whose Lie algebra is $N_{K_0}(\mathfrak{w}) \oplus \mathfrak{a} \oplus \mathfrak{w} \oplus \mathfrak{g}_{2\alpha}$ acts on $\mathbb{C}\mathsf{H}^n$ with cohomogeneity one. We denote by W^{2n-k}_φ the orbit through the origin $o \cong eK$ of this group, where k is its codimension. If $\mathfrak{w}$ is a hyperplane, then $\mathfrak{w}^\perp$ is one-dimensional and, thus, totally real. The corresponding action has exactly one minimal orbit, known as the Lohnherr hypersurface, and the rest of the orbits are equidistant hypersurfaces to it. If $\mathfrak{w}$ has codimension $k > 1$, then there is exactly one singular orbit, known as a Berndt-Brück submanifold, and the rest of the orbits are tubes around it. Any Kähler angle is possible if $n \geq 3$. However, if $\varphi = 0$, this construction reproduces the tubes around a totally geodesic $\mathbb{C}\mathsf{H}^k$, $k \in \{1, \ldots, n\}$, so it is removed from item (6) of Theorem 7 to avoid duplication.

Remark 5 Isoparametric hypersurfaces in $\mathbb{C}\mathsf{H}^n$ have been classified in [33]. It follows from this classification that a hypersurface in $\mathbb{C}\mathsf{H}^n$ is isoparametric if and only if it is an open part of a homogeneous hypersurface or of a tube around the orbit through the origin of the subgroup of $AN \subset \mathsf{SU}_{1,n}$ whose Lie algebra is $\mathfrak{a} \oplus \mathfrak{w} \oplus \mathfrak{g}_{2\alpha}$, where $\mathfrak{w}^\perp$ is a subspace of $\mathfrak{g}_\alpha$ with nonconstant Kähler angle. As a consequence, any isoparametric hypersurface in $\mathbb{C}\mathsf{H}^2$ is homogeneous, but there are infinitely many inhomogeneous examples in $\mathbb{C}\mathsf{H}^n$, $n \geq 3$.

5.4 Homogeneous Hypersurfaces in Quaternionic Hyperbolic Spaces

The classification of cohomogeneity one actions on quaternionic hyperbolic spaces $\mathbb{H}\mathsf{H}^n$ has recently been obtained in [32] by the first two authors and Rodríguez-Vázquez. The corresponding classification of homogeneous hypersurfaces can be read from there.

Theorem 8 (Homogeneous Hypersurfaces in Quaternionic Hyperbolic Spaces) *A homogeneous hypersurface in $\mathbb{H}\mathsf{H}^n$ is congruent to one of the following:*

(1) *A geodesic sphere*
(2) *A tube around a totally geodesic $\mathbb{H}\mathsf{H}^k$ in $\mathbb{H}\mathsf{H}^n$, $k \in \{1, \ldots, n-1\}$*
(3) *A tube around a totally geodesic $\mathbb{C}\mathsf{H}^n$ in $\mathbb{H}\mathsf{H}^n$*
(4) *A horosphere*
(5) *A homogeneous minimal hypersurface P_1 or one of its equidistant hypersurfaces*
(6) *A tube around a homogeneous minimal submanifold $P_\mathfrak{w}$ in $\mathbb{H}\mathsf{H}^n$, where $\mathfrak{w}^\perp$ is a protohomogeneous subspace of $\mathfrak{g}_\alpha$*

Similar to the complex case, tubes around a totally geodesic quaternionic hyperbolic space $\mathbb{H}\mathsf{H}^k$, $k \in \{0, \ldots, n-1\}$, are homogeneous and are congruent to the principal orbits of the action of $\mathsf{Sp}_{1,k} \times \mathsf{Sp}_{n-k}$ on $\mathbb{H}\mathsf{H}^n$. If $k = 0$, we again have geodesic spheres. Tubes around totally geodesic complex hyperbolic spaces $\mathbb{C}\mathsf{H}^n$ in $\mathbb{H}\mathsf{H}^n$ are also homogeneous and correspond to the principal orbits of the action of $\mathsf{SU}_{1,n}$. Although there are more totally geodesic submanifolds of $\mathbb{H}\mathsf{H}^n$, their tubes fail to be homogeneous. The action of N gives rise to a horosphere foliation, all whose orbits are congruent to each other. Examples (5) correspond to the leaves of the solvable foliation. This is constructed, as usual, as the action of the subgroup of $AN \subset \mathsf{Sp}_{1,n}$ whose Lie algebra is $\mathfrak{a} \oplus \mathfrak{w} \oplus \mathfrak{g}_{2\alpha}$, where $\mathfrak{w}$ is a hyperplane in $\mathfrak{g}_\alpha$. This foliation has exactly one minimal leaf, which we have denoted by P_1.

For the rest of the examples, we need to determine all subspaces $\mathfrak{w}$ of $\mathfrak{g}_\alpha \cong \mathbb{H}^{n-1}$ such that $N_{K_0}(\mathfrak{w})$ acts transitively on the unit sphere of $\mathfrak{w}^\perp$ up to congruence by an element of $K_0 \cong \mathsf{Sp}_1\mathsf{Sp}_{1,n-1}$. As we explained before, the subgroup of $K_0AN \subset \mathsf{Sp}_{1,n}$ whose Lie algebra is $N_{K_0}(\mathfrak{w}) \oplus \mathfrak{a} \oplus \mathfrak{w} \oplus \mathfrak{g}_{2\alpha}$ acts on $\mathbb{H}\mathsf{H}^n$ with cohomogeneity one. We will call the subspaces $\mathfrak{w}$ of $\mathfrak{g}_\alpha$ satisfying this condition *protohomogeneous*. In the particular case that $\mathfrak{w}$ is a hyperplane, we recover the solvable foliation, which corresponds to item (5) of Theorem 8.

The space $\mathfrak{g}_\alpha \cong \mathbb{H}^{n-1}$ is a right quaternionic vector space. It can be endowed with a quaternionic structure $\mathfrak{J}$, that is, a vector subspace of $\mathrm{End}_{\mathbb{R}}(\mathbb{H}^{n-1})$ admitting a so-called canonical basis $\{J_1, J_2, J_3\}$ satisfying

$$J_i^2 = -\operatorname{Id}, \quad J_i J_{i+1} = J_{i+2} = -J_{i+1} J_i \quad \text{(indices modulo 3).}$$

For a given subspace $\mathfrak{w}^\perp \subset \mathfrak{g}_\alpha$, each complex structure $J \in \mathfrak{J}$ determines a Kähler angle of a nonzero vector $v \in \mathfrak{w}^\perp$ in the sense we have considered for the complex case. We define the quaternionic Kähler angle of a nonzero $v \in \mathfrak{w}^\perp$ to be the triple $(\varphi_1(v), \varphi_2(v), \varphi_3(v))$ satisfying that there exists a canonical basis $\{J_1, J_2, J_3\}$ such that:

(i) $\varphi_1(v) \le \varphi_2(v) \le \varphi_3(v)$.
(ii) $\varphi_i(v)$ is the Kähler angle of v with respect to J_i, $i \in \{1, 2, 3\}$.
(iii) $\langle \pi_{\mathfrak{w}^\perp} J_i v, \pi_{\mathfrak{w}^\perp} J_j v\rangle = 0$ if $i \ne j$ and where $\pi_{\mathfrak{w}^\perp} \colon \mathfrak{g}_\alpha \to \mathfrak{w}^\perp$ denotes the orthogonal projection onto $\mathfrak{w}^\perp$.
(iv) $\varphi_1(v)$ is minimum, and $\varphi_3(v)$ is maximum among the Kähler angles of v with respect to the complex structures $J \in \mathfrak{J}$.

A probably more telling way of defining the quaternionic Kähler angle is the following. We consider the symmetric bilinear form:

$$L_v \colon \mathfrak{J} \times \mathfrak{J} \to \mathbb{R}, \quad (J, J') \mapsto \langle \pi_{\mathfrak{w}^\perp} J v, \pi_{\mathfrak{w}^\perp} J' v\rangle.$$

Then, the Kähler angle of a nonzero $v \in \mathfrak{w}^\perp$ is the ordered triple $(\varphi_1(v), \varphi_2(v), \varphi_3(v))$ satisfying that the eigenvalues of L_v are precisely $\cos^2(\varphi_i(v))\langle v, v\rangle$. The canonical basis $\{J_1, J_2, J_3\}$ used above to define the quaternionic Kähler angle is precisely a basis that diagonalizes L_v.

If $\mathfrak{w}^\perp$ is protohomogeneous, then $\mathfrak{w}^\perp$ has constant quaternionic Kähler angle. Protohomogeneous subspaces of $\mathbb{H}^n$ have been classified in [32] up to congruence by an element of $\mathsf{Sp}_1\mathsf{Sp}_n$ by making extensive use of the concept of quaternionic Kähler angle. The moduli space $\mathcal{M}_{k,n}$ of nonzero protohomogeneous subspaces of dimension k in $\mathbb{H}^n$, up to congruence in $\mathsf{Sp}_1\mathsf{Sp}_n$, is described in Table 5.

This classification includes well-known examples such as totally real subspaces (precisely those with quaternionic Kähler angle $(\pi/2, \pi/2, \pi/2)$), totally complex subspaces (with quaternionic Kähler angle $(0, \pi/2, \pi/2)$), quaternionic subspaces (with quaternionic Kähler angle $(0, 0, 0)$), subspaces of constant Kähler angle $\varphi \in (0, \pi/2)$ inside a totally complex vector subspace (with quaternionic Kähler

Table 5 Moduli space of protohomogeneous subspaces of dimension k in $\mathbb{H}^n$

$\mathcal{M}_{k,n}$	$k \leq n$	$n < k \leq \frac{4n}{3}$	$\frac{4n}{3} < k \leq 2n$	$k > 2n$
$k \equiv 0 \,(\mathrm{mod}\, 4)$	$(\mathfrak{R}_4^+ \setminus \mathfrak{R}_4^-) \sqcup (\mathfrak{R}_4^- \times \mathbb{Z}_2)$	$\mathfrak{S}$	$\{(0,\varphi,\varphi)\}_{\varphi\in[0,\frac{\pi}{2}]}$	$\{(0,0,0)\}$
$k \equiv 2 \,(\mathrm{mod}\, 4)$	$\{(\varphi,\frac{\pi}{2},\frac{\pi}{2})\}_{\varphi\in[0,\frac{\pi}{2}]}$	$\{(0,\frac{\pi}{2},\frac{\pi}{2})\}$	$\{(0,\frac{\pi}{2},\frac{\pi}{2})\}$	$\emptyset$
$k \neq 3$ odd	$\{(\frac{\pi}{2},\frac{\pi}{2},\frac{\pi}{2})\}$	$\emptyset$	$\emptyset$	$\emptyset$
$k = 3$	$(\mathfrak{R}_3^+ \setminus \mathfrak{R}_3^-) \sqcup (\mathfrak{R}_3^- \times \mathbb{Z}_2)$	$\emptyset$	$\{(\varphi,\varphi,\frac{\pi}{2})\}_{\varphi\in\{0,\frac{\pi}{3}\}}$	$\{(0,0,\frac{\pi}{2})\}$

$$\Lambda = \{(\varphi_1,\varphi_2,\varphi_3) \in [0,\pi/2]^3 : \varphi_1 \leq \varphi_2 \leq \varphi_3\},$$

$$\mathfrak{R}_3^+ = \{(\varphi,\varphi,\pi/2) \in \Lambda : \varphi \in [0,\pi/2]\},$$

$$\mathfrak{R}_3^- = \{(\varphi,\varphi,\pi/2) \in \Lambda : \varphi \in [\pi/3,\pi/2)\},$$

$$\mathfrak{R}_4^+ = \{(\varphi_1,\varphi_2,\varphi_3) \in \Lambda : \cos(\varphi_1) + \cos(\varphi_2) - \cos(\varphi_3) \leq 1\},$$

$$\mathfrak{R}_4^- = \{(\varphi_1,\varphi_2,\varphi_3) \in \Lambda : \cos(\varphi_1) + \cos(\varphi_2) + \cos(\varphi_3) \leq 1,\ \varphi_3 \neq \pi/2\},$$

$$\mathfrak{S} = \{(\varphi_1,\varphi_2,\varphi_3) \in \Lambda : \cos(\varphi_1) + \cos(\varphi_2) + \varepsilon\cos(\varphi_3) = 1,\ \text{for } \varepsilon = 1 \text{ or } \varepsilon = -1\}.$$

angle $(\varphi,\pi/2,\pi/2)$), complexifications of subspaces of constant Kähler angle $\varphi \in (0,\pi/2)$ in a totally complex subspace (with quaternionic Kähler angle $(0,\varphi,\varphi)$), and subspaces of the form $\mathfrak{J}v$, $v \in \mathbb{H}^n$, $v \neq 0$ (with quaternionic Kähler angle $(0,0,\pi/2)$).

However, there are some other nonclassical examples. See [32] for an explicit construction of these subspaces. While two subspaces with different quaternionic Kähler angles cannot be congruent to each other, a remarkable consequence of this classification implies the existence of noncongruent subspaces of $\mathbb{H}^n$ with the same quaternionic Kähler angles. These correspond precisely to the intersections $\mathfrak{R}_3^+ \cap \mathfrak{R}_3^- = \mathfrak{R}_3^-$ and $\mathfrak{R}_4^+ \cap \mathfrak{R}_4^- = \mathfrak{R}_4^-$.

All the examples in Theorem 8 (6) are obtained as tubes around the orbit through the origin $o \cong eK$ of the connected subgroup of $AN \subset G = \mathsf{Sp}_{1,n}$ whose Lie algebra is $\mathfrak{a} \oplus \mathfrak{w} \oplus \mathfrak{g}_{2\alpha}$ and where $\mathfrak{w}^\perp$ is protohomogeneous in $\mathfrak{g}_\alpha \cong \mathbb{H}^{n-1}$. The moduli space $\mathcal{M}_{k,n-1}$ determines the congruence classes of the singular orbits of the corresponding cohomogeneity one actions, which in turn determines the orbit equivalence classes of cohomogeneity one actions on $\mathbb{H}\mathsf{H}^n$.

In order to get a proper classification, we still need to exclude a few classes that intersect with previous items of Theorem 8. If $\mathfrak{w}^\perp$ has quaternionic Kähler angle $(0,0,0)$, then $\mathfrak{w}^\perp$ and also $\mathfrak{w}$ are quaternionic vector subspaces of $\mathfrak{g}_\alpha \cong \mathbb{H}^{n-1}$. In this case, we recover tubes around totally geodesic quaternionic hyperbolic spaces $\mathbb{H}\mathsf{H}^k$, $k \in \{1,\ldots,n-1\}$. As we explained before, we also have to exclude when $\mathfrak{w}$ is a hyperplane, as this gives the solvable foliation.

Remark 6 Consider the connected subgroup of $\mathsf{Sp}_{1,n}$ with Lie algebra $\mathfrak{a} \oplus \mathfrak{w} \oplus \mathfrak{g}_{2\alpha}$, where $\mathfrak{w}$ is an arbitrary proper subspace of $\mathfrak{g}_\alpha$. It follows from [30] that tubes around the orbit through the origin of that group are always isoparametric.

These have constant principal curvatures if and only if $\mathfrak{w}^\perp$ has constant quaternionic Kähler angle. If follows from [32] that by taking direct sums of spaces in both $\mathfrak{R}_4^+$ and $\mathfrak{R}_4^-$ with the same constant quaternionic Kähler angles, we obtain subspaces $\mathfrak{w}^\perp$ that still have constant quaternionic Kähler angle but are not protohomogeneous. This yields examples of isoparametric hypersurfaces with constant principal curvatures in $\mathbb{H}\mathsf{H}^n$, $n \geq 8$, that are not homogeneous.

5.5 *Homogeneous Hypersurfaces in the Cayley Hyperbolic Plane*

Finally, we deal with the Cayley hyperbolic plane $\mathbb{O}\mathsf{H}^2$.

Theorem 9 (Homogeneous Hypersurfaces in the Cayley Hyperbolic Plane) *A homogeneous hypersurface in $\mathbb{O}\mathsf{H}^2$ is congruent to one of the following:*

(1) *A geodesic sphere*
(2) *A tube around a totally geodesic $\mathbb{O}\mathsf{H}^1$*
(3) *A tube around a totally geodesic $\mathbb{H}\mathsf{H}^2$*
(4) *A horosphere*
(5) *A minimal homogeneous hypersurface F_1 or one of its equidistant hypersurfaces*
(6) *A tube around the minimal submanifold F_k of codimension $k \in \{2, 3, 6, 7\}$*
(7) *A tube around the minimal submanifold $F_{4,\varphi}$ of codimension 4 for some $\varphi \in [0, 1]$*

Geodesic spheres are principal orbits of the isotropy action of Spin_9 on $\mathbb{O}\mathsf{H}^2$. Tubes around a totally geodesic $\mathbb{O}\mathsf{H}^1$ on $\mathbb{O}\mathsf{H}^2$ are congruent to the principal orbits of the action of $\mathsf{Spin}^0_{1,8} \subset \mathsf{F}_4^{-20}$, and tubes around a totally geodesic $\mathbb{H}\mathsf{H}^2$ are principal orbits of the action of $\mathsf{Sp}_{1,2}\mathsf{Sp}_1 \subset \mathsf{F}_4^{-20}$. The group N, which is the nilpotent part of the Iwasawa decomposition of F_4^{-20}, gives rise to the horosphere foliation in $\mathbb{O}\mathsf{H}^2$, whose leaves are congruent to each other. Example (5) of Theorem 9 corresponds to the solvable foliation, which is obtained by the action of the subgroup of F_4^{-20} whose Lie algebra is $\mathfrak{a} \oplus \mathfrak{w} \oplus \mathfrak{g}_{2\alpha}$, where $\mathfrak{w}$ is a hyperplane in $\mathfrak{g}_\alpha$. This action has a unique minimal orbit which is denoted by F_1.

Examples (6) and (7) correspond to the nilpotent construction. Berndt and Brück classified in [5] all subspaces $\mathfrak{w}$ of $\mathfrak{g}_\alpha \cong \mathbb{O}$ such that $N_{K_0}(\mathfrak{w})$ acts transitively on the unit sphere of $\mathfrak{w}^\perp$. It turns out that any proper subspace $\mathfrak{w}$ of $\mathfrak{g}_\alpha$ with $\dim \mathfrak{w} \neq 3$ satisfies this condition. Hyperplanes of $\mathfrak{g}_\alpha$ are ones of such spaces, but they correspond to item (5) and produce a foliation. The group $K_0 \cong \mathsf{Spin}_7$ acts on $\mathbb{O} \cong \mathbb{R}^8$ by its irreducible 8-dimensional spin representation. This action induces an action on the Grassmannians $\mathsf{G}_k(\mathbb{R}^8)$ of k-planes in $\mathbb{R}^8$. If $k \neq 4$, this action is transitive, and if $k = 4$, this action is of cohomogeneity one (see the discussion for $\mathbb{O}\mathsf{H}^2$ in [14] and the references therein). This implies that any pair of subspaces of $\mathfrak{g}_\alpha$ of dimension $k \neq 4$ are congruent by an isometry of Spin_7. The singular orbit of the action on $\mathbb{O}\mathsf{H}^2$ of the connected subgroup of F_4^{-20} with Lie algebra $N_{K_0}(\mathfrak{w}) \oplus \mathfrak{a} \oplus \mathfrak{w} \oplus \mathfrak{g}_{2\alpha}$ is denoted by F_k, where $k = \dim \mathfrak{w}^\perp = 8 - \dim \mathfrak{w}$. The moduli space of Spin_7-congruence classes of subspaces of $\mathfrak{g}_\alpha$ of dimension 4 is in one-to-one correspondence with the orbit space $\mathsf{G}_4(\mathbb{R}^8)/\mathsf{Spin}_7 \cong [0, 1]$. The congruence class corresponding to some $\varphi \in [0, 1]$ produces a cohomogeneity one action on $\mathbb{O}\mathsf{H}^2$ whose singular orbit is denoted by $F_{4,\varphi}$.

Remark 7 As in the previous hyperbolic spaces, any tube around the orbit through the origin of the subgroup $S_\mathfrak{w}$ of $AN \subset \mathsf{F}_4^{-20}$ with Lie algebra $\mathfrak{a} \oplus \mathfrak{w} \oplus \mathfrak{g}_{2\alpha}$ is isoparametric. Moreover, in this case, it follows from [30] that each one of these tubes has constant principal curvatures. Thus, for $\dim \mathfrak{w}^\perp = 5$, the corresponding tubes around $S_\mathfrak{w} \cdot o$ are inhomogeneous isoparametric hypersurfaces with constant principal curvatures. If $\dim \mathfrak{w}^\perp = 4$, the constant principal curvatures of the homogeneous tubes around $F_{4,\varphi}$ are independent of φ. Thus, there is an infinite family of noncongruent homogeneous isoparametric hypersurfaces with the same constant principal curvatures counted with multiplicities.

6 Homogeneous Hypersurfaces in Symmetric Spaces of Noncompact Type and Arbitrary Rank

The aim of this section is to provide an overview of the methods of construction and classification of cohomogeneity one actions on symmetric spaces of noncompact type and arbitrary rank. As we commented in the previous section, the classification in rank one is nowadays complete. Although this is not the case for higher rank, there have been recent advances that give us not only some classifications in certain spaces but also, importantly, a panoramic view of the possible types of actions that may arise in any symmetric space of noncompact type.

We will start by explaining four construction techniques that can be regarded as building blocks for the classification problem. These techniques are the construction of codimension one subgroups of the solvable part AN of the Iwasawa decomposition (explained in Sect. 6.1), the actions with a totally geodesic singular orbit (Sect. 6.2), the canonical extension of actions from lower rank symmetric spaces (Sect. 6.3), and the nilpotent construction (Sect. 6.4). Then, in Sect. 6.5, we will report on a structural result that asserts that these four building blocks are enough to construct any cohomogeneity one action on any (not necessarily irreducible) symmetric space of noncompact type.

6.1 Homogeneous Codimension One Foliations

Since any symmetric space of noncompact type M is a Hadamard manifold, any cohomogeneity one action on M has at most one singular orbit. We will explain in this subsection that the case of actions without singular orbit is nowadays well understood.

It follows from the Iwasawa decomposition that the connected solvable subgroup AN of G with Lie algebra $\mathfrak{a}\oplus\mathfrak{n}$ acts freely and transitively on M. Thus, codimension one subgroups of AN give rise to homogeneous codimension one regular foliations on M. Berndt and Tamaru used this in [12] to propose two general methods for constructing cohomogeneity one actions with no singular orbits on a given symmetric space of noncompact type.

The first method produces a regular Riemannian foliation $\mathcal{F}_\ell$ for each one-dimensional subspace ℓ in $\mathfrak{a}$. Define $\mathfrak{h}_\ell$ to be the orthogonal complement of ℓ in $\mathfrak{a} \oplus \mathfrak{n}$, $\mathfrak{h}_\ell = (\mathfrak{a} \ominus \ell) \oplus \mathfrak{n}$. This is a codimension one subalgebra of $\mathfrak{a} \oplus \mathfrak{n}$, so the corresponding connected subgroup H_ℓ of G acts on M with cohomogeneity one and no singular orbits. It turns out that the orbits of this action are congruent to each other. Foliations of M by horospheres (i.e., by the level sets of a Busemann function on M) are a particular type of such a construction [42, Remark 5.4], so we will refer to the $\mathcal{F}_\ell$ as *foliations of horospherical type*.

The second method gives us a foliation $\mathcal{F}_i$ for each simple root $\alpha_i \in \Lambda = \{\alpha_1, \dots, \alpha_r\}$. Let ℓ be a one-dimensional subspace of a simple root space $\mathfrak{g}_{\alpha_i}$. It follows from the properties of root spaces that $\mathfrak{h}_i = \mathfrak{a} \oplus (\mathfrak{n} \ominus \ell)$ is a codimension one subalgebra of $\mathfrak{a} \oplus \mathfrak{n}$, and so, its corresponding connected subgroup H_i of G acts with cohomogeneity one on M. Actions arising in this way have a unique minimal orbit (namely, the orbit through o). We will refer to these $\mathcal{F}_i$ as *foliations of solvable type*.

It was shown in [12] for irreducible M and in [7] for the general case that every cohomogeneity one action on a symmetric space of noncompact type with no singular orbits is orbit equivalent to the action of some H_ℓ or H_i as constructed before. Furthermore, the moduli space of such actions has been studied in [12] and [92]. Two actions of horospherical type $\mathcal{F}_\ell$ and $\mathcal{F}_{\ell'}$ are isometrically congruent precisely whenever there exists an isometry of M that induces a symmetry of the Dynkin diagram of $\mathfrak{g}$ taking ℓ to ℓ'. Something similar happens for the foliations of solvable type: $\mathcal{F}_i$ and $\mathcal{F}_j$ are isometrically congruent if and only if there exists an isometry of M that induces a symmetry of the Dynkin diagram of $\mathfrak{g}$ taking α_i to α_j. In particular, if ℓ and ℓ' are contained in the same root space, they yield congruent foliations. Thus, the moduli space of homogeneous codimension one foliations on a symmetric space of noncompact type up to orbit equivalence is isomorphic to $(\mathbb{R}\mathsf{P}^r \sqcup \{1, \dots, r\})/\mathrm{Aut}(\mathrm{DD}_M)$, where $r = \mathrm{rank}\,(M)$ and $\mathrm{Aut}(\mathrm{DD}_M)$ denotes the subgroup of symmetries of the Dynkin diagram of $\mathfrak{g}$ which are induced by isometries of M.

6.2 Cohomogeneity One Actions with a Totally Geodesic Singular Orbit

Among the cohomogeneity one actions that have a singular orbit, it is natural to first determine those actions whose singular orbit is totally geodesic. Recall that if a cohomogeneity one action on a Euclidean or a real hyperbolic space has a singular orbit, this must be totally geodesic, although this is no longer the case for other hyperbolic spaces, as explained in Sect. 5.

In the article [13], Berndt and Tamaru derived the classification of the totally geodesic submanifolds F of any *irreducible* symmetric space of noncompact type M that arise as singular orbits of cohomogeneity one actions on M, i.e., the totally geodesic submanifolds F such that the tubes around them are homogeneous hypersurfaces. This is basically the only case where the use of duality of symmetric spaces can be applied. However, we recall that one cannot simply analyze case-by-case all possible totally geodesic submanifolds of M, since even nowadays there is no such a classification. Berndt and Tamaru appeal to the use of duality, along with Kollross' classification [61] in the compact setting, as well as Leung's classification [66] of a certain very particular type of totally geodesic submanifolds, called reflective submanifolds. A *reflective submanifold* F of a symmetric space M is a totally geodesic submanifold of M such that the exponential of its normal space at some (and hence all) point, $F^\perp = \exp(\nu_p F)$, is also totally geodesic in M. Recall that, as totally geodesic submanifolds, both F and $F^\perp$ are themselves symmetric spaces.

Berndt and Tamaru proved that F is a totally geodesic singular orbit of a cohomogeneity one action on an irreducible M if and only if one of the following possibilities holds:

(i) F is a reflective submanifold such that $F^\perp$ is a symmetric space of rank one (see [13, Theorem 3.3] for an explicit list).
(ii) F is one of the five possible nonreflective totally geodesic submanifolds related to the exceptional Lie group G_2 appearing in Table 6.

It is important to mention that the lists provided in [13] are given up to congruence in M by isometries of the full isometry group $\mathrm{Isom}(M)$, cf. Problem 3 in Sect. 7.

Let us now assume that M is *reducible*. Put $M = M_1 \times \cdots \times M_s$ for its de Rham decomposition into irreducible symmetric spaces (of noncompact type). For each $i \in \{1, \ldots, s\}$, we write $M_i \cong G_i/K_i$, and hence, $\mathfrak{g} = \mathfrak{g}_1 \oplus \cdots \oplus \mathfrak{g}_s$ is the decomposition of the semisimple Lie algebra $\mathfrak{g}$ into its simple ideals. A fundamental

Table 6 Nonreflective totally geodesic submanifolds related to G_2

M	$\mathsf{SO}_{3,7}/\mathsf{SO}_3 \times \mathsf{SO}_7$	$\mathsf{SO}_7(\mathbb{C})/\mathsf{SO}_7$	$\mathsf{G}_2^2/\mathsf{SO}_4$	$\mathsf{G}_2^{\mathbb{C}}/\mathsf{G}_2$
F	$\mathsf{G}_2^2/\mathsf{SO}_4$	$\mathsf{G}_2^{\mathbb{C}}/\mathsf{G}_2$	$\mathbb{C}\mathsf{H}^2$, $\mathsf{SL}_3(\mathbb{R})/\mathsf{SO}_3$	$\mathsf{SL}_3(\mathbb{C})/\mathsf{SU}_3$

observation made in the recent work [31] is that if a cohomogeneity one action on M with a totally geodesic singular orbit does not split nicely with respect to the previous decompositions (i.e., if it is not orbit equivalent to a product action), then there must exist two homothetic factors M_j and M_k of M of rank one, and the action is orbit equivalent to that of the connected subgroup of G whose Lie algebra is

$$\mathfrak{g}_{j,k,\tau} \oplus \left(\bigoplus_{\substack{i=1 \\ i \neq j,k}} \mathfrak{g}_i \right), \qquad \text{with } \mathfrak{g}_{j,k,\tau} = \{X + \tau X : X \in \mathfrak{g}_j\}, \tag{3}$$

where $\tau \colon \mathfrak{g}_j \to \mathfrak{g}_k$ is a Lie algebra isomorphism. In this case, the singular orbit is also homothetic to M_j and M_k. This result ultimately follows from a classical theorem of Dynkin [43, Theorem 15.1, p. 235] which states that a maximal proper subalgebra of either $\mathfrak{g}$ splits nicely with respect to the decomposition of $\mathfrak{g}$ into simple ideals or it is of the form (3).

All in all, any cohomogeneity one action with a totally geodesic singular orbit on M is determined by one of the actions on an irreducible factor of M listed by Berndt and Tamaru in [13] or by a diagonal action on the product of two homothetic rank one factors of M, as in (3).

6.3 Canonical Extension of Actions on Boundary Components

Consider a subset $\Phi \subset \Lambda$ of simple roots and its associated boundary component B_Φ. Since S_Φ is (up to a covering) the identity component of $\mathrm{Isom}(B_\Phi)$, any isometric action on B_Φ has the same orbits as some connected Lie subgroup H_Φ of S_Φ. Consider the subgroup

$$H_\Phi^\Lambda = H_\Phi A_\Phi N_\Phi$$

of G. Then, H_Φ^Λ acts on M with the same cohomogeneity of the action of H_Φ on B_Φ. Indeed, each H_Φ^Λ-orbit on M, say $H_\Phi^\Lambda \cdot p$, is nothing but the union of all $A_\Phi N_\Phi$-orbits through the points of $H_\Phi \cdot p$. Recall from Sect. 3.4 that all the $A_\Phi N_\Phi$-orbits have the same dimension. We say that H_Φ^Λ is the group obtained by *canonical extension* of H_Φ from the boundary component B_Φ to M. Furthermore, it was proved in [15, Proposition 4.2] that if the actions of two connected subgroups of S_Φ are orbit equivalent on B_Φ by an isometry in S_Φ (equivalently, by an isometry of $\mathrm{Isom}(B_\Phi)^0$), then their canonical extensions are orbit equivalent on M by an element of G.

As boundary components of M are symmetric spaces of noncompact type, it makes sense to study what happens if one applies this procedure twice. Consider the boundary component B_Φ associated with a subset of simple roots $\Phi \subset \Lambda$. Recall that we can naturally identify Φ with a set of simple roots for $\mathfrak{s}_\Phi$. Thus, a boundary component of B_Φ is determined by a subset $\Psi \subset \Phi \subset \Lambda$ and in fact coincides with the boundary component B_Ψ of M associated with Ψ. One gets an inclusion of totally geodesic submanifolds $B_\Psi \subset B_\Phi \subset M$. Let H_Ψ be a connected closed subgroup of S_Ψ acting isometrically on B_Ψ. Then, its canonical extension H_Ψ^Φ is a connected closed subgroup of S_Φ acting isometrically on B_Φ, so we can consider its canonical extension to M, which we denote by $(H_\Psi^\Phi)^\Lambda$. This construction turns out to be the same as directly extending the action of H_Ψ from B_Ψ to the whole M, that is, $(H_\Psi^\Phi)^\Lambda = H_\Psi^\Lambda$ (cf. [31, Lemma 4.2]). Roughly speaking, the composition of canonical extensions is a canonical extension.

Remark 8 The canonical extension method described above admits an interesting version that allows to enlarge submanifolds from boundary components to the whole symmetric space. This procedure preserves important geometric properties such as the constancy of the mean curvature or isoparametricity, as was shown in [35]. More recently, another remarkable extension method of submanifolds and actions in the context of symmetric spaces of noncompact type has been discovered [41]. In this case, the extension does not apply to boundary components, but to certain totally geodesic and flat submanifolds. As a by-product of this method, the first examples of inhomogeneous isoparametric hypersurfaces in any symmetric space of noncompact type and rank higher than two were obtained.

6.4 The Nilpotent Construction Method

Apart from the canonical extension, Berndt and Tamaru proposed in [15] another method for constructing cohomogeneity one actions from the parabolic subgroups of G. Although this procedure was originally formulated for an arbitrary subset of simple roots $\Phi \subset \Lambda$, it will be enough to consider subsets of cardinality $|\Phi| = |\Lambda| - 1$, that is, those giving rise to maximal proper parabolic subgroups of G.

Let $\Phi = \Lambda \setminus \{\alpha_j\}$, for some $\alpha_j \in \Lambda$, and consider the dual vector $H^j \in \mathfrak{a}$ of α_j, defined by $\alpha_i(H^j) = \delta_{ij}$. The subalgebra $\mathfrak{n}_\Phi$ admits a natural gradation $\bigoplus_{\nu\geq 1} \mathfrak{n}_\Phi^\nu$, where $\mathfrak{n}_\Phi^\nu = \bigoplus_{\lambda(H^j)=\nu} \mathfrak{g}_\lambda$. Note that $\lambda(H^j) = \nu$ if and only if λ has coefficient ν in α_j when expressed as a sum of simple roots. Suppose that $\mathfrak{v}$ is a subspace of $\mathfrak{n}_\Phi^1$ of dimension $\dim \mathfrak{v} \geq 2$. Then, $\mathfrak{n}_{\Phi,\mathfrak{v}} = \mathfrak{n}_\Phi \ominus \mathfrak{v}$ is a subalgebra of $\mathfrak{n}_\Phi$. Denote by $N_{\Phi,\mathfrak{v}}$ the corresponding connected Lie subgroup of N_Φ. Assume the following conditions hold:

(NC1) $N_{M_\Phi}(\mathfrak{n}_{\Phi,\mathfrak{v}})$ acts transitively on $B_\Phi = M_\Phi \cdot o$.
(NC2) $N_{K_\Phi}(\mathfrak{n}_{\Phi,\mathfrak{v}}) = N_{K_\Phi}(\mathfrak{v})$ acts transitively on the unit sphere of $\mathfrak{v}$.

Then, the group

$$H_{\Phi,\mathfrak{v}} = N_{L_\Phi}^0(\mathfrak{n}_{\Phi,\mathfrak{v}}) N_{\Phi,\mathfrak{v}} = N_{M_\Phi}^0(\mathfrak{n}_{\Phi,\mathfrak{v}}) A_\Phi N_{\Phi,\mathfrak{v}}$$

acts on M with cohomogeneity one and a singular orbit $H_{\Phi,\mathfrak{v}} \cdot o$. Here, $N^0(\cdot)$ denotes the connected component of the identity of a normalizer. In this case, we say that the action of $H_{\Phi,\mathfrak{v}}$ on M has been obtained by nilpotent construction from the choices of Φ and $\mathfrak{v}$. Moreover, it was proved in [15] that if two subspaces $\mathfrak{v}_1, \mathfrak{v}_2 \subset \mathfrak{n}_\Phi^1$ giving rise to actions by nilpotent construction are conjugate by an element in K_Φ, the actions of the corresponding groups $H_{\Phi,\mathfrak{v}_1}$, $H_{\Phi,\mathfrak{v}_2}$ on M are orbit equivalent (via the same element).

Remark 9 Conditions (NC1)–(NC2) have geometric meaning. Condition (NC1) implies that the orbit $H_{\Phi,\mathfrak{v}} \cdot o$ contains the boundary component B_Φ, and hence its normal space can be identified with $\mathfrak{v}$, i.e., $\nu_o(H_{\Phi,\mathfrak{v}} \cdot o) \cong \mathfrak{v}$. Then, condition (NC2) means that the slice representation of $H_{\Phi,\mathfrak{v}}$ (i.e., the action of the group of differentials of the isometries in $H_{\Phi,\mathfrak{v}}$ on $\nu_o(H_{\Phi,\mathfrak{v}} \cdot o) \cong \mathfrak{v}$) is of cohomogeneity one on the Euclidean space $\nu_o(H_{\Phi,\mathfrak{v}} \cdot o) \cong \mathfrak{v}$, with orbits given by the origin and concentric spheres. Since an isometric action has the same cohomogeneity as its slice representation, we see that both conditions (NC1)–(NC2) imply that $H_{\Phi,\mathfrak{v}}$ acts on M with cohomogeneity one, as claimed above.

Remark 10 Subspaces of $\mathfrak{n}_\Phi^1$ satisfying condition (NC1) (respectively, (NC2)) have been called admissible (resp. protohomogeneous) in [32] and [91]. We observe that if M has rank one, then any proper subset Φ of $\Lambda = \{\alpha_1\}$ is necessarily the empty set, and hence, any proper boundary component is a point. Therefore, the admissibility condition (NC1) is trivially satisfied. Thus, for rank one spaces, the nilpotent construction amounts to the determination of protohomogeneous subspaces. For these spaces, $K_\Phi^0 = K_\emptyset^0 = K_0$, from where one can see that condition (NC2) is equivalent to the definition of protohomogenous subspace given in Sect. 5.

The complete determination of all possible subspaces $\mathfrak{v}$ satisfying conditions (NC1)–(NC2) for a specific symmetric space is usually a very difficult task. Indeed, as commented in Sect. 5, this was even hard in the case of the quaternionic hyperbolic spaces, where condition (NC1) did not play any role.

As before, it is important to determine what happens when one considers an action on a boundary component B_Φ obtained by nilpotent construction, and then one extends it to M. This turns out to be equivalent to an action obtained via nilpotent construction on M. More precisely, let $\alpha_j \in \Phi \subset \Lambda$. Let H_Φ be a subgroup of S_Φ obtained by the nilpotent construction method applied to the symmetric space B_Φ. Then the subgroup H_Φ^Λ of G obtained by canonical extension of the H_Φ-action to M acts on M with the same orbits as the Lie group $H_{\Lambda\setminus\{\alpha_j\},\mathfrak{v}} = N^0_{L_{\Lambda\setminus\{\alpha_j\}}}(\mathfrak{n}_{\Lambda\setminus\{\alpha_j\},\mathfrak{v}})N_{\Lambda\setminus\{\alpha_j\},\mathfrak{v}}$ obtained by nilpotent construction applied to M, for certain subspace $\mathfrak{v}$ of $\mathfrak{n}^1_{\Lambda\setminus\{\alpha_j\}}$ of $\dim \mathfrak{v} \geq 2$. For further details and a proof, see [31, Lemma 4.3].

6.5 *The Classification of Cohomogeneity One Actions*

A general procedure to classify cohomogeneity one actions on a given symmetric space of noncompact type $M \cong G/K$ (not necessarily irreducible) goes as follows. Assume we have a connected Lie subgroup H of G acting on M with cohomogeneity one. If the H-action produces a regular foliation, then the H-action is orbit equivalent to one of the actions described in Sect. 6.1, as explained in that subsection. Thus, let us suppose that the action of H has a singular orbit. The Lie algebra $\mathfrak{h}$ of H is contained in some maximal proper subalgebra $\mathfrak{q}$ of $\mathfrak{g}$. By a result of Mostow [73], there are two possibilities for $\mathfrak{q}$: it is either a maximal proper reductive subalgebra or a maximal proper parabolic subalgebra of $\mathfrak{g}$. Denote by Q the connected subgroup of G with Lie algebra $\mathfrak{q}$. Then:

(a) If $\mathfrak{q}$ is a maximal proper reductive subalgebra of $\mathfrak{g}$, then Q acts with cohomogeneity one and the same orbits as the H-action, one of them being totally geodesic (which is the singular one if M is irreducible and $M \neq \mathbb{R}\mathsf{H}^n$), as shown in [15, Theorem 3.2].
(b) If $\mathfrak{q}$ is a maximal proper parabolic subalgebra of $\mathfrak{g}$, then the H-action is orbit equivalent to an action obtained by canonical extension or by nilpotent construction, as proved by Berndt and Tamaru in [15, Theorem 5.8].

Using this approach, along with a careful analysis of the nilpotent construction, allowed for the classification of the cohomogeneity one actions on several symmetric spaces of noncompact type and rank 2, namely, on

$$\begin{gathered}\mathsf{SL}_3(\mathbb{R})/\mathsf{SO}_3, \quad \mathsf{SL}_3(\mathbb{C})/\mathsf{SU}_3, \quad \mathsf{SL}_3(\mathbb{H})/\mathsf{Sp}_3, \quad \mathsf{SO}_5(\mathbb{C})/\mathsf{SO}_5, \\ \mathsf{G}_2^2/\mathsf{SO}_4, \quad \mathsf{G}_2^{\mathbb{C}}/\mathsf{G}_2, \quad \mathsf{SO}^0_{2,n}/\mathsf{SO}_2\mathsf{SO}_n, \quad \mathsf{SU}_{2,n}/\mathsf{S}(\mathsf{U}_2\mathsf{U}_n).\end{gathered} \tag{4}$$

These classifications were obtained in the series of papers [8, 15, 91].

When trying to implement this approach in spaces of rank greater than 2, it turns out that one can apply a rank reduction procedure. Roughly speaking, if the H-action is orbit equivalent to the canonical extension of some action on a boundary component, we can apply the same procedure as before recursively until we get to an action that can no longer be retrieved by canonical extension. Thus, every cohomogeneity one action with a singular orbit can ultimately be obtained by nilpotent construction or by extending an action of cohomogeneity one with a totally geodesic singular orbit on a boundary component of M. In the latter case, as follows from the discussion in Sect. 6.2, the action being extended is of one of the following two types:

(i) A cohomogeneity one action with a totally geodesic singular orbit on an irreducible boundary component B_Φ of M and hence orbit equivalent to one of the actions classified in [13] in terms of certain reflective submanifolds and some exceptions related to G_2.
(ii) A cohomogeneity one action with a diagonal totally geodesic submanifold on a reducible boundary component $B_{\{\alpha_j,\alpha_k\}} \simeq B_{\{\alpha_j\}} \times B_{\{\alpha_k\}} \cong \mathbb{F}\mathsf{H}^n \times \mathbb{F}\mathsf{H}^n$, given by a connected Lie group with Lie algebra $\mathfrak{s}_{j,k,\tau} = \{X + \tau X : X \in \mathfrak{s}_{\{\alpha_j\}}\}$, where $\tau\colon \mathfrak{s}_{\{\alpha_j\}} \to \mathfrak{s}_{\{\alpha_k\}}$ is a Lie algebra isomorphism between the isometry Lie algebras of both factors of $B_{\{\alpha_j,\alpha_k\}}$.

As a consequence of all the facts sketched above in this section, we have recently obtained the following structural result in [31, Theorem A].

Theorem 10 (Cohomogeneity One Actions on Symmetric Spaces of Noncompact Type) *Let $M \cong G/K$ be a symmetric space of noncompact type, and let H be a connected closed subgroup of G. Then, H acts on M with cohomogeneity one if and only if the H-action is orbit equivalent to one of the following:*

(FH) *An action inducing a regular codimension one foliation of horospherical type*

(FS) *An action inducing a regular codimension one foliation of solvable type*

(CEI) *The canonical extension of a cohomogeneity one action with a totally geodesic singular orbit on an irreducible boundary component*

(CER) *The canonical extension of a cohomogeneity one diagonal action on a reducible boundary component of rank two with two homothetic factors*

(NC) *An action obtained by nilpotent construction*

Remark 11 Cases (CEI) and (NC) in the previous theorem may overlap. Indeed, the nilpotent construction method often produces actions that can be obtained by canonical extension. So far, the only spaces where the nilpotent construction is known to produce actions that cannot be obtained by any other methods are the hyperbolic spaces of nonconstant curvature, $\mathsf{G}_2^2/\mathsf{SO}_4$ and $\mathsf{G}_2^{\mathbb{C}}/\mathsf{G}_2$.

Remark 12 Although the moduli space of cohomogeneity one actions producing regular foliations has been completely determined (see Sect. 6.1), the study of the moduli space of actions with a singular orbit is much more involved. Note that in Sect. 6.3 we have only stated sufficient conditions for two canonical extensions to be orbit equivalent on M. Despite two actions not being orbit equivalent on a boundary component, it could happen that their canonical extensions could be orbit equivalent. It may also happen that two orbit equivalent actions could produce canonical extensions which are not orbit equivalent in M (if the equivalence in the boundary component B_Φ had been obtained by an isometry in $\mathrm{Isom}(B_\Phi) \setminus \mathrm{Isom}(B_\Phi)^0$). Thus, determining the orbit equivalence classes involves additional difficulties (see Problem 3 in Sect. 7).

As an application of Theorem 10, we derived in [31] the classification of cohomogeneity one actions on the family of spaces $\mathsf{SL}_{n+1}(\mathbb{R})/\mathsf{SO}_{n+1}$. We recall that $\mathsf{SL}_{n+1}(\mathbb{R})/\mathsf{SO}_{n+1}$ has rank n. The associated root space decomposition of $\mathfrak{g} = \mathfrak{sl}_{n+1}(\mathbb{R})$ satisfies $\mathfrak{g}_0 = \mathfrak{a}$ and $\dim \mathfrak{g}_\lambda = 1$, for any root $\lambda \in \Sigma$.

Theorem 11 (Cohomogeneity One Actions on $\mathsf{SL}_{n+1}(\mathbb{R})/\mathsf{SO}_{n+1}$) *Let $M \cong \mathsf{SL}_{n+1}(\mathbb{R})/\mathsf{SO}_{n+1}$, $n \geq 1$, and let $\Lambda = \{\alpha_1, \ldots, \alpha_n\}$ be a set of simple roots for $\mathfrak{sl}_{n+1}(\mathbb{R})$ whose Dynkin diagram is*

Table 7 Actions on $\mathsf{SL}_{n+1}(\mathbb{R})/\mathsf{SO}_{n+1}$ obtained by canonical extension

$\mathfrak{h}_\Phi$	Φ	B_Φ	$\operatorname{codim}(H_\Phi^\Lambda \cdot o)$	Comments
$\mathfrak{k}_{\{\alpha_j\}} \cong \mathfrak{so}_2$	$\{\alpha_j\}$	$\mathbb{R}\mathsf{H}^2$	2	$1 \le j \le n$
$\mathfrak{sl}_{k-j+1}(\mathbb{R}) \oplus \mathbb{R}$	$\{\alpha_j, \ldots, \alpha_k\}$	$\mathsf{SL}_{k-j+2}(\mathbb{R})/\mathsf{SO}_{k-j+2}$	$k-j+1$	$1 \le j < k \le n$
$\mathfrak{sp}_2(\mathbb{R})$	$\{\alpha_j, \alpha_{j+1}, \alpha_{j+2}\}$	$\mathsf{SL}_4(\mathbb{R})/\mathsf{SO}_4$	3	$1 \le j \le n-2$
$\mathfrak{s}_{j,k,\tau} \cong \mathfrak{sl}_2(\mathbb{R})$	$\{\alpha_j, \alpha_k\}$	$\mathbb{R}\mathsf{H}^2 \times \mathbb{R}\mathsf{H}^2$	2	$\lvert k-j\rvert > 1$

Any cohomogeneity one action on M is orbit equivalent to one of the following:

(FH) *The action of the connected subgroup of $\mathsf{SL}_{n+1}(\mathbb{R})$ with Lie algebra $(\mathfrak{a} \ominus \ell) \oplus \mathfrak{n}$, for some line ℓ of $\mathfrak{a}$.*

(FS) *The action of the connected subgroup of $\mathsf{SL}_{n+1}(\mathbb{R})$ with Lie algebra $\mathfrak{a} \oplus (\mathfrak{n} \ominus \mathfrak{g}_{\alpha_j})$, for some simple root $\alpha_j \in \Lambda$.*

(CE) *The canonical extension H_Φ^Λ of the action of the connected subgroup H_Φ of $\mathsf{SL}_{n+1}(\mathbb{R})$ on a boundary component B_Φ, for one of the cases in Table 7.*

Theorem 10 can also be used to address the classification problem on reducible symmetric spaces by allowing us to restrict our analysis to the classification problem on each irreducible factor. It turns out that actions of the types (FS), (CEI), and, importantly, (NC) split well with respect to the de Rham decomposition of a reducible symmetric space, so they are product actions. We emphasize that a result analogous to Theorem 12 below is not yet known for compact symmetric spaces (see Problem 5 in Sect. 7).

Theorem 12 (Cohomogeneity One Actions on Reducible Symmetric Spaces of Noncompact Type) *Let M be a symmetric space of noncompact type with de Rham decomposition $M = M_1 \times \cdots \times M_s$, where $M_i = G_i/K_i$, $i = 1, \ldots, s$, and let $G = \prod_{i=1}^s G_i$. Then, a cohomogeneity one action on M is orbit equivalent to one of the following:*

(Prod) *The product action of a subgroup $H_j \times \prod_{\substack{i=1 \\ i \neq j}}^s G_i$ of G, where H_j is a connected Lie subgroup of G_j that acts with cohomogeneity one on the irreducible factor M_j.*

(FH) *The action of the connected subgroup of G with Lie algebra $\mathfrak{h} = (\mathfrak{a} \ominus \ell) \oplus \mathfrak{n}$, for some line ℓ of $\mathfrak{a}$.*

(CER) *The canonical extension of a cohomogeneity one diagonal action on a reducible boundary component of M of rank two with two homothetic factors.*

Theorem 12 can be applied to derive explicit classifications on any product of symmetric spaces of noncompact type for which we already have the complete list of cohomogeneity one actions (namely, all rank one spaces studied in Sect. 5, the rank two spaces in (4), and the spaces $\mathsf{SL}_{n+1}(\mathbb{R})/\mathsf{SO}_{n+1}$). As a very particular instance of these possible applications, we state the following classification of homogeneous hypersurfaces on any finite product of real hyperbolic spaces:

Theorem 13 (Homogeneous Hypersurfaces in Products of Real Hyperbolic Spaces) *A homogeneous hypersurface of* $M = \mathbb{R}\mathsf{H}^{n_1} \times \cdots \times \mathbb{R}\mathsf{H}^{n_r}$ *is congruent to one of the following:*

(FH) *A leaf of a regular codimension one foliation of horospherical type*
(FS) *An extrinsic product* $\mathbb{R}\mathsf{H}^{n_j-1} \times \prod_{i \neq j} \mathbb{R}\mathsf{H}^{n_i}$, *where* $\mathbb{R}\mathsf{H}^{n_j-1}$ *is totally geodesic in* $\mathbb{R}\mathsf{H}^{n_j}$, *or one of its equidistant hypersurfaces*
(CEI) *A tube around the extrinsic product* $\mathbb{R}\mathsf{H}^k \times \prod_{i \neq j} \mathbb{R}\mathsf{H}^{n_i}$, *where* $\mathbb{R}\mathsf{H}^k$ *is totally geodesic in* $\mathbb{R}\mathsf{H}^{n_j}$, *for some* $k \in \{0, \ldots, n_j - 2\}$
(CER) *A tube around the extrinsic product* $\Delta\mathbb{R}\mathsf{H}^{n_j} \times \prod_{i \neq j,k} \mathbb{R}\mathsf{H}^{n_i}$, *where*

$$\Delta\mathbb{R}\mathsf{H}^{n_j} = \{(p, \varphi(p)) : p \in \mathbb{R}\mathsf{H}^{n_j}\}$$

is a totally geodesic real hyperbolic space diagonally embedded in $\mathbb{R}\mathsf{H}^{n_j} \times \mathbb{R}\mathsf{H}^{n_k}$, *for two indices* j, k *with* $n_j = n_k$ *and where* φ *is a homothety between* $\mathbb{R}\mathsf{H}^{n_j}$ *and* $\mathbb{R}\mathsf{H}^{n_k}$

Note that even in the simplest case of a product of two hyperbolic planes, $M = \mathbb{R}\mathsf{H}^2 \times \mathbb{R}\mathsf{H}^2$, the classification of homogeneous hypersurfaces did not seem to be previously known (see [46] for a recent alternative approach via isoparametric hypersurfaces). In this particular case, there are uncountably many cohomogeneity one actions up to orbit equivalence, due to the existence of actions of horospherical type, which are determined by the choice of a line ℓ in the 2-dimensional space $\mathfrak{a}$. Apart from these, there are exactly other three cohomogeneity one actions (up to orbit equivalence) if both factors of M are isometric and exactly five actions otherwise:

(FS) Two of them producing foliations with the totally geodesic codimension one leaf $\mathbb{R}\mathsf{H}^1 \times \mathbb{R}\mathsf{H}^2$ or $\mathbb{R}\mathsf{H}^2 \times \mathbb{R}\mathsf{H}^1$, respectively (being both orbit equivalent if and only if both factors of M are isometric)
(CEI) Other two with the totally geodesic singular orbits $\{o_1\} \times \mathbb{R}\mathsf{H}^2$ or $\mathbb{R}\mathsf{H}^2 \times \{o_2\}$, respectively (again, both orbit equivalent when both factors are isometric)
(CER) The diagonal action of $\mathsf{SO}^0_{1,2} \cong \mathsf{SL}_2(\mathbb{R})$ on M, which has a diagonal totally geodesic $\mathbb{R}\mathsf{H}^2$ as singular orbit

It is interesting to compare this result with the situation in the compact dual of M, namely, the product of two round spheres $\mathbb{S}^2 \times \mathbb{S}^2$. Here, by a result of Urbano [100] (who actually classified isoparametric hypersurfaces in this space), the only homogeneous hypersurfaces are dual analogs to the examples (CEI) and (CER) above. Again, it is important to recall that the generalization of Urbano's classification of homogeneous hypersurfaces for products of several spheres of higher dimensions (i.e., the compact analog of Theorem 13) is still outstanding.

7 Open Problems

We include a list of open problems and questions related to the investigation of homogeneous hypersurfaces in symmetric spaces:

(1) Analyze the nilpotent construction for each symmetric space of noncompact type. This, along with the structural result in Theorem 10, would allow to complete the classification of cohomogeneity one actions in this setting. Due to the difficulty of this problem, we can distinguish two main cases:

   (a) For spaces whose isometry group is a split semisimple Lie group, we expect that the nilpotent construction leads to various linear algebraic problems (each one depending on a certain class of representations) whose solution may be achieved following the lines of the analogous problem for $\mathsf{SL}_n(\mathbb{R})/\mathsf{SO}_n$.
   (b) For the remaining spaces, the linear algebraic problems involved are more complicated, but we expect that the combination of Solonenko's ideas in [91] with the ones used for $\mathsf{SL}_n(\mathbb{R})/\mathsf{SO}_n$ in [31] may eventually lead to a complete classification.

(2) Is there any cohomogeneity one action on a symmetric space of noncompact type, rank at least 2, and of non-(G_2)-type that can be obtained by nilpotent construction but not as a canonical extension?
(3) Investigate the congruence problem of homogeneous hypersurfaces, or, equivalently, determine when two cohomogeneity one actions are orbit equivalent. Whereas for actions of foliation type this problem has already been solved in [92], an eventual positive answer to Question (2) would need a specific (but probably easy) investigation. However, the analysis of the other types of actions seems more difficult. In particular, one would need to address the following issues:

   (a) Given a rank two reducible boundary component $B_\Phi \cong \mathbb{F}\mathsf{H}^n \times \mathbb{F}\mathsf{H}^n$ of M, determine when two different isomorphisms τ_1 and τ_2 between the isometry Lie algebras of the two homothetic factors $\mathbb{F}\mathsf{H}^n$ give rise to orbit equivalent canonical extensions of type (CER).
   (b) Can two orbit equivalent cohomogeneity one actions with totally geodesic singular orbits on an irreducible boundary component produce non-orbit equivalent canonical extensions of type (CEI)? If the answer is affirmative, one would probably have to revisit Berndt and Tamaru's classification of cohomogeneity one actions with totally geodesic singular orbits [13] in order to determine the moduli space of actions up to strong orbit equivalence (i.e., up to orbit equivalence by isometries in the connected component of the identity of the isometry group). This may entail an analysis of a strong congruence problem of Leung's classification of reflective submanifolds [66].

(4) Determine the extrinsic geometry of homogeneous hypersurfaces of symmetric spaces of noncompact type. As an application, one may obtain the classification of homogeneous minimal hypersurfaces.
(5) Leaving aside the noncompact setting, classify cohomogeneity one actions on reducible symmetric spaces of compact type. See [63] for information on this problem.
(6) Initiate the study of homogeneous hypersurfaces of locally symmetric spaces, both of compact and noncompact types.
(7) Derive structure results for cohomogeneity one actions on symmetric spaces of mixed type, including noncompact spaces with Euclidean factors (e.g., $\mathsf{GL}_n^+(\mathbb{R})/\mathsf{SO}_n$).
(8) Obtain characterizations of (certain families of) homogeneous hypersurfaces by (both extrinsic and intrinsic) geometric properties, such as isoparametricity, constancy of principal curvatures, curvature adaptedness, or having an Einstein or Ricci soliton induced metric, cf. [42, 75]. Also, obtaining characterizations of the inhomogeneous isoparametric examples known in most symmetric spaces would be very interesting, in that this would probably entail the introduction of new techniques in submanifold geometry of symmetric spaces. Specifically, although isoparametric hypersurfaces of a product of two real hyperbolic planes turn out to be (open subsets of) homogeneous hypersurfaces [46], for a product of three hyperbolic planes, we know the existence of inhomogeneous examples [41].
(9) In this survey, we assumed actions to be proper and homogeneous submanifolds to be closed and embedded. Under which circumstances and for which ambient spaces can one guarantee that nonproper cohomogeneity one actions have the same orbits as proper cohomogeneity one actions? Can one prove that on a simply connected ambient space there do not exist nonembedded or nonclosed homogeneous hypersurfaces?

Acknowledgments The authors have been supported by the projects PID2019-105138GB-C21/AEI/10.13039/501100011033 (Spain) and ED431F 2020/04 (Xunta de Galicia, Spain). The second and third authors acknowledge support of the Ramón y Cajal program (AEI, Spain) and the FPU program (Ministry of Universities, Spain), respectively.

References

1. Alexandrino, M.M., Bettiol, R.: Lie Groups and Geometric Aspects of Isometric Actions. Springer, Cham (2015)
2. Atiyah M., Berndt, J.: Projective planes, Severi varieties and spheres. In: Surveys in Differential Geometry, vol. VIII, pp. 1–27. Boston, MA (2002); Surv. Differ. Geom., 8, Int. Press, Somerville, MA (2003)
3. Bérard Bergery, L.: Sur de nouvelles variétés riemanniennes d'Einstein. Publ. Inst. É. Cartan **4**, 1–60 (1982)

4. Berndt, J.: Hyperpolar homogeneous foliations on symmetric spaces of noncompact type. In: Proceedings of the 13th International Workshop on Differential Geometry and Related Fields, vol. 13, pp. 37–57. Natl. Inst. Math. Sci. (NIMS), Taejŏn (2009)
5. Berndt, J., Brück, M.: Cohomogeneity one actions on hyperbolic spaces. J. Reine Angew. Math. **541**, 209–235 (2001)
6. Berndt, J., Console, S., Olmos, C.: Submanifolds and Holonomy, 2nd edn. Monographs and Research Notes in Mathematics. CRC Press, Boca Raton, FL (2016)
7. Berndt, J., Díaz-Ramos, J. C., Tamaru, H.: Hyperpolar homogeneous foliations on symmetric spaces of noncompact type. J. Differ. Geom. **86**, 191–235 (2010)
8. Berndt, J., Domínguez-Vázquez, M.: Cohomogeneity one actions on some noncompact symmetric spaces of rank two. Transform. Groups **20**(4), 921–938 (2015)
9. Berndt, J., Olmos, C.: Maximal totally geodesic submanifolds and index of symmetric spaces. J. Differ. Geom. **104**(2), 187–217 (2016)
10. Berndt, J., Olmos, C.: The index conjecture for symmetric spaces. J. Reine Angew. Math. **772**, 187–222 (2021)
11. Berndt, J., Sanmartín-López, V.: Submanifolds with constant principal curvatures in Riemannian symmetric spaces. Commun. Anal. Geom., to appear
12. Berndt, J., Tamaru, H.: Homogeneous codimension one foliations on noncompact symmetric spaces. J. Differ. Geom. **63**, 1–40 (2003)
13. Berndt, J., Tamaru, H.: Cohomogeneity one actions on noncompact symmetric spaces with a totally geodesic singular orbit. Tohoku Math. J. (2) **56**, 163–177 (2004)
14. Berndt, J., Tamaru, H.: Cohomogeneity one actions on noncompact symmetric spaces of rank one. Trans. Am. Math. Soc. **359**(7), 3425–3438 (2007)
15. Berndt, J., Tamaru, H.: Cohomogeneity one actions on symmetric spaces of noncompact type. J. Reine Angew. Math. **683**, 129–159 (2013)
16. Berndt, J., Tricerri, F., Vanhecke, L.: Generalized Heisenberg groups and Damek-Ricci harmonic spaces. Lecture Notes in Mathematics, vol. 1598. Springer, New York (1995)
17. Besse, A. L.: Einstein manifolds. Reprint of the 1987 edition. Classics in Mathematics. Springer, Berlin (2008)
18. Borel, A., Ji, L.: Compactifications of symmetric and locally symmetric spaces. Mathematics: Theory & Applications. Birkhäuser Boston, Boston, MA (2006)
19. Borel, A., Tits, J.: Groupes réductifs. Publ. Sci. IHES **27**, 55–150 (1965)
20. Bredon, G. E.: Introduction to compact transformation groups. Pure and Applied Mathematics, vol. 46. Academic Press, New York-London (1972)
21. Cartan, É.: Sur une classe remarquable d'espaces de Riemann. Bull. Soc. Math. France **54**, 214–264 (1926)
22. Cartan, É.: Familles de surfaces isoparamétriques dans les espaces à courbure constante. Ann. Mat. Pura Appl. **17**, 177–191 (1938)
23. Cartan, É.: Sur des familles remarquables d'hypersurfaces isoparamétriques dans les espaces sphériques. Math. Z. **45**, 334–367 (1939)
24. Cartan, É.: Sur des familles d'hypersurfaces isoparamétriques des espaces sphériques á 5 et á 9 dimensions. Revista Univ. Tucumán **1**, 5–22 (1940)
25. Cecil, T., Ryan, P.: Geometry of hypersurfaces. Springer Monographs in Mathematics. Springer, New York (2015)
26. Chi, Q.-S.: The isoparametric story, a heritage of Élie Cartan. In: Proceedings of the International Consortium of Chinese Mathematicians 2018, pp. 197–260. Int. Press, Boston, MA (2020)
27. Chi, Q.-S.: Isoparametric hypersurfaces with four principal curvatures, IV. J. Differ. Geom. **115**(2), 225–301 (2020)
28. D'Atri, J. E.: Certain isoparametric families of hypersurfaces in symmetric spaces. J. Differ. Geom. **14**(1), 21–40 (1979)
29. Díaz-Ramos, J. C., Domínguez-Vázquez, M.: Inhomogeneous isoparametric hypersurfaces in complex hyperbolic spaces. Math. Z. **271**, 1037–1042 (2012)

30. Díaz-Ramos, J. C., Domínguez-Vázquez, M.: Isoparametric hypersurfaces in Damek-Ricci spaces. Adv. Math. **239**, 1–17 (2013)
31. Díaz-Ramos, J. C, Domínguez-Vázquez, M., Otero, T.: Cohomogeneity one actions on symmetric spaces of noncompact type and higher rank. Adv. Math. **428**, 109165 (2023)
32. Díaz-Ramos, J. C., Domínguez-Vázquez, M., Rodríguez-Vázquez, A.: Homogeneous and inhomogeneous isoparametric hypersurfaces in rank one symmetric spaces. J. Reine Angew. Math. **779**, 189–222 (2021)
33. Díaz-Ramos, J. C., Domínguez-Vázquez, M., Sanmartín-López, V.: Isoparametric hypersurfaces in complex hyperbolic spaces. Adv. Math. **314**, 756–805 (2017)
34. Díaz-Ramos, J. C., Domínguez-Vázquez, M., Sanmartín-López, V.: Submanifold geometry in symmetric spaces of noncompact type. São Paulo J. Math. Sci. **15**(1), 75–110 (2021)
35. Domínguez-Vázquez, M.: Canonical extension of submanifolds in noncompact symmetric spaces. Int. Math. Res. Not. (IMRN) **2015**(22), 12114–12125 (2015)
36. Domínguez-Vázquez, M.: Isoparametric foliations on complex projective spaces. Trans. Am. Math. Soc. **368**(2), 1211–1249 (2016)
37. Domínguez-Vázquez, M., Enciso, A., Peralta-Salas, D.: Solutions to the overdetermined boundary problem for semilinear equations with position-dependent nonlinearities. Adv. Math. **351**, 718–760 (2019)
38. Domínguez-Vázquez, M., González-Álvaro, D., Mouillé, L.: Infinite families of manifolds of positive kth-intermediate Ricci curvature with k small. Math. Ann. **386**, 1979–2014 (2023)
39. Domínguez-Vázquez, M., Gorodski, C.: Polar foliations on quaternionic projective spaces. Tohoku Math. J. (2) **70**(3), 353–375 (2018)
40. Domínguez-Vázquez, M., Manzano, J. M.: Isoparametric surfaces in $\mathbb{E}(\kappa, \tau)$-spaces. Ann. Sc. Norm. Super. Pisa Cl. Sci. (5) **22**, 269–285 (2021)
41. Domínguez-Vázquez, M., Sanmartín-López, V.: Isoparametric hypersurfaces in symmetric spaces of non-compact type and higher rank. arXiv:2109.03850v2
42. Domínguez-Vázquez, M., Sanmartín-López, V., Tamaru, H.: Codimension one Ricci soliton subgroups of solvable Iwasawa groups. J. Math. Pures Appl. **152**, 69–93 (2021)
43. Dynkin, E. B.: Semisimple subalgebras of the semisimple Lie algebras. Am. Math. Soc. Transl. Ser. 2 **6**, 111–244 (1952)
44. Eberlein, P. B.: Geometry of nonpositively curved manifolds. Chicago Lectures in Mathematics. University of Chicago Press, Chicago, IL (1996)
45. Eschenburg, J.-H.: Lectures on symmetric spaces, http://myweb.rz.uni-augsburg.de/~eschenbu/symspace.pdf. Cited 26 April 2023
46. Gao, D., Ma, H., Yao, Z.: On Hypersurfaces of $\mathbb{H}^2 \times \mathbb{H}^2$. arXiv:2209.10467
47. Ge, J., Radeschi, M.: Differentiable classification of 4-manifolds with singular Riemannian foliations. Math. Ann. **363**(1-2), 525–548 (2015)
48. Ge, J., Tang, Z.: Isoparametric functions and exotic spheres. J. Reine Angew. Math. **683**, 161–180 (2013)
49. Ge, J., Tang, Z.: Geometry of isoparametric hypersurfaces in Riemannian manifolds. Asian J. Math. **18**(1), 117–125 (2014)
50. Grove, K., Wilking, B., Ziller, W.: Positively curved cohomogeneity one manifolds and 3-Sasakian geometry. J. Differ. Geom. **78**(1), 33–111 (2008)
51. Harvey, F.R.: Spinors and Calibrations. Academic Press, New York (1990)
52. Harvey, F.R., Lawson, H.B., Jr.: Calibrated geometries. Acta Math. **148**, 47–157 (1982)
53. Helgason, S.: Differential geometry, Lie groups, and symmetric spaces. Corrected reprint of the 1978 original. Graduate Studies in Mathematics, vol. 34. American Mathematical Society, Providence, RI (2001)
54. Helgason, S.: Geometric Analysis on Symmetric Spaces, 2nd edn. Mathematical Surveys and Monographs, vol. 39. American Mathematical Society, Providence, RI (2008)
55. Hsiang, W.-Y., Lawson, H.B., Jr.: Minimal submanifolds of low cohomogeneity. J. Differ. Geom. **5**, 1–38 (1971)
56. Iwata, K.: Classification of compact transformation groups on cohomology quaternion projective spaces with codimension one orbits. Osaka Math. J. **15**(3), 475–508 (1978)

57. Iwata, K.: Compact transformation groups on rational cohomology Cayley projective planes. Tohoku Math. J. (2) **33**(4), 429–442 (1981)
58. Ji, L.: Lectures on locally symmetric spaces and arithmetic groups. Lie groups and automorphic forms, pp. 87–146. AMS/IP Stud. Adv. Math., vol. 37. Amer. Math. Soc., Providence, RI (2006)
59. Kawohl, B.: Symmetrization—or how to prove symmetry of solutions to a PDE. In: Partial Differential Equations (Praha, 1998), pp. 214–229. Chapman & Hall/CRC Res. Notes Math., vol. 406. Chapman & Hall/CRC, Boca Raton, FL (2000)
60. Knapp, A.W.: Lie Groups Beyond an Introduction, 2nd edn. Progress in Mathematics, vol .140. Birkhäuser Boston, Inc., Boston, MA (2002)
61. Kollross, A.: A classification of hyperpolar and cohomogeneity one actions. Trans. Am. Math. Soc. **354**(2), 571–612 (2002)
62. Kollross, A.: Duality of symmetric spaces and polar actions. J. Lie Theory **21**(4), 961–986 (2011)
63. Kollross, A.: Hyperpolar actions on reducible symmetric spaces. Transform. Groups **22**(1), 207–228 (2017)
64. Kollross, A., Rodríguez-Vázquez, A.: Totally geodesic submanifolds in exceptional symmetric spaces. Adv. Math. **418**, 108949 (2023)
65. Lee, J.M.: Introduction to Smooth Manifolds, 2nd edn. Graduate Texts in Mathematics, vol. 218. Springer, New York (2013)
66. Leung, D.S.P.: Reflective submanifolds, III. Congruency of isometric reflective submanifolds and corrigenda to the classification of reflective submanifolds. J. Differ. Geom. **14**(2), 167–177 (1979)
67. Loos, O.: Symmetric Spaces. I: General Theory. W. A. Benjamin, Inc., New York-Amsterdam (1969)
68. Loos, O.: Symmetric Spaces. II: Compact Spaces and Classification. W. A. Benjamin, Inc., New York-Amsterdam (1969)
69. Lorenzo-Naveiro, J. M.: Grupos de transformaciones. Bachelor's thesis, Universidade de Santiago de Compostela (2020)
70. Magnanini, R., Peralta-Salas, D., Sakaguchi, S.: Stationary isothermic surfaces in Euclidean 3-space. Math. Ann. **364**(1-2), 97–124 (2016)
71. Meeks, W.H., III, Mira, P., Pérez, J., Ros, A.: Constant mean curvature spheres in homogeneous three-manifolds. Invent. Math. **224**(1), 147–244 (2021)
72. Michor, P. W.: Topics in differential geometry. Graduate Studies in Mathematics, vol. 93. American Mathematical Society, Providence, RI (2008)
73. Mostow, G.D.: On maximal subgroups of real Lie groups. Ann. Math. (2) **74**, 503–517 (1961)
74. Münzner, H. F.: Isoparametrische Hyperflächen in Sphären. Math. Ann. **251**(1), 57–71 (1980)
75. Nikolayevsky, Y., Park, J.H.: Einstein hypersurfaces in irreducible symmetric spaces. Annali di Matematica **202**, 1719–1751 (2023)
76. O'Neill, B.: Semi-Riemannian geometry. With applications to relativity. Pure and Applied Mathematics, vol. 103. Academic Press, New York (1983)
77. Pelayo, A., Peralta-Salas, D.: A geometric approach to the classification of the equilibrium shapes of self-gravitating fluids. Commun. Math. Phys. **267**(1), 93–115 (2006)
78. Peñate-Moreno, F.: Teoremas de Cartan y Münzner para hipersuperficies isoparamétricas en esferas. Master's thesis, Universidade de Santiago de Compostela (2021)
79. Podestà, F., Thorbergsson, G.: Polar actions on rank-one symmetric spaces. J. Differ. Geom. **53**(1), 131–175 (1999)
80. Qian, C., Tang, Z.: Isoparametric functions on exotic spheres. Adv. Math. **272**, 611–629 (2015)
81. Rodríguez-Vázquez, A.: A nonisoparametric hypersurface with constant principal curvatures. Proc. Am. Math. Soc. **147**, 5417–5420 (2019)
82. Sakaguchi, S.: When are the spatial level surfaces of solutions of diffusion equations invariant with respect to the time variable? J. Anal. Math. **78**, 219–243 (1999)

83. Savo, A.: Heat flow, heat content and the isoparametric property. Math. Ann. **366**(3–4), 1089–1136 (2016)
84. Savo, A.: Geometric rigidity of constant heat flow. Calc. Var. Part. Differ. Equ. **57**(6), Paper No. 156, 26 pp. (2018)
85. Savo, A.: On the heat content functional and its critical domains. Calc. Var. Part. Differ. Equ. **60**(5), Paper No. 167, 28 pp. (2021)
86. Scholze, P.: p-adic geometry. In: Proceedings of the International Congress of Mathematicians—Rio de Janeiro 2018, vol. I. Plenary Lectures, pp. 899–933. World Sci. Publ., Hackensack, NJ (2018)
87. Segre, B.: Famiglie di ipersuperficie isoparametriche negli spazi euclidei ad un qualunque numero di dimensioni. Atti Accad. Naz. Lincei Rend. Cl. Sci. Fis. Mat. Natur. **27**, 203–207 (1938)
88. Serrin, J.: A symmetry problem in potential theory. Arch. Ration. Mech. Anal. **43**, 304–318 (1971)
89. Shklover, V.E.: Schiffer problem and isoparametric hypersurfaces. Rev. Mat. Iberoamericana **16**(3), 529–569 (2000)
90. Siffert, A.: A new structural approach to isoparametric hypersurfaces in spheres. Ann. Global Anal. Geom. **52**(4), 425–456 (2017)
91. Solonenko, I.: Classification of homogeneous hypersurfaces in some noncompact symmetric spaces of rank two. Annali di Matematica (2023). https://doi.org/10.1007/s10231-023-01345-8
92. Solonenko, I.: Homogeneous codimension one foliations on reducible symmetric spaces of noncompact type. arXiv:2112.02189
93. Spivak, M.: A Comprehensive Introduction to Differential Geometry, vol. IV, 2nd edn. Publish or Perish, Inc., Wilmington, DE (1979)
94. Szabó, Z.I.: A short topological proof for the symmetry of 2 point homogeneous spaces. Invent. Math. **106**(1), 61–64 (1991)
95. Takagi, R., Takahashi, T.: On the Principal Curvatures of Homogeneous Hypersurfaces in a Sphere. Differential Geometry (in Honor of Kentaro Yano), pp. 469–481. Kinokuniya, Tokyo (1972)
96. Takagi, R.: On homogeneous real hypersurfaces in a complex projective space. Osaka Math. J. **10**, 495–506 (1973)
97. Tamaru, H.: Parabolic subgroups of semisimple Lie groups and Einstein solvmanifolds. Math. Ann. **351**(1), 51–66 (2011)
98. Thorbergsson, G.: A survey on isoparametric hypersurfaces and their generalizations. Handbook of Differential Geometry, vol. I, pp. 963–995. North-Holland, Amsterdam (2000)
99. Thorbergsson, G.: From isoparametric submanifolds to polar foliations. São Paulo J. Math. Sci. **16**, 459–472 (2022)
100. Urbano, F.: On hypersurfaces of $\mathbb{S}^2 \times \mathbb{S}^2$. Commun. Anal. Geom. **27**(6), 1381–1416 (2019)
101. Venkatesh, A.: Cohomology of arithmetic groups–Fields Medal lecture. In: Proceedings of the International Congress of Mathematicians–Rio de Janeiro 2018, vol. I. Plenary Lectures, pp. 267–300. World Sci. Publ., Hackensack, NJ (2018)
102. Wang, Q. M.: Isoparametric functions on Riemannian manifolds, I. Math. Ann. **277**, 639–646 (1987)
103. Wolf, J. A.: Spaces of Constant Curvature, 6th edn. AMS Chelsea Publishing, Providence, RI (2011)
104. Wu, B.: Isoparametric submanifolds of hyperbolic spaces. Trans. Am. Math. Soc. **331**(2), 609–626 (1992)
105. Ziller, W.: Lie groups, representation theory and symmetric spaces. https://www.math.upenn.edu/~wziller/math650/LieGroupsReps.pdf. Cited 26 April 2023

First Dirichlet Eigenvalue and Exit Time Moments: A Survey

Vicent Gimeno and Ana Hurtado

Abstract In this paper we summarize recent results concerning the connection between the L^1-moment spectrum associated with a domain D in a Riemannian manifold and its Dirichlet spectrum. In particular, we will expose how to obtain estimations of the first Dirichlet eigenvalue of D in a Riemannian manifold with controlled geometry (or in a submanifold of it) via the study of the so-called moment spectrum.

Keywords First Dirichlet eigenvalue · Exit time moment spectrum · Laplace operator · Geodesic ball

1 Introduction

In Riemannian geometry, the study of partial differential equations and differential operators on Riemannian manifolds plays a central role. One of the most celebrated differential operator, used to build up partial differential equations, is the Laplacian. Given a complete Riemannian manifold (M, g), the Laplacian of a C^2-function $f : M \to \mathbb{R}$ is given in local coordinates $(x^1, \cdots, x^n)$ as

$$\Delta f = \frac{1}{\sqrt{\det g}} \sum_{i,j=1}^{n} \frac{\partial}{\partial x^i} \left(\sqrt{\det g}\, g^{ij} \frac{\partial f}{\partial x^j} \right).$$

V. Gimeno (✉)
Departament de Matemàtiques - Intitut de Matemàtiques i Aplicacions de Castelló, Universitat Jaume I, Castelló de la Plana, Spain
e-mail: gimenov@uji.es

A. Hurtado
Departamento de Geometría y Topología and Excellence Research Unit "Modeling Nature" (MNat), Universidad de Granada, Granada, Spain
e-mail: ahurtado@ugr.es

A. Alarcón et al. (eds.), *New Trends in Geometric Analysis*,
RSME Springer Series 10, https://doi.org/10.1007/978-3-031-39916-9_6

With the Laplacian, the classical heat equation, for instance, can be formulated in a complete Riemannian manifold. In particular it can be studied the problem of finding a heat diffusion function $v : [0, T) \times M \to \mathbb{R}$ solution to the following problem:

$$\begin{cases} \frac{\partial v}{\partial t} &= \Delta v, \\ v(0, x) &= v_0(x), \end{cases} \tag{1}$$

where v_0 is a given real function defined on M. It is well-known (see [19]) that when v_0 is a bounded positive continuous function, Eq. (1) has a solution given by

$$v(x, t) = \int_M p_t(x, y) v_0(y) dV(y),$$

where $p_t(x, y)$ is the heat kernel of M. Similarly, the heat problem with Dirichlet boundary condition in a precompact domain $D \subset M$ with smooth boundary has solution

$$v(x, t) = \int_D p_t^D(x, y) v_0(y) dV(y),$$

where $p_t^D(x, y)$ is the Dirichlet heat kernel of D and is given by (see [8], for instance)

$$p_t^D(x, y) = \sum_{i=1}^{\infty} e^{-i\lambda_i(D)t} \phi_i(x)\phi_i(y).$$

Here $\{\phi_i\}$ is a complete orthonormal basis of $L^2(D)$ made up of eigenfunctions for $-\Delta$ with Dirichlet boundary condition on D and $\{\lambda_i(D)\}$ are the corresponding eigenvalues.

Recall that given an open precompact domain D with smooth boundary ∂D, if there exists a C^∞ function f on D, not identically zero, solution of the Dirichlet boundary value problem

$$\begin{aligned} \Delta f + \lambda f &= 0 \text{ on } D \\ f|_{\partial D} &= 0, \end{aligned} \tag{2}$$

for a real number λ, then λ is called an eigenvalue of $-\Delta$ with respect to the Dirichlet boundary condition and f an eigenfunction. In this problem, the eigenvalues are positive with finite multiplicity, and we obtain a discrete increasing sequence of eigenvalues $\{\lambda_i(D)\}$ satisfying

$$0 < \lambda_1(D) < \lambda_2(D) \leq \cdots \quad \text{with} \quad \lambda_i \to \infty \quad \text{when} \quad i \to \infty,$$

with each distinct eigenvalue repeated according with its multiplicity. Then we can construct a complete orthonormal basis of $L^2(D)$ consisting on eigenfunctions $\{\phi_i\}$ with corresponding eigenvalues $\{\lambda_i(D)\}$.

The lowest eigenvalue $\lambda_1(D)$ is called, unsurprisingly, the *first eigenvalue* of D, and it has a positive associated eigenfunction $\phi_1 : D \to \mathbb{R}$. The influence of geometric properties, such as curvature bounds and isoperimetric inequalities, on the first eigenvalue, the spectrum of the Laplacian, or the heat kernel, has been widely studied along the last century. See [8, 18, 19] for a detailed discussion.

On the other hand, with the heat kernel, a (sub)Markov process X_t on M, called *Brownnian motion*, can be constructed with the transition density p_t. The influence of geometric properties on the behavior of the Brownian motion including recurrence, transience, and stochastic completeness has also been largely studied. We refer to [18] for a comprehensive survey on the topic. Furthermore, let $\mathbb{P}^x$, $x \in M$, be the associated family of probability measures weighting those Brownian paths beginning at x. The first exit time τ_D of a bounded domain $D \subset M$ for X_t is

$$\tau_D = \inf\{t \geq 0 : X_t \notin D\}.$$

Hence, by using the expectation operator $\mathbb{E}^x$ with respect to $\mathbb{P}^x$, the *moment spectrum of* D can be defined as the L^1-norm of the k-th moment of τ_D (see [24])

$$\operatorname{mspec}(D) = \{\mathcal{A}_k(D)\}_{k\in\mathbb{N}} = \{\|\mathbb{E}^x[\tau_D^k]\|_{L^1}\}_{k\in\mathbb{N}},$$

with

$$\|\mathbb{E}^x[\tau_D^k]\|_{L^1} = \int_D \mathbb{E}^x[\tau_D^k]dV.$$

In this survey we discuss the relation between the moment spectrum of D and the first eigenvalue $\lambda_1(D)$ of $-\Delta$ with Dirichlet boundary.

In Sect. 3 it is shown that the moment spectrum determines the first eigenvalue of the Laplacian, in the sense that the knowledge of $\operatorname{mspec}(D)$ implies the knowledge of $\lambda_1(D)$. More precisely, in Theorem 1, Theorem 4, and Eq. (18), explicit formulae are provided for the expression of $\lambda_1(D)$ in terms of $\operatorname{mspec}(D)$.

We must remark here that in some cases, it is harder to solve the eigenvalue problem and determine the first eigenvalue than to obtain the whole moment spectrum. For instance, there is no analytic formula for the first eigenvalue of a geodesic ball even in the hyperbolic space of constant sectional curvature of dimension 2.

The problem of finding the moment spectrum can be drastically simplified in the presence of a large group of isometries. In Sect. 2.3, it is shown how to compute the moment spectrum of a geodesic ball in a rotationally symmetric manifold. Hence, the first eigenvalue of a rotationally symmetric geodesic ball can be computed (see Theorem 4 and Eq. (18) again). This computation is used to obtain in Sect. 3 Theorem 7, where several upper bounds for the first eigenvalue of a geodesic ball are obtained in terms only of the area functions of the geodesic spheres in the line of [16]. Finally, in Sect. 4, we explain how to use this bridge between moment spectrum and first eigenvalue to obtain upper and lower bounds for the first eigenvalue from geometric comparisons. These comparisons can be either intrinsic or extrinsic. In the intrinsic case, the first eigenvalue of a domain in a manifold with bounded sectional (or Ricci) curvatures is estimated by the first eigenvalue of a geodesic ball in a suitable model space. In the extrinsic one, we consider domains in a submanifold with controlled mean curvature immersed in an ambient manifold with bounded curvatures.

2 Preliminaries: Poisson Hierarchy, Green Operator, Moment Spectrum, and Model Spaces

2.1 Poisson Hierarchy

Classically, from mathematical physics, given a precompact domain $D \subset \mathbb{R}^2$ with smooth boundary, the *torsional rigidity* $\mathcal{A}_1(D)$ of D is the torque required per unit angle of twist and per unit length when twisting an elastic beam of uniform cross section D; see [1, 31]. The torsional rigidity of D can be calculated as the L_1-norm of the expectation of the first exit time function

$$\mathcal{A}_1(D) = \|\mathbb{E}^x[\tau_D]\|_1 = \int_D \mathbb{E}^x[\tau_D]dV.$$

Given a precompact domain $D \subset M$ with smooth boundary, the mean exit time function

$$E : D \to \mathbb{R}, \quad x \mapsto \mathbb{E}^x[\tau_D],$$

is characterized (see [15]), as the solution of the following second-order PDE, with Dirichlet boundary data

$$\begin{aligned} \Delta E + 1 &= 0, \ \text{ in } D, \\ E|_{\partial D} &= 0. \end{aligned} \tag{3}$$

This characterization of the first moment of the exit time can be extended to any k-th moment. There exists a sequence $\{u_k\}$ of smooth functions such that

$$\mathcal{A}_k(D) = \|\mathbb{E}^x[\tau_D^k]\|_1 = \int_D u_k(x)dV(x).$$

This sequence $\{u_k\}$ is the so-called (see [14]) Poisson hierarchy for D. The functions $\{u_k\}$ of the Poisson hierarchy can be obtained inductively as the solution of the following boundary value problems on D. First we let

$$u_0(x) = 1, \quad \text{for} \quad \text{all} \quad x \in D,$$

and then for $k \geq 1$,

$$\begin{aligned} \Delta u_k + k u_{k-1} &= 0, \text{ on } D, \\ u_k|_{\partial D} &= 0. \end{aligned} \tag{4}$$

We are going to focus our study in $\{\mathcal{A}_k(D)\}_{k=1}^{\infty}$ which is the L^1-moment spectrum of D. But in some parts of this survey, we will also consider the L^p-moment spectrum of D, which can be defined as the following sequence of integrals:

$$\mathcal{A}_{p,k}(D) := \left(\int_D (u_k(x))^p dV \right)^{\frac{1}{p}}, \quad k = 1, 2, \ldots, \infty.$$

2.2 Green Operator

Let D be a bounded open subset with smooth boundary $\partial D \neq \emptyset$ of a Riemannian manifold (M, g). The Green operator $G^D : L_2(D) \to L_2(D)$ is given by

$$G^D(f)(x) = \int_0^{\infty} \int_D p_t^D(x, y) f(y) dV\, dt = \int_D g^D(x, y) f(y) dV,$$

where $p_t^D(x, y)$ is the heat kernel of the operator $-\Delta$ and

$$g^D(x, y) = \int_0^{\infty} p_t^D(x, y) dt$$

is the Green function of D. The Green operator is a bounded self-adjoint operator in $L_2(D)$, and it is the inverse of $-\Delta$. Thus for any $f \in L_2(D)$, there is a unique solution $u = G^D(f)$ to the equation $-\Delta u = f$. Applying G^D to the equation

$$\Delta u + \lambda_i(D) u = 0,$$

the eigenvalue $\lambda_i(D)$ of D is given by $\lambda_i(D) = u / G^D(u)$. This kind of quotients for a suitable function u can be used to obtain estimations of the eigenvalue $\lambda_1(D)$ as we will see in Sect. 3.

In the sequel, we will omit the superscript D in the notation of the Green operator G^D by simplicity in the notation.

2.3 *Radial Green Operator and Poisson Hierarchy for Geodesic Balls of Model Spaces*

The problem of finding $\{u_k\}$ and $\{\mathcal{A}_{p,k}(D)\}$ for a given smoothly bounded precompact domain $D \subset M$ is, in general, hard to solve. This problem can be simplified when certain symmetries are assumed. In particular, the problem can be completely and explicitly solved when the metric tensor of M is rotationally symmetric around a point $p \in M$. We will say that the metric tensor of M is *rotationally symmetric* around p if the metric spheres S_R of radius R centered at p have maximal dimension of the isometry group and, hence, in the case of being simply connected are isometric to some $n-1$-dimensional sphere of constant sectional curvature. Rotationally symmetric metrics can be constructed by using warping products as follows:

Definition 1 (See [17, 18, 30]) A $w-$model M_w^m is a product manifold

$$[0, \Lambda) \times \mathbb{S}_1^{m-1}/\sim, \quad (t_1, \theta_1) \sim (t_2, \theta_2) \quad \text{if} \quad (t_1, \theta_1) = (t_2, \theta_2) \quad \text{or} \quad t_1 = t_2 = 0,$$

endowed with the warping product metric

$$dr \otimes dr + w^2(r) d\theta \otimes d\theta,$$

where $d\theta \otimes d\theta$ is the standard metric of constant sectional curvature 1 in the sphere $\mathbb{S}_1^{m-1}$ and w is the warping function $w : [0, \Lambda) \to \mathbb{R}_+ \cup \{0\}$ with $w(0) = 0$, $w'(0) = 1$, $w^{(k)}(0) = 0$ for all even derivation orders k and $w(r) > 0$ for all $r > 0$. The point $o_w = \pi^{-1}(0)$, where π denotes the projection onto $[0, \Lambda)$, is called the *center point* of the model space. The value Λ is called the *radius of the model space*.

The conditions on w and its derivatives at 0 in the above definition are required to ensure a smooth metric tensor on o_w. Observe that the metric of a model space is rotationally symmetric around the center point o_w, and moreover, if the model space has radius $\Lambda = \infty$, then the center point o_w is a pole of M_w^m.

Remark 1 The simply connected space forms $\mathbb{K}^m(b)$ of constant curvature b can be constructed as $w-$models with warping functions

$$w(r) = w_b(r) = \begin{cases} \frac{1}{\sqrt{b}} \sin(\sqrt{b}\, r) & \text{if } b > 0 \\ r & \text{if } b = 0 \\ \frac{1}{\sqrt{-b}} \sinh(\sqrt{-b}\, r) & \text{if } b < 0, \end{cases} \tag{5}$$

and radius $\Lambda = \infty$ for $b \leq 0$ and $\Lambda = \pi/\sqrt{b}$ for $b \geq 0$. Note that for $b > 0$ the warping function w_b induces a smooth metric tensor on

$$\left[0, \pi/\sqrt{b}\right] \times \mathbb{S}_1^{m-1}/\sim,$$

where $(t_1, \theta_1) \sim (t_2, \theta_2)$ if $(t_1, \theta_1) = (t_2, \theta_2)$, or $t_1 = t_2 = 0$, or $t_1 = t_2 = \pi/\sqrt{b}$. For $b \leq 0$ any center point is a pole.

A complete description of these model spaces can be found in [17, 18, 22, 23, 28]. The sectional curvatures in the radial directions from the center point are determined by the radial function $-\frac{w''}{w}(r(p))$ for any $p \in M_w^m$. Moreover, the mean curvature of the distance sphere of radius R from the center point is a radial function given by

$$\eta_w(R) = \frac{w'(R)}{w(R)} = \ln'(w(R)). \tag{6}$$

The solution to the boundary value problem (4) defined on the geodesic R-ball B_R^w of radius R centered at $o_w \in M_w^m$ is computed in [20], and the functions u_k are given by

$$u_k(r) = k \int_r^R \frac{\int_0^t w^{m-1}(s)\, u_{k-1}(s)\, ds}{w^{m-1}(t)}\, dt. \tag{7}$$

Moreover, by applying the Divergence theorem

$$\mathcal{A}_k(B_R^w) = -\frac{1}{k+1} u'_{k+1}(R) \mathrm{Vol}(S_R^w), \tag{8}$$

where S_R^w is the geodesic R-sphere in M_w^m. Hence, the Poisson hierarchy $\{u_k\}$ and the moment spectrum $\{\mathcal{A}_k(B_R^w)\}$ can be always explicitly computed, and it depends only on the warping function w. Observe moreover that the Poisson hierarchy $\{u_k\}$ consists of radial functions for geodesic balls of a model space. This is in fact a particular case of a broader phenomenon: given an isometry $\phi : M \to M$ such that for a some precompact domain $D \subset M$ with smooth boundary

$$\phi(D) = D, \quad \phi(\partial D) = \partial D,$$

any function $u_k : D \to \mathbb{R}$ satisfying (4) must be invariant under the action of ϕ, i.e.,

$$u_k = u_k \circ \phi. \tag{9}$$

The above equality can be deduced because since ϕ is an isometry, $\Delta(u_k \circ \phi) = (\Delta u_k) \circ \phi$ and therefore $u_k \circ \phi$ satisfies (4) as well. Finally the uniqueness of the solution of (4) implies equality (9).

In a model space M_w^m, the Green operator for radial functions on B_R^w is given by

$$G(u)(r) = \int_r^R \frac{\int_0^t w^{m-1}(s)\, u(s) ds}{w^{m-1}(t)}\, dt, \tag{10}$$

and the moment functions u_k can be written recursively via the Green operator as $u_k(r) = kG(u_{k-1})(r)$. Indeed, in [4] it is proved that for any bounded open subset D with smooth boundary $\partial D \neq \emptyset$ in a Riemannian manifold, the sequence of the Poisson hierarchy for D can be written as

$$u_k = k!G^k(1),$$

where G is the Green operator on D and $G^k = G \circ \cdots \circ G$ k-times.

In this context the Green operator has been used to obtain bounds of $\lambda_1(B_R^w)$ as can be seen in [2–4, 21, 32].

3 First Eigenvalue and Moment Spectrum

In this section we collect some results that show how to obtain estimations of the first Dirichlet eigenvalue of a domain D in terms of its moment spectrum. The relationship between the moment spectrum and the Dirichlet spectrum is given by the equality (see [26] and see [4] for the weighted version)

$$\mathcal{A}_k(D) = k! \sum_{i=1}^{\infty} \frac{a_i^2}{\lambda_i^k}, \tag{11}$$

where $a_i = \int_D \phi_i \, dV$ and $\{\phi_i\}$ is a complete orthonormal basis in $L^2(D)$ of eigenfunctions of $-\Delta$ with associated eigenvalues $\{\lambda_i\}$. In fact, using this formula, in [26] it is proved that:

Theorem 1 *The first Dirichlet eigenvalue $\lambda_1(D)$ of any smoothly bounded precompact domain $D \subseteq M$ can be directly extracted from the corresponding exit time moment spectrum $\{A_k(D)\}_{k=k_0}^{\infty}$ as follows:*

$$\lambda_1(D) = \sup \left\{ \eta \geq 0 : \lim_{k\to\infty} \sup \left(\frac{\eta}{2}\right)^k \frac{\mathcal{A}_k(D)}{\Gamma(n+1)} < \infty \right\}.$$

In [21], Markvorsen, Palmer, and the second author study the problem of determining the first eigenvalue of a geodesic ball B_R^w in a model space M_w^m with pole o_w. For the radial functions $u_k(r)$ defined on $D = B_R^w$ and given by (7), they proved that, since $u_k(r) = kG(u_{k-1}(r))$ for the Green operator G in (10), the functions $\frac{k\,u_{k-1}}{u_k}(r)$ are increasing. Then, applying Barta's inequality

$$\frac{k\,u_{k-1}}{u_k}(o_w) \leq \lambda_1(B_R^w) \leq \frac{k\,\mathcal{A}_{k-1}(B_R^w)}{\mathcal{A}_k(B_R^w)}. \tag{12}$$

For k even the upper bound was generalized in [14] for a bounded open domain D with smooth boundary in a complete Riemannian manifold. When $k = 1$, the above inequalities reads

$$\frac{1}{\int_0^R q_w(t)\,dt} \leq \lambda_1(B_R^w) \leq \frac{\operatorname{vol}(B_R^w)}{\mathcal{A}_1(B_R^w)}, \tag{13}$$

where $q_w(t)$ is the isoperimetric quotient defined by

$$q_w(t) := \frac{\int_0^t w^{m-1}(s)\,ds}{w^{m-1}(t)}. \tag{14}$$

The lower bound was obtained in [7] for geodesic balls in the n-dimensional sphere $\mathbb{S}^n(1)$, and later it was generalized for an arbitrary M_w^m in [3]. Since, in this case,

$$\int_0^R q_w(t)\,dt = \max_{x\in B_R^w} u_1(x),$$

this bound is also related with the following result of Del Grosso-Marchetti proved in [13]; see also [8]:

Proposition 1 *Let $D \subset M$ be a smoothly bounded precompact domain. Then*

$$\lambda_1(D) \geq \frac{1}{\max\limits_{x\in D} u_1(x)}.$$

By using the specific expression of the Green function, it is proved in [4] that in the particular case of geodesic balls in rotationally symmetric manifolds, it can state the following:

Theorem 2 *Let B_R^w be the geodesic ball of radius R in M_w^m. Then,*

$$\max_{x\in B_R^w} u_1(x) = \sum_{i=1}^{\infty} \frac{1}{\lambda_i^{\mathrm{rad}}(B_R^w)}$$

where $\left\{\lambda_i^{\mathrm{rad}}(B_R^w)\right\}_{i=1}^{\infty}$ is the set of eigenvalues associated with radial eigenfunctions.

The upper bound in (12) gives a relation between the first Dirichlet eigenvalue and the torsional rigidity of a geodesic ball in a rotationally symmetric space. For domains in Euclidean space, this is the classical inequality of Pólya's (see [31]; see [33, 34] for further developments), and it was generalized in [14] for a general bounded open domain D in a complete Riemannian manifold. Indeed, they prove that:

Theorem 3 ([14]) *Let $D \subseteq M$ be a bounded open domain with smooth boundary in a complete Riemannian manifold M. Then for every k,*

$$\lambda_1(D) \leq \frac{(k!)^2}{(2k-1)!} \frac{\mathcal{A}_{2k-1}(D)}{\mathcal{A}_k^2(D)} \operatorname{Vol}(D).$$

Relating the quotient before with the variance of the random variable τ_D^k given by the k-power of the first exit time from D, they also obtain that:

Corollary 1 ([14]) *Let D as before. For $k \in \mathbb{N}$, let* $\operatorname{Var}_k(D)$ *be the L^1-norm of the variance of τ_D^k:*

$$\operatorname{Var}_k(D) = \int_D (u_{2k} - u_k)\, dV. \tag{15}$$

Then,

$$\lambda_1(D) \leq \frac{(2k)! - (k!)^2}{(2k-1)!} \frac{\mathcal{A}_{2k-1}(D)}{\operatorname{Var}_k(D)}.$$

It is also established in [14] a lower bound for the first Dirichlet eigenvalue using the quantity a_1^2 appearing in (11) and the L^1-moment spectrum of a bounded open domain D with smooth boundary. Indeed, for all $k \leq 1$,

$$\left(\frac{k!a_1^2}{\mathcal{A}_k(D)}\right)^{1/k} \leq \lambda_1(D). \tag{16}$$

In the previous results, the value of the first eigenvalue of a domain is bounded from above or from below by geometric quantities involving the isoperimetric quotient, the torsional rigidity or the L^1-moment spectrum of the domain D. But the exact value of $\lambda_1(D)$ can be obtained as a limit of some of these estimations. Indeed, the bounds in (12) improves as k increases, and we can obtain better and better estimates of $\lambda_1(B_R^w)$ since the functions u_k can be explicitly computed in this case (see, for instance, [3, 4, 7, 21, 32] for approximations in this line). Moreover, the exact value of $\lambda_1(B_R^w)$ is obtained in the limit:

Theorem 4 ([21]) *Let B_R^w be the geodesic ball of radius R and center the pole o_w in M_w^m. Then,*

$$\lambda_1(B_R^w) = \lim_{k\to\infty} \frac{k\, u_{k-1}(0)}{u_k(0)} = \lim_{k\to\infty} \frac{k\, \mathcal{A}_{k-1}(B_R^w)}{\mathcal{A}_k(B_R^w)},$$

where u_k are the functions defined by (7) and $\{\mathcal{A}_k(B_R^w)\}$ is the L^1-moment spectrum of B_R^w. Moreover, the radial C^2-function $g_\infty(r) := \lim_{k\to\infty} \frac{u_k(r)}{u_k(0)}$ is an eigenfunction of the first eigenvalue.

This result has been generalized in [4] for a bounded open domain D with smooth boundary in a (weighted) Riemannian manifold using the relation (11), and it is also generalized in [12]. By using the expression of the Green operator $G : L^2(D) \to L^2(D)$ in an orthonormal basis in $L_2(D)$ of eigenfunctions of $-\Delta$, and studying the convergence of the sequence $k \mapsto \frac{\|G^k(f)\|_{L^2}}{\|G^{k+1}(f)\|_{L^2}}$, in [4] Bessa, Jorge, and the first author proved the following relation between the Green operator and the spectrum of the Laplacian:

Theorem 5 *Let D be a bounded open subset with smooth boundary $\partial D \neq \emptyset$ in a Riemannian manifold (M^m, g). Let G be the Green operator. Then, for any positive $f \in L^2(D)$,*

$$\lambda_1(D) = \lim_{k\to\infty} \frac{\|G^k(f)\|_{L^2}}{\|G^{k+1}(f)\|_{L^2}}. \tag{17}$$

Indeed, the previous theorem is stated in [4] in the boarder setting of weighted manifolds. Observe, moreover, that using $f = 1$ in the above theorem, we can state that L^2-norm of the Poisson hierarchy is related with the first eigenvalue. In fact,

$$\begin{aligned}\lambda_1(D) &= \lim_{k\to\infty} (k+1)\frac{\|u_k\|_{L^2}}{\|u_{k+1}\|_{L^2}} = \lim_{k\to\infty} (k+1)\frac{\mathcal{A}_{2,k}(D)}{\mathcal{A}_{2,k-1}(D)} \\ &= \lim_{k\to\infty} \frac{k\mathcal{A}_{2,k-1}(D)}{\mathcal{A}_{2,k}(D)}.\end{aligned} \tag{18}$$

More recently Sarrion-Pedralva and the first author used in [16] the above techniques to obtain an upper bound for the first eigenvalue of a geodesic ball B_R centered at a point p of a Riemmanian m-dimensional manifold. The main idea is to build up a model space M_w^m with geodesic spheres of the same volume than the geodesic spheres included in B_R. Then using the first eigenfunction of B_R^w in this model space and the Rayleigh quotient they show that:

Theorem 6 ([16]) *Let (M, g) be a Riemannian manifold, let $p \in M$ with injectivity radius* $\mathrm{inj}(p)$, *let* B_R *be the geodesic ball of radius R centered at p, and let* $\mathrm{A}(t) :=$

$\mathrm{vol}(\mathrm{S}_t)$ *be the area function of the geodesic spheres centered at* p. *If* $R < \mathrm{inj}(p)$, *then the model space* M_w^m *with warping function*

$$w(t) = \left(\frac{\mathrm{vol}(\mathrm{S}_t)}{\mathrm{vol}(\mathbb{S}_1^{m-1})} \right)^{\frac{1}{m-1}} \tag{19}$$

has smooth metric tensor in B_R^w *and*

$$\lambda_1(B_R) \leq \lambda_1(B_R^w). \tag{20}$$

Furthermore, equality (20) *is attained if and only if, for any* $t \in (0, R)$, *the mean curvature pointed inward* $\mathrm{H}_{\mathrm{S}_t}$ *of the geodesic sphere* S_t *of radius* t *centered at* p *is a radial function. Namely, equality* (20) *is attained, if and only if, there exists a smooth function* $h(t)$ *such that*

$$\mathrm{H}_{\mathrm{S}_t} = h(t) \quad \text{for any} \quad 0 < t < R.$$

Since the warping function given in (19) of the model space in the above theorem depends only on the area function of the geodesic spheres, we can rewrite the expression of the radial Green operator in (10) and the Poisson hierarchy of B_R^w in (7) in terms of this area function as follows:

$$G(u)(r) := \int_r^R \frac{\int_0^t A(s)u(s)ds}{A(t)} dt, \quad u_k(r) = k \int_r^R \frac{\int_0^t A(s)\, u_{k-1}(s)\, ds}{A(t)} dt. \tag{21}$$

This allows us to provide the following theorem which makes use of almost every result listed in this survey for the comparison of the first eigenvalue and the Poisson hierarchy for geodesic balls in rotationally symmetric model spaces.

Theorem 7 *Let* (M, g) *be a Riemannian manifold, let* $p \in M$ *with injectivity radius* $\mathrm{inj}(p)$, *and let* B_R *be the geodesic ball of radius* R *centered at* p. *Let us denote by* $\{\mathcal{A}_{1,k}\}_{k=0}^{\infty}$, $\{\mathcal{A}_{2,k}\}_{k=0}^{\infty}$ *and* $\{\mathcal{V}_k\}_{k=0}^{\infty}$ *the following sequences constructed recursively from the area function:*

$$\begin{aligned} \mathcal{A}_{1,k} &:= \textstyle\int_0^R u_k(t)A(t)dt, \\ \mathcal{A}_{2,k} &:= \left(\textstyle\int_0^R u_k^2(t)A(t)dt\right)^{1/2}, \\ \mathcal{V}_k &:= \textstyle\int_0^R \left(u_{2k}(t) - u_k(t)\right) A(t)dt, \end{aligned}$$

with

$$u_k(r) = k \int_r^R \frac{\int_0^t A(s)\, u_{k-1}(s)\, ds}{A(t)} dt, \quad u_0 = 1.$$

If $R < \mathrm{inj}(p)$, *then, for any* $k \geq 1$,

$$\lambda_1(\mathrm{B}_R) \leq \frac{k\mathcal{A}_{1,k-1}}{\mathcal{A}_{1,k}}, \tag{22}$$

$$\lambda_1(\mathrm{B}_R) \leq \frac{(k!)^2}{(2k-1)!}\frac{\mathcal{A}_{1,2k-1}}{\mathcal{A}^2_{1,k}}\mathcal{A}_{1,0}, \tag{23}$$

$$\lambda_1(\mathrm{B}_R) \leq \frac{(2k)!-(k!)^2}{(2k-1)!}\frac{\mathcal{A}_{1,2k-1}}{\mathcal{V}_k}. \tag{24}$$

Moreover,

$$\begin{aligned}\lambda_1(\mathrm{B}_R(p)) &\leq \lim_{k\to\infty}\frac{k\mathcal{A}_{2,k-1}}{\mathcal{A}_{2,k}},\\ \lambda_1(\mathrm{B}_R(p)) &\leq \lim_{k\to\infty}\frac{k\mathcal{A}_{1,k-1}}{\mathcal{A}_{1,k}},\end{aligned} \tag{25}$$

where equality (25) *is attained in if and only if, for any* $t \in (0, R)$, *the mean curvature pointed inward* $\mathrm{H}_{\mathrm{S}_t}$ *of the geodesic sphere* S_t *of radius t centered at p is a radial function.*

Remark 2 Unless in this survey we focus our attention in the first Dirichlet eigenvalue, estimates of the following eigenvalues in terms of the moment spectrum can also be obtained; see [4, 12, 14].

4 Comparison Results for the First Eigenvalue

An important problem in Riemannian geometry is to find good upper and lower estimates of the first Dirichlet eigenvalue of a domain in a Riemannian manifold M. A way to obtain such estimations, in contrast with the techniques used in the above section, is to establish comparison results. If we assume that M has controlled geometry, in the sense that some of its curvatures (or another interesting geometric quantity) are bounded from above or from below by the curvatures of a model space, one can obtain inequalities for the Laplacian of distinguished functions that provide estimations not only for the first Dirichlet eigenvalue but also for other geometric objects which are related with the Laplacian operator. These estimations are usually given in terms of the corresponding quantity for a geodesic ball in the model space that we are using to establish the comparison. Classical results in this direction are Cheng's comparison theorems for the first eigenvalue of a geodesic ball B_R in a manifold with sectional curvatures (respectively, Ricci curvatures) bounded from above (respectively, from below) by a given constant [9, 10]. In this case, $\lambda_1(B_R)$ is bounded by the first eigenvalue of a geodesic ball in a simply connected real space form of constant sectional curvature.

There is a vast literature in this subject; in this section we do not aim to give an exhaustive list of comparison results for the first eigenvalue of the Laplacian, but just to focus our attention in those which are most related to the ones in the previous section.

Let us consider a complete Riemannian manifold (M^m, g) and fix $p \in M$. For any $x \in M - \{p\}$, the sectional curvature of the two-planes $\sigma_x \in T_x M$ that contain the tangent vector to a minimal geodesic from p to x is called the p-radial sectional curvature of M. In a model space M_w^m, the o_w-radial curvatures are determined by the radial function $-w''(r)/w(r)$. In [21] it is established a generalization of Cheng's eigenvalue comparison theorem for Riemannian manifolds with radial sectional curvatures bounded by the curvatures of M_w^m. If g_∞ is the eigenfunction of $\lambda_1(B_R^w)$ given in Theorem 4, then we can obtain a radial function on M by composing g_∞ with the distance function from the point p: $g(r) := g_\infty \circ r$. Comparing the Laplacian in M of the function g with the Laplacian in M_w^m of g_∞ it can prove the following:

Theorem 8 ([21]) *Let B_R be a geodesic ball of a complete Riemannian manifold M^m with a pole p and suppose that the p-radial sectional curvatures of M^m are bounded from below (respectively, from above) by the o_w-radial sectional curvatures of a w-model space M_w^m. Then*

$$\lambda_1(B_R) \leq (\geq) \lambda_1(B_R^w) = \lim_{k\to\infty} \frac{k\mathcal{A}_{k-1}(B_R^w)}{\mathcal{A}_k(B_R^w)}, \tag{26}$$

where B_R^w is the o_w-centered geodesic ball in M_w^m.

An alternative proof of this theorem can be done using the description of $\lambda_1(D)$ given by McDonald and Meyers in Theorem 1 and the isoperimetric type inequalities for the exit time moment spectrum established in [20] that asserts that under the hypothesis of Theorem 8

$$\frac{\mathcal{A}_k\left(\mathrm{B}_R\right)}{\mathrm{vol}\left(\mathrm{S}_R\right)} \geq (\leq) \frac{\mathcal{A}_k\left(\mathrm{B}_R^{\omega}\right)}{\mathrm{vol}\left(\mathrm{S}_R^{\omega}\right)}, \quad \text{for all } k \geq 0. \tag{27}$$

Using this second strategy, if the model manifold M_w^m is *strictly balanced* in the sense that $q_w(r)\eta_w(r) > 1/m$ for all radius r, then equality in (26) for some fixed radius R_0 implies that B_{R_0} and $B_{R_0}^w$ are isometric.

In a similar direction, it is shown in [25] that if a complete Riemannian manifold M^m satisfies the *moment comparison condition with constant curvature space form* $\mathbb{K}^m(b)$, $\mathcal{A}_k(D) \leq \mathcal{A}_k(B)$ for all smoothly bounded domain $D \subset M$ with compact closure and all $k \in \mathbb{N}$, where B is the geodesic ball in $\mathbb{K}^m(b)$ with the same volume of D, then

$$\lambda_1(B) \leq \lambda_1(D).$$

We point out that in [21], the comparison result in Theorem 8 is formulated in a more general context. The authors consider an unbounded complete and closed submanifold P^n with controlled mean curvature in a Riemannian manifold N^m with radial sectional curvatures bounded by a radial function. In this context, an estimation of the first Dirichlet eigenvalue of a extrinsic ball D_R, namely, the intersection of a geodesic ball of the ambient manifold with the submanifold P, is obtained in terms of the first eigenvalue of a geodesic ball in a suitable model space. However the construction of this model space of comparison in terms of the assumptions over P and M is very technical, so for clarity of exposition, we have only stated the corresponding intrinsic version. The details of the general statement can be seen in [21]. As a particular case of these more general statements for a Cartan-Hadamard manifold, applying the techniques in [11] (see also [5]), we have that:

Theorem 9 ([21]) *Let M^m be a Cartan-Hadamard manifold, with sectional curvatures bounded from above by a constant $K_M \leq b \leq 0$. Let $p \in M$ be a pole in M and r the distance function from p. Let $P^n \subseteq M^m$ be a complete and non-compact properly immersed submanifold with*

$$- \langle \nabla r(x), H_P(x) \rangle \leq h(r(x)),$$

for all $x \in P^n$, where H_P is the mean curvature of P^n and $h(r)$ is a radial smooth function. Suppose that

$$(n-1) \cdot \sqrt{-b} \coth(R\sqrt{-b}) \geq n \cdot \sup_{r \in [0,R]} h(r), \tag{28}$$

where we read $\sqrt{-b}\coth(R\sqrt{-b})$ to be $1/R$ when $b = 0$. For any given extrinsic ball $D_R(p)$ in P^n, we then have the following inequality:

$$\lambda_1(D_R) \geq \frac{1}{4}\left((n-1) \cdot \sqrt{-b} \coth(R\sqrt{-b}) \;\; - n \cdot \sup_{r \in [0,R]} h(r) \right)^2 . \tag{29}$$

Notice that when P is a minimal submanifold, we can take as bounding function $h(r(x)) = 0$. In the intrinsic setting, when $P^n = N^m$, $H_P = 0$, and $D_R = B_R$, the classical result of McKean [27] for the fundamental tone of a Cartan-Hadamard manifold is recovered. Moreover, Bessa and Montenegro observe in [5] an improvement of the bound (29) in the intrinsic setting as follows:

Theorem 10 *Under the intrinsic conditions with M^m having sectional curvatures bounded from above by $b \leq 0$*

$$\lambda_1(B_R) \geq \frac{1}{4}\left(\max\left(\frac{m}{R} \, , \; (m-1) \cdot \sqrt{-b} \coth(R\sqrt{-b}) \right) \right)^2 . \tag{30}$$

Comparison results for the first eigenvalue of a domain can be also formulated assuming bounds for the mean curvature of the geodesic spheres of the manifold M instead of bounding the sectional or Ricci curvatures. Given a point $p \in M$, for $R < \operatorname{inj}(p)$, we denote by H_{S_R} the pointed inward mean curvatures of the R-geodesic spheres centered at p in M. Notice that in a model space M_w^m, $H_{S_R^w} = \eta_w(R) = w'(R)/w(R)$. With this weaker hypothesis, in [6] the first eigenvalue of a geodesic ball is compared with the first eigenvalue of a geodesic ball in a model space, and recently, it is shown in [29] the following:

Theorem 11 *Let M^m be a complete manifold, and let $p \in M$ such that* $\operatorname{inj}(p) \leq \operatorname{inj}(o_w)$ *for o_w the center point of a rotationally symmetric space M_w^m. Assume that for $R < \operatorname{inj}(p)$, the mean curvature of the geodesic spheres in M satisfies that*

$$H_{S_t} \geq (\leq) H_{S_t^w} = \frac{w'(t)}{w(t)}, \tag{31}$$

for all $0 < t \leq R$. Then,

$$\lambda_1(B_R) \geq (\leq) \lambda_1(B_R^w). \tag{32}$$

Equality in (32) *implies that*

$$H_{S_t} = H_{S_t^w}, \tag{33}$$

for all $0 < t \leq R$, and then we have the following equalities:

1. *$u_k = u_k^w \circ r_p$ on B_R for all $k \geq 1$, where r_p is the distance function from the point p in M and $\{u_k\}$ and $\{u_k^w\}$ are the Poisson hierarchy for B_R and B_R^w, respectively. Hence, $u_k = u_k^w \circ r_p$ on B_t for all $k \geq 1$ and for all $0 < t \leq R$.*
2. *$\operatorname{vol}(B_t) = \operatorname{vol}(B_t^w)$ and $\operatorname{vol}(S_t) = \operatorname{vol}(S_t^w)$ for all $0 < t \leq R$.*
3. *$\mathcal{A}_k(B_t) = \mathcal{A}_k(B_t^w)$, for all $k \geq 1$ and for all $0 < t \leq R$.*

The proof follows the lines of the proof of Theorem 8, so the authors previously shown that inequality (27) for the exit time moment spectrum is still valid under this weaker hypothesis on the mean curvature of geodesic spheres. In this setting, the equality between the first eigenvalues of the geodesic spheres of the manifold M and the model space M_w^m gives the equality between the mean curvatures of the corresponding geodesic spheres instead of the isometry of the geodesic balls as happens in Theorem 8 where the radial sectional curvatures of M are bounded for the radial sectional curvatures of M_w^m. Observe that this is coherent since bounds on the sectional curvatures of the manifold implies bounds on the mean curvature of its geodesic spheres, but the reciprocal is not true. Indeed, in [6] the authors construct a family of complete smooth metrics on $\mathbb{R}^m$ non-isometric to the constant sectional curvatures b metrics of the simply connected space forms $\mathbb{K}^m(b)$ such that the geodesic balls B_R and $B_R^{w_b}$ have the same first eigenvalue and the geodesic spheres S_t and $S_t^{w_b}$ for $0 < t \leq R$ have the same mean curvatures (see also the examples in [4] and [29]).

Finally, as a consequence of the comparison results obtained in [29] and the study of the equality case in the comparison, it is shown that:

Corollary 2 ([29]) *Under the assumptions of Theorem 11, the following equalities are equivalent:*

1. $\lambda_1(B_R) = \lambda_1(B_R^w)$.
2. $\mathcal{A}_k(B_R) = \mathcal{A}_k(B_R^w)$, *for all* $k \geq 1$.
3. $u_k = u_k^w \circ r_p$ *on* B_R *for all* $k \geq 1$, *where* r_p *is the distance function from the point* p *in* M *and* $\{u_k\}$ *and* $\{u_k^w\}$ *are the Poisson hierarchy for* B_R *and* B_R^w, *respectively.*

Moreover, equality $H_{S_t} = H_{S_t^w}$ *for all* $0 < t \leq R$ *implies any (and hence all) of the equalities* (1), (2), *and* (3).

Acknowledgments The first author was partially supported by the Research grant PID2020-115930GA-100 funded by MCIN/ AEI /10.13039/501100011033, Program of University Jaume I Project UJI-B2021-08, and AICO/2021/252. The second author was partially supported by the research grant PID2020-118180GB-I00 funded by MCIN/AEI/10.13039/501100011033 and by Junta de Andalucía grants A-FQM-441-UGR18 and PY20-00164.

References

1. Bandle, C.: Isoperimetric Inequalities and Applications, vol. 7. Pitman Publishing, Lanham (1980)
2. Barroso, C.S., Bessa, G.P.: A note on the first eigenvalue of spherically symmetric manifolds. Mat. Contemp. **30**, 63–69 (2006). XIV School on Differential Geometry (Portuguese)
3. Barroso, C.S., Bessa, G.P.: Lower bounds for the first Laplacian eigenvalue of geodesic balls of spherically symmetric manifolds. Int. J. Appl. Math. Stat. **6**(D06), 82–86 (2006)
4. Bessa, G.P., Gimeno, V., Jorge, L.P.: Green functions and the Dirichlet spectrum. Rev. Mat. Iberoam. **36**(1), 1–36 (2019)
5. Bessa, G.P., Montenegro, J.F.: Eigenvalue estimates for submanifolds with locally bounded mean curvature. Ann. Global Anal. Geom. **24**(3), 279–290 (2003)
6. Bessa, G.P., Montenegro, J.F.: On Cheng's eigenvalue comparison theorem. In: Mathematical Proceedings of the Cambridge Philosophical Society, vol. 144, no. 3, pp. 673–682. Cambridge University Press, Cambridge (2008)
7. Betz, C., Cámera, G.A., Gzyl, H.: Bounds for the first eigenvalue of a spherical cap. Appl. Math. Optim. **10**(3), 193–202 (1983)
8. Chavel, I.: Eigenvalues in Riemannian Geometry. Academic, Cambridge (1984)
9. Cheng, S.Y.: Eigenfunctions and eigenvalues of Laplacian, Amer. Math. Soc. Proc. Symp. Pure Math. **27**, 185–193 (1975)
10. Cheng, S.Y.: Eigenvalue comparison theorems and its geometric applications. Math. Z. **143**(3), 289–297 (1975)
11. Cheung, L.-F., Leung, P.-F.: Eigenvalue estimates for submanifolds with bounded mean curvature in the hyperbolic space. Math. Z. **236**(3), 525–530 (2001)
12. Colladay, D., Langford, J., McDonald, P.: Comparison results, exit time moments, and eigenvalues on Riemannian manifolds with a lower Ricci curvature bound. J. Geom. Anal. **28**, 3906–3927 (2018)

13. Del Grosso, G., Marchetti, F.: Asymptotic estimates for the principal eigenvalue of the Laplacian in a geodesic ball. Appl. Math. Optim. **10**, 37–50 (1983)
14. Dryden, E.B., Langford, J.J., McDonald, P.: Exit time moments and eigenvalue estimates. Bull. Lond. Math. Soc. **49**, 480–490 (2017)
15. Dynkin, E.B.: Markov Processes, Springer, Berlin (1965)
16. Gimeno, V., Sarrion-Pedralva, E.: First Eigenvalue of the Laplacian of a geodesic ball and area-based symmetrization of its metric tensor. J. Math. Inequal. **16**, 371–391 (2022)
17. Greene, R.E., Wu, H.: Function theory on manifolds which possess a pole. In: Lecture Notes in Mathematics, vol. 699. Springer, Berlin (1979)
18. Grigor'yan, A.: tAnalytic and geometric background of recurrence and non-explosion of the Brownian motion on Riemannian manifolds, Bull. Amer. Math. Soc. (N.S.) **36**(2), 135–249 (1999)
19. Grigor'yan, A.: Heat Kernel and Analysis on Manifolds. AMS/IP Studies in Advanced Mathematics, vol. 47. American Mathematical Society, Providence, International Press, Boston (2009)
20. Hurtado, A., Markvorsen, S., Palmer, V.: Comparison of exit moment spectra for extrinsic metric balls. Potential Anal. **36**(1), 137–153 (2012)
21. Hurtado, A., Markvorsen, S., Palmer, V.: Estimates of the first Dirichlet eigenvalue from exit time moment spectra. Math. Ann. **365**(3–4), 1603–1632 (2016)
22. Markvorsen, S., Palmer, V.: How to obtain transience from bounded radial mean curvature, Trans. Amer. Math. Soc. **357**(9), 3459–3479 (electronic) (2005)
23. Markvorsen, S., Palmer, V.: Torsional rigidity of minimal submanifolds. Proc. Lond. Math. Soc. **93**(1), 253–272 (2006)
24. McDonald, P.: Isoperimetric conditions, Poisson problems, and diffusions in Riemannian manifolds. Potential Anal. **16**(2), 115–138 (2002)
25. McDonald, P.: Exit times, moment problems and comparison theorems. Potential Anal. **38**, 1365–1372 (2013)
26. McDonald, P., Meyers, R.: Dirichlet spectrum and heat content. J. Funct. Anal. **200**(1), 150–159 (2003)
27. McKean, H.P.: An upper bound to the spectrum of Δ on a manifold of negative curvature. J. Differ. Geom. **4**, 359–366 (1970)
28. O'Neill, B.: Semi-Riemannian Geometry with Applications to Relativity. Academic, Cambridge (1983)
29. Palmer, V., Sarrión-Pedralva, E.: First Dirichlet eigenvalue and exit time moment spectra comparisons. Potential Anal. (2023). https://doi.org/10.1007/s11118-022-10058-1
30. Petersen, P.: Riemannian Geometry. Graduate Texts in Mathematics, vol. 171, 2nd edn. Springer, New York, (2006)
31. Pólya, G., Szegö, G.: Isoperimetric Inequalities in Mathematical Physics. Princeton University Press, Princeton (1951)
32. Sato, S.: Barta's inequalities and the first eigenvalue of a cap domain of a 2-sphere. Math. Z. **181**(3), 313–318 (1982)
33. van den Berg, M., Buttazzo, G., Velichkov, B.: Optimization problems involving the first Dirichlet eigenvalue and the torsional rigidity, new trends in shape optimization. In: International Series of Numerical Mathematics, vol. 166, pp. 19–41. Birkhäuser/Springer, Cham (2015)
34. van den Berg, M., Ferone, V., Nitsch, C., Trombetti, C.: On Polya's inequality for torsional rigidity and first Dirichlet eigenvalue. Integr. Equ. Oper. Theory **86**, 579–600 (2016)

Area-Minimizing Horizontal Graphs with Low Regularity in the Sub-Finsler Heisenberg Group $\mathbb{H}^1$

Gianmarco Giovannardi, Julián Pozuelo, and Manuel Ritoré

Abstract In the Heisenberg group $\mathbb{H}^1$, equipped with a left-invariant and not necessarily symmetric norm in the horizontal distribution, we provide examples of entire area-minimizing horizontal graphs which are locally Lipschitz in Euclidean sense. A large number of them fail to have further regularity properties. The examples are obtained by prescribing as singular set a horizontal line or a finite union of horizontal half-lines extending from a given point. We also provide examples of families of area-minimizing cones.

Keywords Sub-Finsler geometry · Perimeter-minimizing sets · Area-minimizing cones · Heisenberg group

1 Introduction

The regularity of perimeter-minimizing sets in sub-Finsler geometry is currently one of the most challenging problems in calculus of variations. A sub-Finsler structure in a Carnot-Carathéodory manifold with a completely non-integrable distribution $\mathcal{H}$ is defined by a smooth norm on $\mathcal{H}$. The case of a Euclidean norm is that of sub-Riemannian geometry. The notion of sub-Riemannian perimeter was introduced by Garofalo and Nhieu [17], while sub-Finsler boundary measures in the first Heisenberg group $\mathbb{H}^1$ were considered by Sánchez [30] and sub-Finsler perimeters in $\mathbb{H}^1$ by Pozuelo and Ritoré [26] and Franceschi et al. [10].

Fine properties of sets of finite perimeter in the Heisenberg groups $\mathbb{H}^n$ were obtained by Franchi et al. [11]. Among others, they obtained a structure result for

G. Giovannardi
Dipartimento di Matematica e Informatica "U. Dini", Universita degli Studi di Firenze, Firenze, Italy
e-mail: gianmarco.giovannardi@unifi.it

J. Pozuelo · M. Ritoré (✉)
Departamento de Geometría y Topología, Universidad de Granada, Granada, Spain
e-mail: pozuelo@ugr.es; ritore@ugr.es

A. Alarcón et al. (eds.), *New Trends in Geometric Analysis*,
RSME Springer Series 10, https://doi.org/10.1007/978-3-031-39916-9_7

the reduced boundary of a set of finite perimeter: except for a set of small spherical Hausdorff dimension, it is the union of $\mathbb{H}$-regular hypersurfaces (i.e., level sets of continuous functions with continuous first derivatives in the horizontal directions); see the Main Theorem in page 486 of [11].

The regularity of sub-Riemannian perimeter-minimizing sets has been investigated by a large number of researchers [1, 3, 5–9, 12, 13, 21, 23, 27–29]. The boundaries of the conjectured solutions to the isoperimetric problem are of class C^2 (see [2]), although there exist examples of area-minimizing horizontal graphs which are merely Euclidean Lipschitz; see [6, 22, 27]. The sub-Riemannian Plateau problem was first considered by Pauls [24]. Afterward, under given Dirichlet conditions on p-convex domains, Cheng et al. [6] proved existence and uniqueness of t-graphs (horizontal graphs of the form $t = u(x, y)$) which are Lipschitz continuous weak solutions of the minimal surface equation in $\mathbb{H}^1$. Later, Pinamonti et al. [25] obtained existence and uniqueness of t-graphs on domains with boundary data satisfying a bounded slope condition, thus showing that Lipschitz regularity is optimal at least in the first Heisenberg group $\mathbb{H}^1$. Capogna et al. [3] established that the intrinsic graph of a Lipschitz continuous function, which is, in addition, a viscosity solution of the sub-Riemannian minimal surface equation in $\mathbb{H}^1$, is of class $C^{1,\alpha}$, with higher regularity in the case of $\mathbb{H}^n$, $n > 1$; see [4]. It was shown in [7] that the regular part of a t-graph of class C^1 with continuous prescribed sub-Riemannian mean curvature in $\mathbb{H}^1$ is foliated by C^2 characteristic curves. Furthermore, in [16] the authors generalized the previous result when the boundary S is a general C^1 surface in a three-dimensional contact sub-Riemannian manifold. Later, Galli in [14] improved the result in [16] only assuming that the boundary S is Euclidean Lipschitz and $\mathbb{H}$-regular. Recently, in [18] the first and third authors extended the result in [14] to the sub-Finsler Heisenberg groups. Up to now, determining the optimal regularity of perimeter-minimizing $\mathbb{H}$-regular hypersurfaces in the Heisenberg group remains an open problem.

Bernstein-type problems for surfaces in $\mathbb{H}^1$ have also received a special attention. The nature of the sub-Riemannian Bernstein problem in the Heisenberg group is completely different from the Euclidean one even for graphs. On the one hand the area functional for t-graphs is convex as in the Euclidean setting. Therefore the critical points of the area are automatically minimizers for the area functional. However, since t-graphs admit singular points where the horizontal gradient vanishes, their classification is not an easy task. Thanks to a deep study of the singular set for C^2 surfaces in $\mathbb{H}^1$, Cheng et al. [5] showed that minimal t-graphs of class C^2 are congruent to a family of surfaces including the hyperbolic paraboloid $u(x, y) = xy$ and the Euclidean planes. Under the hypothesis that the surface is area-stationary, Ritoré and Rosales proved in [28] that the surface must be congruent to a hyperbolic paraboloid or to a Euclidean plane. If we consider the class of Euclidean Lipschitz t-graphs, the previous classification does not hold since there are several examples of area-minimizing surfaces of low regularity; see [27]. The complete classification for C^2 surfaces was established by Hurtado et al. in [21], by showing that a complete, orientable, connected, stable area-stationary surface is congruent either to the hyperbolic paraboloid $u(x, y) = xy$ or to a Euclidean plane.

As in the Euclidean setting, the stability condition is crucial in order to discard some minimal surfaces such as helicoids and catenoids.

On the other hand, the study of the regularity of intrinsic graphs (i.e., Riemannian graphs over vertical planes) is a completely different problem since the area functional for such graphs is not convex. Indeed, Danielli et al. in [8] discovered that the family of graphs

$$u_\alpha(x,t) = \frac{\alpha x t}{1+2\alpha x^2}, \qquad \alpha > 0,$$

are area-stationary but *unstable*. In [22], Monti et al. provided an example of an area-minimizing intrinsic graph of regularity $C^{1/2}(\mathbb{R}^2)$ that is an intrinsic cone. Therefore the Euclidean threshold of dimension $n = 8$ fails in the sub-Riemannian setting. In [1], Barone et al. classified complete C^2 area-stationary intrinsic graphs. Later Danielli et al. in [9] showed that a C^2 complete stable embedded minimal surface in $\mathbb{H}^1$ with empty characteristic set must be a plane. In [15] Galli and Ritoré proved that a complete, oriented, and stable area-stationary C^1 surface without singular points is a vertical plane. Later, Nicolussi Golo and Serra Cassano [23] showed that Euclidean Lipschitz stable area-stationary intrinsic graphs are vertical planes. Recently, Giovannardi and Ritoré [19] showed that in the Heisenberg group $\mathbb{H}^1$ with a sub-Finsler structure, a complete, stable, Euclidean Lipschitz surface without singular points is a vertical plane and Young [32] proved that a ruled area-minimizing entire intrinsic graph in $\mathbb{H}^1$ is a vertical plane by introducing a family of deformations of graphical strips based on variations of a vertical curve.

In this note, we provide examples of entire perimeter-minimizing t-graphs for a fixed but arbitrary left-invariant sub-Finsler structure in the first Heisenberg group $\mathbb{H}^1$. Our examples are inspired by the corresponding sub-Riemannian ones in [27]. Of particular interest are the conical examples, invariant by the non-isotropic dilations of $\mathbb{H}^1$. In the sub-Riemannian case, these examples were investigated in [20] and [27].

The paper is organized in the following way. In Sect. 2 we include some preliminaries. In Theorem 3.1 of Sect. 3 we obtain a necessary and sufficient condition, inspired by Theorem 3.1 in [26], for a surface to be a critical point of the sub-Finsler area. We assume that the surface is piecewise C^2 and composed of pieces meeting in a C^1 way along C^1 curves. This condition will allow us to construct area-minimizing examples in Proposition 4.3 of Sect. 4 and examples with low regularity in Proposition 4.4. The same construction, keeping fixed the angle at one side (and hence at the other one) of the singular line, provides examples of area-minimizing cones; see Corollary 4.5. Finally, in Sect. 5 we exhibit some examples of area-minimizing cones in the spirit of [20]. These examples are obtained in Theorem 5.2 from circular sectors of the area-minimizing cones with one singular half-line obtained in Corollary 4.5.

2 Preliminaries

2.1 The Heisenberg Group

We denote by $\mathbb{H}^1$ the first Heisenberg group: the three-dimensional Euclidean space $\mathbb{R}^3$ with coordinates (x, y, t), endowed with the product $*$ defined by

$$(x, y, t) * (\bar{x}, \bar{y}, \bar{t}) = (x + \bar{x}, y + \bar{y}, t + \bar{t} + \bar{x}y - x\bar{y}).$$

A frame of left-invariant vector fields is given by

$$X = \frac{\partial}{\partial x} + y\frac{\partial}{\partial t}, \qquad Y = \frac{\partial}{\partial y} - x\frac{\partial}{\partial t}, \qquad T = \frac{\partial}{\partial t}.$$

For $p \in \mathbb{H}^1$, the left-translation by p is the diffeomorphism $L_p(q) = p * q$. The horizontal distribution $\mathcal{H}$ is the planar non-integrable one generated by X and Y, which coincides with the kernel of the contact one-form $\omega = dt - ydx + xdy$.

We shall consider on $\mathbb{H}^1$ the left-invariant Riemannian metric $g = \langle \cdot, \cdot \rangle$, so that $\{X, Y, T\}$ is a global orthonormal frame, and let D be the Levi-Civita connection associated with the Riemannian metric g. Setting $J(U) = D_U T$ for any vector field U in $\mathbb{H}^1$, we get $J(X) = Y$, $J(Y) = -X$ and $J(T) = 0$. Therefore $-J^2$ coincides with the identity when restricted to the horizontal distribution. The Riemannian volume of a set E is, up to a constant, the Haar measure of the group and is denoted by $|E|$. The integral of a function f with respect to the Riemannian measure is denoted by $\int f \, d\mathbb{H}^1$.

2.2 Sub-Finsler Norms and Perimeter

Given a convex set $K \subset \mathbb{R}^2$ with $0 \in \text{int}(K)$ and associated asymmetric norm $||\cdot||$ in $\mathbb{R}^2$, we define on $\mathbb{H}^1$ a left-invariant norm $||\cdot||_K$ on the horizontal distribution by means of the equality

$$||fX + gY||_K(p) = ||(f(p), g(p))||,$$

for any $p \in \mathbb{H}^1$. Its dual norm is denoted by $||\cdot||_{K,*}$.

If the boundary of K is of class C^ℓ, for $\ell \geqslant 2$, and the geodesic curvature of ∂K is strictly positive, we say that K is of class C^ℓ_+. When K is of class C^2_+, the outer Gauss map $N_K : \partial K \to \mathbb{S}^1$ is a diffeomorphism and the map

$$\pi_K(fX + gY) = N_K^{-1}\left(\frac{(f, g)}{\sqrt{f^2 + g^2}}\right),$$

defined for non-vanishing horizontal vector fields $U = fX + gY$, satisfies

$$||U||_{K,*} = \langle U, \pi_K(U)\rangle,$$

where $||\cdot||_{K,*}$ is the dual norm of $||\cdot||_K$. See §2.3 in [26].

Definition 2.1 Given a convex body $K \subset \mathbb{R}^2$ containing 0 in its interior, and a measurable set $E \subset \mathbb{H}^1$, its horizontal K-perimeter in an open set $\Omega \subset \mathbb{H}^1$ is

$$P_K(E,\Omega) = \sup\left\{\int_E \operatorname{div}(U)\, d\mathbb{H}^1, U \in \mathcal{H}_0^1(\Omega), ||U||_{K,\infty} \leqslant 1\right\},$$

Here $||U||_{K,\infty} = \sup_{p\in\mathbb{H}^1} ||U_p||_{K_0}$ and $\mathcal{H}_0^1(\Omega)$ is the space of C^1 horizontal vector fields with compact support in Ω. If $\Omega = \mathbb{H}^1$ we write $P_K(E)$ instead of $P_K(E, \mathbb{H}^1)$. When $P_K(E,\Omega)$ is finite, we say that E has finite horizontal K-perimeter in Ω.

Remark 2.2 If E has C^1 boundary ∂E, then its perimeter $P_K(E)$ is equal to the sub-Finsler area A_K of its boundary, defined by

$$A_K(\partial E) = \int_{\partial E} ||N_h||_{K,*} d\sigma.$$

where N_h is the projection on the horizontal distribution $\mathcal{H}$ of the Riemannian normal N with respect to the metric g and $d\sigma$ is the Riemannian measure of ∂E. For more details, see §2.4 in [26].

We will often omit the subscript K to simplify the notation.

3 The First Variation Formula and a Stationary Condition

In this section we present some consequences of the first variation formula. We assume that the Heisenberg group $\mathbb{H}^1$ is endowed with the sub-Finsler structure associated with a convex set K of class C_+^2 with $0 \in \operatorname{int}(K)$. Recall that, given a surface $S \subset \mathbb{H}^1$ of class C^1, its *singular set* S_0 is composed of those points of S where the tangent plane is horizontal. The *regular* part of S is $S \setminus S_0$.

Theorem 3.1 (Theorem 3.1 in [26]) *Let S be an oriented surface of class C^1 such that the regular part $S \setminus S_0$ is of class C^2. Consider a C^2 vector field U with compact support on S, normal component $u = \langle U, N\rangle$, and associated flow $\{\varphi_s\}_{s\in\mathbb{R}}$. Let $\eta = \pi(\nu_h)$, where ν_h is the horizontal unit normal to S. Then we have*

$$\left.\frac{d}{ds}\right|_{s=0} A_K(\varphi_s(S)) = \int_{S\setminus S_0} H_K u\, dS - \int_{S\setminus S_0} \operatorname{div}_S(u\eta^\top)\, dS, \tag{3.1}$$

where div_S *is the Riemannian divergence on* S *and the superscript* $\top$ *indicates the projection over the tangent plane to* S*. The quantity* $H_K = \langle \nabla_Z \pi(\nu_h), Z \rangle$*, for* $Z = -J(\nu_h)$*, is the* K*-mean curvature of* S*.*

Using Theorem 3.1, we can prove the following necessary and sufficient condition for a surface S to be A_K-stationary. When a surface S of class C^1 is divided into two parts S^+, S^- by a singular curve S_0 so that S^+, S^- are of class C^2 up to the boundary, the tangent vectors Z^+, Z^- can be chosen so that they parameterize the characteristic curves (i.e., horizontal curves en the regular part of S) as curves leaving from S_0; see Corollary 3.6 in [5]. In this case $\eta^+ = \pi(\nu_h) = \pi(J(Z^+))$ and $\eta^- = \pi(J(Z^-))$.

Corollary 3.2 *Let* S *be an oriented surface of class* C^1 *such that the singular set* S_0 *is a* C^1 *curve. Assume that* $S \setminus S_0$ *is the union of two surfaces* S^+, S^- *of class* C^2 *meeting along* S_0*. Let* η^+, η^- *the restrictions of* η *to* S^+ *and* S^-*, respectively. Then* S *is area-stationary if and only if*

1. $H_K = 0$*, and*
2. $\eta^+ - \eta^-$ *is tangent to* S_0*.*

In particular, condition $H_K = 0$ *implies that* $S \setminus S_0$ *is foliated by horizontal straight lines.*

Proof We may apply the divergence theorem to the second term in (3.1) to get

$$\frac{d}{ds}\bigg|_{s=0} A_K(\varphi_s(S)) = \int_{S\setminus S_0} H_K u \, dS - \int_{S_0} u \, \langle \xi, (\eta^+ - \eta^-)^\top \rangle \, dS,$$

where ξ is the outer unit normal to S^+ along S_0. Hence the stationary condition is equivalent to $H = 0$ on $S \setminus S_0$ and $\langle \xi, \eta^+ - \eta^- \rangle = 0$. The latter condition is equivalent to that $\eta^+ - \eta^-$ be tangent to S_0.

That $H_K = 0$ implies that $S \setminus S_0$ is foliated by horizontal straight lines was proven in Theorem 3.14 in [26]. □

Since $\nu^+ = J(Z^+)$, $\nu^- = J(Z^-)$, where Z^+ and Z^- are the extensions of the horizontal tangent vectors in S^+, S^-, we have that the second condition in Corollary 3.2 is equivalent to

$$\pi(J(Z^+)) - \pi(J(Z^-)) \text{ is tangent to } S_0. \tag{3.2}$$

So a natural question is, given a C^2_+ convex body K containing 0 in its interior, and a unit vector $v \in \mathbb{S}^1$, can we find a pair of unit vectors Z^+, Z^- such that (3.2) is satisfied? If such vectors exist, how many pairs can we get? The answer follows from the next result.

Lemma 3.3 *Let* K *be a convex body of class* C^2_+ *such that* $0 \in \operatorname{int}(K)$*. Given* $v \in \mathbb{R}^2 \setminus \{0\}$*, let* $L \subset \mathbb{R}^2$ *be the vector line generated by* v*. Then, for any* $u \in \partial K$*, we have the following possibilities:*

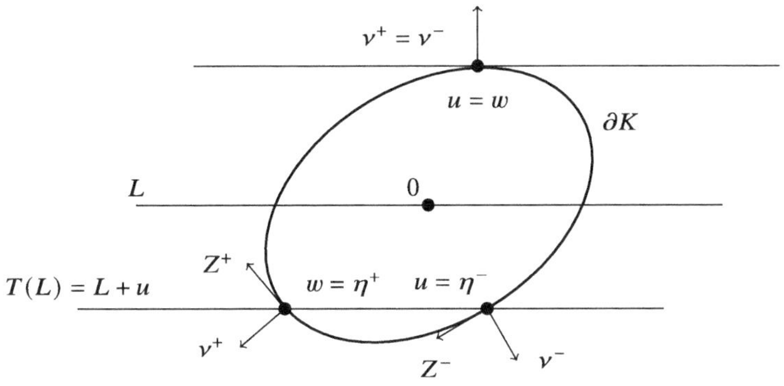

Fig. 1 Geometric construction to obtain $w = \eta^+$ from $u = \eta^-$ so that the stationary condition is satisfied. The case $\nu^+ = \nu^-$ cannot hold

1. The only $w \in \partial K$ such that $w - u \in L$ is $w = u$, or
2. There is only one $w \in \partial K$, $w \neq u$ such that $w - u \in L$.

The first case happens if and only if L is parallel to the support line of K at u.

Proof Let T be the translation in $\mathbb{R}^2$ of vector u. Then $T(L)$ is a line that meets ∂K at u. The line $T(L)$ intersects ∂K only once when L is the supporting line of $T(K)$ at 0; otherwise L intersects ∂K just at another point $w \neq u$ so that $w - u \in L$. □

Remark 3.4 We use Lemma 3.3 to understand the behavior of characteristic curves meeting at a singular point $p \in S_0$. Let Z^+, Z^- be the tangent vectors to the characteristic lines starting from p. Let ν^+, ν^- be the vectors $J(Z^+)$, $J(Z^-)$, and L the line generated by the tangent vector to S_0 at p. The condition that S is stationary implies that $\eta^+ - \eta^- \in L$. If $w = \eta^+$ and $u = \eta^-$ are equal, then $\nu^+ = \nu^-$ are orthogonal to L, which implies that Z^+, Z^- lie in L. This is not possible since characteristic lines meet tranversaly the singular line, again by Corollary 3.6 in [5].

Hence $\eta^+ \neq \eta^-$ and η^+ is uniquely determined from η^- by Lemma 3.3. Obviously the roles of η^+ and η^- are interchangeable (Fig. 1).

4 Examples of Entire K-Perimeter-Minimizing Horizontal Graphs with One Singular Line

Remark 3.4 implies that Z^- can be uniquely determined from Z^+ when S is a stationary surface. Let us see that this result can be refined to provide a smooth dependence of the oriented angle $\angle(v, Z^-)$ in terms of $\angle(v, Z^+)$. We use complex notation for horizontal vectors assuming that the horizontal distribution is positively oriented by $v, J(v)$ for any $v \in \mathcal{H} \setminus \{0\}$.

Lemma 4.1 *Let K be a convex body of class C^2_+ with $0 \in \text{int}(K)$. Consider a unit vector $v \in \mathbb{R}^2$ and let $L \subset \mathbb{R}^2$ be the vector line generated by v. Then, for any $\alpha \in (0, \pi)$, there exists a unique $\beta \in (\pi, 2\pi)$ such that if $Z^+ = ve^{i\alpha}$, $Z^- = ve^{i\beta}$, then $\pi(J(Z^+)) - \pi(J(Z^+))$ belongs to L.*

Moreover the function $\beta : (0, \pi) \to (\pi, 2\pi)$ is of class C^1 with negative derivative.

Proof We change coordinates so that L is the line $y = 0$. We observe that $Z^+ = ve^{i\alpha}$ implies that $J(Z^+) = ve^{i(\alpha+\pi/2)}$. We define $(x, y) : \mathbb{S}^1 \to \partial K$ by

$$(x(\alpha), y(\alpha)) = N_K^{-1}(ve^{i(\alpha+\pi/2)}),$$

where $N_K : \partial K \to \mathbb{S}^1$ is the (outer) Gauss map of ∂K. The functions x, y are C^1 since N_K is C^1. The point $(x(\alpha), y(\alpha))$ is the only one in ∂K such that the clockwise oriented tangent vector to ∂K makes an angle α with the positive direction of the line L. A line parallel to L meets ∂K at a single point only when $\alpha + \pi/2 = \pi/2$ or $\alpha + \pi/2 = 3\pi/2$. Hence, for $\alpha \in (0, \pi)$, there is a unique $\beta \in (\pi, 2\pi)$ such that

$$(x(\beta), y(\beta)) - (x(\alpha), y(\alpha)) \in L.$$

Observe that, for $\alpha \in (0, \pi)$, we have $dy/d\alpha > 0$, and, for $\beta \in (\pi, 2\pi)$, we get $dy/d\beta < 0$. We can use the implicit function theorem (applied to $y(\beta) - y(\alpha)$) to conclude that β is a C^1 function of α. Moreover

$$\frac{d\beta}{d\alpha} = \frac{dy/d\alpha}{dy/d\beta} < 0,$$

as desired. □

Now we give the main construction in this section.

We fix a vector $v \in \mathbb{R}^2 \setminus \{0\}$ and the line $L_v = \{\lambda v : \lambda \in \mathbb{R}\}$. For every $\lambda \in \mathbb{R}$, we consider two half-lines, $r_\lambda^+, r_\lambda^- \subset \mathbb{R}^2$, extending from the point $p = \lambda v \in L_v$ with angles $\alpha(\lambda)$ and $\beta(\lambda)$, respectively. Here $\alpha : \mathbb{R} \to (0, \pi)$ is a non-increasing function, and $\beta(\lambda)$ is the composition of $\alpha(\lambda)$ with the function obtained in Lemma 4.1. Hence $\beta(\lambda)$ is a non-decreasing function. The line L_v can be lifted to the horizontal straight line $R_v = L_v \times \{0\} \subset \mathbb{H}^1$ passing through the point $(0, 0, 0)$, and the half-lines $r_\lambda^\pm$ can be lifted to horizontal half-lines $R_\lambda^\pm$ starting from the point $(\lambda v, 0)$ in the line R_v.

The surface obtained as the union of the half-lines R_λ^+ and R_λ^-, for $\lambda \in \mathbb{R}$, is denoted by $\Sigma_{v,\alpha}$. Since any $R_\lambda^\pm$ is a graph over $r_\lambda^\pm$ and $\bigcup_{\lambda \in \mathbb{R}}(r_\lambda^+ \cup r_\lambda^-)$ covers the xy-plane, we can write the surface $\Sigma_{v,\alpha}$ as the graph of a continuous function $u_\alpha : \mathbb{R}^2 \to \mathbb{R}$. Writing $v = e^{i\alpha_0}$, the surface $\Sigma_{v,\alpha}$ can be parametrized by $\Psi : \mathbb{R}^2 \to \mathbb{R}^3$ as follows:

$$\Psi(\lambda, \mu) = \begin{cases} \left(\lambda e^{i\alpha_0} + \mu e^{i(\alpha_0+\alpha(\lambda))}, -\mu\lambda \sin\alpha(\lambda)\right), & \mu \geqslant 0, \\ \left(\lambda e^{i\alpha_0} + |\mu| e^{i(\alpha_0+\beta(\lambda))}, -|\mu|\lambda \sin\beta(\lambda)\right), & \mu \leqslant 0. \end{cases} \tag{4.1}$$

Example 4.2 A special example to be considered is the sub-Riemannian cone Σ_α, where $\alpha \in (0, \pi)$. The projection of Σ_α to the horizontal plane $t = 0$ is composed of the line $y = 0$ and the half-lines starting from points in $y = 0$ with angles α and $-\alpha$. This cone can be parametrized, for $s \in \mathbb{R}, t \geqslant 0$, by

$$(u, v) \mapsto (u + v\cos\alpha, v\sin\alpha, -uv\sin\alpha)$$

when $y \geqslant 0$, and by

$$(u + v\cos\alpha, -v\sin\alpha, uv\sin\alpha)$$

when $y \leqslant 0$. A straightforward computation implies that Σ_α is the t-graph of the function

$$u_\alpha(x, y) = -xy + \cot\alpha\, y|y|. \tag{4.2}$$

Observe that

$$\lim_{\alpha\to 0} u_\alpha(x, y) = \begin{cases} +\infty, & y > 0, \\ 0, & y = 0, \\ -\infty, & y < 0, \end{cases} \tag{4.3}$$

so that the subgraph of Σ_α converges pointwise locally when $\alpha \to 0$ to a vertical half-space.

The following result provides some properties of u_α when $\alpha(\lambda)$ is a smooth function of λ (Fig. 2).

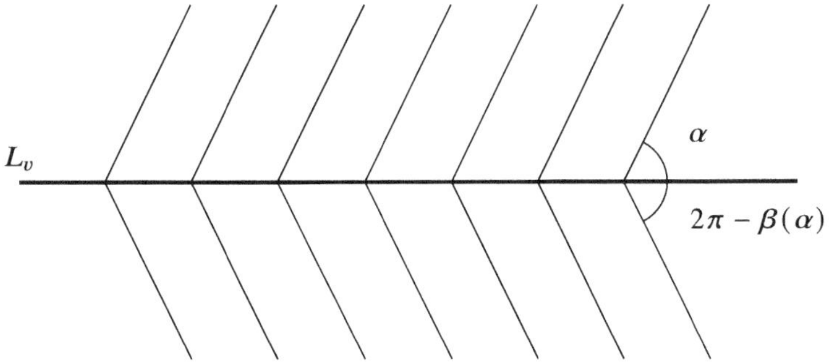

Fig. 2 The planar configuration to obtain the surface $\Sigma_{v,\alpha}$. Here α is a constant function and K is the unit disk D. Such surfaces were called *herringbone surfaces* by Young [31] as they are the union of horizontal rays that branch out of a horizontal line

Proposition 4.3 *Let $\alpha \in C^k(\mathbb{R})$, $k \geqslant 2$, be a non-decreasing function. Then*

(i) u_α *is a* C^k *function in* $\mathbb{R}^2 \setminus L_v$,
(ii) u_α *is merely* $C^{1,1}$ *near* L_v *when* $\beta \neq \alpha + \pi$.
(iii) u_α *is* C^∞ *in any open set* I *of values of* λ *when* $\beta = \alpha + \pi$ *on* I.
(iv) $\Sigma_{v,\alpha}$ *is* K*-perimeter-minimizing when* $\beta = \beta(\alpha)$.
(v) *The projection of the singular set of* $\Sigma_{v,\alpha}$ *to the* xy*-plane is* L_v.

Proof (i), (ii), (iii), and (v) are proven in Lemma 3.1 in [27].

We prove (iv) by a calibration argument. We shall drop the subscript α to simplify the notation. Let E be the subgraph of u and $F \subseteq \mathbb{H}^1$ such that $F = E$ outside a Euclidean ball centered at the origin. Let $P = \{(z,t) : \langle z, v\rangle = 0\}$, $P^1 = \{(z,t) : \langle z, v\rangle > 0\}$ and $P^2 = \{(z,t) : \langle z, v\rangle < 0\}$. We define two vector fields U^1, U^2 on P^1, P^2, respectively, by vertical translations of the vectors $\pi(\nu_E)|_{P^1} = \eta^+$ and $\pi(\nu_E)|_{P^2} = \eta^-$. They are C^2 in the interior of the half-spaces and extend continuously to the boundary plane P. As $\operatorname{div}(U^j)_{(z,t)}$ coincides with the sub-Finsler mean curvature of the translation of $\Sigma_{v,\alpha}$ passing through (z,t) as defined in [26], and these surfaces are foliated by horizontal straight lines in the interior of the half-spaces, by Theorem 3.14 in [26], we get

$$\operatorname{div} U^j = 0 \quad j = 1, 2.$$

Here $\operatorname{div} U$ is the Riemannian divergence of the vector field U. We apply the divergence theorem (Theorem 2.1 in [27]) to get

$$0 = \int_{F\cap P^j\cap B} \operatorname{div} U^j = \int_F \langle U^j, \nu_{P^j\cap B}\rangle |\partial(P^j \cap B)| + \int_{P^j\cap B} \langle U^j, \nu_F\rangle |\partial F|.$$

Let $C = P \cap \bar{B}$. Then, for every $p \in C$, we have $\nu_{P^1\cap B} = J(v)$ is a normal vector to the plane P and $\nu_{P^2\cap B} = -J(v)$, $U^1 = \eta^+$ and $U^2 = \eta^-$. Hence, by Lemma 4.1, we get

$$\langle U^1, \nu_{P^1\cap B}\rangle + \langle U^2, \nu_{P^2\cap B}\rangle = \langle \eta^+ - \eta^-, J(v)\rangle = 0 \quad p \in C.$$

Adding the above integrals, we obtain

$$0 = \sum_{j=1,2} \int_F \langle U^j, \nu_B\rangle d|\partial B| + \int_{B\cap \operatorname{int}(H^j)} \langle U^j, \nu_F\rangle d|\partial F|. \tag{4.4}$$

From the Cauchy-Schwarz inequality and the fact that $|\partial F|$ is a positive measure, we get that

$$\sum_{j=1,2} \int_{B\cap P^j} \langle U^j, \nu_F\rangle d|\partial F| \leqslant P_K(F, B). \tag{4.5}$$

In particular, if we apply the same reasoning to E, equality holds and

$$0 = \sum_{j=1,2} \int_E \langle U^j, \nu_B \rangle d|\partial B| + P_K(E, B). \tag{4.6}$$

From (4.4), (4.5), (4.6), and the fact that $F = E$ in the boundary of B, we get

$$P_K(E, B) \leqslant P_K(F, B),$$

as desired. □

The general properties of $\Sigma_{v,\alpha}$ when α is only continuous are given in the following result.

Proposition 4.4 *Let $\alpha : \mathbb{R} \to \mathbb{R}$ be a continuous and non-decreasing function. Then*

(i) *u_α is locally Lipschitz in Euclidean sense,*
(ii) *E_α is a set of locally finite perimeter in $\mathbb{H}^1$, and*
(iii) *$\Sigma_{v,\alpha}$ is K-perimeter-minimizing in $\mathbb{H}^1$.*

Proof (i) and (ii) are proven in [27], Proposition 3.2. Let

$$\alpha_\varepsilon(x) = \int_{\mathbb{R}} \alpha(y) \delta_\varepsilon(x - y) dy$$

the usual convolution, where δ is a Dirac function and $\delta_\varepsilon = \frac{\delta(x/\varepsilon)}{\varepsilon}$. Then α_ε is a C^∞ non-decreasing function, and α_ε converges uniformly to α on compact sets of $\mathbb{R}$. By Lemma 4.1, $\beta_\varepsilon = \beta(\alpha_\varepsilon)$ is a C^1 non-decreasing function. Since β is C^1 with respect to α, it follows the uniform convergence on compact sets of β_ε to a function $\bar{\beta}$.

Take $F \subset \mathbb{H}^1$ so that $F = E$ outside a Euclidean ball centered at the origin. We follow the arguments of the proof of iv) in Proposition 4.3 and define vector fields $\operatorname{div}(U_\varepsilon^j)$ by translating vertically $\pi(\nu_{E_\varepsilon})$, where E_ε is the subgraph of $\Sigma_{\alpha_\varepsilon}$, to obtain by the divergence theorem

$$\sum_{j=1,2} \int_{B \cap \operatorname{int}(P^i)} \langle U_\varepsilon^j, \nu_{E_\varepsilon} \rangle |\partial E_\varepsilon| = \sum_{j=1,2} \int_{B \cap \operatorname{int}(P^i)} \langle U_\varepsilon^j, \nu_F \rangle |\partial F|.$$

The left-hand side is the K-perimeter of E_ε, while the right-hand side is trivially bounded by the K-perimeter of F. Therefore

$$P_K(E_\varepsilon, B) \leqslant P_K(F, B).$$

Since E_ε converges uniformly in compact sets to E, we obtain the result. □

We study now with some detail the case when $\Sigma_{v,\alpha}$ is a C^∞ surface.

Corollary 4.5 *When α is constant, the surface $\Sigma_{v,\alpha}$ is a K-perimeter-minimizing cone in $\mathbb{H}^1$ of class $C^{1,1}$. The singular set is a horizontal straight line, and the regular part of $\Sigma_{v,\alpha}$ is a C^∞ surface.*

The following extends the already known result that in the sub-Riemannian setting, the surfaces $\Sigma_{v,\pi/2}$ are C^∞.

Lemma 4.6 *Let $v \in \mathbb{R}^2 \setminus \{0\}$ and $\alpha \in (0,\pi)$ be fixed. If K is centrally symmetric with respect to $O = \frac{1}{2}\eta^+ + \frac{1}{2}\eta^-$ then $\beta(\alpha) = \alpha + \pi$, where $\eta^+ = \pi(J(ve^{i\alpha}))$ and $\eta^- = \pi(J(ve^{i\beta}))$.*

Proof Let K be centrally symmetric with respect to O. Then η^- is the symmetric point of η^+. On the other hand, the convex body $K - O$ is symmetric with respect to the origin. Then the dual norm is even and, in particular, $\pi_{K-O}(-\nu^+) = -\pi_{K-O}(\nu^+)$. Now, since a translation takes symmetric points of $K-O$ with respect to the origin to symmetric points of K with respect to O, we get $\nu^- = -\nu^+$. This implies that $\beta(\alpha) = \alpha + \pi$. □

The existence of a convex body K of class C^2_+ such that $0 \in \operatorname{int}(K)$ for which $\Sigma_{v,\alpha}$ is C^∞ is studied in Corollary 4.7 and Proposition 4.8.

Corollary 4.7 *Let $v \in \mathbb{R}^2 \setminus \{0\}$ and $\alpha \in (0,\pi)$ be fixed. Then there exists a convex body K of class C^2_+ with $0 \in \operatorname{int}(K)$ such that $\Sigma_{v,\alpha}$ is C^∞.*

Proof To construct the convex body K, fix a point $p \in \{(x,y) : \langle (x,y), ve^{i\alpha}\rangle > 0\}$ and $O \in J(L) + p \cap L$, where L is the vector line generated by v. Then any K of class C^2_+ centrally symmetric with respect to O containing the origin such that $p \in \partial K$ and $ve^{i\alpha} \perp T_p \partial K$ satisfies the hypothesis of Lemma 4.6, where $\eta^+ = p$ and η^- is the symmetric of η^+ with respect to O. Thus, by (iii) in Proposition 4.3 we get that $\Sigma_{v,\alpha}$ is C^∞. □

Proposition 4.8 *Given a convex body K of class C^2_+ with $0 \in \operatorname{int}(K)$, there exists $v \in \mathbb{R}^2$ such that $\Sigma_{v,\pi/2}$ is C^∞.*

Proof Let p and q be points in K at maximal distance. Then the lines through p and q orthogonal to $q - p$ are support lines to K. Taking $v = q - p$ and setting $p = \eta^+$, we have $q = \eta^-$, while the vectors ν^+ and ν^- are over the line $L(v)$, that is, Z^+ Z^- make angles $\pi/2$ and $3\pi/2$ with $L(v)$. □

For fixed $v \in \mathbb{R}^2$, we define the surface $\Sigma^+_{v,\alpha}$ as the one composed of all the horizontal half-lines R^+_λ and $R^-_\lambda \subseteq \mathbb{R}^2$ extending from the lifting of the point $p = \lambda v \in L_v$, $\lambda \geqslant 0$, to $\mathbb{H}^1$. The surface $\Sigma^+_{v,\alpha}$ has a boundary composed of two horizontal lines and its singular set is the ray $L^+_v = \{\lambda v : \lambda > 0\}$. We present some pictures of such surfaces (Figs. 3, 4 and 5).

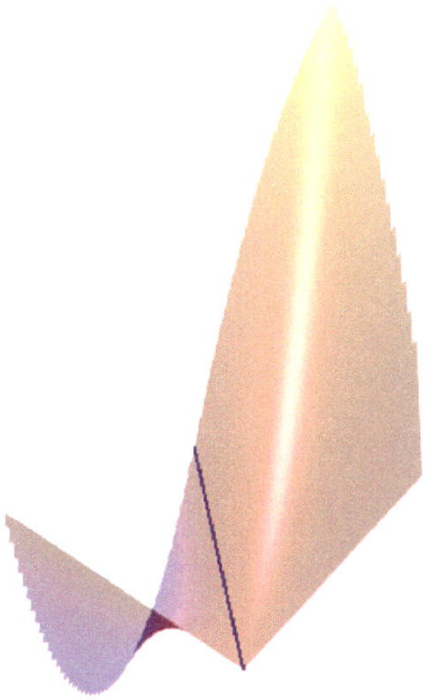

Fig. 3 The surface $\Sigma^+_{\pi/3,\pi/6}$ associated with the norm $||\cdot||_D$, where D is the unit disk. The singular set corresponds to the purple ray of angle $e^{i\pi/3}$

Fig. 4 The surface $\Sigma^+_{\pi/3,\pi/6}$ associated with the p-norm with $p = 1.5$. The left part of the figure coincides with the left part of Fig. 3, while the angle β is bigger. Notice that also the height has increased

5 Area-Minimizing Cones in $\mathbb{H}^1$

We proceed now to construct examples of K-perimeter-minimizing cones in $\mathbb{H}^1$ with an arbitrary finite number of horizontal half-lines meeting at the origin. The building blocks for this construction are liftings of circular sectors of the cones considered in Corollary 4.5.

We first prove the following result.

Fig. 5 The surface $\Sigma^+_{\pi/3,\pi/6}$ with $\beta = \alpha + \pi$. The existence of K is granted by Corollary 4.7

Lemma 5.1 *Let K be a convex body of class C^2_+ such that $0 \in \operatorname{int}(K)$. Let $u, w \in \mathbb{S}^1$, $\theta = \angle(u, w) > 0$. Then there exists $v \in \mathbb{S}^1$ such that the vector line L_v generated by v splits the sector determined by u and w into two sectors of oriented angles α and β such that $\alpha + \beta = \theta$. Moreover, the stationary condition $\pi(J(u)) - \pi_K(J(w)) \in L_v$ is satisfied.*

Proof Let $\nu_u = J(u)$, $\nu_w = J(w)$ and $\eta_u = \pi(\nu_u)$, $\eta_w = \pi(\nu_w)$, $\eta_u \neq \eta_w$ since π is a C^1 diffeomorphism. Thus there exists a unique line $\tilde{L}$ passing through η_u and η_w and $L = \tilde{L} - \eta_u$ is a straight line passing though the origin. Notice that $\tilde{L}$ splits ∂K in two connect open components ∂K_1 and ∂K_2. There exist two points $\eta_1 \in \partial K_1$ and $\eta_2 \in \partial K_2$ such that $L + \eta_1$ (resp. $L + \eta_2$) is the support line at η_1 (resp. η_2). Setting $v_1 = N_{\partial K}(\eta_1)$ and $v_2 = N_{\partial K}(\eta_2)$, we gain that v_i for $i = 1, 2$ is perpendicular to L. Without loss of generality, we set that $-J(v_1)$ belongs to the portion of plane identified by the θ, and $-J(v_2)$ belongs to the portion of plane identified by the $2\pi - \theta$. Then we set $v = -J(v_1)$. Notice that v splits θ in two angles $\beta = \angle(u, v)$, $\alpha = \angle(v, w)$ with $\theta = \alpha + \beta$ and $L = L_v$. □

Now we proceed with the construction inspired by the sub-Riemannian construction in [20]. For $k \geqslant 3$ consider a fixed angle θ_0 and family of positive oriented angles $\theta_1, \ldots, \theta_k$ such that $\theta_1 + \cdots + \theta_k = 2\pi$. Consider the planar vectors $u_0 = (\cos(\theta_0), \sin(\theta_0))$ and

$$u_i = (\cos(\theta_0 + \theta_1 + \cdots + \theta_i), \sin(\theta_0 + \theta_1 + \cdots + \theta_i)), \qquad i = 1, \ldots, k.$$

Observe that $u_k = u_0$. For every $i \in \{1, \dots, k\}$ consider the vectors u_{i-1}, u_i and apply Lemma 5.1 to obtain a family of k vectors v_i in $\mathbb{S}^1$ between u_{i-1} and u_i. We lift the half-lines $L_i = \{\lambda v_i : \lambda \geqslant 0\}$ to horizontal straight lines passing through $(0, 0, 0) \in \mathbb{H}^1$, and we also lift the half-lines

$$\lambda v_i + \{\rho u_{i-1} : \rho \geqslant 0\}, \qquad \lambda v_i + \{\rho u_i : \rho \geqslant 0\},$$

to horizontal straight lines starting from $(\lambda v_i, 0)$. This way we obtain a surface

$$C_K(\theta_0, \theta_1, \dots, \theta_k)$$

with the following properties.

Theorem 5.2 *The surface $C_K(\theta_0, \theta_1, \dots, \theta_k)$ is K-perimeter-minimizing cone which is the graph of a C^1 function.*

Proof $C_K(\theta_0, \theta_1, \dots, \theta_k)$ is a cone by construction. It is an entire graph since it is composed of horizontal lifting of straight half-lines in the xy-plane that covered the whole plane without intersecting themselves transversally. The K-perimeter-minimizing property follows in a similar way to from Proposition 2.4 in [20]. That it is the graph of a C^1 function is proven like in Proposition 3.2(4) in [20]. □

A particular example of area-minimizing cones are those who uses the sub-Riemannian cones C_α restricted to the circular sector with $\theta \in (-\alpha, \alpha)$ as model piece of the cone. Taking $K = D$, $k \geqslant 3$, and the angle $\alpha = \pi/k$, we define

$$C(k) = C_D\Big(\frac{\pi}{k}, \frac{2\pi}{k}, \dots, \frac{2\pi}{k}\Big).$$

Let us denote by u_k the functions in $\mathbb{R}^2$ whose graph is $C(k)$. The behavior when k tends to infinity of u_k in a disk is analyzed in the following result.

Proposition 5.3 *The sequence u_k converge to 0 uniformly on compact subsets of $\mathbb{R}^2$. Moreover, the sub-Riemannian area of u_k converges locally to the sub-Riemannian area of the plane $t = 0$. Moreover the sub-Riemannian area of u_k converges to the one of the plane $t = 0$.*

Proof Since u_k is obtained by collating some rotated copies of u_α, where $\alpha = \pi/k$, we can estimate the height of u_k by the height of u_α. By (4.2), using polar coordinates (r, θ), where $\theta \in [-\alpha, \alpha]$ and $r < r_0$, we get

$$|u_\alpha| \leqslant 2r_0^2 |\sin(\pi/k)|$$

on $D(r_0) = \overline{B}(0, r_0)$. The claim follows since $\lim_{k\to\infty} \sin(\pi/k) = 0$.

The sub-Riemannian area of the graph of u_k over $D(r_0)$ is given by

$$A_D(u_k, r_0) = \int_{D(r_0)} \|\nabla u_k + (-y, x)\| dx dy.$$

Since the sub-Riemannian perimeter is rotationally invariant, we can decompose the above integral as k times the area of the cone C_α in the circular sector with $\theta \in (-\alpha, \alpha)$ and $r < r_0$. By (4.2), it is immediate that

$$\|\nabla u_k(x, y) + (-y, x)\| = 2|y| \sin^{-1}(\alpha).$$

A direct computation shows that

$$A_D(u_k, r_0) = \frac{4\pi r_0^3}{3} \frac{1 - \cos \pi/k}{(\pi/k) \sin \pi/k}.$$

Then $A_D(u_k, r_0)$ tends to $\frac{2\pi r_0^3}{3}$ as $k \to +\infty$ (Figs. 6 and 7). □

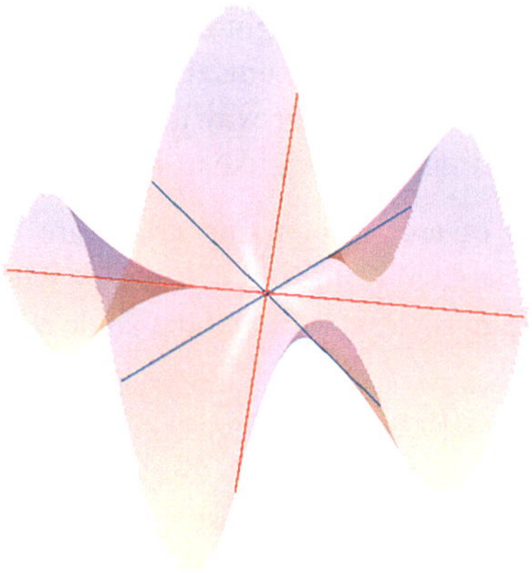

Fig. 6 The cone $C(4)$. The singular set is composed of the red rays of angle $0, \pi/2, \pi, (3\pi)/2$, while the rays of angles $\pi/4$, $(3\pi)/4$, $(5\pi)/4)$, $(7\pi)/4$, where two pieces of the construction meet, are depicted in cyan

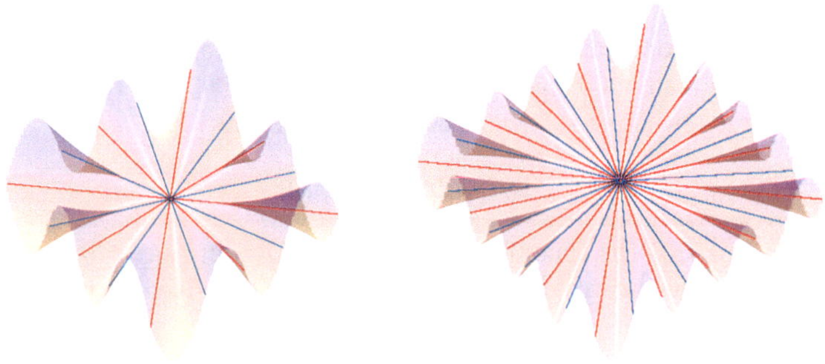

Fig. 7 The cones $C(8)$ and $C(16)$. They are depicted at the same in this Figure and the previous one. As the number of angles increases, the cone produces more oscillations of smaller height

References

1. Barone Adesi, V., Serra Cassano, F., Vittone, D.: The Bernstein problem for intrinsic graphs in Heisenberg groups and calibrations. Calc. Var. Partial Differ. Equ. **30**(1), 17–49 (2007)
2. Capogna, L., Danielli, D., Pauls, S.D., Tyson, J.T.: An Introduction to the Heisenberg Group and the Sub-Riemannian Isoperimetric Problem. Progress in Mathematics, vol. 259. Birkhäuser Verlag, Basel (2007)
3. Capogna, L., Citti, G., Manfredini, M.: Regularity of non-characteristic minimal graphs in the Heisenberg group $\mathbb{H}^1$. Indiana Univ. Math. J. **58**(5), 2115–2160 (2009)
4. Capogna, L., Citti, G., Manfredini, M.: Smoothness of Lipschitz minimal intrinsic graphs in Heisenberg groups $\mathbb{H}^n$, $n > 1$. J. Reine Angew. Math. **648**, 75–110 (2010)
5. Cheng, J.-H., Hwang, J.-F., Malchiodi, A., Yang, P.: Minimal surfaces in pseudohermitian geometry. Ann. Sc. Norm. Super. Pisa Cl. Sci. **4**(1), 129–177 (2005)
6. Cheng, J.-H., Hwang, J.-F., Yang, P.: Existence and uniqueness for p-area minimizers in the Heisenberg group. Math. Ann. **337**(2), 253–293 (2007)
7. Cheng, J.-H., Hwang, J.-F., Yang, P.: Regularity of C^1 smooth surfaces with prescribed p-mean curvature in the Heisenberg group. Math. Ann. **344**(1), 1–35 (2009)
8. Danielli, D., Garofalo, N., Nhieu, D.M.: A notable family of entire intrinsic minimal graphs in the Heisenberg group which are not perimeter minimizing. Amer. J. Math. **130**(2), 317–339 (2008)
9. Danielli, D., Garofalo, N., Nhieu, D.-M., Pauls, S.D.: The Bernstein problem for embedded surfaces in the Heisenberg group $\mathbb{H}^1$. Indiana Univ. Math. J. **59**(2), 563–594 (2010)
10. Franceschi, V., Monti, R., Righini, A., Sigalotti, M.: The isoperimetric problem for regular and crystalline norms in $\mathbb{H}^1$. J. Geom. Anal. **33**(1), 8, 40 (2023)
11. Franchi, B., Serapioni, R., Serra Cassano, F.: Rectifiability and perimeter in the Heisenberg group. Math. Ann. **321**(3), 479–531 (2001)
12. Galli, M.: First and second variation formulae for the sub-Riemannian area in three-dimensional pseudo-Hermitian manifolds. Calc. Var. Partial Differ. Equ. **47**(1–2), 117–157 (2013)
13. Galli, M.: On the classification of complete area-stationary and stable surfaces in the subriemannian Sol manifold. Pacific J. Math. **271**(1), 143–157 (2014)
14. Galli, M.: The regularity of Euclidean Lipschitz boundaries with prescribed mean curvature in three-dimensional contact sub-Riemannian manifolds. Nonlinear Anal. **136**, 40–50 (2016)
15. Galli, M., Ritoré, M.: Area-stationary and stable surfaces of class C^1 in the sub-Riemannian Heisenberg group $\mathbb{H}^1$. Adv. Math. **285**, 737–765 (2015)
16. Galli, M., Ritoré, M.: Regularity of C^1 surfaces with prescribed mean curvature in three-dimensional contact sub-Riemannian manifolds. Calc. Var. Partial Differ. Equ. **54**(3), 2503–2516 (2015)
17. Garofalo, N., Nhieu, D.-M.: Isoperimetric and Sobolev inequalities for Carnot-Carathéodory spaces and the existence of minimal surfaces. Comm. Pure Appl. Math. **49**(10), 1081–1144 (1996)
18. Giovannardi, G., Ritoré, M.: Regularity of Lipschitz boundaries with prescribed sub-Finsler mean curvature in the Heisenberg group $\mathbb{H}^1$. J. Differ. Equ. **302**, 474–495 (2021)
19. Giovannardi, G., Ritoré, M.: The Bernstein problem for (X, Y)-Lipschitz surfaces in three-dimensional sub-Finsler Heisenberg groups (2021). arXiv e-prints, page arXiv:2105.02179
20. Golo, S.N., Ritoré, M.: Area-minimizing cones in the Heisenberg group H. Ann. Fenn. Math. **46**(2), 945–956 (2021)
21. Hurtado, A., Ritoré, M., Rosales, C.: The classification of complete stable area-stationary surfaces in the Heisenberg group $\mathbb{H}^1$. Adv. Math. **224**(2), 561–600 (2010)
22. Monti, R., Serra Cassano, F., Vittone, D.: A negative answer to the Bernstein problem for intrinsic graphs in the Heisenberg group. Boll. Unione Mat. Ital. **1**(3), 709–727 (2008)
23. Nicolussi, S., Serra Cassano, F.: The Bernstein problem for Lipschitz intrinsic graphs in the Heisenberg group. Calc. Var. Partial Differ. Equ. **58**(4), 141, 28 (2019)

24. Pauls, S.D.: Minimal surfaces in the Heisenberg group. Geom. Dedicata **104**, 201–231 (2004)
25. Pinamonti, A., Serra Cassano, F., Treu, G., Vittone, D.: BV minimizers of the area functional in the Heisenberg group under the bounded slope condition. Ann. Sc. Norm. Super. Pisa Cl. Sci. **14**(3), 907–935 (2015)
26. Pozuelo, J., Ritoré, M.: Pansu-Wulff shapes in $\mathbb{H}^1$. Adv. Calc. Var. **16**(1), 69–98 (2023)
27. Ritoré, M.: Examples of area-minimizing surfaces in the sub-Riemannian Heisenberg group $\mathbb{H}^1$ with low regularity. Calc. Var. Partial Differ. Equ. **34**(2), 179–192 (2009)
28. Ritoré, M., Rosales, C.: Area-stationary surfaces in the Heisenberg group $\mathbb{H}^1$. Adv. Math. **219**(2), 633–671 (2008)
29. Rosales, C.: Complete stable CMC surfaces with empty singular set in Sasakian sub-Riemannian 3-manifolds. Calc. Var. Partial Differ. Equ. **43**(3–4), 311–345 (2012)
30. Sánchez, A.P.: A Theory of Sub-Finsler Area in the Heisenberg Group. PhD Thesis, Tutfs University (2017)
31. Young, R.: Harmonic intrinsic graphs in the Heisenberg group. Ann. Sc. Norm. Super. Pisa Cl. Sci. (To Appear). https://doi.org/10.2422/2036-2145.202105_054
32. Young, R.: Area-minimizing ruled graphs and the Bernstein problem in the Heisenberg group. Calc. Var. Partial Differ. Equ. **61**(4), 142, 32 (2022)

On the Double Soul Conjecture

David González-Álvaro and Luis Guijarro

Abstract A conjecture proposed by Karsten Grove (Geometry of, and via, symmetries. In: Conformal, Riemannian and Lagrangian Geometry (Knoxville, TN, 2000). University Lecture Series, vol. 27, pp. 31–53. American Mathematical Society, Providence, 2002) asks whether every closed simply connected Riemannian manifold with nonnegative sectional curvature can be written as the union of two disk bundles along their boundaries. If true, the conjecture would give a description of nonnegative curvature and would allow to attack several of the other main conjectures in the area. In this note, we discuss the context of this conjecture, provide some of the reasons for its possible validity, and discuss several related results.

Keywords Nonnegative sectional curvature · Soul theorem

1 Introduction

The geometrical and topological structure of open (non-compact complete without boundary) Riemannian manifolds of nonnegative sectional curvature is considerably well understood, thanks to the soul theorem of Cheeger and Gromoll in 1972 [7]. They showed that such a space X contains a totally convex and totally geodesic closed (compact without boundary) submanifold S (called the soul) for which the total space of its normal bundle in X is diffeomorphic to X.

As a consequence of the soul theorem, the classification of *all* open manifolds of nonnegative curvature reduces to the following two tasks:

D. González-Álvaro (✉)
ETSI de Caminos, Canales y Puertos, Universidad Politécnica de Madrid, Madrid, Spain
e-mail: david.gonzalez.alvaro@upm.es

L. Guijarro
Universidad Autónoma de Madrid and ICMAT-CSIC, Madrid, Spain
e-mail: luis.guijarro@uam.es

A. Alarcón et al. (eds.), *New Trends in Geometric Analysis*,
RSME Springer Series 10, https://doi.org/10.1007/978-3-031-39916-9_8

- Classify *all* closed Riemannian manifolds of nonnegative sectional curvature,
- Classify which vector bundles over each of the spaces in the previous item admit a metric of nonnegative curvature.

Unfortunately, little is known about either of the items above. Regarding the first item, and in contrast to the non-compact case, both the geometrical and topological structure of closed nonnegatively curved manifolds are poorly understood; see Sect. 2 for a brief review of the current knowledge. In this survey we discuss a conjecture which asks about their geometrical structure. For convenience we use the following definition, which does not involve any Riemannian metric nor curvature condition.

Definition
A closed manifold is said to admit a double disk bundle decomposition if it can be written as the union $D(E_1) \cup D(E_2)$ of two disk bundles $D(E_i) \to S_i$ glued together along their common boundary $\partial D(E_1) = \partial D(E_2)$ by a diffeomorphism.

Double disk bundle decompositions arise in several geometrical situations; see Sect. 3 for some background. In particular, disk bundles and double disk bundle decompositions play an important role in the context of nonnegative curvature, as we next explain.

On the one hand, the arguments of Cheeger and Gromoll in the proof of the soul theorem can be adapted to the situation where the manifold X is compact with non-empty totally geodesic boundary [6, Remark in p. 626]. In this case, X similarly contains a totally convex and totally geodesic closed submanifold S (the soul) such that X is diffeomorphic to the total space of the disk bundle associated with the normal bundle of $S \subset X$.

On the other hand, many closed manifolds of nonnegative curvature have been constructed as the gluing of two disk bundles. Karsten Grove conjectured in [21] that *all* closed manifolds of nonnegative curvature should carry such a decomposition.

Double Soul Conjecture (Karsten Grove)
A closed simply connected Riemannian manifold with nonnegative sectional curvature carries a double disk bundle decomposition.

Despite the fact that the number of existing methods to construct nonnegatively curved manifolds is very small, it is already unknown whether all the resulting spaces satisfy the conjecture. More precisely, it is an open question whether an arbitrary biquotient carries a double disk bundle decomposition. This indicates how challenging would be to tackle the conjecture in the general case.

Moreover, if the conjecture were true, it is not clear how the double disk bundle decomposition would be necessarily related to the metric. Some evidence for this is that the base space of each disk bundle would not be a totally geodesic submanifold; see Sect. 4. This is in contrast to the soul theorem, where the soul *is* a totally geodesic submanifold of the ambient space.

After a first version of this survey was finished, Jason DeVito made available a preprint [13] where he studies a generalized version of the double soul conjecture in which there is no restriction for the fundamental group of the manifolds. In that paper, DeVito shows that there exist *non-simply connected* nonnegatively curved closed manifolds which admit no double disk bundle decomposition.

Organization of the Paper In Sects. 2 and 3 we give some background on nonnegatively curved manifolds and on double disk bundle decompositions, respectively. In Sect. 4 we discuss the most obvious difficulties for a potential proof in the general case. Section 5 reviews previous constructions of nonnegatively curved manifolds as double disk bundles. Section 6 examines some of the metric features, under ideal conditions, of a nonnegatively curved metric obtained by gluing of two disc bundles. In Sect. 7 we compare existing results on double disk bundles and on nonnegatively curved manifolds. In Sect. 8 we study which biquotients satisfy the conjecture. Section 9 discusses the relation between the double soul conjecture and the Bott-Grove-Halperin conjecture.

2 Context of Nonnegatively Curved Manifolds

In this section, we briefly review the state of the art for manifolds of nonnegative curvature. For further information, we refer to the surveys [22, 41, 42], which contain references to all the results that we mention here.

Manifolds of nonnegative curvature lie within the broader context of spaces with lower curvature bounds. These include manifolds with other constraints on the sectional, Ricci, or scalar curvature, but also singular spaces like Alexandrov spaces or RCD spaces [4, 17]. Lower curvature bounds arise in natural geometrical situations like taking Gromov-Hausdorff limits or isometric quotients.

Probably the most classical curvature bound is positive sectional curvature. Its investigation has seen important advances since the last century, including various sphere theorems, the classification of homogeneous positively curved manifolds, the construction of new inhomogeneous examples, or structural results under additional (often symmetry) assumptions. However, many questions remain unanswered, and weaker curvature conditions have been investigated, one of them being precisely nonnegative sectional curvature. One particularly interesting question is whether a nonnegatively curved metric can be deformed to a positively curved metric. This is known to be false in general by Synge's theorem, but the question remains open in the simply connected case.

The study of nonnegatively curved manifolds became very active in the 1970s with seminal works like the soul theorem of Cheeger and Gromoll, the construction by Gromoll and Meyer of the first nonnegatively curved metric on an exotic sphere, or Cheeger's construction of nonnegatively curved metrics on the connected sum of two compact rank one symmetric spaces.

Unfortunately, not much is yet known about the class of manifolds admitting a metric of nonnegative curvature. Both the number of existing methods to construct such manifolds and the number of known topological obstructions to admit such a metric are small. Let us review them.

The existing examples of closed manifolds with nonnegative sectional curvature have been constructed in one of the two following ways:

- as the gluing of two adequate disk bundles along their boundary, or
- as biquotients.

The spaces in the first item evidently satisfy the double soul conjecture, and the concrete constructions shall be outlined in Sect. 5. As for biquotients, their definition as well as a discussion on whether they satisfy the conjecture is contained in Sect. 8.

The power of these two methods may be illustrated by the following facts:

- every dimension ≥ 6 is known to support infinitely many closed simply connected nonnegatively curved manifolds of pairwise distinct homotopy type,
- the only dimensions where some rational (non-standard) sphere is known to carry a nonnegatively curved metric are 5 and $4n - 3$ with $n \geq 2$,
- while in dimension 7 all exotic spheres are known to carry a nonnegatively curved metric (thanks to the work of Gromoll and Meyer in 1974, Grove and Ziller in 2000 [25], and Goette, Kerin and Shankar in 2020 [18]), it is unknown whether any other exotic sphere of dimension $\neq 7$ admits such a metric.

Moving to topological obstructions, essentially the only one specific to closed simply connected nonnegatively curved manifolds M is Gromov's Betti Number Theorem. It states that, for any field of coefficients F, the total Betti number $\sum_i \operatorname{rank} H^i(M, F)$ of any n-dimensional manifold M is bounded above by a constant depending only on n.

Gromov's theorem can be used to rule out the existence of nonnegatively curved metrics on many simply connected manifolds. For example, for any nonnegatively curved manifold M not being a homotopy sphere, there is some integer k such that the connected sum $M\sharp \ldots \sharp M$ of k copies of M does not admit a nonnegatively curved metric. This implies that surgery constructions will not preserve nonnegative curvature, in general.

3 Context of Double Disk Bundles

In this section we review some classical situations and results where double disk bundles appear. More recent and specific results will appear in the following sections.

The most visual double disk bundle decomposition is that of a sphere S^n as the union of two disks $D^n \cup D^n$ (hemispheres) glued along their boundary (the equator).

In Milnor's discovery of several *seven* exotic spheres, he described them as double disk bundles in two ways [35, Chapter 30, p. 377]: as $(S^3 \times D^4) \cup_{\phi_1} (S^3 \times D^4)$ and as $D^7 \cup_{\phi_2} D^7$, for certain self-diffeomorphisms ϕ_1 and ϕ_2 of S^3 and S^6, respectively.

The construction of gluings of the form $(S^{n-1} \times D^n) \cup_\phi (S^{n-1} \times D^n)$ for self-diffeomorphisms ϕ of $S^{n-1} \times S^{n-1}$ is studied in [3, I.7]. Depending on ϕ, the resulting manifolds may have the homology of S^{2n-1} or $S^{n-1} \times S^n$.

We remark that each of the disk bundles can be different, and, moreover, the corresponding fiber (i.e., the disk) can be of different dimension. For example, the sphere S^n can be written as $(S^p \times D^{q+1}) \cup (D^{p+1} \times S^q)$; see the Introduction of [14].

Besides the explicit constructions above, double disk bundles arise in quite a variety of geometrical situations. It is the case of a subclass of cohomogeneity one manifolds, i.e., manifolds M endowed with a Lie group action G such that the orbit space M/G is 1-dimensional. When M/G is an interval $[-1, 1]$ (or equivalently when M is closed and its fundamental group is finite), the space M automatically carries a very concrete double disk bundle decomposition in terms of the isotropy groups of the action; see [25, Section 1] for details. This gives an abundant source of manifolds carrying double disk bundles decompositions, e.g. all compact rank one symmetric spaces.

We remark that the class of closed manifolds admitting a double disk bundle decomposition is strictly larger than that of closed cohomogeneity one manifolds. In the simply connected case, the example of lowest dimension is the connected sum $\mathbb{C}P^2 \sharp \mathbb{C}P^2$. This space is a double disk bundle (see, e.g., [6]) but is known to admit no cohomogeneity one action (see [28]).

Double disk bundle decompositions arise in other contexts of Riemannian Geometry such as Dupin hypersurfaces, isoparametric hypersurfaces, or certain singular Riemannian foliations. We refer to the introduction of [14] for a discussion of these situations and for the corresponding references.

For later convenience, we recall the following terminology for some simple cases of double disk bundles:

- The *double* of a disk bundle $D(E) \to S$ is the gluing $D(E) \cup D(E)$ via the identity map on the boundary $\partial D(E)$.
- A *twisted double* of a disk bundle $D(E) \to S$ is the gluing $D(E) \cup_\phi D(E)$ via a self-diffeomorphism ϕ of the boundary.

These notions can be generalized to the situation where the disk bundle $D(E)$ is replaced by any compact manifold W with non-empty boundary ∂W; see [35, Chapter 30, p. 377]. In this more general context, a remarkable theorem of Barden and Smale states that every closed odd-dimensional manifold M is a twisted double $M = W \cup_\phi W$; see [35, Chapter 30, p. 377]. In contrast, it has been proved in [14, Corollary C] that, among all (infinitely many) simply connected closed 5-dimensional manifolds, only four of them (S^5, $S^3 \times S^2$, the Wu-manifold $SU(3)/SO(3)$ and the non-trivial bundle $S^3 \hat{\times} S^2$) admit a double disk bundle decomposition.

4 Difficulties for a Proof: Absence of Totally Geodesic Hypersurfaces and Souls

As already anticipated in the introduction, an important difficulty toward a proof for the double soul conjecture is that a potential double disk bundle decomposition of a closed nonnegatively curved manifold does not seem to be related to the metric in any obvious way. Here we review some basic observations in this direction.

The natural strategy to prove the conjecture for a closed manifold would be to find an adequate hypersurface diving the manifold in two halves and then to apply an adapted version of the soul theorem to each half. This would yield the desired double disk bundle decomposition.

A special situation where this strategy can be applied is when the manifold contains a totally geodesic hypersurface with trivial normal bundle. In this case, the proof of Cheeger and Gromoll applies in a straightforward way to each half, as observed by Cheeger in [6, Remark in p. 626].

However, it is quite uncommon for a nonnegatively curved manifold to possess a totally geodesic hypersurface. Already among spaces with a large isometry group, which naturally tend to have totally geodesic submanifolds, one can find many examples with no totally geodesic hypersurfaces. For example, round spheres are the only irreducible simply connected compact symmetric spaces admitting totally geodesic hypersurfaces [8] (we refer to [34] for a complete classification of homogeneous spaces admitting totally geodesic hypersurfaces and for references to previous works). Thus, if a nonnegatively curved manifold has a double disk bundle decomposition, the properties of the hypersurface along which the disk bundles are glued are unclear.

In any case, and coming back to a potential strategy for a proof of the conjecture, suppose you have overcome the difficulty of finding an adequate hypersurface separating the manifold into two halves. Then, it is not obvious how the soul theorem should be applied to each half. More precisely, the soul theorem in its standard form would yield the existence of totally geodesic submanifolds in M: one in each half (namely, its soul), each of which would be the base of a disk bundle whose gluing would recover the diffeomorphism type of M. Unfortunately, nonnegatively curved manifolds do not have two totally geodesic “souls” in general. Let us be more precise.

Take any of the known examples with (not only nonnegative but) positive curvature and slightly perturb the metric. Then, on the one hand, one retains positive curvature (since it is an open condition) and, in particular, nonnegative curvature. On the other hand, such a perturbed metric will have no totally geodesic submanifolds of dimension ≥ 2 in general; see [33]. This implies that if the manifold admitted a double disk bundle decomposition with the base S_i of each disk bundle being totally geodesic for the perturbed metric, then $\dim S_i$ would be ≤ 1. However, not all the known positively curved examples carry such a decomposition. As we have been informed by Jason DeVito, the methods from the article [14] can be used to show that the Wallach homogeneous space $SU(3)/T^2$ does not admit a double disk bundle decomposition with base spaces of dimension ≤ 1. Thus, a potential proof for the double soul conjecture should deal with two submanifolds playing the role of the soul of each half, but being not necessarily totally geodesic.

5 Construction of Examples as Double Disk Bundles

An obvious indication that the conjecture might hold is that many of the known examples of closed manifolds of nonnegative curvature have been constructed as the gluing of two adequate disk bundles along their boundary, as already mentioned in Sect. 2. Such constructions are known to be doable only in some rather special situations. And even when such a construction is possible, it is often a very hard task to determine the diffeomorphism type of the resulting manifold out of the geometric information of each disk bundle. We next review the history of the known constructions.

The first gluing in this context was due to Cheeger [6]. He considered compact rank one symmetric spaces $\mathbb{K}P^n$ with a disk D of the same dimension removed, where $\mathbb{K}$ is any of $\mathbb{R}$, $\mathbb{C}$, $\mathbb{H}$, or Ca. He used the description of $\mathbb{K}P^n \setminus D$ as the normal disk bundle of the natural inclusion $\mathbb{K}P^{n-1} \subset \mathbb{K}P^n$ to construct a nonnegatively curved metric which on the boundary is isometric to the round metric on the sphere. As a consequence, he obtained metrics of nonnegative curvature on the connected sum of any two compact rank one symmetric spaces of the same dimension.

Cheeger already observed in [6, Remark in p. 626] that some of his gluings, namely, $\mathbb{C}P^2\sharp\mathbb{C}P^2$ and $\mathbb{C}P^2\sharp-\mathbb{C}P^2$, are not diffeomorphic to any homogeneous space. In 2002, Totaro [39] determined which of the Cheeger manifolds were diffeomorphic to biquotients. It turned out that most of them were, except for a few cases which are not even homotopy equivalent to any biquotient.

In 1998 [26], the second author of the present survey showed that any nonnegatively curved metric g on an open manifold X with soul S can be deformed to a nonnegatively curved metric $\bar{g}$ which splits as a metric product $N \times [a, \infty)$ outside some tubular neighborhood of S, where N is diffeomorphic to the unit normal bundle of S in X.

This metric deformation has an immediate consequence for the construction of closed nonnegatively curved manifolds. If $D(S)$ denotes the normal disc bundle of S in X, then there is a nonnegatively curved metric on the double $D(S) \cup_{\mathrm{Id}} D(S)$ of $D(S)$. The same construction works more generally for the twisted doubles $D(S) \cup_\phi D(S)$ glued along any isometry ϕ of the boundary of $D(S)$. In order to apply this method, two tasks need to be done:

- construct nonnegatively curved metrics on open manifolds (which by the soul theorem must be total spaces of vector bundles over closed nonnegatively curved manifolds), and
- determine the diffeomorphism type of the (twisted) doubles induced by the spaces of the previous item.

The first item has been investigated for vector bundles over homogeneous spaces, biquotients, and cohomogeneity one manifolds; see [20] and the references therein. The second item seems too hard to treat in certain generality, but it could produce interesting manifolds in some particular cases.

The next construction came in the year 2000. Grove and Ziller [25] showed that any cohomogeneity one G-manifold M with singular orbits of codimension 2 admits a G-invariant metric of nonnegative curvature. By the O'Neill Formula for Riemannian submersions, the quotient M/L by any subgroup $L < G$ acting freely on M also carries a nonnegatively curved metric (see also [41, Theorem 2.8]). While M is foliated by homogeneous spaces, M/L is foliated by biquotients.

Again, it seems unmanageable to determine all diffeomorphism types of the manifolds belonging to these classes. However, Grove and Ziller discovered very interesting manifolds within them and hence carrying nonnegatively curved metrics: all of the ten Milnor 7-spheres and more generally all S^3-bundles over S^4, all principal $S^3 \times S^3$-bundles over S^4, or all of the four homotopy $\mathbb{R}P^5$'s. Only recently, Goette, Kerin, and Shankar discovered in [18] larger families of ten manifolds and seven manifolds, generalizing those of Grove and Ziller. The 7-dimensional family is specially interesting as it realizes all exotic seven spheres, as well as infinitely many spaces with the cohomology ring of some S^3-bundle over S^4 but of different homotopy type [19].

We finally mention other works dealing with gluings and nonnegative curvature. Schwachhöfer and Tapp studied in [36] generalizations of the metric construction of Grove and Ziller. Torres constructed nonnegatively curved metrics on various non-simply connected manifolds in dimension 4 [38]. This last construction has been extended to other dimensions by Torres and the second author in [27] to produce nonnegatively curved non-simply connected manifolds in all dimensions ≥ 4, including examples in dimensions 4 (resp. 5) that are not even homotopy equivalent (resp. homeomorphic) to any biquotient or cohomogeneity one manifold.

6 The Perfect Nonnegatively Curved Glued Metric and Some of Its Properties

In [26], there is a procedure to change the metric of an open nonnegatively curved Riemannian manifold X to another with a cylindrical end. Thus, the double of X carries a nonnegatively curved metric, although more interesting gluings are possible. This has been extended in [27] to a procedure that allows, in certain cases, to construct nonnegatively curved metrics in the gluing of two possibly different disc bundles D_1 and D_2. Specifically, it is enough to have nonnegatively curved metrics on each side, with isometric boundaries and with the normal exponential map of the soul in each side being a diffeomorphism for the disc bundle of the right radius. This gives some motivation to consider the metric structure of such gluings. Since it falls far from the reach of this survey, we just make a few simple observations here.

We start with two disc bundles with metrics of nonnegative sectional curvature

$$D^{k_1} \to N_1 \to B_1, \qquad D^{k_2} \to N_2 \to B_2,$$

such that, for $i = 1, 2$,

1. ∂N_i are convex hypersurfaces of N_i and isometric for the induced metrics;
2. N_i has soul isometric to B_i;
3. N_i is a tubular neighborhood of B_i of some radius r_i;
4. $\exp : \nu_{r_i}(B_i) \to N_i$ is a diffeomorphism.

Then we can use the construction in [26] (see specific details in [27]) to modify the metric on each N_i so that together to all the above properties we also have that, outside of a tubular neighborhood of radius $r_i/2$, N_i is isometric to a metric product $\partial N_i \times [r_i/2, r_i]$. Since we assumed from the beginning that $\partial N_1 \simeq \partial N_2$, we can use any isometry $f : \partial N_1 \to \partial N_2$ to construct a metric g in the gluing

$$M := N_1 \cup_f N_2.$$

We call g a *perfect glued metric* and examine now some of its properties.

- B_1 and B_2 are metric dual sets; by this we mean that, if

$$r_1 + r_2 > \max\{\text{diam } B_1, \text{diam } B_2\}, \tag{1}$$

  then

$$B_2 = \max \text{dist}_{B_1}, \quad B_1 = \max \text{dist}_{B_2}$$

  Observe that the construction of the perfect glued metric allows us to increase the values of r_1 and r_2 arbitrarily so that the above restriction (1) is satisfied.
- We have that

$$\exp_1 : \nu_{r_1+r_2}(B_1) \to M \setminus B_2$$

  is a diffeomorphism, with a similar statement for B_2.
- B_1 is the cut locus of B_2 and vice versa.
- Let $p \in M \setminus (B_1 \cup B_2)$, and let

$$v = \frac{\exp_1^{-1}(p)}{\|\exp_1^{-1}(p)\|} \in \nu^1(B_1), \qquad w = \frac{\exp_2^{-1}(p)}{\|\exp_2^{-1}(p)\|} \in \nu^1(B_2). \tag{2}$$

  Then the geodesics with initial conditions v and w satisfy

$$\gamma_v(t) = \gamma_w(r_1 + r_2 - t). \tag{3}$$

- The set at distance $r_0 := \frac{r_1+r_2}{2}$ from either B_1 or B_2 is a closed totally geodesic hypersurface that we denote by H.
- There are Riemannian submersions $\pi_i : H \to B_i$ such that, if we denote by S_{r_0} the normal sphere bundle of radius r_0,

$$\pi_i \circ \exp_i : S_{r_0}(B_i) \to B_i$$

is the bundle projection map.
- Associated with the submersions π_i, there are vector bundle decompositions into horizontal and vertical vectors

$$TH = \mathcal{H}_1 \oplus \mathcal{V}_1 = \mathcal{H}_2 \oplus \mathcal{V}_2,$$

that extend trivially (after adding the line bundle generated by minimal geodesics connecting B_1 to B_2) to the whole of $M \setminus \{ B_1, B_2 \}$.
- For each direction $v \in \mathcal{H}_{1,p} \cap \mathcal{H}_{2,p}$, there is a totally geodesic immersed flat $\phi : \mathbb{R} \times [0, r_1+r_2] \to M$. This follows from Perelman's rigidity theorem applied to each half of M, namely, each direction in $\mathcal{H}_1$ provides a half flat in N_1 and similarly with $\mathcal{H}_2$, N_2. The condition $v \in \mathcal{H}_{1,p} \cap \mathcal{H}_{2,p}$ and the uniqueness of a geodesic from its initial conditions implies that these flats match smoothly at the glued part.
- There is a fixed point free involution $a : H \to H$: for any $p \in H$, consider the minimal geodesic normal to B_1 passing through p, and follow it until it hits H again. More precisely, since H lies at distance r_0 from both B_1 and B_2, we define

$$a : H \to H, \qquad a(p) = \exp_1(3 \exp_1^{-1}(p)).$$

If $a(p) = p$ for some $p \in H$, then for v, w as in (2), we would get

$$q = \exp_2(r_0 \cdot w) = \exp_2(-r_0 \cdot w),$$

contradicting that $\exp_2$ is a diffeomorphism in $\nu_{r_1+r_2}(B_2)$.

The map a does not need to preserve orientation or be homotopic to the identity; this can be seen easily by considering the gluing of two copies of $B \times D^\ell$ by the identity map. a will then agree with the product of the identity in B and the antipodal map in ∂D^ℓ.
- The construction of $a : H \to H$ can be used to show that there is an extension

$$a : M \setminus \{ B_1, B_2 \} \to M \setminus \{ B_1, B_2 \}$$

that is still a diffeomorphism and interchanges the two halves separated by H.
- By changing the number 3 in the definition of the map a by some other odd integer, we get a family of involutions

$$a_k : H \to H, \qquad a_k(p) = \exp_1((2k + 1) \exp_1^{-1}(p)).$$

It is clear from the construction that

$$a_k \circ a_\ell = a_{k-\ell}, \qquad \text{for all } k, \ell \in \mathbb{Z},$$

It would be interesting to examine the topological dynamics of this collection of maps.

- Let $p \in H$, and denote by $F_1(p)$, $F_2(p)$ the fibers through p of π_1 and π_2 respectively. Then

$$a_k(F_1(p) \cap F_2(p)) = F_1(a_k(p)) \cap F_2(a_k(p)), \qquad \text{for all } k \in \mathbb{Z}.$$

- $\mathcal{V}_1 \cap \mathcal{V}_2$ is the tangent space to $F_1(p) \cap F_2(p)$.

If we allow taking r_1 and r_2, sufficiently large, we can obtain nonnegatively curved metrics on M with some special properties. Without loss of generality, we will assume that $r_1 = r_2 = r$.

- There are metrics g_r in M with $\lambda_1(M) \to 0$ as $r \to \infty$, where $\lambda_1(M)$ is the first eigenvalue of the Laplacian. To see this, observe that the Cheeger constant of M (defined as

$$h(M) = \inf_E \frac{A(E)}{\min V(A), V(B)},$$

where E is a smooth hypersurface dividing M into two pieces A and B), will tend to zero when $r_1 = r_2 = r \to \infty$, since in that case, for $E = H$,

$$A(H) = \text{ constant}, \quad V(A), V(B) \simeq C \cdot r$$

for some constant $C > 0$. Then Buser's inequality [5] provides $\lambda_1(M) \leq 10h(M)^2 \to 0$.
- By rescaling the metric g_r by $1/r$, we obtain metrics g_r in M collapsing to the interval $[0, 1]$ in the Gromov-Hausdorff distance as $r \to \infty$.

7 Intertwining Results from Double Disk Bundles and Nonnegative Curvature

In this section we discuss a question which is actually more general than the conjecture: which smooth manifolds (i.e., not necessarily endowed with any Riemannian metric) admit a double disk bundle decomposition?

While this question seems unmanageable in general, there exist very interesting results in some particular cases. Even more interesting to us, they are very related to other existing results in the context of nonnegative curvature.

The first results to review deal with low dimensions. There exist complete classifications of closed simply connected manifolds which admit a double disk bundle decomposition in dimensions 4 and 5, due to Ge and Radeschi [16] and to DeVito et al. [14], respectively. The lists consist of S^4, $\mathbb{C}P^2$, $S^2 \times S^2$, and both $\mathbb{C}P^2 \sharp \pm \mathbb{C}P^2$ and of S^5, $S^2 \times S^3$, $SU(3)/SO(2)$ and $S^2 \hat{\times} S^3$ (the non-trivial S^3-bundle over S^2), respectively. It is remarkable that these lists coincide, respectively, in each dimension, with those of the known simply connected closed nonnegatively curved manifolds, which in turn coincide with the classification [9] of closed simply connected biquotients.

DeVito et al. [14] moreover obtained a partial classification in dimension 6. They showed that the only 6-dimensional manifolds with vanishing second Betti number $b_2 = 0$ which admit a double disk bundle decomposition are S^6 and $S^3 \times S^3$. Again, the only known examples of nonnegatively curved closed simply connected manifolds with $b_2 = 0$ are S^6 and $S^3 \times S^3$.

In arbitrary dimensions, it seems hard to determine whether an arbitrary manifold admits a double disk bundle decomposition. There are, however, certain situations where one can ensure such a decomposition.

One such situation has been studied in the recent thesis [32] by Andrew Lutz, supervised by Jason DeVito and Krishnan Shankar. Its Theorem 0.0.1 states that a manifold $M = (S^{n_1} \times \cdots \times S^{n_r})/T^k$ constructed as the quotient of a product of spheres under a torus linear action carries a (very specific) double disk bundle decomposition. All of these spaces carry nonnegatively curved submersion metrics from the product of round metrics on the total space. Hence Lutz's result, combined with the topological results from [40] (see Proposition 2.1 therein and the first paragraph in page 231), yields examples of double disk bundles carrying nonnegatively curved metrics in infinitely many homotopy types in each odd dimension ≥ 7 and in each even dimension ≥ 14.

Instead of imposing conditions on the diffeomorphism type of the manifold, one can do so on the metric. Spindeler investigated *fixed point homogeneous* metrics, i.e., such that the action of the isometry group G on the corresponding manifold M has nonempty fixed point set M^G and the equality $\dim M/G - \dim M^G = 1$ holds, where $\dim M/G$ denotes the dimension of the regular part of orbit space M/G. One of the consequences of Spindeler's work is that fixed point homogeneous nonnegatively curved manifolds carry double disk bundle decompositions [37, Theorem 4.1].

DeVito, Galaz-García, and Kerin derived further sufficient conditions for a closed (not necessarily Riemannian) manifold M to admit a double disk bundle decomposition; see [14, Proposition 3.1]. These conditions cover in particular all known examples of manifolds with positive sectional curvature; see [14, Theorem 3.3]. They also cover some (but not all) biquotients, as explained in Sect. 8.

One of the observations in [14, Proposition 3.1] is that the total space of a fiber bundle over a double disk bundle is itself a double disk bundle. This method can be iterated to produce many examples. In particular, and when allowing for an infinite fundamental group, it can be used to construct examples of double disk bundles that admit no metric of nonnegative curvature, as follows. Since any infranilmanifold is itself an iterated principal circle bundle [1], we immediately obtain that every nilmanifold is a double disk bundle. However, a metric of nonnegative sectional curvature on a nilmanifold needs to be flat, since its universal cover would need to be the Euclidean space (see Theorem 9.1 in [7]). Therefore, most nilmanifolds are double disc bundles, but will never admit a nonnegatively curved metric.

We finally mention the recent preprint of DeVito [12], where he studies the question of which rational spheres admit *linear* double disk bundle decompositions (i.e., each disk bundle extends to a vector bundle). He proves that a rational sphere M of even dimension ≥ 6 admits a linear double disk bundle decomposition if and only if M is a homotopy sphere. For $n = 4$, the result of Ge and Radeschi above implies that the only rational sphere admitting a double disk bundle decomposition is actually S^4. Parallelly, as mentioned in Sect. 2, there is no example of an even-dimensional rational sphere (other than the standard sphere S^{2n}) with a nonnegatively curved metric.

DeVito's result should be compared to the situation in dimensions 5 and $4n - 1$ with $n \geq 2$, where there are rational spheres which are not homotopy spheres but still admit a (linear) double disk bundle decomposition. Examples are given by the Wu manifold $SU(3)$ or the unit tangent bundle $T^1 S^{2n}$, $n \geq 2$. Moreover, these spaces are homogeneous and hence carry nonnegatively curved metrics (and hence provide the examples mentioned in Sect. 2). As far as we know, there are no analogous results in dimensions of the form $4n + 1$ with $n \geq 2$.

8 Proof for Biquotients

As mentioned in Sect. 2, biquotients constitute one of the two known sources for the construction of nonnegatively curved manifolds. Here we recall their definition, and afterward we discuss their relation to the conjecture.

A (closed) biquotient is defined as the quotient of a (closed) homogeneous space G/H (with G compact) under the free (isometric) action of a subgroup $K < G$. The corresponding biquotient is commonly denoted by $K\backslash G/H$. Evidently, the class of biquotients contains that of homogeneous spaces and, in particular, that of Lie groups. Moreover, these inclusions are strict. For the first one, the Gromoll-Meyer exotic sphere is by construction a biquotient, while the only homotopy spheres that admit homogeneous structures are the standard ones. For the second inclusion, recall that the only sphere diffeomorphic to a Lie group is S^3.

All biquotients carry a metric of nonnegative curvature inherited from a bi-invariant metric on G. Coming back to the double soul conjecture, it is currently an open question whether it holds for all biquotients. Rather surprisingly, it is already unknown for the subclass of Lie groups: none of the sufficient conditions explained in Sects. 3 or 7 for a manifold to admit a double disk bundle decomposition seem to apply to the exceptional Lie groups E_7 or E_8. In particular, Kollross [31] showed that E_7 and E_8 are the only simple Lie groups which do not admit cohomogeneity one isometric actions with respect to a bi-invariant metric.

On the positive side, using Kollross' result, it is observed in [14, Proposition 3.2] that any compact, connected Lie group G which is not isomorphic to a finite quotient of a product $\prod G_i$ with all G_i being E_7 or E_8 admits a double disk bundle decomposition.

Further positive results can be derived in low dimensions, due to the existing classifications of homogeneous spaces and biquotients. Lutz proved in his thesis [32, Theorem 0.0.2] that all known closed simply connected nonnegatively curved manifolds of dimension ≤6 carry a (very specific) double disk bundle decomposition. This result hence covers all biquotients up to dimension 6 (see also [15] for dimensions 4 and 5). In this survey, we extend this particular case and make the following observation.

Proposition 1 *All closed simply connected biquotients of dimension ≤7 satisfy the double soul conjecture.*

For homogeneous spaces, we can improve the conclusion further.

Proposition 2 *All closed simply connected homogeneous spaces of dimension ≤10 satisfy the double soul conjecture.*

Proof of Propositions 1 and 2 The statements above follow from combining the existing classifications of the manifolds under consideration with the known sufficient conditions from [14, Proposition 3.1(d)] for a manifold to admit a double disk bundle decomposition.

Closed simply connected homogeneous spaces of dimension ≤9 were classified in Klaus' thesis [30], and those of dimension 10 can be extracted from the work of Böhm and Kerr [2]. The list consists of the following diffeomorphism types:

- compact rank one symmetric spaces;
- the following single examples in each dimension:
  - the Wu-manifold $SU(3)/SO(3)$,
  - the Grassmannian $SO(5)/(SO(3)SO(2))$ and the Wallach space $SU(3)/T^2$,
  - the Berger space $B^7 = SO(5)/SO(3)$ and the unit tangent bundle $T^1S^4 = SO(5)/SO(3)$ (obviously, the inclusions $SO(3) < SO(5)$ are different),
  - the group $SU(3)$, the Grassmannian $SU(4)/S(U(2)U(2))$, the Wolf space $G_2/SO(4)$ and the flag $Sp(2)/T^2$,
  - the Grassmannian $SU(4)/SO(4)$,
  - the group $Sp(2)$, the Grassmannian $SO(7)/(SO(5)SO(2))$, the projectivized tangent bundle $P_{\mathbb{C}}T\mathbb{C}P^6 = SU(4)/S(U(1)U(1)U(2))$ and $G_2/U(2)$ (which is a rational $\mathbb{C}P^5$ [29]);
- products of spaces in the previous two items;
- circle bundles over spaces in the previous three items.

It can be easily checked in the literature that all the spaces in the first two items admit either cohomogeneity one actions or fibrations over cohomogeneity one spaces; hence, all carry double disk bundle decompositions. These can be lifted to produce double disk bundle decompositions on the spaces in the last two items.

The classification of closed simply connected biquotients of dimension ≤7 was obtained by DeVito in [10]. We will not review those of dimension ≤6 since they have been covered in Lutz's thesis [32]. We thus focus in dimension 7. Apart from the homogeneous spaces listed above, the following diffeomorphism types arise in the biquotient case:

- quotients $M = (S^{n_1} \times \cdots \times S^{n_r})/T^k$ of a product of spheres under a torus linear action;
- (not necessarily positively curved) Eschenburg spaces, i.e., circle quotients of $SU(3)$;
- quotients of $(SU(3)/SO(3)) \times SU(2)$ by a circle action.

The spaces in the first item are covered in [32, Theorem 0.0.1]. Regarding Eschenburg spaces, it is not known whether all of them carry cohomogeneity one actions or fiber over any other manifold. However, as observed in the proof of [14, Theorem 3.3], each of the free circle actions on $SU(3)$ is equivalent to some subaction of a cohomogeneity one action on $SU(3)$; hence the quotient carries a double disk bundle decomposition by DeVito et al. [14, Proposition 3.1(c)]. For the spaces in the third item, observe that the product action by $SU(3) \times T^2$ on $(SU(3)/SO(3)) \times SU(2)$ is of cohomogeneity one. Since $SO(3)$ is maximal in $SU(3)$, the possible circle actions on $(SU(3)/SO(3)) \times SU(2)$ are, up to conjugacy, subactions of $SU(3) \times T^2$. Again, the corresponding quotients carry a double disk bundle decomposition by DeVito et al. [14, Proposition 3.1(c)]. □

We remark that a potential proof to extend Proposition 2 to dimension 11, if true, would not be as straightforward. One reason is that none of the known sufficient conditions for a manifold to admit a double disk bundle decomposition seem to apply to the space $G_2/SO(3)$. Here $SO(3)$ is a maximal subgroup of G_2, and hence there are no homogeneous fibrations associated with intermediate subgroups. Moreover, $G_2/SO(3)$ (which is a rational 11-sphere, see [29]) does not admit any cohomogeneity one action, as follows from the recent classification of cohomogeneity one actions on rational spheres by DeVito [11].

9 Relation to the Bott-Grove-Halperin Conjecture

The most prominent conjecture in the field of nonnegative curvature is the Bott-Grove-Halperin conjecture. It suggests, in its simpler form, that a closed simply connected nonnegatively curved should be rationally elliptic. This is a topological condition which can be characterized as follows: a closed simply connected manifold M is *rationally elliptic* if only a finite number of its homotopy groups have infinite order; otherwise it is called *rationally hyperbolic*. This conjecture was discussed in the literature for the first time by Grove and Halperin [23, Conjecture 1.5], where they attribute it to Bott. This conjecture, nowadays called the Bott-Grove-Halperin conjecture, if true, would vastly improve our understanding of nonnegatively curved manifolds.

Rationally elliptic manifolds have been studied from the point of view of rational homotopy theory, and they are topologically very rigid. For example, their Euler characteristic is nonnegative. Moreover, the total Betti number $\sum_i \operatorname{rank} H^i(M, \mathbb{Q})$ of a rationally elliptic n-dimensional manifold M is bounded above by 2^n. This should be compared to Gromov's Betti number Theorem stated in Sect. 2, which also gives un upper bound for $\sum_i \operatorname{rank} H^i(M, \mathbb{Q})$. However, Gromov's bound is vastly greater than 2^n.

In contrast to the situation for the double soul conjecture, the Bott-Grove-Halperin conjecture is known to hold for all existing examples of nonnegatively curved manifolds. In the case of biquotients $K\backslash G/H$, it follows from the long exact homotopy sequence associated with the fibration $(K \times H) \to G \to K\backslash G/H$, since Lie groups are known to be rationally elliptic. In the case of the nonnegatively curved examples constructed as the gluing of two disk bundles, rational ellipticity can be proven by using the methods of Grove and Halperin from [24].

In view of the Bott-Grove-Halperin conjecture and the double soul conjecture, it is natural to compare the class of rationally elliptic manifolds and the class of manifolds admitting a double disk bundle decomposition. This has been done in detail in [14], and they obtained structural results in low dimensions. We highlight the following results, which illustrate the fact that the two classes do not coincide:

- In each dimension 5 and 6, there are infinitely many rationally elliptic manifolds which admit no double disk bundle decomposition.
- In each dimension $n \geq 6$, there are manifolds admitting a double disk bundle composition which are rationally hyperbolic. Examples are $S^n \times (\mathbb{C}P^2\sharp\ldots\sharp\mathbb{C}P^2)$ for any $n \geq 2$ and ≥ 3 copies of $\mathbb{C}P^2$, as well as $S^n \times ((S^2 \times S^4)\sharp(S^2 \times S^4))$ for any $n \geq 2$. The latter is specially interesting as it has the same Betti numbers as a rationally elliptic manifold, namely, $S^n \times S^2 \times \mathbb{C}P^2$.

Acknowledgments We thank Anand Dessai and Andreas Kollross for informing us about the references [35] and [8, 34], respectively. We also thank Jason DeVito, Fernando Galaz-García, Karsten Grove, and Rafael Torres for comments on a preliminary version of this survey. We are particularly grateful to Jason DeVito for useful conversations which in particular led to Propositions 1 and 2.

This work was partially supported by the PID2021-124195NB-C32 and by the Severo Ochoa Programme for Centers of Excellence in R&D, CEX2019-000904-S (MICINN).

References

1. Belegradek, I.: Iterated circle bundles and infranilmanifolds. Osaka J. Math. **57**(1), 165–168 (2020)
2. Böhm, C., Kerr, M.M.: Low-dimensional homogeneous Einstein manifolds. Trans. Amer. Math. Soc. **358**(4), 1455–1468 (2006)
3. Bredon, G.E.: Introduction to Compact Transformation Groups. Pure and Applied Mathematics, vol. 46. Academic, New York (1972)
4. Burago, D., Burago, Y., Ivanov, S.: A Course in Metric Geometry. Graduate Studies in Mathematics, vol. 33. American Mathematical Society, Providence (2001)

5. Buser, P.: A note on the isoperimetric constant. Ann. Sci. École Norm. Sup. **15**(2), 213–230 (1982)
6. Cheeger, J.: Some examples of manifolds of nonnegative curvature. J. Differ. Geometry **8**, 623–628 (1973)
7. Cheeger, J., Gromoll, D.: On the structure of complete manifolds of nonnegative curvature. Ann. Math. **96**(2), 413–443 (1972)
8. Chen, B.-Y., Nagano, T.: Totally geodesic submanifolds of symmetric spaces. II. Duke Math. J. **45**(2), 405–425 (1978)
9. DeVito, J.: The classification of compact simply connected biquotients in dimensions 4 and 5. Differ. Geom. Appl. **34**, 128–138 (2014)
10. DeVito, J.: The classification of compact simply connected biquotients in dimension 6 and 7. Math. Ann. **368**(3–4), 1493–1541 (2017)
11. DeVito, J.: Cohomogeneity one manifolds with singly generated rational cohomology II (2021). Preprint arXiv:2108.10919
12. DeVito, J.. Rational spheres and double disk bundles (2021). Preprint arXiv:2105.02150
13. DeVito, J.: Counterexamples to the non-simply connected Double Soul Conjecture (2023). Preprint arXiv:2301.05675
14. DeVito, J., Galaz-García, F.: Kerin, M.: Manifolds that admit a double disk-bundle decomposition. To appear in the Indiana University Journal of Mathematics
15. Galaz-García, F., Kerin, M.: Cohomogeneity-two torus actions on non-negatively curved manifolds of low dimension. Math. Z. **276**(1–2), 133–152 (2014)
16. Ge, J., Radeschi, M.: Differentiable classification of 4-manifolds with singular Riemannian foliations. Math. Ann. **363**(1–2), 525–548 (2015)
17. Gigli, N., Pasqualetto, E.: Lectures on Nonsmooth Differential Geometry. SISSA Springer Series, vol. 2. Springer, Cham (2020)
18. Goette, S., Kerin, M., Shankar, K.: Highly connected 7-manifolds and non-negative sectional curvature. Ann. of Math. **191**(3), 829–892 (2020)
19. Goette, S., Kerin, M., Shankar, K.: Highly connected 7-manifolds, the linking form and non-negative curvature (2020). Preprint arXiv:2003.04907
20. González-Álvaro, D., Zibrowius, M.: The stable converse soul question for positively curved homogeneous spaces. J. Differ. Geom. **119**(2), 261–307 (2021)
21. Grove, K.: Geometry of, and via, symmetries. In: Conformal, Riemannian and Lagrangian Geometry (Knoxville, TN, 2000). University Lecture Series, vol. 27, pp. 31–53. American Mathematical Society, Providence (2002)
22. Grove, K.: A panoramic glimpse of manifolds with sectional curvature bounded from below. Translated from Algebra i Analiz **29**(1), 7–48 (2017). St. Petersburg Math. J. **29**(1), 3–31 (2018)
23. Grove, K.; Halperin, S.: Contributions of rational homotopy theory to global problems in geometry. Inst. Hautes Études Sci. Publ. Math. **56**, 171–177 (1982/1983)
24. Grove, K., Halperin, S.: Dupin hypersurfaces, group actions and the double mapping cylinder. J. Differ. Geom. **26**(3), 429–459 (1987)
25. Grove, K., Ziller, W.: Curvature and symmetry of Milnor spheres. Ann. Math. **152**(1), 331–367 (2000)
26. Guijarro, L.: Improving the metric in an open manifold with nonnegative curvature. Proc. Amer. Math. Soc. **126**(5), 1541–1545 (1998)
27. Guijarro, L., Torres, R.: Gluing constructions of Riemannian manifolds with nonnegative sectional curvature. In preparation (2023)
28. Hoelscher, C.A.: Classification of cohomogeneity one manifolds in low dimensions. Pacific J. Math. **246**(1), 129–185 (2010)
29. Kapovitch, V., Ziller, W.: Biquotients with singly generated rational cohomology. Geom. Dedicata **104**, 149–160 (2004). See the Updated 2017 Version https://arxiv.org/pdf/math/0210231.pdf
30. Klaus, S.: Einfach-zusammenhängende Kompakte Homogene Räume bis zur Dimension Neun, diploma Thesis, Johannes Gutenberg Universität Mainz (1988)

31. Kollross, A.: Low cohomogeneity and polar actions on exceptional compact Lie groups. Transform. Groups **14**(2), 387–415 (2009)
32. Lutz, A.: Classification of codimension one biquotient foliations in low dimensions. Ph.D. Thesis at the University of Oklahoma (2021)
33. Murphy, T.: Wilhelm, Frederick random manifolds have no totally geodesic submanifolds. Michigan Math. J. **68**(2), 323–335 (2019)
34. Nikolayevsky, Y.: Totally geodesic hypersurfaces of homogeneous spaces. Israel J. Math. **207**(1), 361–375 (2015)
35. Ranicki, A.: High-Dimensional Knot Theory. Algebraic Surgery in Codimension 2. With an Appendix by Elmar Winkelnkemper. Springer Monographs in Mathematics. Springer, New York (1998).
36. Schwachhöfer, L.J., Tapp, K: Cohomogeneity one disk bundles with normal homogeneous collars. Proc. Lond. Math. Soc. **99**(3), 609–632 (2009)
37. Spindeler, W.: Nonnegatively curved quotient spaces with boundary. Bol. Soc. Mat. Mex. **26**(2), 719–747 (2020)
38. Torres, R.: An orbit space of a nonlinear involution of $S^2 \times S^2$ with nonnegative sectional curvature. Proc. Amer. Math. Soc. **147**(8), 3523–3532 (2019)
39. Totaro, B.: Cheeger manifolds and the classification of biquotients. J. Differ. Geom. **61**(3), 397–451 (2002)
40. Wang, M.Y., Ziller, W.: Einstein metrics on principal torus bundles. J. Differ. Geom. **31**(1), 215–248 (1990)
41. Wilking, B.: Nonnegatively and positively curved manifolds. In: Surveys in Differential Geometry. vol. XI, pp. 25–62. International Press, Somerville (2007)
42. Ziller, W.: Examples of Riemannian manifolds with non-negative sectional curvature. In: Surveys in Differential Geometry. Vol. XI, pp. 63–102. International Press, Somerville (2007)

Consequences and Extensions of the Brunn-Minkowski Theorem

María A. Hernández Cifre, Eduardo Lucas, Francisco Marín Sola, and Jesús Yepes Nicolás

Abstract In this work we study some extensions and consequences of the fundamental Brunn-Minkowski inequality, using two different approaches: on one hand we deal with the so-called Grünbaum inequality, a beautiful consequence of the Brunn-Minkowski theorem which asserts, roughly speaking, that any hyperplane passing through the centroid divides any compact convex set into two not too small parts; on the other hand we study discrete versions of the Brunn-Minkowski inequality for the *lattice point enumerator*, that is, the functional counting how many points with integer coordinates are contained in a bounded set.

Keywords Brunn-Minkowski inequality · Lattice point enumerator · Grünbaum inequality

1 Introduction

As usual, we write $\mathbb{R}^n$ to represent the n-dimensional Euclidean space, endowed with the (Euclidean) inner product $\langle\cdot,\cdot\rangle$. One of the cornerstones of convex geometry is the Brunn-Minkowski inequality, which, in its classical form, provides a relation between the notions of Minkowski addition (of compact sets) and volume.

Theorem 1 *Let $K, L \subset \mathbb{R}^n$ be non-empty compact sets. Then, for all $\lambda \in (0,1)$,*

$$\mathrm{vol}\big((1-\lambda)K + \lambda L\big)^{1/n} \geq (1-\lambda)\mathrm{vol}(K)^{1/n} + \lambda \mathrm{vol}(L)^{1/n}, \tag{1}$$

with equality for some $\lambda \in (0,1)$, when $\mathrm{vol}(K)\mathrm{vol}(L) > 0$, if and only if K and L are homothetic compact convex sets.

M. A. Hernández Cifre · E. Lucas · F. Marín Sola · J. Yepes Nicolás (✉)
Dpto. de Matemáticas, Universidad de Murcia, Murcia, Spain
e-mail: mhcifre@um.es; eduardo.lucas@um.es; francisco.marin7@um.es; jesus.yepes@um.es

A. Alarcón et al. (eds.), *New Trends in Geometric Analysis*,
RSME Springer Series 10, https://doi.org/10.1007/978-3-031-39916-9_9

Here $\mathrm{vol}(\cdot)$ denotes the n-dimensional Lebesgue measure, and $+$ is used for the *Minkowski addition*, i.e., $A + B = \{a + b : a \in A, b \in B\}$ for any non-empty sets $A, B \subset \mathbb{R}^n$. Moreover, λA represents the set $\{\lambda a : a \in A\}$, for $\lambda \geq 0$.

Despite its apparent simplicity, the Brunn-Minkowski inequality is one of the most powerful results in convex geometry and beyond: for instance, its equivalent analytic version (the Prékopa-Leindler inequality; see, e.g., [12, Theorem 8.14]) and the fact that the compactness assumption can be weakened to Lebesgue measurability (see [20]) have allowed it to move to much wider fields. It implies very important inequalities such as the isoperimetric and Urysohn inequalities (see, e.g., [31, page 382]), and it has been the starting point for new developments like the L_p-Brunn-Minkowski theory (see, e.g., [21, 22]) or a reverse Brunn-Minkowski inequality (see, e.g., [27]), among many others. It would not be possible to collect here all references regarding equivalent versions, applications, and/or generalizations of the Brunn-Minkowski inequality. For extensive and beautiful surveys on them, we refer to [2, 7].

The classical *Brunn concavity principle* (see, e.g., [25, Theorem 12.2.1]) is one of the abovementioned equivalent versions of the Brunn-Minkowski inequality. It asserts that, for any non-empty compact and convex set $K \subset \mathbb{R}^n$ and a hyperplane H, the cross-sectional volume function $f : H^{\perp} \longrightarrow \mathbb{R}_{\geq 0}$ defined by

$$f(x) = \mathrm{vol}_{n-1}\big(K \cap (x + H)\big)^{1/(n-1)}$$

is concave; here $H^{\perp}$ represents the orthogonal complement of H. Moreover, in the following, we will denote by $M|H$ the orthogonal projection of a subset $M \subset \mathbb{R}^n$ onto H.

This result is the key fact in the classical proof of a celebrated theorem by Grünbaum [13]. In order to state it, we need further notation: for any compact set $K \subset \mathbb{R}^n$ with non-empty interior, we write $\mathrm{g}(K)$ to represent its *centroid*, i.e., the affine-covariant point

$$\mathrm{g}(K) := \frac{1}{\mathrm{vol}(K)} \int_K x \, \mathrm{d}x.$$

Moreover, given $u \in \mathbb{S}^{n-1}$, we write $H_u := \big\{x \in \mathbb{R}^n : \langle x, u\rangle = 0\big\}$ and $H_u^- := \big\{x \in \mathbb{R}^n : \langle x, u\rangle \leq 0\big\}$ to denote the (vector) hyperplane orthogonal to u and the corresponding closed halfspace with u as outer normal unit vector. Finally, we will say that K is a cone in the direction u if K is the convex hull of $\{x\} \cup \big(K \cap (y + H_u)\big)$, for some $x, y \in \mathbb{R}^n$.

Theorem 2 (Grünbaum) *Let $K \subset \mathbb{R}^n$ be a compact convex set, with non-empty interior, having its centroid at the origin. Then*

$$\frac{\mathrm{vol}(K \cap H_u^-)}{\mathrm{vol}(K)} \geq \left(\frac{n}{n+1}\right)^n \tag{2}$$

for all $u \in \mathbb{S}^{n-1}$. *Equality holds, for some* $u \in \mathbb{S}^{n-1}$, *if and only if* K *is a cone in the direction* u.

In the last years Grünbaum's result has been extended to the case of sections [6, 28] and projections [33] of compact convex sets and has been even generalized to the analytic setting of log-concave functions [26] and p-concave functions for $p > 0$ [28] (we refer the reader to [7] for more information on log-concave and p-concave functions). Moreover, it has been also extended to the case of compact sets with a p-concave cross-sectional volume function [23], for $p \geq 0$.

The original proof of Theorem 2 relies on exploiting the Brunn concavity principle to compare both the volume of the compact convex set K and of $K \cap H_u^-$ with those of a suitable cone C in the direction $u \in \mathbb{S}^{n-1}$ and $C \cap H_u^-$, respectively. In this paper we show, on the one hand, how one can derive Grünbaum's result as a direct application of the Brunn-Minkowski theorem (Theorem 1). Furthermore, the characterization of the equality given in Theorem 2 now will follow from the equality case of Theorem 1.

On the other hand, we devote this work to exploring discrete versions of the Brunn-Minkowski inequality. Nowadays there is a growing interest for studying discrete analogues of classical (continuous) results, which can be carried out from two points of view: either considering finite subsets $A, B \subset \mathbb{Z}^n$ of integer points and measuring with the cardinality $|\cdot|$ or working with compact sets $K, L \subset \mathbb{R}^n$ and using the so-called lattice point enumerator as measure, that is, $\mathrm{G}_n(K) = |K \cap \mathbb{Z}^n|$.

Regarding the cardinality, and besides the simple and classical inequality

$$|A + B| \geq |A| + |B| - 1 \tag{3}$$

for finite $A, B \subset \mathbb{Z}^n$, Gardner and Gronchi obtained in [8] a beautiful and powerful discrete Brunn-Minkowski inequality: they proved that if A, B are finite subsets of the integer lattice $\mathbb{Z}^n$, with dimension $\dim B = n$, then

$$|A + B| \geq \left|D^B_{|A|} + D^B_{|B|}\right|.$$

Here, for any $m \in \mathbb{N}$, D^B_m is a B-*initial segment*, i.e., the set of the first m points of $\mathbb{Z}^n_{\geq 0} = \left\{x \in \mathbb{Z}^n : x_i \geq 0\right\}$ in the so-called B-order, which is a particular order defined on $\mathbb{Z}^n_{\geq 0}$ depending only on the cardinality of B. They also derive some inequalities that improve previous results obtained by Ruzsa in [29, 30]. For a proper definition and a deep study of it, we refer the reader to [8].

Recently [9, 16, 19], different discrete analogues of the Brunn-Minkowski inequality have been obtained for the cardinality, including the case of its classical form (1): in [16] it is shown that if $A, B \subset \mathbb{Z}^n$ are non-empty finite sets, then

$$\left|\bar{A} + B\right|^{1/n} \geq |A|^{1/n} + |B|^{1/n},$$

where $\bar{A}$ is a suitably defined extension of A not depending on B.

In this paper we will focus on investigating discrete Brunn-Minkowski type inequalities for the lattice point enumerator and will present the more recent advances in this respect.

2 Deriving Grünbaum's Inequality as a Consequence of the Brunn-Minkowski Theorem

Before showing Theorem 2, we need to introduce some notation. Given a compact convex set $K \subset \mathbb{R}^n$ with non-empty interior, and a vector $u \in \mathbb{S}^{n-1}$, we denote by $K_u(t) = K \cap (tu + H_u)$ and by $K_u^-(t) = K \cap (tu + H_u^-)$, for any $t \in \mathbb{R}$. Furthermore, we observe that if K has centroid at the origin, then, using Fubini's theorem, we get

$$0 = \int_K \langle x, u\rangle \, \mathrm{d}x = \int_a^b t \operatorname{vol}_{n-1}\big(K_u(t)\big) \, \mathrm{d}t, \tag{4}$$

where $a, b \in \mathbb{R}$ are such that $K|H_u^{\perp} = [au, bu]$ (here, as usual, by $[x, y]$ we denote the segment with endpoints $x, y \in \mathbb{R}^n$).

Now we are in a position to prove Theorem 2. We will follow here the approach used in [24] to derive the functional version of Grünbaum's inequality.

***Proof** (of **Theorem** 2)* Let $u \in \mathbb{S}^{n-1}$ be fixed and assume that $K|H_u^{\perp} = [au, bu]$ for some $a, b \in \mathbb{R}$ with $a < b$. First we observe that since K is a compact convex set with interior points, we have that $\operatorname{vol}_{n-1}\big(K_u(t)\big) > 0$ for all $t \in (a, b)$ and so the condition (4) yields $a < 0 < b$. In particular we have $\operatorname{vol}_{n-1}\big(K_u(0)\big) > 0$.

On the one hand, from the convexity of K, we get

$$K_u^-\big((1-\lambda)t_1 + \lambda t_2\big) \supset (1-\lambda)K_u^-(t_1) + \lambda K_u^-(t_2)$$

for all $t_1, t_2 \in [a, b]$ and all $\lambda \in [0, 1]$. Then the Brunn-Minkowski inequality (1) applied to the equation above implies that $\operatorname{vol}\big(K_u^-(\cdot)\big)^{1/n}$ is a concave function on $[a, b]$, and further we have $\operatorname{vol}\big(K_u^-(t)\big) = 0$ for all $t \leq a$ and $\operatorname{vol}\big(K_u^-(t)\big) = \operatorname{vol}(K)$ for all $t \geq b$.

On the other hand, since $\operatorname{vol}_{n-1}\big(K_u(\cdot)\big)$ is continuous in (a, b) (due to the fact that every concave function is continuous in the interior of its domain and $\operatorname{vol}_{n-1}\big(K_u(\cdot)\big)^{1/(n-1)}$ is so), from the fundamental theorem of calculus and Fubini's theorem, we have that

$$\frac{\mathrm{d}}{\mathrm{d}t}\operatorname{vol}\big(K_u^-(t)\big) = \operatorname{vol}_{n-1}\big(K_u(t)\big) \tag{5}$$

for all $t \in (a, b)$. Thus $\operatorname{vol}\bigl(K_u^-(\cdot)\bigr)^{1/n}$ is concave and differentiable on (a, b), and then its tangent at $t = 0$, which is given by the function $h : \mathbb{R} \longrightarrow \mathbb{R}$ defined by

$$h(t) = \frac{1}{n}\operatorname{vol}\bigl(K_u^-(0)\bigr)^{1/n}(mt + n)$$

for

$$m = \frac{\operatorname{vol}_{n-1}\bigl(K_u(0)\bigr)}{\operatorname{vol}\bigl(K_u^-(0)\bigr)} > 0,$$

lies above its graph. Then $0 \leq \operatorname{vol}\bigl(K_u^-(t)\bigr)^{1/n} \leq h(t)$ for all $t \in [a, b]$, and further, taking into account that h is negative on $(-\infty, -n/m)$ and $\operatorname{vol}\bigl(K_u^-(t)\bigr)^{1/n}$ is constant for all $t \geq b$, we have

$$\operatorname{vol}\bigl(K_u^-(t)\bigr)^{1/n} \leq h(t) \qquad \text{for all } t \in \Bigl[-\frac{n}{m}, \infty\Bigr). \tag{6}$$

Moreover, applying integration by parts (jointly with (5)) and using (4), we get

$$\int_a^b \operatorname{vol}\bigl(K_u^-(t)\bigr)\,\mathrm{d}t = b\operatorname{vol}(K) - \int_a^b t\operatorname{vol}_{n-1}\bigl(K_u(t)\bigr)\,\mathrm{d}t = b\operatorname{vol}(K). \tag{7}$$

Hence, noticing on one hand that $\operatorname{vol}\bigl(K_u^-(\cdot)\bigr)$ is strictly increasing on $[a, b]$ and that $\operatorname{vol}\bigl(K_u^-(t)\bigr) = \operatorname{vol}(K)$ for all $t \geq b$ on the other hand, by (7) and (6), we have

$$\begin{aligned}
b\operatorname{vol}(K) &= \int_a^b \operatorname{vol}\bigl(K_u^-(t)\bigr)\,\mathrm{d}t = \int_{-n/m}^b \operatorname{vol}\bigl(K_u^-(t)\bigr)\,\mathrm{d}t \\
&= \int_{-n/m}^{1/m} \operatorname{vol}\bigl(K_u^-(t)\bigr)\,\mathrm{d}t + \int_{1/m}^b \operatorname{vol}\bigl(K_u^-(t)\bigr)\,\mathrm{d}t \\
&\leq \int_{-n/m}^{1/m} h(t)^n\,\mathrm{d}t + \Bigl(b - \frac{1}{m}\Bigr)\operatorname{vol}(K) \\
&= \frac{\operatorname{vol}\bigl(K_u^-(0)\bigr)}{m}\Bigl(\frac{n+1}{n}\Bigr)^n + \Bigl(b - \frac{1}{m}\Bigr)\operatorname{vol}(K).
\end{aligned}$$

Therefore

$$\operatorname{vol}\bigl(K \cap H_u^-\bigr) = \operatorname{vol}\bigl(K_u^-(0)\bigr) \geq \Bigl(\frac{n}{n+1}\Bigr)^n \operatorname{vol}(K),$$

and so (2) follows. Furthermore, equality holds, for such a fixed vector $u \in \mathbb{S}^{n-1}$, if and only if

$$\mathrm{vol}\big(K_u^-(t)\big) = h(t)^n \tag{8}$$

for all $t \in [a, b]$, with $a = -n/m$ and $b = 1/m$.

First, we assume that the above conditions hold (for such $u \in \mathbb{S}^{n-1}$ fixed). Hence $\mathrm{vol}\big(K_u^-(\cdot)\big)^{1/n}$ is affine on $[a, b]$, which implies, from the equality case of the Brunn-Minkowski theorem (see Theorem 1), that $K_u^-(t_1)$ and $K_u^-(t_2)$ are homothetic for all $t_1, t_2 \in [a, b]$. Then, for every $t \in [a, b]$, we have $K_u^-(t) = r(t)K_u^-(b) + y_t = r(t)K + y_t$ for some $r(t) \geq 0$ and some $y_t \in \mathbb{R}^n$, from where we further get

$$K_u(t) = r(t)K_u(b) + y_t \tag{9}$$

for all $t \in [a, b]$. Moreover, for the suitable constants $A, B > 0$, we have

$$\begin{aligned} \mathrm{vol}(K)r(t)^n &= \mathrm{vol}\big(r(t)K + y_t\big) = \mathrm{vol}\big(K_u^-(t)\big) = h(t)^n = A(mt+n)^n \\ &= B(t-a)^n, \end{aligned}$$

where in the last equality, we have used that $a = -n/m$. Thus, $r(t) = C(t-a)$ for some $C > 0$ and since $r(b) = 1$, we get

$$r(t) = \frac{t-a}{b-a} \tag{10}$$

for all $t \in [a, b]$. Now, for every fixed $t \in [a, b]$, if we set $\lambda = (b-t)/(b-a) \in [0, 1]$ then $t = (1-\lambda)b + \lambda a$ and so, from the convexity of K, we have

$$K_u(t) \supset \left(\frac{t-a}{b-a}\right) K_u(b) + \left(\frac{b-t}{b-a}\right) K_u(a). \tag{11}$$

Since $K_u(a) = r(a)K_u(b) + y_a = y_a$, taking volumes in (11) and using (9) and (10), we obtain that

$$\begin{aligned} \mathrm{vol}_{n-1}\big(K_u(t)\big) &\geq \mathrm{vol}_{n-1}\left[\left(\frac{t-a}{b-a}\right) K_u(b) + \left(\frac{b-t}{b-a}\right) K_u(a)\right] \\ &= \left(\frac{t-a}{b-a}\right)^{n-1} \mathrm{vol}_{n-1}\big(K_u(b)\big) = \mathrm{vol}_{n-1}\big(r(t)K_u(b)\big) \\ &= \mathrm{vol}_{n-1}\big(K_u(t)\big), \end{aligned}$$

and thus (11) holds with equality, for all $t \in [a, b]$. We conclude that K is the convex hull of $K_u(b)$ and the point $K_u(a)$, that is, K is a cone in the direction u.

Finally, if K is a cone in a direction $u \in \mathbb{S}^{n-1}$, then $K_u(t) = r(t)K_u(b) + y_t$ (cf. (11)), where r is given by (10) and y_t is the point $(1 - r(t))K_u(a)$. So, using the well-known formula for the volume of a cone, we get

$$\frac{\mathrm{vol}(K \cap H_u^-)}{\mathrm{vol}(K)} = \frac{-a\,\mathrm{vol}_{n-1}\big(K_u(0)\big)}{(b-a)\mathrm{vol}_{n-1}\big(K_u(b)\big)} = \left(\frac{-a}{b-a}\right)^n, \tag{12}$$

where in the last equality, we have used that

$$\mathrm{vol}_{n-1}\big(K_u(0)\big) = r(0)^{n-1}\mathrm{vol}_{n-1}\big(K_u(b)\big) = \left(\frac{-a}{b-a}\right)^{n-1} \mathrm{vol}_{n-1}\big(K_u(b)\big).$$

Now, (4) implies that

$$0 = \int_a^b t\, r(t)^{n-1}\,\mathrm{d}t = (b-a)\frac{nb+a}{n(n+1)},$$

which is equivalent to $nb = -a$. Replacing the latter in (12), we conclude that (2) indeed holds with equality when K is a cone in the direction u. This finishes the proof.

Given a non-negative measurable function with finite positive integral, its centroid is the point defined by

$$\mathrm{g}(f) := \frac{1}{\int_{\mathbb{R}^n} f(x)\,\mathrm{d}x}\int_{\mathbb{R}^n} x f(x)\,\mathrm{d}x.$$

In [24] (see also the references therein), it is shown that one can obtain the functional analogue of Grünbaum's inequality (2) by exploiting the functional counterpart of the Brunn-Minkowski inequality, the so-called *Borell-Brascamp-Lieb inequality* (see [3] and [4]). More precisely, given a p-concave function $f : \mathbb{R}^n \longrightarrow \mathbb{R}_{\geq 0}$ for some $p \in [0, \infty]$ with compact support and centroid at the origin, and any hyperplane H, one has

$$\int_{H^-} f(x)\,\mathrm{d}x \geq \left(\frac{np+1}{(n+1)p+1}\right)^{(np+1)/p} \int_{\mathbb{R}^n} f(x)\,\mathrm{d}x. \tag{13}$$

As usual, if $p = 0$ or $p = \infty$, the constant appearing in the right-hand side of the above inequality is the value that is obtained "by continuity," that is, the limit as $p \to 0^+$ or $p \to \infty$, respectively. We notice that Grünbaum's inequality (2) is then recovered from (13) by taking f the characteristic function of the n dimensional compact convex set K with centroid at the origin, which is ∞-concave.

3 Discrete Brunn-Minkowski Type Inequalities for the Lattice Point Enumerator

We note that the known discrete Brunn-Minkowski inequalities for the cardinality in its classical form involve the Minkowski addition of two finite subsets $A, B \subset \mathbb{Z}^n$, but not its *convex combination*. Indeed, if one aims to get a discrete analog of (1), one should observe the following: for any pair of non-empty finite sets $A, B \subset \mathbb{R}^n$, using (3) and the convexity of the function $t \mapsto t^n$ for $t \geq 0$, one gets

$$\begin{aligned}\bigl|(1-\lambda)A+\lambda B\bigr| &\geq \bigl|(1-\lambda)A\bigr|+|\lambda B|-1=|A|+|B|-1\\ &=(1-\lambda)|A|+\lambda|B|+\lambda|A|+(1-\lambda)|B|-1\\ &\geq(1-\lambda)|A|+\lambda|B|\geq\Bigl((1-\lambda)|A|^{1/n}+\lambda|B|^{1/n}\Bigr)^n;\end{aligned}$$

this inequality is however meaningless from a geometric point of view, because while the quantities $|A|$, $|B|$ on the right-hand side are reduced by the factors $1-\lambda$ and λ, the sets $(1-\lambda)A$ and λB on the left-hand side have the same cardinality as A and B, respectively.

So, one needs to involve a way of "counting points" for which dilations affect, and a perfect candidate for this is the lattice point enumerator G_n (for compact subsets of $\mathbb{R}^n$). However, and as in the case of the cardinality, one cannot expect to obtain a discrete Brunn-Minkowski inequality in the classical form for the lattice point enumerator, namely, the relation

$$\mathrm{G}_n\bigl((1-\lambda)K+\lambda L\bigr)^{1/n}\geq(1-\lambda)\mathrm{G}_n(K)^{1/n}+\lambda\mathrm{G}_n(L)^{1/n}$$

is in general not true. In fact, just taking $K=\{0\}$ and the cube $L=[0,m]^n$ with $m\in\mathbb{N}$ odd, then it is

$$\mathrm{G}_n\left(\frac{1}{2}K+\frac{1}{2}L\right)^{1/n}=\frac{m+1}{2}<\frac{m+2}{2}=\frac{1}{2}\mathrm{G}_n(K)^{1/n}+\frac{1}{2}\mathrm{G}_n(L)^{1/n}.$$

So, the question arises what is the "best" way to define a set M, for given compact sets $K, L\subset\mathbb{R}^n$, such that $(1-\lambda)K+\lambda L\subset M$ and

$$\mathrm{G}_n(M)^{1/n}\geq(1-\lambda)\mathrm{G}_n(K)^{1/n}+\lambda\mathrm{G}_n(L)^{1/n}$$

holds *for all* $\lambda\in(0,1)$.

In [19] the authors answered this question by proving that if $K, L\subset\mathbb{R}^n$ are non-empty bounded sets and $\lambda\in(0,1)$, then

$$\mathrm{G}_n\bigl((1-\lambda)K+\lambda L+(-1,1)^n\bigr)^{1/n}\geq(1-\lambda)\mathrm{G}_n(K)^{1/n}+\lambda\mathrm{G}_n(L)^{1/n},$$

the inequality being sharp. Furthermore, the cube cannot be reduced in the latter inequality, and it implies the classical Brunn-Minkowski inequality (1) for bounded convex sets.

The latter inequality was obtained as a direct consequence of a (more general) functional discrete inequality: indeed, the authors proved a discrete version of the Borell-Brascamp-Lieb inequality. Furthermore, they showed that their discrete Borell-Brascamp-Lieb type inequality implies the classical functional one (under mild assumptions on the functions there involved), which makes it a powerful result in the field.

Here we provide a new proof of the above Brunn-Minkowski type inequality for the lattice point enumerator, using a completely geometrical approach, and we show that it implies the continuous version in the more general case of Jordan measurable sets. More precisely, we show the following result.

Theorem 3 *Let $K, L \subset \mathbb{R}^n$ be non-empty bounded sets and let $\lambda \in (0, 1)$. Then*

$$\mathrm{G}_n\big((1-\lambda)K + \lambda L + (-1, 1)^n\big)^{1/n} \geq (1-\lambda)\mathrm{G}_n(K)^{1/n} + \lambda \mathrm{G}_n(L)^{1/n}. \tag{14}$$

Moreover, it implies the Brunn-Minkowski inequality (1) *for bounded Jordan measurable sets.*

Before showing this result, we need some additional notation and an auxiliary property: we will represent by B_0 the n-dimensional Euclidean open unit ball, and we denote by $\mathrm{cl}\, M$ the closure of a set $M \subset \mathbb{R}^n$.

The proof of the theorem relies on the following relations between the volume and the lattice point enumerator of a non-empty bounded measurable set $A \subset \mathbb{R}^n$:

$$\begin{aligned} \mathrm{G}_n(A) &\leq \mathrm{vol}\left(A + \left(-\tfrac{1}{2}, \tfrac{1}{2}\right)^n\right), \\ \mathrm{vol}(A) &\leq \mathrm{G}_n\left(A + \left(-\tfrac{1}{2}, \tfrac{1}{2}\right)^n\right). \end{aligned} \tag{15}$$

The first inequality can be found in [10, (3.3)], whereas the second one is gathered in [11, p. 877] (see also [1]).

Moreover, if A is further Jordan measurable, it is a well-known fact that, roughly speaking, the volume and the lattice point enumerator are equivalent when A is large enough, i.e.,

$$\lim_{r \to \infty} \frac{\mathrm{G}_n(rA)}{r^n} = \mathrm{vol}(A) \tag{16}$$

(see, e.g., [12, Formula (3), p.120]). Furthermore, the following property holds.

Lemma 1 *Let $A \subset \mathbb{R}^n$ be a non-empty bounded Jordan measurable set and let $M \subset \mathbb{R}^n$ be a non-empty bounded set containing the origin. Then*

$$\lim_{r\to\infty} \frac{\mathrm{G}_n(rA+M)}{r^n} = \mathrm{vol}(A). \tag{17}$$

Proof Given $m \in \mathbb{N}$, it follows that for any $r > 0$ large enough one has $(1/r)M \subset (1/m)B_0$ and thus

$$\begin{aligned} \mathrm{vol}(A) &= \lim_{r\to\infty} \frac{\mathrm{G}_n(rA)}{r^n} \leq \liminf_{r\to\infty} \frac{\mathrm{G}_n(rA+M)}{r^n} \leq \limsup_{r\to\infty} \frac{\mathrm{G}_n(rA+M)}{r^n} \\ &\leq \limsup_{r\to\infty} \frac{\mathrm{G}_n\Big(r\big(\mathrm{cl}\,A + \frac{1}{m}B_0\big)\Big)}{r^n} \leq \lim_{r\to\infty} \frac{\mathrm{G}_n\Big(r\big(F_m + \frac{2}{m}B_0\big)\Big)}{r^n}, \end{aligned} \tag{18}$$

where F_m is some finite subset of $\mathrm{cl}\,A$ such that $\mathrm{cl}\,A \subset F_m + (1/m)B_0$ (which exists from the compactness of $\mathrm{cl}\,A$). Now, since $F_m + (2/m)B_0$ is a finite union of open balls and so it is Jordan measurable, we have, by (16), that

$$\lim_{r\to\infty} \frac{\mathrm{G}_n\Big(r\big(F_m + \frac{2}{m}B_0\big)\Big)}{r^n} = \mathrm{vol}\left(F_m + \frac{2}{m}B_0\right) \leq \mathrm{vol}\left(\mathrm{cl}\,A + \frac{2}{m}B_0\right). \tag{19}$$

Moreover, since $\mathrm{cl}\,A$ is compact, a standard straightforward computation shows that

$$\mathrm{cl}\,A = \bigcap_{m=1}^{\infty} \left(\mathrm{cl}\,A + \frac{2}{m}B_0\right).$$

Since the boundary of A has null Lebesgue measure (because A is Jordan measurable), the latter identity together with the fact that

$$\mathrm{vol}\left(\bigcap_{m=1}^{\infty} \left(\mathrm{cl}\,A + \frac{2}{m}B_0\right)\right) = \lim_{m\to\infty} \mathrm{vol}\left(\mathrm{cl}\,A + \frac{2}{m}B_0\right),$$

which holds because $\mathrm{cl}\,A + (2/m)B_0$ is a decreasing sequence (see, e.g., [5, Proposition 1.2.5 (b)]) allows us to deduce that

$$\mathrm{vol}(A) = \mathrm{vol}(\mathrm{cl}\,A) = \lim_{m\to\infty} \mathrm{vol}\left(\mathrm{cl}\,A + \frac{2}{m}B_n\right).$$

Therefore, since m was arbitrary in (18) and (19), (17) holds.

Now we are in a position to prove Theorem 3.

***Proof (of Theorem** 3)* Noticing that $M + (-1/2, 1/2)^n$ is open (and thus measurable) for any non-empty subset $M \subset \mathbb{R}^n$, from (1) and (15), we get

$$\begin{aligned}
\mathrm{G}_n\big((1-\lambda)K + \lambda L + (-1,1)^n\big)^{1/n} &\geq \mathrm{vol}\left((1-\lambda)K + \lambda L + \left(-\tfrac{1}{2},\tfrac{1}{2}\right)^n\right)^{1/n} \\
&= \mathrm{vol}\left((1-\lambda)\left[K + \left(-\tfrac{1}{2},\tfrac{1}{2}\right)^n\right] + \lambda\left[L + \left(-\tfrac{1}{2},\tfrac{1}{2}\right)^n\right]\right)^{1/n} \\
&\geq (1-\lambda)\mathrm{vol}\left(K + \left(-\tfrac{1}{2},\tfrac{1}{2}\right)^n\right)^{1/n} + \lambda \mathrm{vol}\left(L + \left(-\tfrac{1}{2},\tfrac{1}{2}\right)^n\right)^{1/n} \\
&\geq (1-\lambda)\mathrm{G}_n(K)^{1/n} + \lambda \mathrm{G}_n(L)^{1/n}.
\end{aligned}$$

In order to conclude the proof, we show that (14) implies (1) when K and L are non-empty bounded Jordan measurable sets. Then, using (14), (16), and Lemma 1, we get

$$\begin{aligned}
&(1-\lambda)\mathrm{vol}(K)^{1/n} + \lambda \mathrm{vol}(L)^{1/n} \\
&\quad = (1-\lambda)\left(\lim_{r\to\infty} \frac{\mathrm{G}_n(rK)}{r^n}\right)^{1/n} + \lambda\left(\lim_{r\to\infty} \frac{\mathrm{G}_n(rL)}{r^n}\right)^{1/n} \\
&\quad = \lim_{r\to\infty} \frac{(1-\lambda)\mathrm{G}_n(rK)^{1/n} + \lambda \mathrm{G}_n(rL)^{1/n}}{r} \\
&\quad \leq \lim_{r\to\infty} \frac{\mathrm{G}_n\big((1-\lambda)(rK) + \lambda(rL) + (-1,1)^n\big)^{1/n}}{r} \\
&\quad = \left(\lim_{r\to\infty} \frac{\mathrm{G}_n\Big(r\big((1-\lambda)K + \lambda L\big) + (-1,1)^n\Big)}{r^n}\right)^{1/n} \\
&\quad = \mathrm{vol}\big((1-\lambda)K + \lambda L\big)^{1/n},
\end{aligned}$$

as desired.

Other discrete analogues of the Brunn-Minkowski inequality for the lattice point enumerator can be found in [14, 18, 19, 32]. We conclude by highlighting the following nice result obtained by Halikias, Klartag, and Slomka in [14]: for non-empty bounded sets $K, L \subset \mathbb{R}^n$, one has

$$\mathrm{G}_n\left(\frac{K+L}{2} + (-1,0]^n\right)\mathrm{G}_n\left(\frac{K+L}{2} + [0,1)^n\right) \geq \mathrm{G}_n(K)\mathrm{G}_n(L),$$

which yields the discrete multiplicative Brunn-Minkowski type inequality

$$\mathrm{G}_n\left(\frac{K+L}{2}+[0,1]^n\right) \geq \sqrt{\mathrm{G}_n(K)\mathrm{G}_n(L)}.$$

In this line, in [19] it is also shown that

$$\mathrm{G}_n\left(\frac{K+L}{2}+[0,1]^n\right)^{1/n} \geq \frac{\mathrm{G}_n(K)^{1/n}+\mathrm{G}_n(L)^{1/n}}{2}, \tag{20}$$

provided that K, L contain some integer point. More recently, some other discrete Brunn-Minkowski type inequalities have been considered. We emphasize some extensions of (14) to the L_p setting, both in the case of $p \geq 1$ [17] and for $p \in [0, 1)$ [15].

Acknowledgments This research is part of the grant PID2021-124157NB-I00, funded by MCIN/AEI/10.13039/501100011033/ "ERDF A way of making Europe." It is also supported by "Ayudas a proyectos para el desarrollo de investigación científica y técnica por grupos competitivos," included in the Programa Regional de Fomento de la Investigación Científica y Técnica (Plan de Actuación 2022) of the Fundación Séneca-Agencia de Ciencia y Tecnología de la Región de Murcia, REF. 21899/PI/22.

References

1. Alonso-Gutiérrez, D., Lucas, E., Yepes Nicolás, J.: On Rogers-Shephard type inequalities for the lattice point enumerator. Commun. Contemp. Math. (2022). https://doi.org/10.1142/S0219199722500225
2. Barthe, F.: Autour de l'inégalité de Brunn-Minkowski. Ann. Fac. Sci. Toulouse: Math. Sér. 6, Tome **12**(2), 127–178 (2003)
3. Borell, C.: Convex set functions in d-space. Period. Math. Hungar. **6**, 111–136 (1975)
4. Brascamp, H.J., Lieb, E.H.: On extensions of the Brunn-Minkowski and Prékopa-Leindler theorems, including inequalities for log concave functions and with an application to the diffusion equation. J. Func. Anal. **22**(4), 366–389 (1976)
5. Cohn, D.L.: Measure Theory, 2nd revised edn. Birkhäuser/Springer, New York (2013)
6. Fradelizi, M., Meyer, M., Yaskin, V.: On the volume of sections of a convex body by cones. Proc. Amer. Math. Soc. **145**(7), 3153–3164 (2017)
7. Gardner, R.J.: The Brunn-Minkowski inequality. Bull. Am. Math. Soc. **39**(3), 355–405 (2002)
8. Gardner, R.J., Gronchi, P.: A Brunn-Minkowski inequality for the integer lattice. Trans. Am. Math. Soc. **353**(10), 3995–4024 (2001)
9. Green, B., Tao, T.: Compressions, convex geometry and the Freiman-Bilu theorem. Q. J. Math. **57**(4), 495–504 (2006)
10. Gritzmann, P., Wills, J.M.: Lattice points. In: Gruber, P.M., Wills, J.M. (eds.) Handbook of Convex Geometry, pp. 765–798. North-Holland, Amsterdam (1993)
11. Gritzmann, P., Wills, J.M.: Finite packing and covering. In: Gruber, P.M., Wills, J.M. (eds.) Handbook of Convex Geometry, pp. 861–898. North-Holland, Amsterdam (1993)
12. Gruber, P. M.: Convex and Discrete Geometry. Springer, Berlin (2007)
13. Grünbaum, B.: Partitions of mass-distributions and of convex bodies by hyperplanes. Pac. J. Math. **10**, 1257–1261 (1960)

14. Halikias, D., Klartag B., Slomka, B.A.: Discrete variants of Brunn-Minkowski type inequalities. Ann. Fac. Sci. Toulouse: Math. Sér. 6, Tome **30**(2), 267–279 (2021)
15. Hernández Cifre, M.A., Lucas, E.: On discrete log-Brunn-Minkowski type inequalities. SIAM J. Discrete Math. **36**(3), 1748–1760 (2022)
16. Hernández Cifre, M.A., Iglesias, D., Yepes Nicolás, J.: On a discrete Brunn-Minkowski type inequality. SIAM J. Discrete Math. **32**, 1840–1856 (2018)
17. Hernández Cifre, M.A., Lucas, E., Yepes Nicolás, J.: On discrete L_p Brunn-Minkowski type inequalities. Rev. Real Acad. Cienc. Exactas Fis. Nat. Ser. A-Mat. **116**, Article 164 (2022)
18. Iglesias, D., Lucas, E., Yepes Nicolás, J.: On discrete Brunn-Minkowski and isoperimetric type inequalities. Discrete Math. **345**(1), 112640 (2022)
19. Iglesias, D., Yepes Nicolás, J., Zvavitch, A.: Brunn-Minkowski type inequalities for the lattice point enumerator. Adv. Math. **370**, 107193 (2020)
20. Lusternik, L. A.: Die Brunn-Minkowskische Ungleichung für beliebige messbare Mengen. C. R. (Dokl.) Acad. Sci. URSS **8**, 55–58 (1935)
21. Lutwak, E.: The Brunn-Minkowski-Firey theory, I: Mixed volumes and the Minkowski problem. J. Diff. Geom. **38**, 131–150 (1993)
22. Lutwak, E.: The Brunn-Minkowski-Firey theory, II: Affine and geominimal surface areas. Adv. Math. **118**, 244–294 (1996)
23. Marín Sola, F., Yepes Nicolás, J.: On Grünbaum type inequalities. J. Geom. Anal. **31**(10), 9981–9995 (2021)
24. Marín Sola, F., Yepes Nicolás, J.: A simple proof of the functional version of Grünbaum's inequality. Submitted
25. Matoušek, J.: Lectures on discrete geometry. Graduate Texts in Mathematics, vol. 212. Springer, New York (2002)
26. Meyer, M., Nazarov, F., Ryabogin D., Yaskin, V.: Grünbaum-type inequality for log-concave functions. Bull. Lond. Math. Soc. **50**(4), 745–752 (2018)
27. Milman, V.D.: Inégalité de Brunn-Minkowski inverse et applications à la théorie locale des espaces normés (An inverse form of the Brunn-Minkowski inequality, with applications to the local theory of normed spaces). C. R. Acad. Sci. Paris Ser. I Math. **302**(1), 25–28 (1986)
28. Myroshnychenko, S., Stephen, M., Zhang, N.: Grünbaum's inequality for sections. J. Funct. Anal. **275**(9), 2516–2537 (2018)
29. Ruzsa, I.Z.: Sum of sets in several dimensions. Combinatorica **14**, 485–490 (1994)
30. Ruzsa, I.Z.: Sets of sums and commutative graphs. Studia Sci. Math. Hungar. **30**, 127–148 (1995)
31. Schneider, R.: Convex bodies: the Brunn-Minkowski theory. 2nd expanded ed. Encyclopedia of Mathematics and Its Applications, vol. 151. Cambridge University Press, Cambridge (2014)
32. Slomka, B.A.: A Remark on discrete Brunn-Minkowski type inequalities via transportation of measure. Available via ArXiv. https://arxiv.org/abs/2008.00738
33. Stephen, M., Zhang, N.: Grünbaum's inequality for projections. J. Funct. Anal. **272**(6), 2628–2640 (2017)

An Account on Links Between Finsler and Lorentz Geometries for Riemannian Geometers

Miguel Ángel Javaloyes, Enrique Pendás-Recondo, and Miguel Sánchez

Abstract Some links between Lorentz and Finsler geometries have been developed in the last years, with applications even to the Riemannian case. Our purpose is to give a brief description of them, which may serve as an introduction to recent references. As a motivating example, we start with Zermelo navigation problem, where its known Finslerian description permits a Lorentzian picture which allows for a full geometric understanding of the original problem. Then, we develop some issues including (a) the accurate description of the Lorentzian causality using Finsler elements, (b) the non-singular description of some Finsler elements (such as Kropina metrics or complete extensions of Randers ones with constant flag curvature), (c) the natural relation between the Lorentzian causal boundary and the Gromov and Busemann ones in the Finsler setting, and (d) practical applications to the propagation of waves and firefronts.

Keywords Zermelo navigation · Randers and Kropina metrics · Wind Finsler metrics · Stationary and SSTK spacetimes · Causality · Gromov and Busemann compactifications · Causal boundary · Lorentz-Finsler metrics · Huygens principle · Wildfire models

1 Introduction

Lorentz and Finsler geometries are two quite different extensions of the Riemannian one, which may serve as an arena to test and, eventually, to extend powerful Riemannian methods. Typically, Finsler and Lorentz metrics appear when either

M. Á. Javaloyes · E. Pendás-Recondo (✉)
Departamento de Matemáticas, Universidad de Murcia, Murcia, Spain
e-mail: majava@um.es; e.pendasrecondo@um.es

M. Sánchez
Departamento de Geometría y Topología, Facultad de Ciencias & IMAG (Centro de Excelencia María de Maeztu), Universidad de Granada, Granada, Spain
e-mail: sanchezm@ugr.es

A. Alarcón et al. (eds.), *New Trends in Geometric Analysis*,
RSME Springer Series 10, https://doi.org/10.1007/978-3-031-39916-9_10

the anisotropies of the space or the relativistic inclusion of time lead to replace the (Riemannian) infinitesimal Euclidean scalar products by the infinitesimal models of those geometries (i.e., possibly non-reversible norms or Lorentzian scalar products, respectively). The unification of both extensions in a single Lorentz-Finsler geometry has been considered by researchers interested in certain generalizations of general relativity, which have received a strong impulse recently [33, 70]. However, at a less speculative level, some links between the Finsler and Lorentz settings appear naturally, as well as their unification in a Lorentz-Finsler one, enhancing the techniques and results in both geometries and multiplying their applications even at a "real world" level [50, 51]. The purpose of the present article is to give a non-technical survey about some links between both geometries. The style is adapted to readers with a background in Riemannian geometry and interest in geometric analysis, and it may motivate the study of long references such as [23, 40].

In a nutshell, (conformal) Lorentzian geometry can be applied at a non-relativistic level in order to describe an object or wave that propagates at a finite maximum velocity $v_{\max}$. This velocity will resemble the speed of light in relativity, as it allows one to construct lightcones by using the velocities of objects moving at maximum speed. Then, the (conformally invariant) relativistic stuff about lightlike directions and causality can be reinterpreted for the description. The possible variation of $v_{\max}$ with the point and time can be directly incorporated in this picture. When the maximum velocity varies not only with these elements but also with the direction, then Finsler geometry comes into play, and the most general description leads to Lorentz-Finsler metrics (which are endowed with anisotropic lightcones).

However, there are relevant cases where a direction-dependent function $v_{\max}$ matches with a classical Lorentz metric. Noticeably, this happens for the Zermelo navigation problem of, say, a zeppelin whose velocity is affected by the wind. Indeed, the anisotropy of $v_{\max}$ with respect to observers on Earth is due to the direction of the wind; thus, it will disappear for observers comoving with it.

As will be stressed in Sects. 3–5, this stimulates the development of notions relative to geodesics, distances, and boundaries in each one of the three settings, Riemannian, Finslerian, and Lorentzian, taking into account possible applications in the last two. Moreover, Lorentz-Finsler methods become applicable to practical purposes, such as the monitoring of the front of a wave in an anisotropic medium or a wildfire, which is emphasized in Sect. 6.

Here, we begin by considering the aforementioned Zermelo navigation problem in Sect. 2, which will serve as a motivation for the remainder. This shows a first relation between three variational problems: (a) Zermelo minimization of the arrival time for trajectories between two points, (b) minimization of a (non-symmetric) distance by geodesics in the class of Finsler metrics of Randers type, and (c) the relativistic Fermat principle for light rays in the class of standard stationary spacetimes. We emphasize that the spacetime viewpoint allows one to remove two usual constraints in the Finslerian approach of Zermelo problem: (1) mild wind, overcome here by using wind Finsler structures and SSTK spacetimes in Sect. 2.2.2, and (2) time-independence, overcome by using non-stationary Lorentz metrics.

In Sect. 3, geometric applications for relativistic spacetimes are obtained. First, we give a brief explanation about the role of Cauchy hypersurfaces for Einstein equations in Sect. 3.2 and the meaning of causality conditions for spacetimes in Sect. 3.3. Then, these elements are characterized in terms of Finslerian ones, namely, a Randers metric in the case of a standard stationary spacetime in Sect. 3.4 and a wind Riemannian structure for SSTK ones in Sect. 3.5. Notice that the former spacetimes are an extension of the latter, allowing a description of settled black holes beyond their (Killing) horizons.

In Sect. 4, we will explore the Finslerian applications of Lorentzian geometry. In some cases, the stationary spacetime viewpoint has suggested some results related to geodesics that turn out to hold for arbitrary Finsler metrics. But it is especially interesting the case of singular Finsler metrics such as Kropina or, more generally, the wind Riemannian ones. The viewpoint of Lorentzian geometry desingularizes the problem allowing for a better understanding of these geometries. Moreover, the wind Riemannian structures provide the natural full setting to understand the classification of Randers metrics of constant flag curvature, thus revisiting the landmark obtained in [5].

In Sect. 5 we deepen in the geometric applications by considering boundaries. It is worth pointing out that, in the Riemannian setting, the Gromov and Busemann compactification have been widely studied since the 1970s, the latter in cases such as Hadamard manifolds or CAT(0) spaces. However, their systematic development for (possibly incomplete, non-reversible) Finslerian metrics had to wait until a specific motivation came from Lorentz geometry. In relativity, the causal boundary introduced by Geroch, Kronheimer, and Penrose had been computed in a limited number of cases. The above links between Finsler and Lorentz metrics prove that, for stationary spacetimes, the causal boundary corresponds to a general type of Busemann boundaries of a non-reversible Finsler metric which include the (forward and backward) Cauchy ones. These boundaries, as well as their relation with the Gromov one, are now well understood and explained here.

In Sect. 6, we go further both in the generality of the setting and the practical applications. Focusing on the propagation of a wave whose velocities depend on the time and direction, we show that the Lorentz-Finsler setting provides a neat geometrical picture of the evolution of the wavefront. In fact, the computations reduce to solving the ODE system given by the geodesic equations of a specific Lorentz-Finsler metric. Moreover, this approach can be applied in real-world models to obtain the evolution of any physical phenomenon that satisfies Huygens' principle, wildfires being the paradigmatic example.

2 A Motivating Example: Zermelo Navigation Problem

The classical Zermelo navigation problem was proposed by the German mathematician Ernst Zermelo in his 1931 paper [82] as follows:

> In an unbounded plane where the wind distribution is given by a vector field as a function of position and time, a ship moves with constant velocity relative to the surrounding air mass. How must the ship be steered in order to come from a starting point to a given goal in the shortest time?

E. Zermelo himself solved the problem using calculus of variations, reducing the problem to solve the so-called Zermelo's equation

$$\frac{d\theta}{dt} = \sin^2(\theta)\frac{\partial W}{\partial x^1} + \sin(\theta)\cos(\theta)\left(\frac{\partial W}{\partial x^1} - \frac{\partial W}{\partial x^2}\right) - \cos^2(\theta)\frac{\partial W}{\partial x^2},$$

where x^1, x^2 are the coordinates of $\mathbb{R}^2$, θ is the angle of the trajectory of the ship with the x^1-axis, and W is the variable wind which depends on time and position (see [26, Eq. 459.8]). In the 1930s, this problem received the attention of some very well-known mathematicians such as Levi-Civita, Von Mises, and Manià [59, 60, 77] and became one of the classical problems in the calculus of variations (see [26]). Zermelo problem can also be solved using optimal control theory (see the classical book [16] or [12, 14, 15, 73] for recent developments), but our interest will focus on more geometrical methods, namely, the use of Finsler geometry to solve the problem. This will be possible whenever the wind is time-independent and its contribution to the velocity does not exceed that provided by the engine.

2.1 The Case of Mild Time-Independent Wind

2.1.1 Basic Finsler Setup

Recall that a Finsler metric in a manifold M is defined as a non-negative function $F : TM \to [0, +\infty)$, being TM the tangent bundle of M, which is smooth away from the zero section, positive homogeneous of degree one, namely, $F(\mu v) = \mu F(v)$ for all $v \in TM$ and $\mu > 0$, and such that for every $v \in TM \setminus \mathbf{0}$, the symmetric bilinear form $g_v : T_{\pi(v)}M \times T_{\pi(v)}M \to \mathbb{R}$ is positive definite, where $\pi : TM \to M$ is the canonical projection and g_v is defined as

$$g_v(u, w) = \frac{1}{2}\frac{\partial^2}{\partial t \partial s}F(v + su + tw)^2|_{t=s=0}, \tag{1}$$

for all $u, w \in T_{\pi(v)}M$. A very important element of a Finsler metric is its indicatrix $\Sigma = \{v \in TM : F(v) = 1\}$. Indeed, the indicatrix determines completely the Finsler metric, and the positive definiteness of g_v is equivalent to having an indicatrix with a positive definite second fundamental form in $T_{\pi(v)}M$, thus, of positive sectional curvature, with respect to any Euclidean scalar product in $T_{\pi(v)}M$. So, at each tangent space $T_{\pi(v)}M$, the indicatrix yields the hypersurface $\Sigma \cap T_{\pi(v)}M$, which is a positively curved smooth sphere enclosing the zero vector in its (bounded) interior region; see, for example, [52, Theorem 2.14]. Any hypersurface with these properties can be regarded as the unit sphere of a (non-necessarily symmetric) norm. Indeed, given Σ, the Finsler metric is determined as follows: for every $v \in TM \setminus \mathbf{0}$, there exists a unique $\mu(v) > 0$ such that $\mu(v)v \in \Sigma$. Then the Finsler metric is obtained as $F(v) = 1/\mu(v)$. Given a Finsler manifold (M, F), it is possible to define the length of any piecewise smooth curve $c : [a, b] \to M$ as

$$\ell_F(c) = \int_a^b F(\dot{c}(s))ds.$$

Observe that this length is independent of positive reparametrizations of the curve, but when one changes the orientation of c, the length could change. This length leads to a non-necessarily symmetric distance $d_F : M \times M \to [0, +\infty)$ defined as the infimum

$$d_F(x_0, y_0) = \inf_{c \in C_{x_0,y_0}(M)} \ell_F(c), \tag{2}$$

where $C_{x_0,y_0}(M)$ is the subset of piecewise smooth curves between x_0 and y_0. Then, *pregeodesics* are defined as the curves that locally minimize the length functional, namely, a small enough piece of a pregeodesic has length equal to the distance between the endpoints of that piece. Moreover, a pregeodesic will be said a *geodesic* if in addition it is an affine reparametrization of an arc-parametrized pregeodesic; we will assume that the domain of each geodesic is an *inextendible* interval $I \subset \mathbb{R}$, except if otherwise specified. A Finsler manifold is *forward (resp. backward) complete* if the domain of its geodesics is always unbounded from above (resp. below). It is possible to define two types of balls for every $r > 0$ and $x_0 \in M$:

$$B_F^+(x_0, r) = \{y \in M : d_F(x_0, y) < r\}, \quad \text{forward ball}, \tag{3}$$

$$B_F^-(x_0, r) = \{y \in M : d_F(y, x_0) < r\}, \quad \text{backward ball}. \tag{4}$$

2.1.2 Classical Finslerian Solution to Zermelo Problem

Let M be a manifold, $x_0, y_0 \in M$, and let us try to minimize the arrival time for a moving object from x_0 to y_0 in the following situation, which is more general than Zermelo's. Assume that its velocity is prescribed at every oriented direction (such a velocity can also be interpreted as the maximum permitted velocity for the object). More precisely, for each $u \in TM \setminus \{\mathbf{0}\}$, its oriented direction is $[u] = \{\lambda u : \lambda > 0\}$ and the prescribed velocity $v_0([u]) = \mu u$ for some $\mu > 0$. Formally, v_0 becomes a section of the bundle $TM \to SM$, where SM is the sphere bundle of all the oriented directions. Further assume that at every $x \in M$, the set of prescribed velocities therein $\{v_0([u]) : \pi(u) = x\}$ is a positively curved smooth hypersurface of T_xM, diffeomorphic to a sphere and enclosing the zero vector. Observe that, as explained above, this hypersurface Σ is the indicatrix of a Finsler metric Z and, by construction, $Z \circ v_0 \equiv 1$. As we will see later, this is what happens in Zermelo problem when the wind is mild.

Assume that $[a, b] \ni s \mapsto c(s) \in M$ is any smooth curve from x_0 to y_0 with non-vanishing velocity. When the moving object follows the trajectory determined by c at the prescribed velocity, then one has a time reparametrization $[0, T] \ni t \mapsto t(s) \in [a, b]$ such that

$$v_0([\dot{c}(s)]) = \frac{d(c \circ s)}{dt}(t(s)) = \dot{s}(t(s))\dot{c}(s) = \frac{1}{\dot{t}(s)}\dot{c}(s) \tag{5}$$

and, as a consequence, $\dot{t}(s) = Z(\dot{c}(s))$ (as $Z(v_0([\dot{c}(s)])) = 1$). The elapsed time by the object is then

$$T = t(b) - t(a) = \int_a^b \dot{t}(s)ds = \int_a^b Z(\dot{c}(s))ds,$$

that is, it is the length of c computed with the Finsler metric Z. Therefore, in order to find the trajectories that minimize the elapsed time, we will have to find the minimizing geodesics of Z from x_0 to y_0, so that the time T in the previous formula is equal to $d_Z(x_0, y_0)$ for d_Z as in (2).

Observe that the time-independent Zermelo problem with mild wind is a particular case of this situation. In such case, the subset of velocities is the translation of a sphere with the wind W, and this translated sphere still encloses the 0 vector. Let us denote by g_R the Riemannian metric having as a unit sphere the velocities without wind and assume that the wind is never stronger than this velocity, namely, $g_R(W, W) < 1$. The pair (g_R, W) is usually called the navigation data of Zermelo problem. Then the translated sphere is determined by those $v \in TM \setminus \{\mathbf{0}\}$ such that

$$g_R\left(\frac{v}{Z(v)} - W, \frac{v}{Z(v)} - W\right) = 1.$$

Solving an equation of second order, we finally deduce that

$$Z(v) = \sqrt{\frac{1}{\lambda}g_R(v, v) + \frac{1}{\lambda^2}g_R(W, v)^2} - \frac{1}{\lambda}g_R(W, v), \quad \text{where} \quad \lambda{=}1{-}g_R(W, W). \tag{6}$$

This is a metric of Randers type, namely, of the form

$$F(v) = \sqrt{\tilde{h}(v, v)} + \tilde{\omega}(v), \tag{7}$$

where $\tilde{h}$ is a Riemannian metric and $\tilde{\omega}$ is a one-form on M such that $\|\tilde{\omega}\|_{\tilde{h}} < 1$ everywhere. The condition on $\tilde{\omega}$ guarantees that F is positive away from the zero section. Moreover, it turns out that it also implies that the fundamental tensor (1) is positive definite (see [4, Sect. 11.1] or [52, Corollary 4.17]). It was shown in [5] that the family of Randers metrics is the same as Zermelo metrics in (6), namely, all the metrics as in (7) admit an expression as in (6) for some navigation data (g_R, W) which is unique. Observe that the original Zermelo problem was formulated for a Euclidean metric g_R, whereas in [5] it was generalized to Riemannian manifolds. A further generalization to Finsler metrics can be found in [74, Sect. 3] (see also [56] for Zermelo problem in pseudo-Riemannian and pseudo-Finsler metrics).

2.1.3 Solution Using a Stationary Spacetime: Fermat Principle

Here, we anticipate the use of some simple notions on spacetimes and causality; the unfamiliarized reader can look at Sect. 3.1 first. A standard stationary spacetime is a Lorentzian manifold $(\mathbb{R} \times M, g)$ such that

$$g((\tau, v), (\tau, v)) = -\Lambda\tau^2 + 2\omega(v)\tau + g_0(v, v), \tag{8}$$

where $(\tau, v) \in \mathbb{R} \times T_{x_0}M \equiv T_{(t_0,x_0)}(\mathbb{R} \times M)$, with $(t_0, x_0) \in \mathbb{R} \times M$, $\Lambda : M \to \mathbb{R}$ smooth and positive, ω a one-form and g_0 a Riemannian metric, both on M. Observe that ∂_t is a Killing field of g which is timelike, and we will assume that $(\mathbb{R} \times M, g)$ is time-oriented by ∂_t.

In this setting, consider the problem of "traveling" from a point ("event") $(0, x_0)$ to a vertical line $\mathbb{R} \ni s \to (s, y_0)$ ("stationary observer at y_0"), so that the increase in the t coordinate is minimum. In order to model the meaning of traveling, one considers (future-directed) timelike or lightlike curves departing from the event but, as the latter curves will "move faster" (at the prescribed "speed of light"), we will restrict ourselves to the space of (smooth, future-directed) lightlike curves from the event to the stationary observer. Lightlike pregeodesics present known extremization properties which imply that light rays are the unique local minimizers of the coordinate t.

Remark 1 This underlies the *relativistic Fermat principle*, namely, the lightlike geodesics joining the event and the observer are the critical points for the arrival time functional in the aforementioned space of lightlike curves. This Fermat principle has been present in general relativity from the very beginning. The first version is due to Hermann Weyl [79] for static spacetimes shortly after the irruption of Einstein field equations. In the following, we will use the extended version for stationary spacetimes developed by Levi-Civita [58]. One of the key points of this version is that the isometries of the Killing field allow for a reduction of the problem to a hypersurface which is not necessarily orthogonal to the stationary observers, and therefore, it is not its natural restspace. As a consequence, the velocity of the light rays measured using this hypersurface is not isotropic and provides the indicatrix of a Finsler metric.

As a lightlike curve $\gamma = (t, x) : [a, b] \to \mathbb{R} \times M$ satisfies that

$$-\Lambda\dot{t}^2 + 2\omega(\dot{x})\dot{t} + g_0(\dot{x}, \dot{x}) = 0,$$

it turns out that

$$t(b) = \int_a^b \left(\frac{1}{\Lambda}\omega(\dot{x}) + \sqrt{\frac{1}{\Lambda^2}\omega(\dot{x})^2 + \frac{1}{\Lambda}g_0(\dot{x}, \dot{x})} \right) ds,$$

namely, the arrival time coincides with the length of the Finsler metric of Randers type

$$F(v) = \frac{1}{\Lambda}\omega(v) + \sqrt{\frac{1}{\Lambda^2}\omega(v)^2 + \frac{1}{\Lambda}g_0(v, v)}. \tag{9}$$

Thus, using the relativistic Fermat principle, one obtains the following characterization of lightlike geodesics (see, for example, [21, Proposition 4.1]).

Proposition 1 *A lightlike curve* $(t, x) : [a, b] \to \mathbb{R}\times M$ *is a geodesic of* $(\mathbb{R}\times M, g)$ *if and only if* x *is a geodesic of* (M, F) *up to parametrization. Moreover, in this case, if* (t, x) *is parametrized with the time-coordinate, then* x *is an* F*-unit geodesic.*

The Fermat metric (9) is very similar to the Zermelo metric (6). Indeed, both families provide the same class of metrics.

Proposition 2 *The Fermat metric* (9) *associated with a standard stationary spacetime as in* (8) *is the Zermelo metric with navigation data* (g_R, W) *determined by*

$$\omega = -g_0(W, \cdot), \quad g_R = \frac{1}{\Lambda + \|\omega\|_0^2} g_0. \tag{10}$$

Conversely, a Zermelo metric with navigation data (g_R, W) *on* M *is the Fermat metric of a standard stationary spacetime* $(\mathbb{R} \times M, g)$ *as in* (8) *determined by*

$$g_0 = g_R, \quad \omega = -g_0(W, \cdot) \quad \text{and} \quad \Lambda = 1 - g_R(W, W). \tag{11}$$

Moreover, this correspondence between standard stationary spacetimes and Zermelo data is one-to-one if we assume that $\Lambda + \|\omega\|_0^2 = 1$, *which fixes an element in the conformal class of* $(\mathbb{R} \times M, g)$.

To give an idea of the proof, observe that (10) implies

$$1 - g_R(W, W) = 1 - \frac{1}{\Lambda + \|\omega\|_0^2} g_0(W, W) = 1 - \frac{\|\omega\|_0^2}{\Lambda + \|\omega\|_0^2} = \frac{\Lambda}{\Lambda + \|\omega\|_0^2}, \tag{12}$$

and putting all this together, one gets straightforwardly that the Zermelo metric with data given in (10) is just the Fermat metric in (9) (see [13, Proposition 3.1] and [43] for the equivalence between Fermat and Zermelo metrics). Observe also that the Fermat metric is conformally invariant, namely, if we multiply g by a positive function ϕ on M, then the associated Fermat metric is preserved. Using this fact, it is possible to choose an element of the conformal class of g normalized in some sense. Indeed, given $(\mathbb{R} \times M, \tilde{g} = \phi g)$ with $\phi = 1/(\Lambda + \|\omega\|_0^2)$, its Fermat metric satisfies that $\tilde{\Lambda} + \|\tilde{\omega}\|_0^2 = 1$, where $\tilde{\Lambda}$, $\tilde{\omega}$, and $\tilde{g}_0$ are the data of $\tilde{g}$ and the norm is computed with $\tilde{g}_0$. In this case, it follows from (12) that $\tilde{\Lambda} := \phi\Lambda = 1 - g_R(W, W)$.

Remark 2 Proposition 2 makes apparent the following non-relativistic interpretation of the Fermat metric. The indicatrix of the Fermat metric is the subset of the velocities of light rays measured by the stationary observers in the tangent space to the slices $\{t_0\} \times M$, which is the translation of a Riemannian sphere. Clearly, this interpretation is not relativistic (all relativistic observers measure the same *speed of light*!). Indeed, a relativistic observer would use their space at rest, which is orthogonal to their timelike direction. This space is infinitesimal and can be identified with the tangent space to the slices only when $\omega = 0$.

It is worth pointing out that the stationary spacetime associated with a Zermelo metric can be interpreted as a sort of analogue gravity (see [6, Sect. 2] and [35]).

2.2 *The Case of Arbitrary Time-Independent Wind*

2.2.1 Emergence of Wind Riemannian Structures

Let us assume now that in the navigation data (g_R, W), the wind W is arbitrary, namely, let us remove the constraint $g_R(W, W) < 1$. This means that we will translate the spheres of g_R with a vector which is possibly not contained in the sphere. The set of all the translated spheres will form a smooth hypersurface $\Sigma \subset TM$ which generalizes the indicatrix of a Finsler metric and is called a *wind Riemannian structure*.

So, each translated sphere Σ_x plays the role of an indicatrix at $x \in M$, which might not enclose the zero in its interior region; thus the regions with $g_R(W, W) > 1$, $g_R(W, W) = 1$ or $g_R(W, W) < 1$ are called of *strong, critical*, or *mild wind*, respectively. In particular, when the wind is strong, one has some *admissible directions* at each $x \in M$, defined as the oriented directions $[u]$ which intersect the indicatrix; they form a closed solid cone in T_xM. Those $[u]$ in the interior of the cone intersect twice the indicatrix and, so, provide naturally two velocities $v_+([u])$, $v_-([u])$ (the latter contained in the segment between the origin and the former); both velocities merge in a single one for the admissible directions in the boundary (see Fig. 1).

In the region of strong wind, Zermelo problem for $x_0, y_0 \in M$ splits into several ones. First, determine whether there exists an admissible curve from x_0 to y_0. If this is the case, prescribe either v_+ or v_- for the moving object and find extremal curves for the arrival time. In the case of v_+, it is natural to wonder for a (local or global) minimizer of the arrival time, and one properly has the previous Zermelo problem. Indeed, such a problem makes sense for curves which may cross regions of strong,

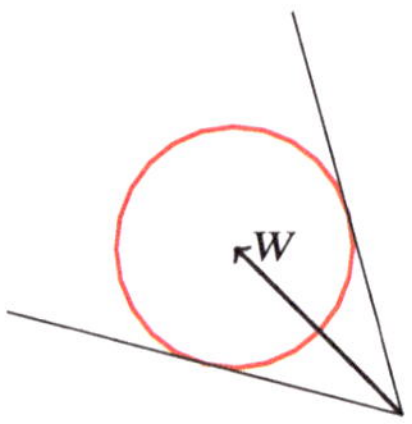

Fig. 1 In the presence of a strong wind W with $g_R(W, W) > 1$, the only directions allowed for the moving object are those which intersect the red sphere

critical, and mild wind, and v_+ will match smoothly with v_0 on the whole M. In the case of prescribing v_-, the natural problem would be to find *maximizers* of the arrival time entirely contained in the region of strong wind. This case would occur when the object tries to delay the arrival at y_0 as much as possible by making its engine power go against the direction of the wind. Technically, this makes sense because the portion of the indicatrix corresponding to v_- is concave (as happens for the unit timelike directions of a Lorentz metric).

These problems were first studied in [23]. Formally, previous ideas of the mild case work similarly to obtain a Finsler metric which measures the elapsed time along an admissible trajectory. In the strong wind case, we obtain two metrics:

$$Z(v) =\frac{1}{\lambda}(\sqrt{\lambda g_R(v,v) + g_R(W,v)^2} - g_R(W,v)), \tag{13}$$

$$Z_l(v) =\frac{1}{\lambda}(-\sqrt{\lambda g_R(v,v) + g_R(W,v)^2} - g_R(W,v)), \tag{14}$$

being defined both of them in

$$A_l = \{v \in TM : \lambda < 0,\ \lambda g_R(v,v) + g_R(W,v)^2 > 0,\ -g_R(W,v) < 0\}.$$

Observe that if we want to compute the minimizing solutions of Zermelo problem, then we have to consider Z, which corresponds to v_+ above. The above expressions do not hold in the critical case $g_R(W, W) = 1$, but it is possible to obtain expressions valid for arbitrary winds multiplying by the conjugate in (13), (14), i.e.,

$$Z(v) =\frac{g_R(v,v)}{\sqrt{\lambda g_R(v,v) + g_R(W,v)^2} + g_R(W,v)}, \tag{15}$$

$$Z_l(v) =\frac{-g_R(v,v)}{\sqrt{\lambda g_R(v,v) + g_R(W,v)^2} - g_R(W,v)}. \tag{16}$$

Observe that now Z is defined for an arbitrary W in the domain

$$A = \{v \in TM \setminus \mathbf{0} : \lambda > 0 \text{ or } \lambda g_R(v,v) + g_R(W,v)^2 > 0,\ g_R(W,v) > 0\}$$

and, in the case of critical wind, the metric Z is of Kropina type, namely, the quotient of a Riemannian metric by a one-form:[1]

$$Z(v) = \frac{g_R(v,v)}{2g_R(W,v)}.$$

In particular, the domain A coincides with $TM \setminus \mathbf{0}$ when $\lambda > 0$ (the region of mild wind), it is the half-space $g_R(W, v) > 0$ in the tangent bundle of the region

$$M_{crit} = \{x \in M : \lambda = 0\}$$

(the region of critical wind) and a conic region in

$$M_l = \{x \in M : \lambda < 0\}$$

(the region of strong wind). On M_l, the domain A coincides with the timelike vectors of the Lorentzian metric $-h$ on M_l,[2] where

$$h(v, v) = \lambda g_R(v, v) + g_R(W, v)^2 \tag{17}$$

in the half-space $h(W, v) < 0$. Moreover, the domain of Z_l is $A_l = A \cap TM_l$. It turns out that in the region $M_{crit} \cup M_l$, the metric Z is conic, namely, at every $x \in M_{crit} \cup M_l$, it is not defined in the whole TM but in the conic region $A_x = A \cap T_xM$. Moreover, it is positive homogeneous of degree one, and its fundamental tensor (1) is positive definite. The metric Z can be extended to the boundary of A_x, but this extension is not smooth, and its fundamental tensor cannot be extended to the boundary. However, Z is a classical Finsler metric in the region $M \setminus \{M_{crit} \cup M_l\}$. On the other hand, the metric Z_l is always *conic*,[3] and its fundamental tensor has index $n - 1$. Moreover, it can be extended to the boundary of A_x, this extension coincides with that of Z, but again its fundamental tensor does not admit such extension (see [23, Sects. 3.3 to 3.5]).

Summing up, Zermelo problem is modeled in terms of a wind Riemannian structure, whose Finslerian description retains some elements of the mild wind case. However, it also includes new ingredients (Kropina metric, F_l with concave indicatrix) which become complicated and apparently singular. Next, the spacetime viewpoint will simplify the picture giving an elegant solution.

---

[1] These metrics are well-known in the Finsler literature since the original article [57].

[2] Notice that, whenever $\lambda < 0$, h has signature $(+, -, \ldots, -)$. Anyway, h is well defined as a signature changing metric on the whole M, so that it can be used to determine the admissible curves between any two points $x_0, y_0 \in M$. Comparing with $\tilde{h}$ in (7) (which was defined for $\lambda > 0$), one has $\tilde{h} = h/\lambda^2$.

[3] See [52] for a full development of this condition.

2.2.2 Solution Using an SSTK Spacetime

The process to solve Zermelo problem for mild wind using a spacetime developed in Sect. 2.1.3 can be extended for a general wind just by considering Lorentzian metrics as in (9) with an arbitrary Λ (not necessarily positive) satisfying $\Lambda + \|\omega\|_0^2 > 0$; this condition must be imposed because it is equivalent to the Lorentzian character of (8). Following [23], this class of spacetimes is called *SSTK (standard with a space-transverse Killing vector field)* spacetimes. Observe that the points $x \in M$ with $\Lambda(x) < 0$ correspond to vertical lines $s \to (s, x) \in \mathbb{R} \times M$ which are spacelike in $(\mathbb{R} \times M, g)$, and then ∂_t is not timelike therein. It is still consistent though to consider the time-orientation determined by $dt > 0$, namely, a timelike vector (τ, v) is future-directed if and only if $\tau > 0$. It is worth pointing out that the usual relativistic Fermat principle does not apply to such a vertical line, as it is not necessarily timelike; however, as shown in [23, Theorem 7.4], this principle can be extended for arrival curves of arbitrary causal character. Thus, the solution of Zermelo problem in this setting can be described as follows:

> Given $x_0, y_0 \in M$, $x_0 \neq y_0$ and the Zermelo data (g_R, W), construct an SSTK spacetime with g_0, ω and Λ given as in (11). An admissible curve $c : [0, T] \to M$ from x_0 to y_0 with prescribed velocity in the wind Riemannian structure Σ is a critical trajectory for the arrival time T if and only if the future-directed lightlike curve $[0, T] \ni t \mapsto (t, c(t)) \in \mathbb{R} \times M$ is a lightlike pregeodesic of the spacetime.
>
> In this case, the arrival time T of c is globally minimizing (resp. maximizing) if and only if $(T, y_0) \in \mathbb{R} \times M$ is the first (resp. last) t of the observer $\{(t, y_0), t \in \mathbb{R}\}$ at y_0 reached by future-directed causal curves starting at $(0, x_0)$.

We can also recover c as a unit speed geodesic for F or F_l (depending whether it is locally minimizing or maximizing) by repeating the process of Sect. 2.1.3. Indeed, it follows that a lightlike curve $\gamma = (t, x) : [a, b] \to \mathbb{R} \times M$ is a lightlike pregeodesic if and only if its projection x is either a pregeodesic of one of the conic metrics

$$F(v) = \frac{g_0(v, v)}{-\omega(v) + \sqrt{\omega(v)^2 + \Lambda g_0(v, v)}}, \tag{18}$$

$$F_l(v) = \frac{-g_0(v, v)}{\omega(v) + \sqrt{\omega(v)^2 + \Lambda g_0(v, v)}}, \tag{19}$$

or it is a suitable curve with velocity in the closure of the conic domain constructed from the lightlike pregeodesics of h [23, Theorem 5.5].

As in the previous subsection, now the Fermat metric (9) splits into two, and one has to multiply by the conjugate to obtain the metrics (18) (defined on the whole M) and (19) (only on M_l); compare with those in (13)–(16). This splitting is naturally interpreted now, because there are two future-directed lightlike vectors which project onto each $v \in TM_l$ (see Fig. 2). Extending Proposition 2, these two metrics can be identified, respectively, with those in (15) and (16).

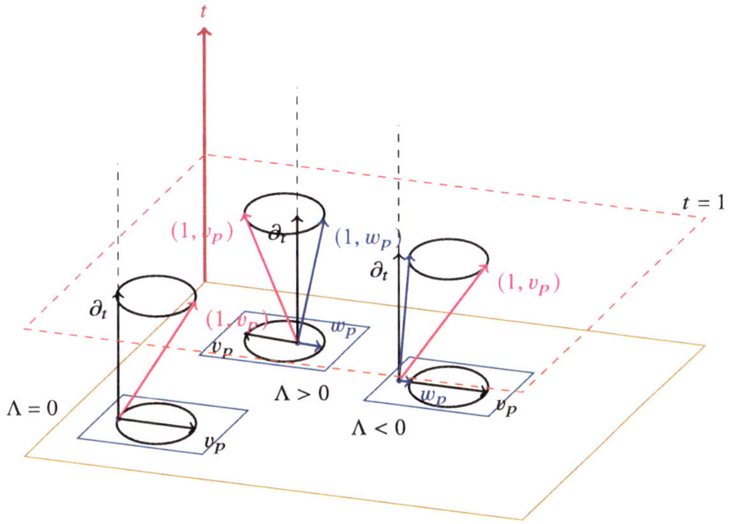

Fig. 2 We show how to obtain the wind Riemannian structure associated with the SSTK spacetime by intersecting the lightlicone with the slice $t = 1$ in the three different cases: $\Lambda = 0$ (critical wind), $\Lambda > 0$ (mild wind), and $\Lambda < 0$ (strong wind). Observe that in the first two cases, v_p is the projection of a unique lightlike vector, while in the last one there are exactly two lightlike vectors projecting and pointing out in the same direction as v_p

Summing up, the spacetime viewpoint yields a full solution of Zermelo problem which permits a unified description of the solutions and allows one to recover the (apparently singular) Finslerian description in the spacelike part. In particular, the lightlike geodesics in the spacetime, which solve the problem, can be characterized as follows [23, Theorem 5.5].

Theorem 1 *A lightlike curve* $(t, x) : [a, b] \to \mathbb{R} \times M$ *is a geodesic of* $(\mathbb{R} \times M, g)$ *if and only if it lies in one of the following three cases: (a)* $\dot{x}$ *is entirely contained in*

$$A = \{v \in TM \setminus \mathbf{0} : \text{ either } \Lambda > 0 \text{ or } \Lambda g_0(v, v) + \omega(v)^2 > 0,\ \omega(v) < 0\} \qquad (20)$$

and, in this case, x *is a pregeodesic of either* F *or* F_l*, (b)* x *is constantly equal to some* x_0 *with* $d\Lambda(x_0) = 0$*, or (c)* x *is a suitable curve with* $\dot{x}$ *in the closure of* A.[4]

[4] It is worth emphasizing that this last case appears in the same footing as the others from the spacetime viewpoint. However, in the Finslerian description, it corresponds with the limit of the geodesics of F and F_l (indeed, as a limit, $F(\dot{x}) = F_l(\dot{x}) \equiv 1$). Moreover, x might start at the region of M_l, arrive at the region of critical wind M_{crit}, and come back to M_l. On M_l, x becomes a pregeodesic of $\Lambda g_0 + \omega \otimes \omega$ consistent with (17) (see the case (iii) (b) in the aforementioned [23, Theorem 5.5]).

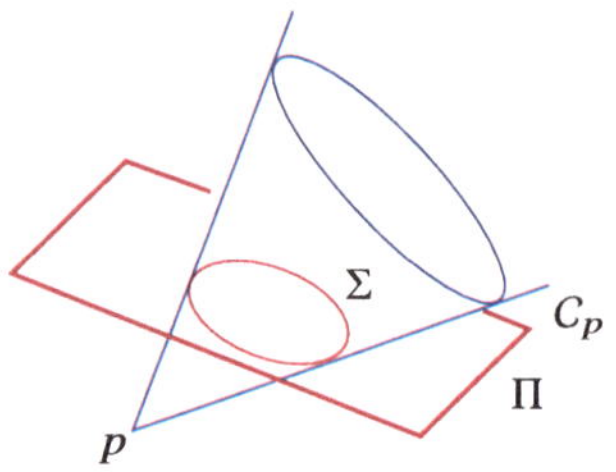

Fig. 3 The cone structure C_p is intersected with an affine hyperplane Π in a strongly convex hypersurface Σ of Π

2.3 The Time-Dependent Case

The Zermelo problem in a manifold M with a time-dependent wind can be handled by using time-dependent Finsler metrics. This was done by Manià [60] as well as by Markvorsen using frozen metrics [62]. Here, we will develop a spacetime picture as a natural extension of our framework. Indeed, we will consider again a more general problem, namely, we will assume that the velocity is prescribed at every direction, but, now, this prescription may have a dependence on time. Thus, one has a smooth hypersurface $\Sigma_{(t_0,x_0)}$ of $T_{x_0}M$ diffeomorphic to a sphere and positively curved at each $t_0 \in \mathbb{R}$, and all these hypersurfaces vary smoothly with (t_0, x_0), providing a smooth submanifold[5] of $T(\mathbb{R} \times M)$ with codimension 2.

Observe that in the non-relativistic spacetime $\mathbb{R} \times M$, being the coordinate $\mathbb{R}$ the (absolute) time, a curve $s \to (t(s), x(s))$ corresponds to a curve with velocity $v(s) = \frac{\dot{x}(s)}{\dot{t}(s)} \in T_{x(s)}M$ at the instant s (recall (5)). This means that if the velocities at each time t_0 are prescribed by the hypersurface $\Sigma_{t_0} = \cup_{x_0 \in M} \Sigma_{(t_0,x_0)} \subset TM$, then the vectors $(\tau, v) \in T_{(t_0,x_0)}(\mathbb{R} \times M)$ tangent to the allowed curves (traveling at the prescribed velocities) must satisfy that $v/\tau \in \Sigma_{t_0}$ (with $\tau > 0$). It turns out that these allowed tangent vectors determine a cone structure C (see [55, Definition 2.7]), namely, the smooth embedded hypersurface $C \subset T(\mathbb{R} \times M) \setminus \mathbf{0}$ given by

$$C_{(t_0,x_0)} = \{\tau(1, v) \in T_{(t_0,x_0)}(\mathbb{R} \times M) : \tau > 0,\ v \in \Sigma_{(t_0,x_0)}\},$$

where $C_{(t_0,x_0)} = C \cap T_{(t_0,x_0)}(\mathbb{R} \times M)$ for every $(t_0, x_0) \in \mathbb{R} \times M$. The requirements on each $\Sigma_{(t_0,x_0)}$ become equivalent to the *strong convexity* of C [55, Proposition 2.26], that is, when $C_{(t_0,x_0)}$ is intersected by an affine hyperplane of $H \subset T_{(t_0,x_0)}(\mathbb{R} \times M)$ which is not tangent to $C_{(t_0,x_0)}$, then $H \cap C_{(t_0,x_0)}$ is a strongly convex hypersurface in the hyperplane (in fact, a positively curved sphere); see Fig. 3.

When the natural vector field ∂_t is timelike for C (i.e., it lies in the interior of the solid cone bounded by C at each point), the easiest way to describe C is using a Lorentz-Finsler metric G in $\mathbb{R} \times M$ constructed as follows. Let F_{t_0} be the Finsler

[5] See [23, Definition 2.8 and Example 2.16] for a subtlety about smoothness applicable here.

metric in M whose indicatrix coincides with Σ_{t_0}, that is,

$$\Sigma_{t_0} = \{v \in TM : F_{t_0}(v) = 1\}.$$

Then, define G at every point (t_0, x_0) as

$$G(\tau, v) = \tau^2 - F_{t_0}(v)^2, \quad \forall(\tau, v) \in T_{(t_0,x_0)}(\mathbb{R} \times M). \tag{21}$$

Observe that G is a Lorentz-Finsler metric[6] with fundamental tensor of Lorentzian type (replace F^2 with G in (1)). It turns out that the cone structure C is given by the lightlike vectors of G (with $\tau > 0$). Indeed, $G(\tau, v) = 0$ if and only if $\tau^2 = F_{t_0}(v)^2$, which is equivalent to $v/\tau \in \Sigma_{t_0}$, whenever $\tau > 0$.

Once G is obtained, the admissible trajectories between an event $(0, x_0)$ and an "observer" $\mathbb{R} \ni t \to (t, y_0)$ at a different point $y_0 \in M$ are lightlike curves of $(\mathbb{R} \times M, G)$. The Fermat principle can also be extended to the Finslerian relativistic case (see, [69]). Then, it follows that the solutions of the time-dependent case are provided by the projections x on M of the lightlike pregeodesics (t, x) of $(\mathbb{R} \times M, G)$ going from the event to the observer, with $\dot{t} > 0$.

Finally, the case when ∂_t is not timelike can be handled by using an auxiliary timelike vector field X which would represent comoving observers; see [50, Remark 6.1].

3 Relativistic Applications

3.1 Basic Lorentz Setup

Recall first some basic notions related to relativistic spacetimes following [8, 67], which are standard for readers with background in Riemannian geometry. Let g be a Lorentz metric on an $(n+1)$ manifold $\mathcal{M}$ with signature $(-, +, \ldots, +)$, which implies that each connected component of $\mathcal{M}$ will be either non-compact or of zero Euler characteristic. A non-zero tangent vector $v \in T\mathcal{M}$ is called *timelike*, *lightlike*, or *spacelike* if $g(v, v) < 0$, $g(v, v) = 0$ or $g(v, v) > 0$, respectively; *causal* vectors are the timelike or lightlike ones and *null* vectors the lightlike or zero ones.[7] It is obvious that any conformal Lorentz metric $g^* = \Omega g$ (for some function Ω) will have the same cones as g and, remarkably, the converse is true, that is, the lightlike vectors determine the conformal class of the metric (see, for example, [8, Lemma 2.1]).

[6] In the sense of [55, Sect. 3.2] (thus, positive and two-homogeneous) up to the fact that it may be non-smooth on ∂_t. This is not relevant here, because we are only interested in the directions of C (anyway, G can be smoothen on ∂_t; see [55, Theorem 5.6].)

[7] The convention for the zero vector depends on the reference, we follow [65].

$(\mathcal{M}, g)$ is called a *spacetime* when $\mathcal{M}$ is connected, and one assumes that a time-orientation has been chosen. The latter is a continuous choice of one of the two cones determined by the causal vectors at each tangent space $T_p\mathcal{M}$, $p \in \mathcal{M}$; each chosen cone will be called *future*, while the non-chosen one will be *past*. The properties of time-orientations are similar to those of usual orientations on $\mathcal{M}$ (including the existence of a time-orientable double covering), but both notions are independent. In spacetimes, the names of *future* or *past-directed* are applied directly to causal vectors; moreover, all the related names (timelike, future- or past-directed, etc.) are transferred directly to (smooth) curves taking into account the character of the velocity. Submanifolds are called spacelike, timelike, or lightlike depending on whether the induced metric has index 0, 1 or it is degenerate, respectively.

A major topic of research in mathematical relativity with interest in geometric analysis is the *initial value problem* for Einstein equations, which is posed on *Cauchy hypersurfaces*; See, for example, [29, 31, 66, 72]. Next, we will describe briefly the setting and, then, some applications of the aforementioned geometric links with Finsler manifolds.

3.2 The Initial Value Problem and Cauchy Hypersurfaces

The Einstein equation for a spacetime $(\mathcal{M}, g)$, $n+1 > 2$, is

$$\mathrm{Ric} + \frac{1}{2}\,\mathrm{Scal}\, g + \Lambda_c g = T, \tag{22}$$

where Ric and Scal denote, respectively, the Ricci tensor and scalar curvature of g, $\Lambda_c \in \mathbb{R}$ is the *cosmological constant*, and T is the *stress-energy tensor* up to a constant (which depends on units and provides the suitable coupling between matter and geometry). Roughly speaking, the initial value problem will consist in specifying initial data $(h, \mathcal{A})$ on some 3-manifold S and finding a spacetime $(\mathcal{M}, g)$ satisfying (22), where S is embedded as a spacelike hypersurface with metric h and second fundamental form $\mathcal{A}$. In particular, the initial data must satisfy the constraints associated with the Gauss and Codazzi equations, which turn out essential for the analysis of solutions. In general, one should also specify initial data on S for T as well as additional equations to be satisfied for the specific type of matter modeled by T (so that (22) remains as a hyperbolic system of equations in g and T). However, the latter are not required for the vacuum case $T = 0$ (which may be a good approximation for modeling the empty space outside a star). It is worth pointing out that, when $T = 0$, (semi-)Riemannian Schur's lemma reduces (22) to the equation of Einstein manifolds $\mathrm{Ric} = ag$, $a \in \mathbb{R}$ and, in the case of $\Lambda_c = 0$, to the Ricci flat case $\mathrm{Ric} = 0$.

Global existence and uniqueness of solutions for the vacuum case and other T's have been proved, starting at the classical results by Choquet-Bruhat and Geroch [28, 30, 41]. A posteriori, the initial data manifold $S \subset \mathcal{M}$ will be a (smooth, spacelike) *Cauchy hypersurface* of the solution $(\mathcal{M}, g)$, i.e., S is crossed exactly once by any inextendible causal curve.[8] The uniqueness of the crossing point is related to the existence of solutions (otherwise, data on a point of S might influence on data on points of S reached by a future-directed causal curve), and the existence of the crossing point is related to the uniqueness of the solutions. Indeed, the existence of a *unique* maximal solution is ensured *assuming that S remains Cauchy*, but different extensions might exist if this assumption is dropped (such a possibility is related to the so-called strong cosmic censorship hypothesis; see, for example, [78, Sect. 12.1]).

Summing up, a posteriori, the solution $(\mathcal{M}, g)$ to the initial value problem of the Einstein equation will admit a Cauchy hypersurface S, and $(\mathcal{M}, g)$ can be regarded as the domain of dependence of the data on S. Thus, it arises the natural question of determining the class of spacetimes admitting such a hypersurface as well as to determine when a given spacelike hypersurface is Cauchy, which will be adressed in the next two subsections.

3.3 Global Hyperbolicity and Causality of Spacetimes

Recall first the following basic elements of causality for a spacetime $(\mathcal{M}, g)$ (see [65] for a complete study). For $p, q \in \mathcal{M}$, we say that p lies in the chronological (resp. strict causal) past of q if there exists a future-directed timelike (resp. causal) curve from p to q; in this case, we write $p \ll q$ (resp. $p < q$). The chronological (resp. causal) future of p is defined as

$$I^+(p) = \{q \in \mathcal{M} : p \ll q\} \quad (\text{resp. } J^+(p) = \{q \in \mathcal{M} : p < q\} \cup \{p\}), \tag{23}$$

then, one defines analogously the chronological and causal pasts $I^-(p), J^-(p)$. Following [11], a spacetime is called *globally hyperbolic* when

$$p \not< p \quad \text{and} \quad J^+(p) \cap J^-(q) \text{ is compact} \tag{24}$$

for all $p, q \in \mathcal{M}$. The first condition means that the spacetime does not contain any causal loop, that is, $(\mathcal{M}, g)$ is *causal*, and the second one that no observer can see the sudden appearance or disappearance of a particle from the spacetime, i.e., $(\mathcal{M}, g)$ does not contain *naked singularities*. Taking into account Geroch's [41] and the further developments in [9, 10], the following characterization holds.

Theorem 2 *A spacetime is globally hyperbolic if and only if it admits a (smooth, spacelike) Cauchy hypersurface. In this case, there exists some onto smooth map*

[8] Sometimes Cauchy hypersurfaces and the initial value problem are allowed to be more general so that non-smooth data or data on degenerate hypersurfaces are permitted, but we will not go into these issues.

$\tau : \mathcal{M} \to \mathbb{R}$ such that the spacetime splits as an orthogonal product

$$\mathcal{M} \cong \mathbb{R} \times M, \quad g = -\Lambda d\tau^2 + g_t,$$

where $\Lambda > 0$ is a function on $\mathcal{M}$, g_t is a Riemannian metric on each slice $\{t\} \times M$ varying smoothly with t, and each slice becomes a Cauchy hypersurface in $\mathcal{M}$.[9] *Such a τ is called a Cauchy temporal function. Moreover, any Cauchy hypersurface S can be regarded as a level of some Cauchy temporal function, obtaining then an orthogonal splitting as above such that $S \cong \{0\} \times M$.*

A generalization of globally hyperbolic spacetimes are the *causally simple* ones, where the second hypothesis in (24) is weakened into the condition: $J^+(p)$, $J^-(p)$ are closed for all p, q. For these spacetimes, no splitting as in Theorem 2 can be ensured; however, they are *stably causal*, that is, they admit a *temporal function* τ. This means that $\tau : \mathcal{M} \to \mathbb{R}$ is onto with grad(τ) timelike and past-directed (thus, τ grows strictly on any future-directed causal curve and the slices $\tau =$ constant are spacelike). A simple example of these properties is the following (see [8, Theorem 3.67] and Theorem 4).

Proposition 3 *Let (M, g_R) be a Riemannian manifold and consider the natural product spacetime $(\mathcal{M} = \mathbb{R} \times M, g = -d\tau^2 + g_R)$. Then, the canonical projection $\tau : \mathbb{R} \times M \to \mathbb{R}$ is a temporal function and*
(A) $(\mathcal{M}, g)$ is causally simple if and only if g_R is convex (i.e., each $p, q \in M$ can be joined by a minimizing geodesic).

(B) The following properties are equivalent:

- *(B1) $(\mathcal{M}, g)$ is globally hyperbolic.*
- *(B2) The slices $\{t\} \times M$ are Cauchy hypersurfaces (i.e., τ is Cauchy temporal).*
- *(B3) g_R is complete.*

Remark 3 Taking into account the conformal invariance of causal properties, this simple result is applicable to some interesting classes of spacetimes which can be written as warped products.

In particular, the *standard static* spacetimes are written as $\mathcal{M} = \mathbb{R} \times M, g = -\Lambda d\tau^2 + g_0$ where g_0 is Riemannian and $\Lambda > 0$ depends only on the M part (this is a subclass of the standard stationary spacetimes already introduced, which will be developed next). Clearly g is conformal to a product with $g_R = g_0/\Lambda$.

Another conformal class are the so-called *Generalized Robertson Walker (GRW)* spacetimes [2], $\mathcal{M} = I \times M, g = -dt^2 + f^2 g_R$, where $I \subset \mathbb{R}$ is an interval and $f > 0$ depends only on the I part. Putting $d\tau = dt/f$, the metric g is conformal to a product spacetime on $J \times M$, where $J \subset \mathbb{R}$ is another interval computable from $\int dt/f(t)$. It is easy to check that the causal properties in Proposition 3 remain valid

---

[9] The natural time-orientation chosen here and in the remainder is the one such that the timelike vector ∂_τ becomes future-pointing.

if $\mathbb{R} \times M$ is replaced by $J \times M$, in contrast with the properties of the boundaries (see Sect. 5).

3.4 Finsler Applications to Stationary Spacetimes

Next, our aim is to obtain a result which extends Proposition 3 to the standard stationary spacetimes $(\mathcal{M} = \mathbb{R} \times M, g)$ introduced in (8).

3.4.1 Background on the (Non-symmetric) Finslerian Distance d_F

In order to state our main results, let us summarize first some metric properties valid for any Finsler metric F following [22]. Recall from (2) that F provides a (non-necessarily symmetric) distance d_F as well as the corresponding forward and backward balls $B_F^{\pm}(p, r)$ in (3) and (4). These distances yield a natural notion of *forward and backward Cauchy sequences* and, then, of *forward and backward completeness*. Such notions are nicely interpreted taking into account the *reversed Finsler metric*, $\tilde{F}(v) := F(-v)$ for all $v \in TM$. Indeed, the latter clearly satisfies $B_{\tilde{F}}^{+}(p, r) = B_F^{-}(p, r)$. Moreover, if γ is a geodesic for F, then its reversed parametrization (i.e., $\tilde{\gamma}(s) := \gamma(-s)$ for all s) becomes a geodesic for $\tilde{F}$, and so, one speaks of *forward (resp. backward) geodesic completeness* when all the inextendible geodesics have an upper (resp. lower) unbounded domain. A version of Hopf-Rinow theorem and some further computations show the following.

Theorem 3 *Let F be any Finsler metric on M.*

(A) The following conditions are equivalent (see, for example, [4, Theorem 6.6.1]):

- *(A1) forward (resp. backward) completeness of d_F,*
- *(A2) compactness of all its closed forward (resp. backward) balls,*
- *(A3) forward (resp. backward) geodesic completeness of F.*

(B) Let $d_F^s(p, q) := (d_F(p, q)+d_F(q, p))/2$ for all $p, q \in M$, then d_F^s is a (usual) distance on M, called the symmetrized distance, and the following conditions are equivalent (see [22, Proposition 2.2, Theorem 5.2]):

- *(B1) compactness of all the closed balls for the symmetrized distance d_F^s,*
- *(B2) precompactness of all the intersections $B_F^{+}(p, r) \cap B_F^{-}(q, r')$, $p, q \in M$, $r, r' > 0$,*
- *(B3) the last property putting $p = q, r = r'$.*

In this case, d_F^s is complete, and F (and so $\tilde{F}$) is convex (i.e., every two points are connected by a minimizing geodesic). Moreover, the above properties hold if either d_F or $d_{\tilde{F}}$ is complete.

Remark 4

(1) The symmetrized distance d_F^s is not associated with any length space, in general. This underlies the fact that the converse of the last assertion in the proposition does not hold, that is, there are even Randers examples with d_F^s complete but non-compact closed balls (see [22, Example 2.3]).
(2) It is easy to check that the forward completeness of F does not imply the backward one. In the particular case of a Fermat metric (9), the forward or backward completeness of F implies the completeness of $-\frac{1}{\Lambda}\tilde{g} = \frac{1}{\Lambda^2}\omega^2 + \frac{1}{\Lambda}g_0$, but the converse does not hold (see [22, Proposition 5.2, Example 2.3]).

3.4.2 Finslerian Description of Causality

It is not difficult to realize that the chronological and causal futures and pasts in a standard stationary spacetime, $I^{\pm}(p_0)$, $J^{\pm}(p_0)$, $p_0 = (t_0, x_0) \in \mathbb{R} \times M$, can be described in terms of the balls for the distance d_F associated with the Fermat metric F defined in (9) as follows (see [22, Proposition 4.2]):

$$I^+(t_0, x_0) = \cup_{s>0}\{(t_0 + s) \times B_F^+(x_0, r)\}.$$

Remark 5 This can be seen as a consequence of a further interpretation of the Zermelo problem described in Sect. 2 (see especially Remark 2). We can consider that the prescribed velocities are the top speeds that a moving object can reach. Accordingly, the trajectories in the spacetime will be causal curves, being lightlike when a certain path is traveled at the top speed. Moreover, by the causal properties, a curve is a solution of Zermelo problem if and only if it remains in $J^+(t_0, x_0) \setminus I^+(t_0, x_0)$, which implies that it is a lightlike pregeodesic and its projection is a minimizing pregeodesic of F of length $t_1 - t_0$ (recall Proposition 1). As in Riemannian geometry, the geodesics of F are always locally minimizing, and the lightlike geodesics departing from p lie initially in $J^+(p) \setminus I^+(p)$.

These are the keys for the description of causality using the Fermat metric F in (9), allowing one to prove the following result (see [22, Theorems 4.3 and 4.4]).

Theorem 4 *For any standard stationary spacetime* $(\mathcal{M}, g)$ *with associated Fermat metric* F *(as in* (8), (9) *above), the canonical projection* $t : \mathbb{R} \times M \to \mathbb{R}$ *is a temporal function and*

(A) The following properties are equivalent:

(A1) $(\mathcal{M}, g)$ *is causally simple.*
(A2) F *is convex (i.e., each* $p, q \in M$ *can be joined by a minimizing* F*-geodesic).*

(B) The following properties are equivalent:

(B1) $(\mathcal{M}, g)$ *is globally hyperbolic.*

(B2) the intersections $B_F^+(p,r) \cap B_F^-(q,r')$, $p, q \in M, r, r' > 0$ *are precompact.*

(C) The following properties are equivalent:

(C1) The slices $\{t_0\} \times M$ *are Cauchy hypersurfaces (so, t is Cauchy temporal).*
(C2) The Fermat metric F is forward and backward complete.

The comparison between Proposition 3 and Theorem 4 shows that the case of standard stationary spacetimes is much subtler than the case of product spacetimes. Much of this subtlety comes from the following fact. Both cases are presented with a natural slicing by spacelike hypersurfaces which are invariant by the timelike Killing vector field ∂_t. In the product case, this slicing is priviledged, as it corresponds with the integral hypersurfaces of the distribution $\partial_t^\perp$, orthogonal to ∂_t. In the standard stationary spacetime, however, the slices are not especially priviledged, and one can always express the spacetime as a standard stationary one in many different ways, as emphasized next.

Any function $f : M \to \mathbb{R}$ which yields a spacelike graph $S^f = \{(f(x), x), x \in M\} \subset \mathcal{M}$ generates a new splitting of the spacetime $(\mathcal{M}, g)$ as a standard stationary one such that $\mathcal{M} \cong \mathbb{R} \times S^f$ (just moving S^f with the flow of ∂_t). The Fermat metric associated with this new splitting is $F^f := F - df$ (see [22, Proposition 5.9]). The characterization of the causality conditions in Theorem 4 together with the last observation about the relation between Fermat metrics associated with different slices of the standard stationary spacetime has some striking consequences for Randers metrics.

Corollary 1 *Let* (M, F) *be a Randers manifold (recall* (7)*) and assume that the intersections* $B_F^+(p,r) \cap B^-(p,r)$ *are precompact for all* $p \in M$ *and* $r > 0$. *Then* F *is convex, and there exists a function* $f : M \to \mathbb{R}$ *such that* $F^f = F - df$ *is a complete Randers metric on* M.

Both properties follow from well-known causal properties. Indeed, as stated in Theorem 4, the precompactness of the intersection of the balls of the Fermat metric is equivalent to the global hyperbolicity of the spacetime, which implies the causal simplicity, and then again, by Theorem 4, the convexity of the Fermat metric. Moreover, as stated in Theorem 2, if $(\mathbb{R} \times M, g)$ is globally hyperbolic, then it admits a smooth spacelike Cauchy hypersurface, and the Fermat metric associated with this slice can be expressed as $F^f = F - df$ for a certain $f : M \to \mathbb{R}$, and it is complete (see [22, Theorem 5.10]).

Remark 6 The previous corollary holds for any Finsler metric. This can be proved either directly, as it was done by Matveev in [63], or generalizing Theorem 4 to stationary Finsler spacetimes as in [18, 19] (and, then, using the generalization of the existence of a smooth Cauchy spacelike hypersurface for cone structures, firstly obtained by Fathi and Siconolfi in [37]).

Remark 7 Theorem 4 can be used to get some different types of results for standard stationary spacetimes:

(i) As a first consequence, one can obtain some multiplicity results for lightlike geodesics (periodic or between two fixed points) of globally hyperbolic stationary spacetimes with arbitrary big arrival times. This follows using multiplicity results for Finsler metrics (see [13, 20, 21]). Moreover, using the auxiliary product spacetime $((\mathbb{R} \times M) \times \mathbb{R}, g + dr^2)$, one can also obtain multiplicity results for timelike geodesics with prescribed proper time between a point and a vertical line of $(\mathbb{R} \times M, g)$. Finally, using some results on convex boundaries for Finsler metrics in [7], it follows the existence of lightlike or timelike geodesics with prescribed proper time for stationary spacetimes with suitable boundaries (see [24]).

(ii) Theorem 4 can also be generalized to pre-Randers metrics, namely, Randers metrics as in (7) with arbitrary ω, this time with some subtleties as the slices $\{t_0\} \times M$ do not have to be necessarily spacelike (see [47, Sect. 3.2]). This more general case has applications to the study of magnetic geodesics and also to strengthen results about the existence of t-periodic lightlike geodesics of stationary-complete spacetimes (see [47, Corollary 5.12]).

(iii) Observe that the Finsler metric F^f in Corollary 1 has the same pregeodesics as F. Moreover, their distances are straightly related having the same triangular function $T(x, y, z) = d(x, y) + d(y, z) - d(x, z)$. As a further interplay between Lorentz and Finsler geometries, the maps which preserve this triangular function are called *almost isometries*, and they can be identified with conformal maps of the spacetime that preserve ∂_t (see [48, Proposition 4.7] for details). Moreover, the almost isometries have been characterized in terms of the semi-Lipschitz functions of (M, F) (see [17, 32]).

3.5 Application to General SSTK Spacetimes

The previous setting can be extended to any SSTK spacetime, as proven in [23]. This is more technical, and only the main ingredients will be sketched. For the basic notions on SSTK spacetimes, recall Sect. 2.2.2.

Remark 8 In relativity, SSTK spacetimes are applicable to describe stationary black holes.[10] Indeed, the so-called Killing horizons can be written as degenerate hypersurfaces of an SSTK spacetime determined by $\Lambda = 0$ (see Fig. 4). Typically, this notion appears under assumptions on asymptotic flatness (so that one can think of Lorentz-Minkowski observers at infinity), and Λ becomes positive asymptotically. Thus, the open subset $\Lambda > 0$ is expected to have a connected part which

[10] See [78] for background, especially Chap. 12 and Sects. 12.3, 12.5.

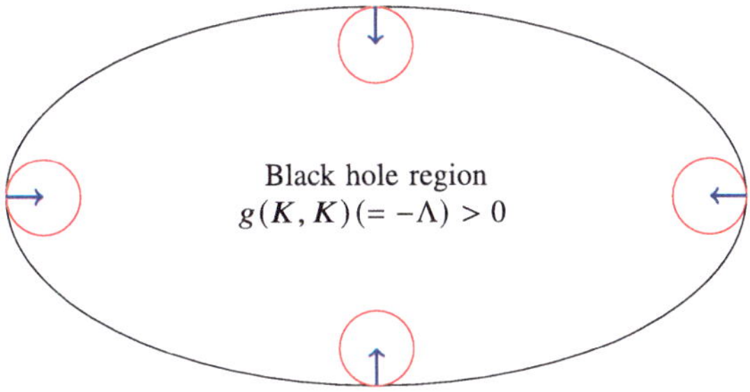

Fig. 4 Killing horizon determined by the region $\Lambda = 0$ in a slice $\{t_0\} \times M$. The blue arrows correspond to the wind W in the critical region for Zermelo data as in Sect. 2.2.1

includes the asymptotic region. This would be the region outside the Killing horizon, and (each connected component of) its boundary would be a Killing horizon.

In Kerr spacetime, the so-called ergosphere is its natural Killing horizon. In the case of Schwarzschild spacetime, such a horizon is equal to its event horizon, informally, the hypersurface that, once crossed by an observer coming from infinity, cannot be crossed again (a spacetime is expected to be globally hyperbolic outside its event horizon). Indeed, Schwarzschild spacetime admits a Killing vector field K which is timelike outside this horizon and makes it a static spacetime there (thus, this Killing K becomes priviledged because its orthogonal distribution is integrable; recall Remark 3). The static description fails in the horizon, but Schwarzschild spacetime can be described as an SSTK therein (being written the region outside the horizon as a stationary spacetime).

3.5.1 The Particular Case $\Lambda \geq 0$

Observe that when $\Lambda \geq 0$, the metric F in (18) is always of Randers type as in (7) (when $\Lambda > 0$) or of Kropina type $F(v) = -g_0(v, v)/(2\omega(v))$ (when $\Lambda = 0$). Accordingly, we will say that F is a Randers-Kropina metric in this case. A noticeable subtlety is that the domain of F (and, so, the velocities of the admissible curves joining two points) is restricted in the Kropina case to the half space $\omega < 0$ (recall (20)). So, the *F-separation* d_F, defined formally as a Finslerian distance, will yield $d_F(x, y) = \infty$ if there is no admissible curve from x to y. Even though not all the statements in Hopf-Rinow theorem hold now, d_F is fairly well-behaved.[11] Indeed, regarding d_F as the F-separation, one can ensure (see [23, Theorem 4.9]):

> *The characterizations of causality and Cauchy hypersurfaces of a standard stationary spacetime in Theorem 4 remain valid for any SSTK spacetime with $\Lambda \geq 0$.*

[11] In particular, it is continuous outside the diagonal [23, Theorem 4.5] (this property does not hold for any conic Finsler metric [52, Proposition 3.9]).

3.5.2 The General Case

Recall that when $\Lambda < 0$, there are two pseudo-Finsler[12] metrics F and F_l (see (18) and (19)) both defined in the conic subset

$$A_l = \{v \in TM : \Lambda < 0,\ \omega(v)^2 + \Lambda g_0(v, v) > 0,\ -\omega(v) > 0\},$$

and all the lightlike geodesics of $(\mathbb{R} \times M, g)$ project into pregeodesics of F or F_l or into lightlike pregeodesics of $\omega^2 + \Lambda g_0$ (which matches (17) up to a sign; see Theorem 1). Then a precise description of the causality and Cauchy hypersurfaces of SSTK spacetimes in terms of its associated wind Riemannian structure is available; see [23, Theorem 5.9]. Even though it is subtler than the previous cases,[13] to determine when the slices $t =$ constant are Cauchy hypersurfaces is simple. Indeed, only the conic metric F (which is defined on the whole spacetime) becomes relevant. Therefore, as emphasized in [53], the extension $\bar{F}$ of F to all the projections of future-directed causal vectors is continuous and permits an extension $d_{\bar{F}}$ of the F-separation on the whole M working as in Sect. 3.5.1 (see [53, Sects. 3.1.1 and 3.2.1]). Then the Cauchy completeness of $d_{\bar{F}}$ becomes naturally equivalent to the completeness of the wind Riemannian structure of the SSTK spacetime [53, Theorem 3.23] and, as a consequence [53, Corollary 4.1]:

For any SSTK spacetime, the slices $t =$ constant are Cauchy if and only if the extended distance $d_{\bar{F}}$ is (forward and backward) complete.

4 Finsler Applications

Next, some applications of the Lorentz-Finsler approach to the Finslerian setting will be gathered. We have already pointed out some of them. Indeed, one of the first consequences was to highlight the importance of the compactness of the symmetrized closed balls, or equivalently, the precompactness of the intersection of forward and backward balls. This condition is equivalent to the global hyperbolicity of the stationary spacetime associated with a Randers metric. Moreover, it also implies the geodesic connectedness of (M, F) (recall Theorem 3) and the existence of a complete Finsler metric $F^f = F - df$ for a certain function $f : M \to \mathbb{R}$ with the same pregeodesics as F (recall Corollary 1 and Remark 6).

---

[12] Here pseudo-Finsler means that the fundamental tensor (1) is non-degenerate, and the domain is conic, not necessarily the whole TM; in particular, so are the Finsler metric F and the Lorentz-Finsler one F_l.

[13] Among the subtleties appearing here, it is worth mentioning the step *causal continuity* (i.e., the spacetime is causal and the causal future and past vary continuously with the point) in the ladder of spacetimes. This is more restrictive than stable causality, and it is satisfied by all the standard stationary ones, as well as whenever $\Lambda \geq 0$; however, it is not satisfied by all the SSTK spacetimes.

The subtlety of this property was stressed in [53, Example 2.11] by exhibiting a Randers manifold (M, R) satisfying as follows: (M, R) *does* not *have compact closed symmetrized balls but its universal covering* $(\hat{M}, \hat{R})$ *does*. This follows by applying Theorem 4 to a globally hyperbolic stationary spacetime previously constructed by Harris [46, Example 3.4(b)], which admitted a non-globally hyperbolic quotient by isometries preserving the timelike Killing field.

A further application of stationary spacetimes to Randers metrics is a result about the smoothness of the distance to (or from) a closed subset. As it was shown in [22, Theorem 5.12 and Corollary 5.13], the subset of points where this distance is not smooth must have negligible n-Hausdorff measure.

4.1 Randers-Kropina Metrics

The applications of spacetimes to Finsler geometry can be even more powerful when one considers Randers-Kropina metrics or, more generally, wind Riemannian structures. This is because these metrics have certain type of singularities in the boundary of the domain, which disappear from the spacetime viewpoint. First, we have a direct extension to some properties for Finsler metrics stated above (see [23, Proposition 6.6 and Corollary 6.11]):

Theorem 5 *Let* (M, F) *be a Randers-Kropina manifold (recall Sect. 3.5.1) such that all the intersections of its backward and forward balls are precompact. Then*

(i) If $d_F(p, q) < +\infty$, *(or equivalently, there exists an admissible curve between* p *and* q*), then there is a minimizing geodesic joining them.*
(ii) There exists a smooth function $f : M \to \mathbb{R}$ *such that* $F - df$ *is a geodesically complete Randers-Kropina metric.*

For arbitrary wind Riemannian structures, it is possible to give a version of part (i) of the last theorem with a more technical definition of forward and backward balls (see [23, Definition 2.26] and [23, Proposition 6.6]). The extension of part (ii) is not straightforward. Again, the precompacteness of the intersections of the forward and backward balls imply global hyperbolicity and, thus, the existence of a Cauchy hypersurface S. However, as the integral curves of the Killing field $K = \partial_t$ may be spacelike, they can cross S more than once or not at all (thus, an SSTK splitting on S would not be obtained).

In [27, Theorem 1.5 (B)] the authors prove geodesic connectivity for the Kropina manifold $(M, g/\omega)$ with M compact and $\omega \wedge d\omega \neq 0$. This follows from part (i) above by observing that $\omega \wedge d\omega \neq 0$ implies that every two points are connected by an admissible curve. Moreover, stationary spacetimes have also been used to obtain multiplicity results for Kropina geodesics between two points in a compact manifold (see [25]).

4.2 *Classification of Randers and Wind Riemannian Spaceforms*

Let us recall that flag curvature is a fundamental invariant of a Finsler manifold which measures how geodesics with initial velocity in a certain plane spread apart. Flag curvature is the generalization of sectional curvature to Finsler geometry, and the problem of classifying the constant flag curvature Finsler manifolds remains as one of the biggest challenges. This classification has only been achieved for some families of Finsler metrics such as Randers and Kropina [5, 80]. Let us recall that Zermelo problem was essential to obtain the final classification of Randers spaceforms in the celebrated paper [5]. Indeed, it turns out that a Randers manifold has constant flag curvature if and only if its Zermelo data (g_R, W) is given by a Riemannian metric g_R with constant sectional curvature and a homothety W of g_R. Differently from the Riemannian counterpart, many of these Randers spaceforms are not geodesically complete. This anomaly disappears when one drops the Randers restriction $g_R(W, W) < 1$ and, thus, considers the wider family of wind Riemannian structures. Recall that there is a natural notion of geodesic intrinsic to such a structure Σ [23, Definition 2.35] and, then, of geodesic completeness, the latter implying the geodesic completeness of both its conic Finsler metric F and the Lorentz Finsler one F_l [23, Corollary 5.6 and Fig. 10]. Moreover, as the fundamental tensors of F and F_l are computable and non-degenerate away from the boundary of A_l, their flag curvatures are well-defined; so, Σ is called of constant flag curvature when the flag curvatures of both F and F_l are equal to the same constant. Then, the techniques used to prove the Randers classification can be extended to these structures to obtain the following (see [54, Theorem 3.12]).

Theorem 6 *The complete simply connected wind Riemannian structures with constant flag curvature lie in one of the following two exclusive cases, determined by the Zermelo data* (g_R, W)*:*

(i) (M, g_R) *is a model space of constant curvature and* W *is any of its Killing vector fields.*
(ii) (M, g_R) *is isometric to* $\mathbb{R}^n$ *and* W *is a properly homothetic (non-Killing) vector field.*

Moreover, the inextensible simply connected Randers (resp. Kropina) manifolds with constant flag curvature are the maximal simply connected open subsets of the previous wind Riemannian structures where the wind is mild (resp. critical).

Notice how, in the last assertion, the assumption of completeness must be replaced by the inextensibility of the Randers or Kropina metric as a manifold of this same type. We can go further and give a characterization in terms only of the conic metric F defined on the whole M. Indeed, it is natural to rename it as the *Zermelo metric* Z for the data (g_R, W) (given explicitly in (15)). At the end of Sect. 3.5.2, we saw that such a Z determines univocally an extended distance $d_{\bar{Z}}$. Moreover, the Cauchy completeness of the latter is equivalent to the completeness of the wind Riemannian structure (see [53, Theorem 3.23]), thus yielding the following.

Corollary 2 *A Zermelo metric Z with complete extended distance $d_{\tilde{Z}}$ on a simple connected manifold M has constant flag curvature $k \in \mathbb{R}$ if and only if its Zermelo data (g_R, W) satisfies either (i) or (ii) in Theorem 6.*

5 Interplay Finsler/Lorentz for Boundaries

Let (M, g_R) be a Riemannian manifold. Its elementary Cauchy boundary $\partial_C M$ provides a completion M_C, and, when g_R is complete, its Gromov boundary $\partial_G M$ (see [44]) provides a compactification M_G. In the case of Hadamard manifolds, this compactification agrees with the previous one by Eberlein and O'Neill (see [36]), which was introduced in a very different way by using Busemann functions associated with rays.[14] Being the main properties of these boundaries well established since the beginning of the 1980s, natural questions about the relation among them, as well as its extension to (possibly non-reversible) Finslerian metrics, had remained dormant. However, an additional motivation for their study came from the links with the Lorentzian setting.

Roughly, the link with the Riemannian case appears when one computes the causal boundary of a product $\mathbb{R} \times M, g = -dt^2 + g_R$. In a natural way, the computation of this boundary leads to consider a sequential compactification M_B of (M, g_R) in terms of Busemann-type functions, thus extending Eberlein and O'Neill compactification. This compactification includes the Cauchy completion M_C in a natural but subtle sense, and it is also related to Gromov's M_G.

Amazingly, M_B and M_G are equivalent except in some rather pathological cases, which also correspond with known pathologies of the causal boundary. All this can be extended to the Finslerian setting by providing a natural link with the causal boundary of standard stationary spacetimes (eventually extendible to Finslerian spacetimes). However, the possible non-reversibility of the Finslerian metric introduces subtleties even at the level of the Cauchy boundary. The systematic analysis of all these issues was carried out in [40], to be followed next.

[14] Hadamard manifolds (which are complete, simply connected and with non-positive sectional curvature) become diffeomorphic to $\mathbb{R}^n$ by Cartan-Hadamard theorem. Eberlein and O'Neill's boundary becomes a topological sphere S^{n-1} located at infinity. Indeed, the Busemann functions yield a quotient in the space of rays so that each function can be regarded as a direction at infinity; then, the set of all these directions can be regarded as the sphere S^{n-1}.

5.1 Gromov Compactification for Incomplete Finslerian Manifolds

Let us start considering a metric space (M, d) associated with a connected Riemannian manifold (M, g_R), or, with more generality, M can satisfy just to be connected, locally compact and second countable, while the distance d is just derived from a length space.[15] In particular, the metric space associated with any *reversible* Finsler manifold is included now and, later, we will refer to the Finsler case when taking into account non-reversibility.

5.1.1 The Symmetric d Case

The Cauchy completion M_C of such a space is standard, and we emphasize that M_C may be non-locally compact (thus, it will not lie under the general hypothesis for (M, d) above). Indeed, it is not difficult to construct a bidimensional Riemannian example starting at a variation of the comb space

$$M = ((0, \infty) \times \{0\}) \cup (K \times (0, 1)) \subset \mathbb{R}^2, \quad \text{where} \quad K = \{1/m : m \in \mathbb{N}\}. \tag{25}$$

Here, $(0, 0)$ is identifiable to a point of $\partial_C M$ which does not have any compact neighborhood in M_C.

Let us construct the Gromov compactification of (M, d) without the usual assumption on completeness for d. Consider the space of all the 1-Lipschitz functions $\mathcal{L}_1(M, d)$. For each $x \in M$, the function $M \ni y \mapsto d_x(y) := d(x, y)$ belongs to $\mathcal{L}_1(M, d)$. Moreover, d_x as well as any function $d_x + C$, where C is a constant, determines univocally x. Thus, M can be identified with a subset of the quotient space $\mathcal{L}_1(M, d)/\mathbb{R}$ under the relation of equivalence:

$$f \sim f' \Leftrightarrow f - f' = C \in \mathbb{R}, \quad \text{where} \quad f, f' \in \mathcal{L}_1(M, d). \tag{26}$$

Definition 1 The Gromov completion M_G of (M, d) is the closure of M in $\mathcal{L}_1(M, g)/\mathbb{R}$ (with, say, the uniform convergence on compact sets topology).

It is not difficult to check that $\mathcal{L}_1(M, g)/\mathbb{R}$ is compact and, moreover (see [40, Theorem 4.12 and Corollary 4.13]):

Proposition 4 *M_G is a compact metrizable space and M_C is naturally included in it. The inclusion $M_C \hookrightarrow M_G$ is continuous, and it is an embedding if and only if M_C is locally compact; moreover, $M \hookrightarrow M_G$ is a dense embedding.*

---

[15] Notice that, for a smooth curve c in such a space, the role of its Riemannian norm $g_R(\dot{c}(s), \dot{c}(s))$ at each s is played by the local dilatation there; moreover, notions such as geodesic or cut point have a natural sense (see [45, Chap. 1] for background).

To check that the inclusion of M_C in the example (25) is not an embedding, notice that the sequence $\{(1/n, 1/2)\}_n$ converges to $(0, 0)$ in M_G, but it is not convergent in M_C. This suggests some subtleties for this boundary. Indeed, Gromov's boundary $\partial_G M := M_G \setminus M$ is divided into a *Cauchy-Gromov* boundary $\partial_{CG} M$, whose points are the limits of bounded sequences in M, and a *proper* Gromov boundary $\partial_{\mathcal{G}} M$ containing limits of unbounded sequences. Clearly, $\partial_{CG} M$ contains the Cauchy boundary $\partial_C M$; however it may contain more points if M_C is not locally compact.

Example 1 Modify the space in (25) by adding an upper half line:

$$M = ((0, \infty) \times \{0, 1\}) \cup (K \times (0, 1)) \subset \mathbb{R}^2, \quad \text{where} \quad K = \{1/m : m \in \mathbb{N}\}.$$

Now, for each $y \in (0, 1)$ the sequence $\{(1/n, y)\}_n$ converges to a distinct limit, so that $\{0\} \times [0, 1]$ can be regarded as a subset of $\partial_{CG} M$. We emphasize that *no boundary point* $(0, y)$, $y \in (0, 1)$*, can be an endpoint of a curve starting at* M.

5.1.2 The Non-symmetric *d* Case

When considering the Finsler case, d will mean the non-necessarily symmetric distance defined in (2), and the following subtleties must be taken into account. Following [81], a map $d : M \times M \to \mathbb{R}$ is called a *generalized distance* when it satisfies, for all $x, y, z \in M$: (a1) $d(x, y) \geq 0$, (a2) $d(x, y) = d(y, x) = 0$ if and only if $x = y$, (a3) $d(x, z) \leq d(x, y) + d(y, z)$, and (a4) given a sequence $\{x_m\}_m \subset M$ and $x \in M$, then $\lim_{m \to \infty} d(x_m, x) = 0$ if and only if $\lim_{m \to \infty} d(x, x_m) = 0$; when the hypothesis (a4) is dropped, then d is called a *quasidistance*. A generalized distance d gives two notions of Cauchy sequences (forward and backward) and, then, forward and backward Cauchy completions M_C^+, M_C^-, respectively.[16] Moreover, the distance d^s obtained by symmetrizing d provides another completion M^s, and the corresponding boundaries satisfy $\partial_C^s M = \partial_C^+ M \cap \partial_C^- M$ in a natural way. The continuous extension of d to such a completion is only a quasidistance; indeed, the topologies generated by the forward and backward balls are not equivalent and M_C^+ is only a T_0 space. Notice that, in our previous study of Finsler metrics, we considered a non-symmetric distance which is, indeed, a generalized distance, and all the assertions above apply [40, Chap. 3].

In order to consider the Gromov completion for a Finsler manifold, notice that there are two non-symmetric notions of 1-Lipschitzian:

$$\mathcal{L}_1^+(M, d) = \{f \text{ smooth} : f(y) - f(x) \leq d(x, y)\},$$
$$\mathcal{L}_1^-(M, d) = \{f \text{ smooth} : f(x) - f(y) \leq d(x, y)\}.$$

[16] There is a non-equivalent way to define forward (and backward) Cauchy sequences; however, it would yield the same forward Cauchy completion (see [40, Sect. 3.2.2]).

Accordingly, there are two Gromov compactifications $M_G^{\pm}$,[17] namely, M_G^+ is the closure of M in $\mathcal{L}_1^+(M, d)/\mathbb{R}$.

Remark 9 The inclusion $M_C^+ \hookrightarrow M_G^+$ is subtler than in Proposition 4, as it satisfies now (see [40, Corollary 5.25]):

(A) it is continuous if and only if the backward balls generate a finer topology on M_C^+ than the forward balls,
(B) it is an embedding when (B1) M_C^+ is locally compact (as in the Riemannian case) and (B2) the extension of d to M_C^+ is a generalized distance.

5.2 The Causal Boundary of a Spacetime

For spacetimes, there are two conformally invariant boundaries which are applied to general relevant classes of spacetimes. The first one is the so-called *conformal boundary*, introduced by Penrose in [68], which underlies notions such as *asymptotic flatness*, and it is widely used in relativity. Essentially, the idea is to find a suitable open conformal embedding of the spacetime in a bigger one and, then, to regard its topological boundary as the conformal one. The second one is the *causal boundary*, firstly introduced by Geroch, Kronheimer and Penrose [42], but later redefined several times (we will refer to the last one, [39]). This boundary $\partial_c \mathcal{M}$ is defined in an intrinsic way for any spacetime under the weak condition of being *strongly causal*.[18] There are general conditions which ensure that these boundaries agree (as well as counterexamples otherwise), see [39, Sect. 4 and Appendix].[19]

Here, we are interested in the causal boundary $\partial_c \mathcal{M}$, which will be described very briefly now, and we refer to [39] for exhaustive details and references. To construct it, one starts defining a *terminal indecomposable past* (resp. future) set P, or *TIP* (resp. F, *TIF*) for short, as the chronological past (resp. future) of any inextendible future-directed (resp. past-directed) timelike curve γ, i.e., $P = I^-(\gamma)$ (resp. $F = I^+(\gamma)$). The set of all the TIPs (resp TIFs) is the *future* (resp. *past*) causal boundary $\partial_c^+ \mathcal{M}$ (resp. $\partial_c^- \mathcal{M}$). To construct $\partial_c \mathcal{M}$ one has to take into account that a TIP and a TIF might represent intuitively the same boundary point (see Fig. 5). So, one introduces the *Szabados relation*: $P \sim F$ iff, on the one hand, P is included in the common past of F (i.e., $(\cap_{x \in F} I^-(x) \supset P)$ and P is maximal among the TIPs satisfying this property, and on the other hand, the dual property holds for F. Then $\partial_c \mathcal{M}$ is composed of $\partial_c^+ \mathcal{M} \cup \partial_c^- \mathcal{M}$ up to the pairings introduced by $\sim$. Even

[17] Its consistency relies on a non-symmetric version of Arzela theorem [40, Theorem 5.12].

[18] Intuitively, it does not admit "almost closed" causal curves. A formalization of this property is that each point $p \in M$ has a neighborhood U such that any inextendible causal curve starting at U will leave U at some point so that it will not return to U.

[19] Noticeably, they agree in the class of globally hyperbolic spacetimes-with-timelike-boundary, see [1].

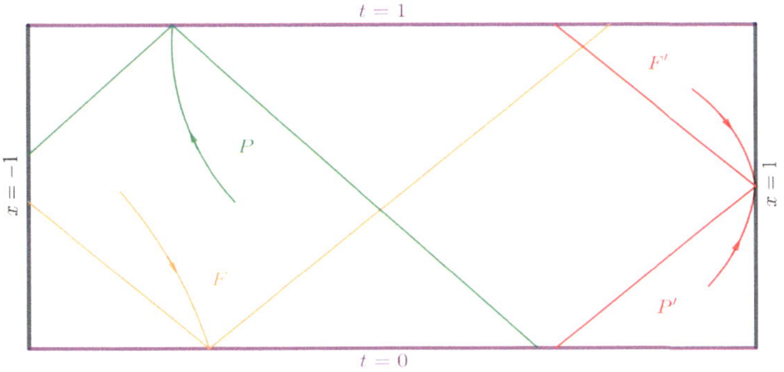

Fig. 5 Let $\mathcal{M} = (0, 1) \times (-1, 1)$ (in Lorentz-Minkowski spacetime). Green P (resp. orange F) corresponds with a TIP (resp. TIF) which represents a point in the causal boundary; intuitively, this point is identifiable with a boundary point in the line $t = 1$ (resp. $t = 0$). Red P', F' are a TIP and a TIF that intuitively represent the same boundary point at $x = 1$. They are paired by Szabados relation, yielding a single point of $\partial_c\mathcal{M}$

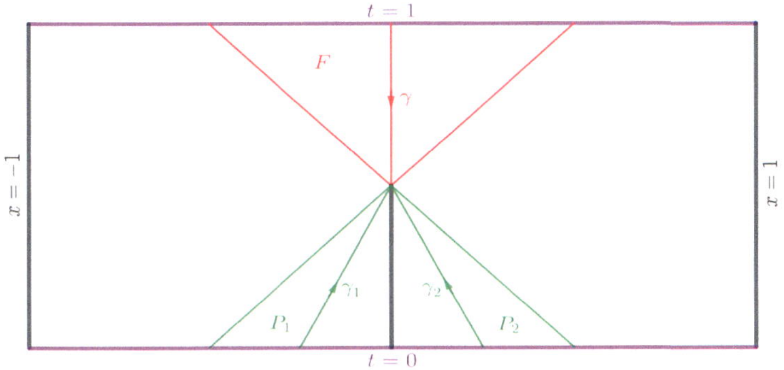

Fig. 6 Modify the example in Fig. 5 by removing a segment in the t axis, $\mathcal{M}' := \mathcal{M} \setminus \{(t, 0) : t \leq 1/2\}$. The TIF $F = I^+(\gamma)$, with $\gamma : (0, 1/2) \ni t \mapsto (1 - t, 0)$ is Szabados related with each one of the TIPs P_1, P_2, where $P_i = I^-(\gamma_i)$, with $\gamma_i : (0, 1/2) \ni t \mapsto (t, (-1)^i ((1/4) - t/2))$. Then, $P_1 \sim F$ and $P_2 \sim F$. Therefore, (P_1, F) and (P_2, F) are two distinct points of $\partial_c\mathcal{M}$ (they are not Hausdorff separated by the chronological topology)

though this is a neat definition, examples such as Fig. 6 show that the pairings may be non-trivial. In what follows, all the elements of $\partial_c\mathcal{M}$ will be regarded as a pair (P, F) with the convention that $F = \emptyset$ (resp. $P = \emptyset$) when P (resp. F) is unpaired.

Remark 10

(1) Globally hyperbolic spacetimes are characterized as the strongly causal ones whose $\partial_c\mathcal{M}$ is composed only of unrelated pairs $(P, \emptyset), (\emptyset, F)$, see [39, Theorem 3.29].
(2) The chronological relation $\ll$ in $\mathcal{M}$ (introduced around (23)) admits a natural extension $\overline{\ll}$ to $\partial_c M$, namely, $(P, F)\overline{\ll}(P', F')$ whenever $F \cap P' \neq \emptyset$.

The most natural topology for $\partial_c M$ is the so-called *chronological topology*. We will not go into the details of this topology, but just point out the following two important features, in relation to the general Busemann completion M_B of any Finslerian manifold M to be described below:

1. The topology of M_B is inspired by the chronological topology of the causal completion of a standard stationary spacetime $\mathcal{M} = \mathbb{R} \times M$. In particular, $\partial_c \mathcal{M}$ will be described directly from the Busemann boundary $\partial_B M$.
2. When this topology is Hausdorff, M_B is identifiable to M_G. Otherwise, the non-Hausdorff property of M_B will be related to the appearance of somewhat pathological properties of M_G, as the one emphasized in Example 1 (see Theorem 7 below).

This second item supports the relevance of the Busemann completion even in a purely Riemannian setting, and the first one supports the previously defined causal boundary.

5.3 A New Busemann Boundary

Next, we will construct a general Busemann boundary which generalizes Eberlein and O'Neill's, following [40, Sects. 4.2 and 5.2]. We will start at the symmetric case with a length metric space (M, d) as in the case of Gromov's, but one can consider a connected Riemannian manifold (M, g_R) to be more specific.

In the standard approach, one assumes the completeness of d, considers a ray c (a half unit geodesic with no cut locus), and defines its Busemann function b_c as $b_c(x_0) := \lim_{t\to\infty}(t - d(x_0, c(t)))$ for all $x_0 \in M$. However, we will drop completeness and admit (generalized) Busemann functions for more general curves, namely,

$$b_c(x_0) = \lim_{t\to\Omega}(t - d(x_0, c(t))) \quad \text{for any } c : [0, \Omega) \to M \text{ with } F(\dot{c}) \leq 1. \tag{27}$$

Easily, if b_c is ∞ at some $x_0 \in M$, then $b_c \equiv \infty$, and $B(M)$ will denote the set of all the finite Busemann functions on M. $B(M)$ will be regarded as a topological space with the *chronological topology* defined by means of a limit operator L. Specifically, given $\{f_m\}_m \subset B(M)$, the subset $L(\{f_m\}_m) \subset B(M)$ is defined by

$$f \in L(\{f_m\}_m) \Leftrightarrow \begin{cases} (a)\ f \leq \liminf_m f_m \quad \text{and} \\ (b)\ \forall g \in B(M) \text{ with } f \leq g \leq \limsup_m f_m, \text{ it is } g = f. \end{cases}$$

Then, the topology is defined by declaring that a subset $C \subset B(M)$ is *closed* if and only if $L(\sigma) \subset C$ for any sequence $\sigma = \{f_m\}_m \subset C$.

Definition 2 As a pointset, the Busemann completion M_B of (M, d) is $M_B := B(M)/\mathbb{R}$ (this is the quotient by an additive constant as in (26)), endowed with the quotient of the chronological topology on $B(M)$.

Notice that $B(M) \subset \mathcal{L}_1(M, g)$ and, thus, naturally $M_B \subset M_G$. However, the inclusion may be non-continuous (indeed, the topology of M_B is always coarser than that of M_G). Observe also that one can consider $M_C \subset M_B$, as the points in the Cauchy completion correspond to two Busemann functions for curves with $\Omega < \infty$. Indeed, one has the disjoint union $M_B = M \cup \partial_C M \cup \partial_{\mathcal{B}} M$, where $\partial_{\mathcal{B}} M$ contains the classes of Busemann functions with $\Omega = \infty$.

Theorem 7 *[40, Sect. 5.2] The following properties hold:*

(A) M_B is sequentially compact.
(B) M_B is a T_1 topological space (and points in the boundary may be non-T_2 related).
(C) $M_C \hookrightarrow M_B \hookrightarrow M_G$ but the topology of M_B is coarser than the others.
(D) They are equivalent:

(D1) $M_B = M_G$ as pointsets.
(D2) M_B is Hausdorff.
(D3) No sequence contained in M converges to two distinct points in $\partial_B M$.
(D4) $M_B \hookrightarrow M_G$ is an embedding.
(D5) $M_B \hookrightarrow M_G$ is a homeomorphism.

Recall that, from (D), M_G contains "extra points" when M_B and M_G are not naturally equivalent (recall Example 1 and see also [40, Theorem 5.39 and Remark 5.41]).

In the Finslerian case, take into account that, now, each curve c yields two Busemann functions, depending on the ordering of the arguments of d in (27). So, as in the case of Gromov's, we have two Busemann completions $M_B^{\pm}$ depending on that ordering. Then, as in the Riemannian case, one has canonical inclusions $M_C^{\pm} \hookrightarrow M_B^{\pm} \hookrightarrow M_G^{\pm}$, and Theorem 7 is extended naturally to this case, [40, Theorem 5.39]. In particular, the topology of M_B^+ is coarser than those of M_C^+ and M_G^+, M_B^+ is identifiable to M_G^+ if and only if M_B^+ is Hausdorff, and the Hausdorffness of M_B^+ is independent of that of M_B^-. Moreover, one also has the disjoint unions:

$$M_B^+ = M \cup \partial_C^+ M \cup \partial_{\mathcal{B}}^+ M, \quad M_B^- = M \cup \partial_C^- M \cup \partial_{\mathcal{B}}^- M.$$

5.4 *The Causal Boundary of Stationary Spacetimes*

Let us describe $\partial_c \mathcal{M}$ for a standard stationary spacetime $\mathcal{M} = \mathbb{R} \times M$ as in (8) by using the Busemann completions $M_B^{\pm}$ of the Finsler manifold (M, F), where F is the Fermat metric in (9). The generalized distance d associated with F will be denoted here d^+ (as it will be related to the future causal boundary $\partial_c\mathcal{M}^+$) and its reversed one d^-. Observe that the interpretation of the Fermat metric provides the following characterization of the chronological relation $\ll$ (recall (23)):

$$(t_0, x_0) \ll (t_1, x_1) \Leftrightarrow d^+(x_0, x_1) < t_1 - t_0.$$

Let us see how Busemann functions appear when one computes the TIPs for $\partial^+\mathcal{M}$. Let γ be any future-directed timelike curve, and let $P = I^-(\gamma)$. If γ is future inextendible, then it yields a TIP. If γ is continuously extendible to a point $p \in \mathcal{M}$, then P is equal to $I^-(p)$, and it is called a PIP, *proper indecomposable past set.* PIPs (and analogously PIF's) permit to identify $\mathcal{M}$ in the future causal completion $\mathcal{M}_c^+ = \mathcal{M} \cup \partial_c^+\mathcal{M}$ and, then, in the causal completion $\mathcal{M}_c = \mathcal{M} \cup \partial_c\mathcal{M}$. Parametrizing γ with the t coordinate of $\mathbb{R} \times M$, we have

$$\gamma(t) = (t, c(t)), \quad t \in [\alpha, \Omega), F(\dot{c}) < 1,$$

and, then,

$$\begin{aligned} P &= \{(t_0, x_0) \in \mathcal{M} : (t_0, x_0) \ll \gamma(t) \text{ for some } t \in [\alpha, \Omega)\} \\ &= \{(t_0, x_0) \in \mathcal{M} : t_0 < t - d^+(x_0, c(t)) \text{ for some } t \in [\alpha, \Omega)\} \\ &= \{(t_0, x_0) \in \mathcal{M} : t_0 < \lim_{t\to\Omega}(t - d^+(x_0, c(t)))\} \\ &= \{(t_0, x_0) \in V : t_0 < b_c^+(x_0)\}, \end{aligned}$$

where $b_c^+(x_0) = \lim_{t\to\Omega}(t - d^+(x_0, c(t)))$ is the forward Busemann function of c in (M, d^+). So, the set of Busemann functions $B^+(M)$ for (M, d^+) satisfies[20]

$$\mathcal{M}_c^+(= \{\text{TIPs and PIPs on } \mathcal{M}\}) \equiv B^+(M) \cup \{b_c \equiv \infty\},$$

where the PIPs correspond to converging c (thus, necessarily, $\Omega < \infty$) and the TIPs to non-converging c, including the case $b_c \equiv \infty$ (this can be obtained with a curve of type $\gamma(t) = (t, x_0), t \geq 0$, for any $x_0 \in M$). Notice that in the construction of $\mathcal{M}_c^+$, no quotient in the set of Busemann functions is carried out. Next, let us describe briefly $\partial_c\mathcal{M}$. We will restrict to its pointset and chronological structures and refer to [40, Chap. 5] for the topological structure and full details.

5.4.1 The Static Case

Recall that in the static case, $\mathcal{M}$ can be regarded as a product $\mathbb{R} \times M$ with F equal to a Riemannian metric by using conformal invariance. Essentially, $\partial_c\mathcal{M}$ becomes

[20] A subtlety is that the Busemann functions which can be constructed with the restriction $F(\dot{c}) \leq 1$ coincide with those constructed with $F(\dot{c}) < 1$.

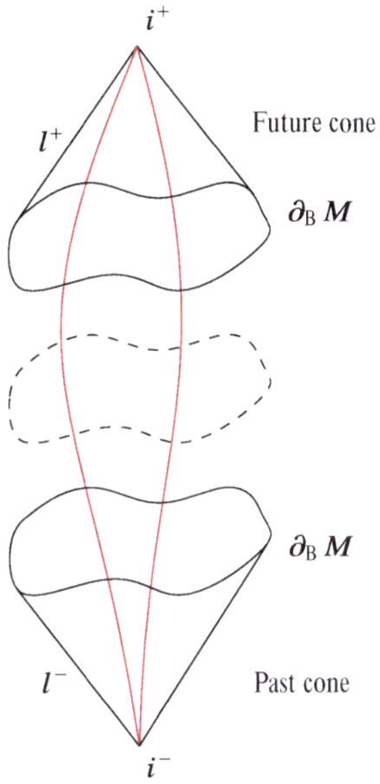

Fig. 7 The causal boundary of a static spacetime. The red lines joining i_- and i_+ correspond to points in $\partial_C M$

a double cone with some lines connecting its apexes, constructed as follows (see Fig. 7):

1. Two points $i^+ = (M, \emptyset)$, $i^- = (\emptyset, M)$ (which correspond to $b_c \equiv \infty$). They are the apexes of a symmetric double cone (invariant by $t \mapsto -t$) on $\partial_B M$.
2. Two horismotic lines (i.e., locally horismotic with no cut points) for each point of $\partial_{\mathcal{B}} M$, one of them l^+ ending at i^+ and the other l^- starting at i^-. This means that the line l^+ is composed of boundary points of type $(P, \emptyset)$ such that (a) l^+ is totally ordered by the relation of inclusion for the first factor (i.e., l^+ is *locally horismotic*) ending at $P = \mathcal{M}$, and (b) no two points in l^+ are related by the extended chronological relation $\overline{\ll}$ defined in Remark 10 (2) (i.e., l^+ has *no cut points*).
3. A timelike line connecting i^-, i^+ for each point of the Cauchy boundary $\partial_C M$. Such a line is a continuous curve in $\mathcal{M}_c$ totally ordered by the extended chronological relation $\overline{\ll}$. These are the only points in $\partial_c \mathcal{M}$ with non-trivial pairings (P, F).

Consistently with Remark 10 (1), the spacetime is globally hyperbolic if the timelike lines do not exist, that is, if $\partial_C M = \emptyset$.

5.4.2 The General Stationary Case

For the sake of simplicity, we will assume that the causal completion $\mathcal{M}_c$ is *simple as a pointset*, that is, each TIP P and each TIF F determines unambiguously a point of $\partial_c \mathcal{M}$ (thus, the situation in Fig. 6 cannot occur); simple sufficient hypotheses to ensure this property are, for example, $M_B^{\pm}$ being Hausdorff or d^+ being extendible to the Cauchy boundary $\partial_C^+ M$ as a generalized distance; see [40, Fig. 6.2].

Then, the three elements in the static picture must be modified allowing more generality, and a fourth ingredient appears, namely, now:

1. $i^+ = (M, \emptyset), i^- = (\emptyset, M)$ are regarded as apexes for two (non-symmetric) cones on $\partial_B^+ M$, $\partial_B^- M$, respectively.
2. Each horismotic line l^+ (resp. l^-) appears for a point in $\partial_{\mathcal{B}}^+ M$ (resp. $\partial_{\mathcal{B}}^- M$).
3. The timelike lines connecting i^+ and i^- appear for each point in the Cauchy boundary for the symmetrized distance, $\partial_C^s M$.
4. The points in $\partial_C^{\pm} M \backslash \partial_C^s M$ determine locally horismotic lines (eventually starting at i^-, ending at i^+, or both).

Recall that the subtle last possibility cannot occur when $\partial_C^{\pm} M = \partial_C^s M$, in particular, when d^+ extends to $\partial_C^+ M$ as a generalized distance [40, Proposition 3.28].

6 Lorentz-Finsler Metrics and Practical Applications

In this section we will take a look at some real-world situations where the interplay between Lorentz and Finsler geometries appears naturally. Specifically, we will focus on the wave propagation: Lorentz metrics and the spacetime viewpoint are essential when considering rheonomic (i.e., time-dependent) waves, whereas Finsler metrics effectively model the anisotropic (i.e., direction-dependent) case. The combination of both cases leads naturally to Lorentz-Finsler metrics. Interestingly, the applications can be generalized to any physical phenomenon that satisfies Huygens' principle, such as wildfires or seismic waves.

6.1 Anisotropic Wave Propagation and Huygens' Principle

Recall from Sect. 2.3 that the (time and direction-dependent) velocities of a moving object can be effectively described by the indicatrix Σ_t of a time-dependent Finsler metric F_t on $\{t\} \times M$. Then the trajectories of the object are given in the (globally hyperbolic[21]) spacetime $\mathcal{M} = \mathbb{R} \times M$ by the lightlike curves of the Lorentz-Finsler metric $G = dt^2 - F_t^2$ introduced in (21),[22] since (considering t-parametrized curves)

$$\gamma(t) = (t, x(t)) \text{ lightlike} \Leftrightarrow G(\dot{\gamma}(t)) = 0 \Leftrightarrow x(t)\ F_t\text{-unit} \Leftrightarrow \dot{x}(t) \in \Sigma_t.$$

---

[21] This is not a restriction in any realistic situation (see [50, Remark 3.2]).

[22] Note that all the usual concepts about causality can be directly translated from the Lorentzian case to the Lorentz-Finsler metrics and the more general setting of cone structures (see, for example, [55, 64]). However, due to the non-reversibility of the Finsler metrics, usually only future directions are considered.

Consider now the following variation of Zermelo problem: instead of finding the fastest trajectory for a moving object between two prescribed points, we seek to determine the evolution of an anisotropic wave starting from $\mathcal{S} \subset M$ with velocities given by Σ_t. We will restrict ourselves to the "mild wind" case, i.e., Σ_t encloses the zero section,[23] so that the wave propagates over M in all directions.

Among all the possible trajectories of the wave (i.e., lightlike curves departing from $\mathcal{S}$), we are interested in those that generate the wavefront. If $\mathcal{S} = \{p\}$, namely, the wave starts from a single point $p \in M$, then all the spacetime points that can be reached by the wave are given by the causal future $J^+(p)$; accordingly, all the spatial points the wave passes through are the projection of $J^+(p)$ on M. Following this reasoning, $\partial J^+(p)$ provides the outermost points reached by the wave, and therefore, $\partial J^+(p) \cap (\{t_0\} \times M)$ is the wavefront at each time $t_0 > 0$. Observe that $J^+(p)$ will be closed due to the global hyperbolicity of G. In the more general case where $\mathcal{S}$ is a compact hypersurface of M,[24] playing the role of the initial wavefront, we can apply Huygens' principle: each point of the front behaves as an independent source of the wave, and thus

$$\mathrm{Front}(t_1) = \partial\left(\cup_{p\in \mathrm{Front}(t_0)} J^+(p)\right) \cap (\{t_1\} \times M) = \partial J^+(\mathrm{Front}(t_0)) \cap (\{t_1\} \times M),$$

for any $t_1 > t_0 \geq 0$. In particular, if we put $t_0 = 0$, then the wavefront at any time $t_1 > 0$ is given by $\partial J^+(\mathcal{S}) \cap (\{t_1\} \times M)$. So, the next step is to determine the trajectories that make up $\partial J^+(\mathcal{S})$ (see Fig. 8).

6.2 *Solution in Terms of Lorentz-Finsler Geodesics*

So far, we know that a wave trajectory remaining in the wavefront must satisfy two conditions: it must be lightlike, and it must be entirely contained in $\partial J^+(\mathcal{S})$. The next result states some crucial properties of this type of curves (see [50, Sect. 4] and [49, Sect. 5]).

Proposition 5 *Let $\gamma : [0, t_0] \to \mathcal{M}$ be a (t-parametrized) causal curve entirely contained in $\partial J^+(\mathcal{S})$. Then:*

(i) γ is a lightlike pregeodesic of $(\mathcal{M}, G)$ departing G-orthogonally from $\mathcal{S}$, i.e., $g^G_{\dot\gamma(0)}(\dot\gamma(0), v) = 0$ for all $v \in T_{\gamma(0)}\mathcal{S}$, being g^G the fundamental tensor of G. In fact, $\partial J^+(\mathcal{S})$ admits a unique foliation by such pregeodesics.

(ii) γ is time-minimizing: for any $p_0 = (t_0, x_0) \in \mathrm{Im}(\gamma)$, γ is the first causal curve from $\mathcal{S}$ to arrive at the vertical line $t \mapsto (t, x_0)$.

[23] Alternatively and more generally, we can consider ∂_t as an observer's vector field co-moving with the medium in which the wave propagates, as suggested at the end of Sect. 2.

[24] For simplicity, $\mathcal{S}$ will be assumed to be a hypersurface of M, although the results we present here can be generalized to any submanifold (see [50]).

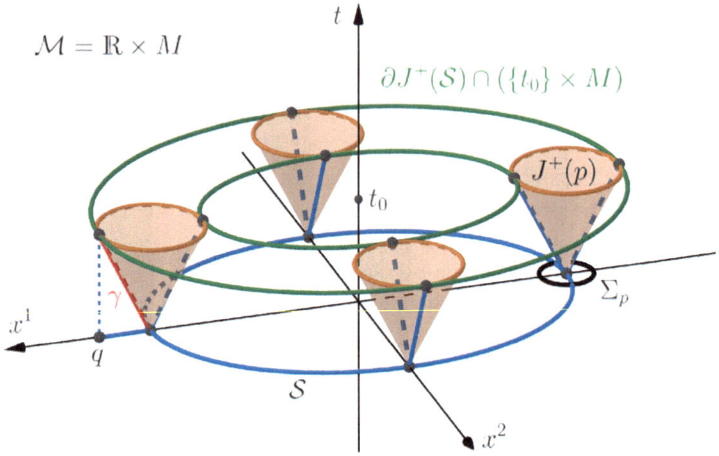

Fig. 8 A simple representation of the wave evolution in dimension $n + 1 = 3$. The indicatrix Σ_p determines the velocity of the wave for each direction and the causal future $J^+(p)$ shows the region in the spacetime that can be reached by the wave starting at p. The envelope of all these chronological futures, $\partial(\cup_{p\in\mathcal{S}} J^+(p)) = \partial J^+(\mathcal{S})$, is generated by the first-arriving trajectories of the wave, so that the intersection with $\{t_0\} \times M$ is the wavefront at any $t_0 > 0$ (when $\mathcal{S}$ is a hypersurface of M there are always two wavefronts). A wave trajectory such as γ is a lightlike pregeodesic G-orthogonal to $\mathcal{S}$ and minimizes the propagation time from $\mathcal{S}$ to q

Essentially, the wavefront is made up of lightlike pregeodesics that are time-minimizing in the sense that no other causal curve from $\mathcal{S}$ arrives earlier at each of their points. In other words, these trajectories solve Zermelo problem from $\mathcal{S}$ to any of their points.

Observe that at each point of $\mathcal{S}$ there are exactly two lightlike G-orthogonal directions: one points to the exterior of $\mathcal{S}$ and the other to the interior. Therefore, there are two wavefronts: one heading outward and the other inward (see Fig. 8). From now on we will focus on the one going outward, since it is usually the most interesting from a practical viewpoint.[25] Consider then the *wavemap*, defined as

$$\begin{aligned} f\colon [0,\infty) \times \mathcal{S} &\longrightarrow \mathcal{M} \\ (t,s) &\longmapsto f(t,s) = (t, x(t,s)), \end{aligned}$$

where, for each $s \in \mathcal{S}$, $t \mapsto f(t,s)$ is the t-parametrized pregeodesic of G whose initial velocity is the unique lightlike vector G-orthogonal to $\mathcal{S}$ and pointing outward. The curve $t \mapsto f(t, s_0)$ represents the *spacetime trajectory* of the wave from $s_0 \in \mathcal{S}$, being its projection $t \mapsto x(t,s)$ the corresponding *spatial trajectory*.

[25] Nevertheless, the results we present here directly apply to the other wavefront simply by replacing "outward" with "inward."

Working in coordinates $\{x^0 := t, x^1, \ldots, x^n\}$, the following result characterizes the wavemap in terms of the ODE system for the t-parametrized lightlike pregeodesics of G (see [50, Theorem 4.11]).

Theorem 8 *For each $s_0 \in \mathcal{S}$, the wavemap $f(t, s_0) = (t, x^1(t, s_0), \ldots, x^n(t, s_0))$ is given by the following ODE system:*

$$\ddot{x}^k = \sum_{i,j=0}^{n} \left(-\gamma^k_{ij}(\dot{f})\dot{x}^i\dot{x}^j + \gamma^0_{ij}(\dot{f})\dot{x}^i\dot{x}^j\dot{x}^k \right), \quad k = 1, \ldots, n, \tag{28}$$

along with the initial conditions:

- $f(0, s_0) = s_0 \in \mathcal{S}$, *and*
- $\dot{f}(0, s_0)$ *is lightlike, G-orthogonal to $\mathcal{S}$ and pointing outward,*

where $\dot{f} = \dot{f}(t, s_0) = (1, \dot{x}^1(t, s_0), \ldots, \dot{x}^n(t, s_0))$ denotes the velocity (tangent vector) of the curve $t \mapsto f(t, s_0)$ and γ^k_{ij} are the formal Christoffel symbols of G, defined as

$$\gamma^k_{ij}(v) := \frac{1}{2}\sum_{r=0}^{n} g^{kr}(v)\left(\frac{\partial g_{rj}}{\partial x^i}(v) + \frac{\partial g_{ri}}{\partial x^j}(v) - \frac{\partial g_{ij}}{\partial x^r}(v) \right), \quad i, j, k = 0, \ldots, n,$$

with $g_{ij}(v) := g^G_v(\partial_{x^i}, \partial_{x^j})$, for any lightlike $v \in T\mathcal{M}$.

Remark 11 In the time-independent anisotropic case (namely, $F_t = F$ is time-independent), t-parametrized lightlike pregeodesics of G project onto unit speed geodesics of F (recall Proposition 1 for the Randers case). Namely, the spatial trajectories of the wave are F-geodesics F-orthogonal to $\mathcal{S}$. Moreover, the propagation time is given by the F-length: if $\gamma : [0, t_0] \to M$ is a spatial trajectory of the wave, then t_0 coincides with the length of γ computed with F.

6.3 Cut Points and Determination of the Wavefront

As long as a spacetime trajectory remains in $\partial J^+(\mathcal{S})$, it keeps providing a point in the wavefront. However, this may not be the case for all t. We define the *(null) cut function* as

$$\begin{aligned} \mathrm{cut}\colon \mathcal{S} &\longrightarrow [0, \infty] \\ s &\longmapsto \mathrm{cut}(s) := \mathrm{Max}\{t : f(t, s) \in \partial J^+(\mathcal{S})\}, \end{aligned}$$

which also relates to the property of being time-minimizing. Specifically:

- If $t \leq \mathrm{cut}(s_0)$, then $f(t, s_0) \in \partial J^+(\mathcal{S})$ is a point of the wavefront and the corresponding spacetime trajectory is time-minimizing.
- If $t > \mathrm{cut}(s_0)$, then $f(t, s_0) \in I^+(\mathcal{S})$ is not in the wavefront, so there exists another lightlike pregeodesic G-orthogonal to $\mathcal{S}$ and contained in $\partial J^+(\mathcal{S})$ that arrives earlier at the same spatial point $x(t, s_0) \in M$ (recall that $J^+(\mathcal{S})$ is closed).

We call $\mathrm{cut}(s)$ and $(\mathrm{cut}(s), f(\mathrm{cut}(s), s))$ the *cut instant* and *cut point* of the corresponding trajectory, respectively. Cut points are interesting from a practical viewpoint because they mark regions where different wave trajectories converge. More precisely (see [51, Proposition A.1]):

Proposition 6 *Let $(t_0, p_0) \in \mathcal{M}$ be the cut point of $\gamma : t \mapsto f(t, s_0)$. Then, at least one of the following properties holds: (a) (t_0, p_0) is the first intersection point of γ with another spacetime trajectory of the wave, or (b) (t_0, p_0) is the first focal point of $\mathcal{S}$ along γ (it is possible for both conditions to hold simultaneously).*

In some situations, the detection of these points becomes crucial. In the case of wildfires, for example, cut points determine possible crossovers of the fire, which can become extremely dangerous both because of the increased heat intensity and because they may leave behind regions completely surrounded by the fire (see [51, Sect. 4.2]).

Focusing now on the determination of the wavefront, observe that if $t_0 < \mathrm{cut}(s)$ for all $s \in S$, then the curve $s \mapsto f(t_0, s)$ exactly coincides with the wavefront at $t = t_0$, i.e., $\mathrm{Im}(f(t_0, s)) = \partial J^+(\mathcal{S}) \cap (\{t_0\} \times M)$. This can be guaranteed at least for a small time, since there always exists some $\varepsilon > 0$ such that $\mathrm{cut}(s) > \varepsilon$ for all $s \in \mathcal{S}$ (see [50, Theorem 4.8]). In general though, the wavefront will be given by all the trajectories which have not arrived yet at their cut points:

$$\partial J^+(\mathcal{S}) = \{f(t, s) : t \leq \mathrm{cut}(s), s \in \mathcal{S}\}.$$

So, in order to obtain the wavefront, we have to compute the wavemap through the ODE system (28), but anytime a spacetime trajectory reaches its cut point, it should be discarded, as subsequent points no longer lie on the wavefront. This is not demanding from a computational viewpoint, since each trajectory is independent from the others and those located beyond their cut points can be removed (or simply ignored) with no harm to the overall computation.

6.4 The Case of Wildfires

The theoretical setting we have presented in this section can be applied to model the propagation of any physical phenomena that behaves as a wave, i.e., that satisfies Huygens' principle. One of the most interesting examples is the case of wildfires (see [50, 51, 61, 62]). Consider a wildfire that spreads over a surface $\hat{M} \subset \mathbb{R}^3$. We can select a (global) coordinate chart $(M, \hat{z}^{-1})$, where $\hat{z}$ is the graph

$$\begin{aligned} \hat{z}\colon M \subset \mathbb{R}^2 &\longrightarrow \hat{M} \subset \mathbb{R}^3 \\ (x, y) &\longmapsto \hat{z}(x, y) := (x, y, z(x, y)), \end{aligned}$$

and consider, as above, the spacetime $(\mathcal{M}, G = dt^2 - F_t^2)$, where the indicatrix of F_t at each $p = (t, x) \in \mathcal{M}$ provides the velocity of the fire for every direction. The Finslerian nature of the model in this specific case is absolutely essential, as several physical effects (mainly the slope and the wind) cause the propagation of the fire to be anisotropic.

If we know F_t, i.e., we know the velocity of the fire at each point, time, and direction, then solving (28) we obtain the evolution of the fire over time. The aim of a wildifire model is therefore to provide such F_t.

In the isotropic case (without slope and wind), the indicatrix Σ_t is a sphere whose radius depends on the fuel and metereological conditions and may vary from one point to another (and over time) due to the change of vegetation, soil, moisture, etc. In order to include the isotropy caused by the wind, the most straightforward approximation is to consider that the wind displaces and deforms the sphere to an ellipse with a certain eccentricity depending on the wind strength. This approximation has been widely used since the experimental results by Anderson [3] and the subsequent PDE system developed by Richards [71] for the wavefront of a wildfire with an elliptical growth.[26] Richards' equations are still used nowadays by fire growth simulators such as FARSITE [38] and Prometheus [76], which even extend the elliptical approximation to the isotropy caused by the slope, i.e., the wildfire becomes a displaced ellipse in the upward direction (since the fire moves faster upward than downward).

From a Finslerian viewpoint, the elliptical model translates into the metric F_t being of Randers type. Markvorsen was the first to propose the use of these metrics for wildfire modeling, transforming Richards' PDE into an ODE (the geodesic equations of the Randers metric; see [61]) and even developing a rheonomic Lagrangian viewpoint to include the time dependence (see [62]). For further developments using Randers metrics, see [34]. Our work including Lorentz-Finsler metrics completes this theoretical framework and provides a full geometrical picture of the evolution of the wildfire in the most general situation. Specifically, there are mainly two important advantages of working with the Lorentz-Finsler setting over the classical elliptical one:

- *Flexibility*: the infinitesimal growth of the wildfire is not restricted to be elliptical and can adopt any other (strongly convex) pattern. In particular, the effect of the wind and the slope can be qualitatively different. For example, in [51] we

[26] Richards' equations are equivalent to the ODE system (28) when Σ_t is an ellipse (see [50, Sect. 5.2]).

have developed a specific model where the wind induces a sort of double semi-elliptical growth, which had already been pointed out by Anderson as the best experimental fitting in [3], while the slope generates the indicatrix of a reverse Matsumoto metric.[27]
- *Efficiency*: computationally speaking, solving an ODE is in general more efficient than solving a PDE. In addition, cut points represent a problem from the PDE viewpoint, since the firefront computed at an instant of time depends on the previously obtained, and therefore they must be corrected every time there is a crossover. This process has to be implemented through algorithms that are usually expensive in time and computing power (see, for example, [38] and references therein). In comparison, this problem is greatly simplified in the ODE case, where we only need to remove the trajectories that reach their cut points, without even affecting the computation of the firefront (see [51]).

Acknowledgments MAJ and EPR were partially supported by the projects PGC2018-097046-B-I00 and PID2021-124157NB-I00, funded by MCIN/AEI/10.13039/501100011033/ "ERDF A way of making Europe" and also by Ayudas a proyectos para el desarrollo de investigación científica y técnica por grupos competitivos (Comunidad Autónoma de la Región de Murcia), included in the Programa Regional de Fomento de la Investigación Científica y Técnica (Plan de Actuación 2022) of the Fundación Séneca-Agencia de Ciencia y Tecnología de la Región de Murcia, REF. 21899/PI/22. EPR and MS were partially supported by the project PID2020-116126GB-I00 funded by MCIN/AEI/10.13039/501100011033 and P20-01391 (PAIDI 2020, Junta de Andalucía), as well as the framework IMAG-María de Maeztu grant CEX2020-001105-M/AEI/10.13039/501100011033. EPR was also supported by Ayudas para la Formación de Profesorado Universitario (FPU) from the Spanish Government.

References

1. Aké, L., Flores, J.L., Sánchez, M.: Structure of globally hyperbolic spacetimes-with-timelike-boundary. Rev. Mat. Iberoam. **37**(1), 45–94 (2021)
2. Alías, L.J., Romero, A., Sánchez, M.: Uniqueness of complete spacelike hypersurfaces of constant mean curvature in generalized Robertson-Walker spacetimes. Gen. Relativ. Gravitation **27**(1), 71–84 (1995)
3. Anderson, H.E.: Predicting wind-driven wild land fire size and shape. Res. Pap. INT-305, USDA Forest Service, Intermountain Forest and Range Experiment Station, Ogden (1983)
4. Bao, D., Chern, S.-S., Shen, Z.: An Introduction to Riemann-Finsler geometry. In: Graduate Texts in Mathematics, vol. 200. Springer, New York (2000)
5. Bao, D., Robles, C., Shen, Z.: Zermelo navigation on Riemannian manifolds. J. Differ. Geom. **66**, 377–435 (2004)
6. Barceló, C., Liberati, S., Visser, M.: Analogue gravity. Living Rev. Relativ. **14**, 3 (2011)
7. Bartolo, R., Caponio, E., Germinario, A.V., Sánchez, M.: Convex domains of Finsler and Riemannian manifolds. Calc. Var. Partial Differ. Equ. **40**, 335–356 (2011)

[27] In its usual version, Matsumoto metrics effectively measure the travel time for a walker on a slope, favoring the downward direction (see [75]).

8. Beem, J.K., Ehrlich, P.E., Easley, K.L.: Global Lorentzian geometry. In: Monographs and Textbooks in Pure and Applied Mathematics, vol. 202, 2nd ed. Marcel Dekker, Inc., New York (1996)
9. Bernal, A.N., Sánchez, M.: Smoothness of time functions and the metric splitting of globally hyperbolic spacetimes. Comm. Math. Phys. **257**(1), 43–50 (2005)
10. Bernal, A.N., Sánchez, M.: Further results on the smoothability of Cauchy hypersurfaces and Cauchy time functions. Lett. Math. Phys. **77**(2), 183–197 (2006)
11. Bernal, A.N., Sánchez, M.: Globally hyperbolic spacetimes can be defined as 'causal' instead of 'strongly causal'. Classical Quantum Gravity **24**(3), 745–749 (2007)
12. Biferale, L., Bonaccorso, F., Buzzicotti, M., Clark Di Leoni, P., Gustavsson, K.: Zermelo's problem: optimal point-to-point navigation in 2D turbulent flows using reinforcement learning. Chaos **29**(10), 103138 (2019), 13 pp.
13. Biliotti, L., Javaloyes, M.A.: t-periodic light rays in conformally stationary spacetimes via Finsler geometry. Houston J. Math. **37**, 127–146 (2011)
14. Bonnard, B., Cots, O., Wembe, B.: A Zermelo navigation problem with a vortex singularity. ESAIM Control Optim. Calc. Var. **27**(suppl. S10), 37 (2021)
15. Bonnard, B., Cots, O., Gergaud, J., Wembe, B.: Abnormal geodesics in 2D-Zermelo navigation problems in the case of revolution and the fan shape of the small time balls. Systems Control Lett. **161**, 105140 (2022)
16. Bryson, Jr., A.E., Ho, Y.C.: Applied optimal control. Optimization, estimation, and control. Revised printing. Hemisphere Publishing Corp., Washington, distributed by Halsted Press [John Wiley & Sons, Inc.], New York (1975)
17. Cabello, J., Jaramillo, J.: A functional representation of almost isometries. J. Math. Annal. Appl. **445**(2), 1243–1257 (2017)
18. Caponio, E., Stancarone, G.: Standard static Finsler spacetimes. Int. J. Geom. Methods Mod. Phys. **13**, 1650040 (2016)
19. Caponio, E., Stancarone, G.: On Finsler spacetimes with a timelike Killing vector field. Classical Quantum Gravity **35**, 085007 (2018)
20. Caponio, E., Javaloyes, M.A., Masiello, A.: Morse theory of causal geodesics in a stationary spacetime. Ann. Inst. H. Poincaré Anal. Non Linéaire **27**, 857–876 (2010)
21. Caponio, E., Javaloyes, M.A., Masiello, A.: On the energy functional on Finsler manifolds and applications to stationary spacetimes. Math. Ann. **351**, 365–392 (2011)
22. Caponio, E., Javaloyes, M.A., Sánchez, M.: On the interplay between Lorentzian causality and Finsler metrics of Randers type. Rev. Mat. Iberoam. **27**(3), 919–952 (2011)
23. Caponio, E., Javaloyes, M.A., Sánchez, M.: Wind Finslerian structures: from Zermelo's navigation to the causality of spacetimes. To appear in Mem. Amer. Math. Soc. ArXiv e-prints, arXiv:1407.5494 [math.DG] (2014)
24. Caponio, E., Germinario, A.V., Sánchez, M.: Convex regions of stationary spacetimes and Randers spaces. Applications to lensing and asymptotic flatness. J. Geom. Anal. **26**(2), 791–836 (2016)
25. Caponio, E., Giannoni, F., Masiello, A., Suhr, S.: Connecting and closed geodesics of a Kropina metric. Adv. Nonlin. Stud. **21**, 683–695 (2021)
26. Carathéodory, C.: Calculus of Variations and Partial Differential Equations of the First Order. Holden-Day Inc., San Francisco (1967)
27. Cheng, J.-H., Marugame, T., Matveev, V., Montgomery, R.: Chains in CR geometry as geodesics of a Kropina metric. Adv. Math. **350**, 973–999 (2019)
28. Choquet-Bruhat, Y.: The Cauchy problem. In: Gravitation: An Introduction to Current Research. Wiley, New York (1962)
29. Choquet-Bruhat, Y.: Introduction to General Relativity, Black Holes, and Cosmology. Oxford University Press, Oxford (2015)
30. Choquet-Bruhat, Y., Geroch, R.: Global aspects of the Cauchy problem in general relativity. Commun. Math. Phys. **14**, 329–335 (1969)
31. Chrusciel, P.T., Friedrich, H. (eds.): The Einstein Equations and the Large Scale Behavior of Gravitational Fields. Birkhäuser, Berlin (2004)

32. Daniilidis, A., Jaramillo, J., Venegas, F.: Smooth semi-Lipschitz functions and almost isometries between Finsler manifolds. J. Func. Anal. **279**(8), 108662 (2020)
33. Danilo, M., Torri, C., Pfeifer, C., Voicu, N. (eds.) Beyond Riemannian Geometry in Classical and Quantum Gravity. Special Issue of Universe (2022). ISSN 2218-1997
34. Dehkordi, H.R.: Applications of Randers geodesics for wildfire spread modelling. Appl. Math. Model. **106**, 45–59 (2022)
35. Dehkordi, H.R., Saa, A.: Huygens' envelope principle in Finsler spaces and analogue gravity. Classical Quantum Gravity **36**(8), 085008 (2019)
36. Eberlein, P., O'Neill, B.: Visibility manifolds. Pacific J. Math. **46**, 45–109 (1973)
37. Fathi, A., Siconolfi, A.: On smooth time functions. Math. Proc. Camb. Philos. Soc. **152**(2), 303–339 (2012)
38. Finney, M.A.: FARSITE: Fire Area Simulator-model development and evaluation. Res. Pap. RMRS-RP-4, USDA Forest Service, Rocky Mountain Research Station, Ogden, 1998 (revised 2004)
39. Flores, J.L., Herrera, J., Sánchez, M.: On the final definition of the causal boundary and its relation with the conformal boundary. Adv. Theor. Math. Phys. **15**, 991–1057 (2011)
40. Flores, J.L., Herrera, J., Sánchez, M.: Gromov, Cauchy and causal boundaries for Riemannian, Finslerian and Lorentzian manifolds. Mem. Amer. Math. Soc. **226**, 1064 (2013)
41. Geroch, R.: Domain of dependence. J. Math. Phys. **11**, 437–449 (1970)
42. Geroch, R.P., Kronheimer, E.H., Penrose, R.: Ideal points in spacetime. Proc. Roy. Soc. Lond. A **237**, 545–567 (1972)
43. Gibbons, G., Herdeiro, C., Warnick, C., Werner, M.: Stationary metrics and optical Zermelo-Randers-Finsler geometry. Phys. Rev. D **79**, 044022 (2009)
44. Gromov, M.: Hyperbolic manifolds, groups and actions. In: Riemann Surfaces and Related Topics: Proceedings of the 1978 Stony Brook Conference (State Univ. New York, Stony Brook, 1978). Annals of Mathematics Studies, vol. 97. Princeton University Press, Princeton (1981)
45. Gromov, M.: Metric structures for Riemannian and non-Riemannian spaces. Progress in Mathematics, vol. 152, Birkhäuser Boston Inc., Boston (1999)
46. Harris, S.: Static- and stationary-complete spacetimes: algebraic and causal structures. Classical Quantum Gravity **32**, 135026 (2015)
47. Herrera, J., Javaloyes, M.A.: Stationary-complete spacetimes with non-standard splittings and pre-Randers metrics. J. Geom. Phys. **163**, 104120 (2021)
48. Javaloyes, M.A., Lichtenfelz, L., Piccione, P.: Almost isometries of non-reversible metrics with applications to stationary spacetimes. J. Geom. Phys. **89**, 38–49 (2015)
49. Javaloyes, M.A., Pendás-Recondo, E.: Lightlike hypersurfaces and time-minimizing geodesics in cone structures. In: A.L. Albujer et al. (eds.), Developments in Lorentzian Geometry, Springer Proceedings in Mathematics & Statistics, vol. 389. Springer Nature Switzerland AG, Cham (2022)
50. Javaloyes, M.A., Pendás-Recondo, E., Sánchez, M.: Applications of cone structures to the anisotropic rheonomic Huygens' principle. Nonlin. Anal. **209**, 112337 (2021)
51. Javaloyes, M.A., Pendás-Recondo, E., Sánchez, M.: A general model for wildfire propagation with wind and slope. SIAM J. Appl. Algebra Geom. **7**(2), 414–439 (2023)
52. Javaloyes, M.A., Sánchez, M.: On the definition and examples of Finsler metrics. Ann. Sc. Norm. Super. Pisa Cl. Sci. (5) **13**(3), 813–858 (2014)
53. Javaloyes, M.A., Sánchez. M.: Some criteria for wind Riemannian completeness and existence of Cauchy hypersurfaces. In: Lorentzian Geometry and Related Topics, Springer Proceedings in Mathematics & Statistics, vol. 211. Springer, Cham (2017)
54. Javaloyes, M.A., Sánchez, M.: Wind Riemannian spaceforms and Randers-Kropina metrics of constant flag curvature. Eur. J. Math. **3**, 1225–1244 (2017)
55. Javaloyes, M.A., Sánchez, M.: On the definition and examples of cones and Finsler spacetimes. RACSAM **114**, 30 (2020)
56. Javaloyes, M.A., Vitório, H.: Some properties of Zermelo navigation in pseudo-Finsler metrics under an arbitrary wind. Houston J. Math. **44**(4), 1147–1179 (2018)

57. Kropina, V.K.: Projective two-dimensional Finsler spaces with special metric (Russian). Trudy Sem. Vektor. Tenzor. Anal. **11**, 277–292 (1961)
58. Levi-Civita, T.: The Absolute Differential Calculus. Blackie & Son Limited, London and Glasgow (1927)
59. Levi-Civita, T.: Über Zermelo's Luftfahrtproblem. Z. Angew. Math. Mech. **11**, 314–322 (1931)
60. Manià, B.: Sopra un problema di navigazione di Zermelo. Math. Ann. **113**, 584–599 (1937)
61. Markvorsen, S.: A Finsler geodesic spray paradigm for wildfire spread modelling. Nonlin. Anal. RWA **28**, 208–228 (2016)
62. Markvorsen, S.: Geodesic sprays and frozen metrics in rheonomic Lagrange manifolds. ArXiv e-prints, arXiv:1708.07350 [math.DG] (2017)
63. Matveev, V.S.: Can we make a Finsler metric complete by a trivial projective change? In: Recent Trends in Lorentzian geometry, Springer Proceedings in Mathematics & Statistics, vol. 26. Springer, New York (2013)
64. Minguzzi, E.: Causality theory for closed cone structures with applications. Rev. Math. Phys. **31**(5), 1930001 (2019)
65. Minguzzi, E., Sánchez, M.: The causal hierarchy of spacetimes. Recent Developments in Pseudo-Riemannian Geometry. ESI Lect. Math. Phys., pp. 299–358. Eur. Math. Soc., Zürich, (2008)
66. Natário, J.: An Introduction to Mathematical Relativity. Latin American Mathematics Series. Springer, Cham (2021)
67. O'Neill, B.: Semi-Riemannian Geometry. Pure and Applied Mathematics, vol. 103. Academic Press, Inc., New York (1983)
68. Penrose, R.: Conformal treatment of infinity. In: Relativité, Groupes et Topologie (Lectures, Les Houches, 1963 Summer School of Theoret. Phys., Univ. Grenoble), Gordon and Breach, New York, 1964. Reprinted in: Gen. Relativity Gravitation **43**, 901–922 (2011)
69. Perlick, V.: Fermat principle in Finsler spacetimes. Gen. Relativity Gravitation **38**(2), 365–380 (2006)
70. Perlick, V. (ed.) Finsler Modification of Classical General Relativity. Special Issue of Universe (2020). ISSN 2218-1997
71. Richards, G.D.: Elliptical growth model of forest fire fronts and its numerical solution. Int. J. Numer. Methods Engrg. **30**(6), 1163–1179 (1990)
72. Ringström, H.: The Cauchy Problem in General Relativity. ESI Lectures in Mathematics and Physics, European Mathematical Society (EMS), Zürich (2009)
73. Serres, U.: On Zermelo-like problems: Gauss-Bonnet inequality and E. Hopf theorem. J. Dyn. Control Syst. **15**(1), 99–131 (2009)
74. Shen, Z.: Finsler metrics with $\mathbf{K} = 0$ and $\mathbf{S} = 0$. Canad. J. Math. **55**, 112–132 (2003)
75. Shimada, H., Sabau, S.V.: Introduction to Matsumoto metric. Nonlin. Anal. **63**, 165–168 (2005)
76. Tymstra, C., Bryce, R.W., Wotton, B.M., Taylor, S.W., Armitage, O.B.: Development and structure of Prometheus: the Canadian Wildland Fire Growth Simulation Model. Inf. Rep. NOR-X-417, Nat. Resour. Can., Can. For. Serv., North. For. Cent., Edmonton (2010)
77. von Mises, R.: Zum Navigationsproblem der Luftfahrt. Z. Angew. Math. Mech. **11**, 373–381 (1931)
78. Wald, R.M.: General Relativity. The University of Chicago Press, Chicago (1984)
79. Weyl, H.: Zur gravitationstheorie. Ann. Phys. **54**, 117–145 (1917)
80. Yoshikawa, R., Sabau, S.V.: Kropina metrics and Zermelo navigation on Riemannian manifolds. Geom. Dedicata **171**, 119–148 (2014)
81. Zaustinsky, E.M.: Spaces with non-symmetric distance. Mem. Am. Math. Soc. **34**, 1–91 (1959)
82. Zermelo, E.: Über das Navigationsproblem bei ruhender oder veränderlicher Windverteilung. Z. Angew. Math. Mech. **11**, 114–124 (1931)

Geometric and Architectural Aspects of the Singular Minimal Surface Equation

Rafael López

Abstract A singular minimal surface is the mathematical model of a dome suspended by its own weight. We review the main properties of these surfaces when the geometry is cylindrical and rotational, showing its relation with the constructions of the Spanish architect Antonio Gaudí.

Keywords Singular minimal surface · Center of gravity · Construction of domes · Stability

1 Physical Motivation and the Variational Problem

If is well-known that if the shape of a hanging flexible chain suspended from its endpoints is the catenary, it is not the way how to describe the shape of a bounded piece Σ of homogeneous flexible cloth (a two-dimensional surface) hanging from its boundary curve and suspended by its weight. In a position of static equilibrium, the tension forces exerted on the surface Σ only have tangential components. Therefore, if we turn Σ upside down, these forces become only internal compressive forces. In consequence these surfaces are models of domes and cupolas, resulting in reduced collapse. For this reason, these surfaces are the two-dimensional analogues of the catenary.

The catenary, as a mathematical model of an arch, is the curve

$$y(x) = \frac{1}{c}\cosh(cx + d), \quad c, d \in \mathbb{R}, c > 0. \tag{1}$$

The Spanish architect Antonio Gaudí (1852–1926) used catenary arches in the construction of vaults in halls and corridors (Fig. 1, left). To be precise, Gaudí repeated the inverted shape of a catenary along the direction of the corridor.

R. López (✉)
Department of Geometry and Topology, University of Granada, Granada, Spain
e-mail: rcamino@ugr.es

A. Alarcón et al. (eds.), *New Trends in Geometric Analysis*,
RSME Springer Series 10, https://doi.org/10.1007/978-3-031-39916-9_11

Fig. 1 Left: Corridor in the Colegio Teresiano, Barcelona [59]. Right: Funicular models of the church of Colonia Güell, Barcelona [60]

From the geometric viewpoint, the corridor is a cylindrical surface generated by a catenary.

We point out that Gaudí suspected that the ideal shape of a dome suspended by its weight was not "trivial" as the catenaries for arches. For that reason, Gaudí opted to use funicular models [55] of skeletons made by threads and put small sand bags hanging from them as shown in Fig. 1, right. Moving the vertices of the skeleton and with different weights, he obtained models of domes that were employed later in some of his constructions, as, for example, in the Sagrada Familia (Barcelona).

Coming back to the two-dimensional analogue of the catenary, the problem of the shape of a dome had attracted the interest of mathematicians such as Beltrami, Germain, Lagrange, and Jellet [4, 22, 24, 27, 33, 58]. Using techniques of calculus of variations, Poisson finally derived the equation that governs the shape of a hanged surface [52, p. 185]. We give these arguments. Let (x, y, z) denote the canonical coordinates of Euclidean space $(\mathbb{R}^3, \langle, \rangle)$, where z indicates the opposite direction of the gravity acceleration and $\langle, \rangle$ stands for the Euclidean metric of $\mathbb{R}^3$. First, we need to precise the initial data of the problem. In the one-dimensional case, we have a chain of given length and suspended from its endpoints, and in the two-dimensional case we prescribe a value for the area of the surface and the boundary curve of the surface.

Let Σ be a compact surface made of a flexible, homogeneous material of density σ per unit area. Suppose that the area of Σ is $A_0 > 0$ and its boundary is a closed curve $\Gamma \subset \mathbb{R}^3$. Let $\{\Sigma_t : t \in (-\epsilon, \epsilon)\}$ be a uniparametric smooth family of compact surfaces with boundary Γ and area A_0 and viewed as a variation of Σ, that is,

$\Sigma_0 = \Sigma$. The energy acting to all these surfaces is the weight, and we ask when Σ has the least energy. The gravity is measured with respect to a fix horizontal plane, for instance, the plane of equation $z = 0$. It is equivalent to say that Σ has minimum energy to have the lowest position of the center of gravity among all surfaces Σ_t. As it is usual in the calculus of variations, we will ask for something less than the minimum of the energy, and we only require that the energy has a critical point at Σ_0. On the other hand, since our problem is local, we can assume that all Σ_t are graphs over a bounded domain $\Omega \subset \mathbb{R}^2$ of the xy-coordinate plane.

Let Σ be a graph $z = u(x, y)$ defined on Ω, where $u_{|\partial\Omega} : \partial\Omega \to \mathbb{R}$ parametrizes the curve Γ. The height of the center of gravity of Σ is

$$\frac{1}{A_0}\int_\Omega \sigma\, u\sqrt{1+u_x^2+u_y^2}\,dxdy = \frac{\sigma}{A_0}\int_\Omega u\sqrt{1+|Du|^2}\,dxdy.$$

To simplify the arguments, we will assume that $A_0 = 1$ and $\sigma = 1$. The surfaces Σ_t are now represented by $u(x, y) + th(x, y)$, where $h\colon \Omega \to \mathbb{R}$ is a smooth function with $h_{|\partial\Omega} = 0$. Using Lagrange multipliers, the energy is the functional

$$J[u] = \int_\Omega u\sqrt{1+|Du|^2}\,dxdy + \lambda\int_\Omega \sqrt{1+|Du|^2}\,dxdy, \tag{2}$$

where $\lambda \in \mathbb{R}$. An extremal u of J satisfies

$$\left.\frac{d}{dt}\right|_{t=0} J[u+th] = 0$$

for all smooth function $h\colon \Omega \to \mathbb{R}$ with $h = 0$ on $\partial\Omega$. The Lagrangian is

$$L(x, y, u, p, q) = (u+\lambda)\sqrt{1+p^2+q^2}.$$

Using the divergence theorem, the boundary condition $h = 0$ on $\partial\Omega$, and the Fundamental Lemma of calculus of variations, u is an extrema if and only if

$$\frac{\partial L}{\partial u} - \left(\frac{\partial L}{\partial p}\right)_x - \left(\frac{\partial L}{\partial q}\right)_y = 0.$$

A straightforward computation leads to the Euler-Lagrange equation

$$\left(\frac{u_x}{\sqrt{1+|Du|^2}}\right)_x + \left(\frac{u_y}{\sqrt{1+|Du|^2}}\right)_y = \frac{1}{(u+\lambda)\sqrt{1+|Du|^2}}.$$

By moving Σ vertically, we can assume that $\lambda = 0$, hence

$$\mathrm{div}\frac{Du}{\sqrt{1+|Du|^2}} = \frac{1}{u\sqrt{1+|Du|^2}}. \tag{3}$$

It gives the following geometric interpretation. The left-hand side is the twice of the mean curvature H of the surface $z = u(x, y)$. Regarding the right-hand side, if Σ is parametrized by $X(x, y) = (x, y, u(x, y))$, then the unit normal vector field N of Σ is

$$N = \frac{1}{\sqrt{1 + |Du|^2}}(-u_x, -u_y, 1).$$

Thus, if $\mathbf{v} = (0, 0, 1)$ is the unit vertical vector on the z-axis, then $u = \langle X, \mathbf{v}\rangle$ and $1/\sqrt{1 + |Du|^2} = \langle N, \mathbf{v}\rangle$ which are the terms that appear in (3). Without further ado, we give the following definition.

Definition 1 Let $\mathbf{v} \in \mathbb{R}^3$ be a unit vector. A surface $\Sigma \subset \mathbb{R}^3$ is said to be a singular minimal surface with respect to $\mathbf{v}$ if it satisfies

$$2H(p) = \frac{\langle N(p), \mathbf{v}\rangle}{\langle p, \mathbf{v}\rangle}, \quad p \in \Sigma. \tag{4}$$

With this definition, a singular minimal surface is the ideal model of a dome where the only force acting on the surface is the gravity. For this reason, we can borrow the words of the German architect Frei Otto asserting that a singular minimal surface is the model of a "perfect roof" [48]. The name of singular minimal surface was coined by Dierkes in [14] because Eq. (3) degenerates at $u = 0$. See also [5, 8, 12].

Remark 1 The previous arguments of calculus of variations are still valid in the one-dimensional case. If $\gamma(x) = (x, u(x))$ is a planar curve, then (3) is

$$\left(\frac{u'}{\sqrt{1 + u'^2}}\right)' = \frac{1}{u\sqrt{1 + u'^2}}, \quad u' = \frac{du}{dx}.$$

Simplifying,

$$\frac{u''}{1 + u'^2} = \frac{1}{u}, \tag{5}$$

whose solution is the catenary (1).

We are assuming in Definition 1 that the surface does not intersect the vector plane orthogonal to $\mathbf{v}$, and, without loss of generality, we will suppose in this paper that Σ is included in the halfspace $\langle p, \mathbf{v}\rangle > 0$. Once we have the mathematical model of a dome, the next step is to find examples of singular minimal surfaces. The abundance or not of examples will show us the richness or not of the theory we are developing. A simple, and at the same time interesting, way is considering surfaces constructed by rigid motions of curves. To be precise, a Darboux surface is a surface parametrized by $\Psi(s, t) = A(t) \cdot \gamma(s) + \beta(t)$, where γ and β are two spatial curves of $\mathbb{R}^3$ and $A(t)$ is an orthogonal matrix [11, Livre I]. In this paper, we focus on two particular cases of Darboux surfaces: cylindrical surfaces ($A(t)$ is the identity

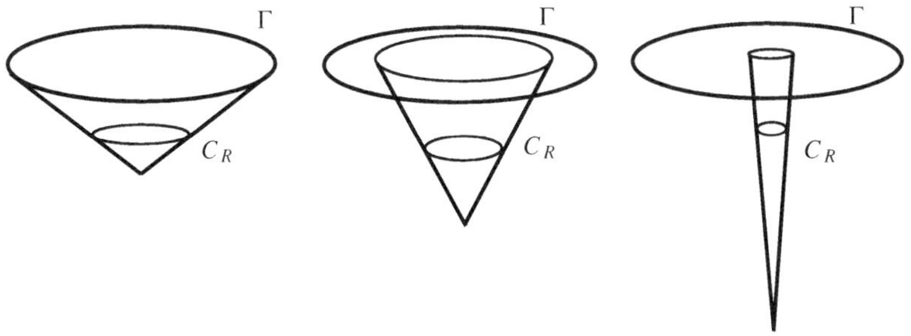

Fig. 2 An example of a uniparametric family of surfaces $\{S_R : R \in (0, 1]\}$, all with the same area and boundary curve, but the height of the centers of gravity of S_R go to $-\infty$ as $R \to 0$

and $\gamma(s)$ is a straight line) and rotational surfaces ($A(t)$ is a uniparametric group of rotations and β is constant). See [37] for other examples of singular minimal surfaces, such as translation surfaces and helicoidal surfaces.

We finish this section with two observations. First, the Euler-Lagrange equation (3) (or (4)) is the only approach until the first order to the problem of minimization of the energy (2). If we are finding minimizers, it is natural to impose that the second derivative of the energy is non-negative. This, in mathematical language, is equivalent to ask for stable surfaces. This problem will be discussed in Sect. 6 for a particular geometry of the surface. The second observation is that, in general, the problem of finding minimizers for given data (area and boundary) is open in all its generality. The next example, motivated by Nitsche, shows surfaces with prescribed boundary and area whose centers of gravity can be as lower as one desires [45]. In the plane P of equation $z = 0$, we prescribe a circle Γ of radius 1 and an area $A_0 = 2\pi$. For each $0 < R \leq 1$, we construct the surface $S_R = \Omega_R \cup C_R$, where $\Omega_R \subset P$ is the annulus $R^2 \leq x^2 + y^2 \leq 1$ and C_R is the cone $u(r) = -h(R-r)/R$ in polar coordinates, $0 \leq r \leq R$, where $h = \sqrt{2 + 1/R^2}$: see Fig. 2. With this choice of h, all surfaces S_R share the same area A_0 and the same boundary Γ. Since the center of gravity of a cone (in our case, C_R) lies on one third of its boundary plane, the height h_s of the center of gravity of S_R is

$$h_s = -\frac{1 + R^2}{6R}\sqrt{1 + 2R^2},$$

which goes to $-\infty$ as $R \to 0$.

This paper is organized as follows. In Sect. 2 we classify all cylindrical singular minimal surfaces. In Sect. 3 we prove that the singular minimal surfaces are minimal surfaces of $\mathbb{R}^3$ when the space $\mathbb{R}^3$ is endowed with a metric conformal to the Euclidean one. Section 4 is devoted to the classification of rotational singular minimal surfaces with special attention to surfaces intersecting orthogonally the rotation axis. These surfaces, called rotational tectums, are the models of rotational domes. In architecture, the literature on the shape of domes is great, see for example [19, 26, 46, 50, 53, 57]. Hence, in Sect. 5 we will compare the center of gravity of rotational tectums with catenary rotation surfaces and paraboloids, showing that although these last two surfaces are not singular minimal surfaces, the positions of their centers of gravity are very close to the ideal model. In Sect. 6 we give a Plateau-Rayleigh result of instability, proving that long cylindrical singular minimal surfaces are not stable. Finally, in Sect. 7, we show that the geometry of a compact singular minimal surface is partially determined by its boundary curve, which imposes restrictions to the existence and the shape of the surface that spans.

2 Cylindrical Singular Minimal Surfaces

It is natural to ask if the corridors constructed by Gaudí using catenary vaults are examples of singular minimal surfaces. From the mathematical viewpoint, we are asking which are the singular minimal surfaces of cylindrical type.

Theorem 1 *The cylindrical singular minimal surfaces are planes parallel to* $\mathbf{v}$*, or the rulings are orthogonal to* $\mathbf{v}$ *and the generating curve is a catenary.*

Proof Let Σ be a cylindrical surface parametrized by $X(s,t) = \gamma(s) + t\mathbf{w}$, $t \in \mathbb{R}$, where $\gamma = \gamma(s)$, $s \in I \subset \mathbb{R}$, is a curve parametrized by arc-length included in a plane orthogonal to $\mathbf{w}$. Then $2H = \kappa$, where κ is the curvature of γ. Since $N = \gamma' \times \mathbf{w}$, then (4) is

$$\kappa(s) = \frac{\langle \gamma'(s) \times \mathbf{w}, \mathbf{v}\rangle}{\langle \gamma(s), \mathbf{v}\rangle + t\langle \mathbf{w}, \mathbf{v}\rangle}.$$

This equation can be written as a polynomial on the variable t, $P_0(s) + P_1(s)t = 0$ for all $s \in I$, where

$$P_0 = \kappa(s)\langle \gamma(s), \mathbf{v}\rangle - \langle \gamma'(s) \times \mathbf{w}, \mathbf{v}\rangle,$$
$$P_1 = \kappa(s)\langle \mathbf{w}, \mathbf{v}\rangle = 0.$$

If κ is constantly 0, then γ is a straight line, Σ is a plane, and from $P_0 = 0$, we have $\langle N, \mathbf{v}\rangle = 0$. This implies that Σ is a plane parallel to $\mathbf{v}$. If γ is not a straight line, then $P_1 = 0$ implies that $\mathbf{w}$ is orthogonal to $\mathbf{v}$, proving that the rulings are orthogonal to $\mathbf{v}$. After a rigid motion of $\mathbb{R}^3$, we assume that $\mathbf{v} = (0, 0, 1)$, $\mathbf{w} = (0, 1, 0)$, and that the curve γ is contained in the xz-coordinate plane. If γ is the graph of $z = z(x)$, then $P_0 = 0$ is

$$\kappa = \frac{1}{z\sqrt{1 + z'^2}}.$$

Writing κ in terms of $z(x)$, we conclude that

$$\frac{z''}{(1 + z'^2)^{3/2}} = \frac{1}{z\sqrt{1 + z'^2}},$$

which coincides with (5). This shows that γ is a catenary, proving the result. □

As a consequence of this theorem, the corridors built by Gaudí using catenary arches are, actually, singular minimal surfaces, and its construction of corridors follow the ideal model of the Definition 1. With historical perspectives, it surprisingly means that Gaudí used singular minimal surfaces without knowing them. Although one can suspect that if the catenary is the model of the shape of an arch (and Gaudí knew that), then a corridor should be constructed by repeated catenaries; this evidence is now supported by Theorem 1.

For future discussions, we need to generalize the concept of singular minimal surfaces as follows. By replacing the energy functional (2) by

$$J_\alpha[u]=\int_\Omega u^\alpha\sqrt{1+|Du|^2}\,dxdy+\lambda\int_\Omega\sqrt{1+|Du|^2}\,dxdy,$$

where $\alpha\in\mathbb{R}$ is a constant, and following the same arguments, the Euler-Lagrange associated to this energy is

$$\operatorname{div}\frac{Du}{\sqrt{1+|Du|^2}}=\frac{\alpha}{u\sqrt{1+|Du|^2}}.$$

A surface $\Sigma\subset\mathbb{R}^3$ is called an α-*singular minimal surface* if

$$2H(p)=\alpha\,\frac{\langle N(p),\mathbf{v}\rangle}{\langle p,\mathbf{v}\rangle},\qquad p\in\Sigma.$$

If $\alpha=0$, we obtain a minimal surface ($H=0$). Another interesting case is $\alpha=-2$ because Σ is a minimal surface in three-dimensional hyperbolic space when this space is viewed in the upper halfspace model. If we now restrict our attention on α-singular minimal surfaces of cylindrical type, then Theorem 1 follows being valid, and the generating curve $u=u(x)$ satisfies

$$\frac{u''}{1+u'^2}=\frac{\alpha}{u}. \tag{6}$$

Planar curves $u=u(x)$ satisfying (6) will be called α-*catenaries*. Multiplying by u' and integrating, there exists a constant $c>0$ such that

$$u'=\pm\sqrt{c^2u^{2\alpha}-1}. \tag{7}$$

Explicit solutions are obtained for some values of α. For example, if $\alpha=1$, then u is a catenary; if $\alpha=-1$, then u is a half-circle; and if $\alpha=1/2$, then u is a parabola. The next proposition shows the main properties of the α-catenaries [37]: see Fig. 3.

Proposition 1 *Let u be a solution of (6). Then u has a unique critical point, which we can assume at $x=0$. Moreover, u is symmetric about the y-axis. Let $I=(-a,a)$ be the maximal domain of u.*

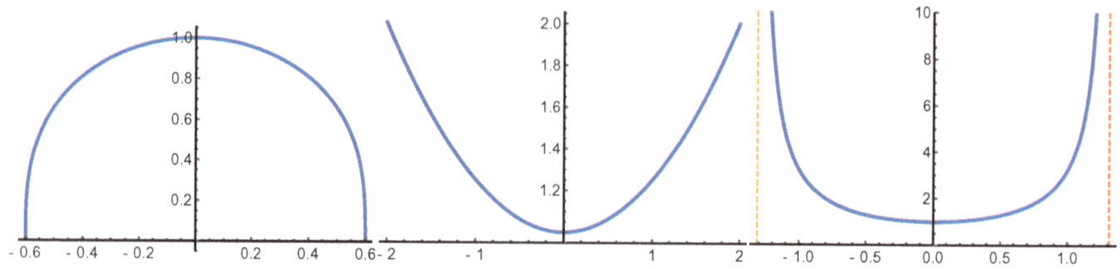

Fig. 3 Examples of α-catenaries. Case $\alpha = -2$ (left), $\alpha = 0.5$ (middle), and $\alpha = 2$ (right), where the curve is asymptotic to the vertical lines of equation $x = \pm R$, $R \approx 1.311$

1. *Case* $\alpha > 0$. *The function u is convex, and the function u has a minimum at* $x = 0$. *The domain of u is* $\mathbb{R}$ *if* $\alpha \in (0, 1]$, *and it is a bounded interval if* $\alpha > 1$. *In both cases,* $\lim_{x\to\pm a} u(x) = \infty$. *In particular, if* $\alpha > 1$, *u is asymptotic to the vertical lines* $x = \pm a$.
2. *Case* $\alpha < 0$. *The function u is concave, and the function u has a maximum at* $x = 0$. *The domain of u is a bounded interval and* $\lim_{x\to\pm a} u(x) = 0$.

3 Singular Minimal Surfaces as Minimal Surfaces in a Conformal Space

We prove that singular minimal surfaces are minimal surfaces in a certain Riemannian manifold. Without loss of generality, we assume $e_3 = (0, 0, 1)$ as the vector $\mathbf{v}$ in (4). In the halfspace $\mathbb{R}^3_+$, define the conformal metric $\widetilde{g} = z\langle , \rangle$.

Theorem 2 *If* Σ *is a surface of* $\mathbb{R}^3_+$, *then* Σ *is a singular minimal surface if and only if* Σ *is a minimal surface of* $(\mathbb{R}^3_+, \widetilde{g})$.

Proof Let $\widetilde{\nabla}$ and $\overline{\nabla}$ be the Levi-Civita connections for $\widetilde{g}$ and $\langle , \rangle$, respectively. The relation between both connections is

$$\widetilde{\nabla}_U V = \overline{\nabla}_U V + \frac{1}{2z}\left(U(z)V + V(z)U - \langle U, V\rangle\overline{\nabla}z\right) \tag{8}$$

for tangent vector fields U and V [6]. There is also a relation between the mean curvatures H and $\widetilde{H}$ of Σ with the induced metric from the Euclidean one and from $\widetilde{g}$, respectively. If A and $\widetilde{A}$ are the corresponding second fundamental forms, then

$$\widetilde{A}(U, V) = z^{1/2}\left(A(U, V) - \frac{1}{2z}\langle\overline{\nabla}z, N\rangle\langle U, V\rangle\right).$$

Hence

$$\widetilde{H} = z^{-1/2}\left(H - \frac{1}{2z}\langle\overline{\nabla}z, N\rangle\right) = z^{-1/2}\left(H - \frac{\langle N, e_3\rangle}{2z}\right). \tag{9}$$

Thus $\widetilde{H} = 0$ if and only Σ is a singular minimal surface. □

The result of Theorem 2 has similarities in the context of weighted manifolds of Gromov [25]. If $\psi \in C^\infty(\mathbb{R}^3)$, consider a density function e^ψ to measure the weighted area and volume. If dA and dV are the elements of area and volume of $\mathbb{R}^3$ with the Euclidean metric $\langle , \rangle$, the weighted area and volume elements are defined by

$$d\mathcal{A}_\psi = e^\psi \, dA, \quad d\mathcal{V}_\psi = e^\psi \, dV.$$

This is not equivalent to multiply the metric by the factor e^ψ because if it were, the area and the volume change by different powers of e^ψ. The formulas of the first variation of weighted area and volume of a compact surface Σ in $\mathbb{R}^3$ are

$$\mathcal{A}'_\psi(0) = -\int_\Sigma u\left(2H - \langle\overline{\nabla}\psi, N\rangle\right) d\mathcal{A}_\psi, \quad \mathcal{V}'_\psi(0) = \int_\Sigma u \, d\mathcal{A}_\psi, \tag{10}$$

where u is the normal component of the variational vector field. We refer to [3, 9] for details. The function

$$H_\psi = H - \frac{1}{2}\langle\overline{\nabla}\psi, N\rangle$$

is called the *weighted mean curvature* of Σ. In particular, from (10), Σ is a critical point of the weighted area functional $\mathcal{A}_\psi$ for any variation preserving the boundary of Σ if and only if $H_\psi = 0$. Now we rediscover the concept of singular minimal surface. Let Σ be a surface of $\mathbb{R}^3_+$ and consider the particular function

$$\psi(x, y, z) = \log z.$$

Since $\overline{\nabla}\psi = e_3/z$, we have $H_\psi = H - \langle N, e_3\rangle/(2z)$. Notice that the weighted mean curvature H_ψ does not coincide with the mean curvature $\widetilde{H}$ given in (9). However, we have the following equivalences.

Theorem 3 *Let Σ be a surface of $\mathbb{R}^3_+$. Then the following statements are all equivalent:*

1. *Σ is a singular minimal surface.*
2. *$H_\psi = 0$ for the density $e^\psi = z$.*
3. *$\widetilde{H} = 0$, that is, Σ is a minimal surface of $(\mathbb{R}^3_+, \widetilde{g})$.*

Not only the singular minimal surfaces are minimal surfaces in $(\mathbb{R}^3_+, \widetilde{g})$, but the geodesics of this space also have an interpretation in terms of singular minimal surfaces, more exactly, with the concept of α-catenaries defined in the previous section. First we calculate the geodesics of $(\mathbb{R}^3_+, \widetilde{g})$. Let $\gamma(s) = (x(s), y(s), z(s))$ be a curve parametrized by arc-length with respect to the metric $\widetilde{g}$. In particular,

$$1 = z(s)\left(x'(s)^2 + y'(s)^2 + z'(s)^2\right). \tag{11}$$

Then γ is a geodesic in $(\mathbb{R}^3_+, \widetilde{g})$ if and only if $\widetilde{\nabla}_{\gamma'}\gamma' = 0$. From (8), this is equivalent to

$$\overline{\nabla}_{\gamma'}\gamma' = -\frac{1}{2z}\left(2\langle\gamma', e_3\rangle\gamma' - |\gamma'|^2 e_3\right).$$

This leads to the ordinary differential equations

$$\begin{aligned} x''(s) &= -\frac{z'(s)}{z(s)}x'(s), \\ y''(s) &= -\frac{z'(s)}{z(s)}y'(s), \\ z''(s) &= -\frac{1}{2z(s)}\left(z'(s)^2 - x'(s)^2 - y'(s)^2\right) \end{aligned} \tag{12}$$

If $x(s)$ and $y(s)$ are both constant functions, then the third equation of (12) implies that γ is a vertical straight line. Assume now that both functions are not constant. Integrating the first two equations of (12) gives

$$x'(s) = \frac{a_1}{z(s)}, \quad y'(s) = \frac{a_2}{z(s)}, \tag{13}$$

for some nonzero real numbers $a_1, a_2 \neq 0$. Substituting into (11),

$$1 = \frac{a_1^2 + a_2^2}{z(s)} + z(s)z'(s)^2. \tag{14}$$

Let $m = a_1^2 + a_2^2$. Due to the third equation of (12), the function $z = z(s)$ cannot be constant. Thus

$$z' = \pm\frac{\sqrt{z-m}}{z}. \tag{15}$$

The third equation of (12) is now

$$z'' = \frac{2m - z}{2z^3}. \tag{16}$$

which is equivalent to (14). A first consequence is that the geodesics are planar curves.

Proposition 2 *The geodesics of* $(\mathbb{R}^3_+, \widetilde{g})$ *are contained in vertical planes.*

Proof The result is true if the geodesic is a vertical straight line. Otherwise, and from (13), the tangent vector $\gamma'(s)$ is

$$\gamma'(s) = \frac{1}{z(s)}(a_1, a_2, 0) + z'(s)e_3.$$

This implies that $\gamma'(s)$ is contained in the plane determined by the vectors $(a_1, a_2, 0)$ and e_3 for all $s \in I$. In particular, this plane is vertical, proving the result. □

Theorem 4 *The geodesics of* $(\mathbb{R}^3_+, \widetilde{g})$ *are* $1/2$*-catenaries and vice versa.*

Proof It is immediate that vertical geodesics are generating curves of singular minimal surfaces of cylindrical type for any α and vice versa.

Let γ be a geodesic of $(\mathbb{R}^3_+, \widetilde{g})$ and suppose that γ is not a vertical straight line. We calculate the curvature κ_γ of γ. After a rotation about e_3, if necessary, let write $\gamma(s) = (x(s), 0, z(s))$, or simply $\gamma(s) = (x(s), z(s))$ viewed γ as a planar curve. Then

$$\kappa_\gamma(s) = \frac{x'(s)z''(s) - z'(s)x''(s)}{|\gamma'(s)|^3}.$$

From (13), $\gamma' = (a_1 z^{-1}, z')$, $a_1 \neq 0$, and $\gamma'' = (-a_1 z^{-2} z', z'')$. We know that $|\gamma'|^3 = z^{-3/2}$. Because $m = a_1^2$, using (15) and (16), we obtain

$$\kappa_\gamma(s) = \frac{a_1}{2z(s)^{3/2}}. \tag{17}$$

On the other hand, we calculate the curvature of the solution $u_{1/2}$ of (6) for $\alpha = 1/2$:

$$\kappa_{u_{1/2}}(t) = \frac{u''_{1/2}(t)}{(1 + u'_{1/2}(t)^2)^{3/2}}.$$

Using the expression of $u''_{1/2}(t)$ in (6) together that of $u'_{1/2}(t)$ in (7),

$$\kappa_{u_{1/2}}(t) = \frac{1}{2\sqrt{c}\, u_{1/2}(t)^{3/2}}.$$

Comparing this identity with (17), we deduce that, up to a constant, κ_γ coincides with $\kappa_{u_{1/2}}$. Using the classical theorem of the local theory of planar curves, we conclude that γ and the graph of $u_{1/2}$ coincide up to a dilation of $\mathbb{R}^3_+$ from the origin and a horizontal translation of $\mathbb{R}^3_+$, which proves the theorem. □

This result extends as follows. If $\widetilde{g}_\alpha$ is the conformal metric $\widetilde{g}_\alpha = z^\alpha \langle , \rangle$, then the geodesics of $(\mathbb{R}^3, \widetilde{g}_\alpha)$ are $\alpha/2$-catenaries and vice versa [17].

4 Rotational Singular Minimal Surfaces

We classify all surfaces of revolution that are singular minimal surfaces. First we point out an observation about the construction of rotational domes. The reader can think that if the catenary is the one-dimensional solution of the suspended surface, it is enough to rotate a vertical catenary with respect to its rotation axis to obtain a model of a rotational dome. However, this surface, which we will call a *catenary rotation surface*, is not a singular minimal surface. It is more interesting to point out that Gaudí never thought that catenary rotation surfaces were models of domes, despite Gaudí used catenary arches in roofs of corridors, hence the efforts of Gaudí to find these shapes of domes from his funicular models. We will revisit the catenary rotation surfaces in the next section, but it should be noted that these surfaces have other interesting mathematical properties [29, 30].

Suppose that Σ is a singular minimal surface which it is also a surface of revolution about an axis L. A first question is the relation between L and the vector $\mathbf{v}$ of (4). Initially, there is not a priori condition between them. However, it is natural to think that L is parallel to the direction of the force of the gravity that motivated the definition of singular minimal surfaces. If, in addition, we are thinking in the construction of rotational domes, one is inclined to think that the rotation axis must be vertical, that is, parallel to the direction of the gravity. However, we have a surprising result.

Theorem 5 *Let Σ be a surface of revolution whose rotation axis is L. If Σ satisfies (4), then L is parallel to $\mathbf{v}$, or L is contained in the orthogonal vector plane to $\mathbf{v}$.*

Proof After a rigid motion of $\mathbb{R}^3$, we can assume that L is the z-axis. If the generating curve is parametrized by $r \mapsto (r, 0, u(r))$, where $u = u(r)$ is a smooth function and $r \in I \subset \mathbb{R}^+$, then Σ parametrizes as

$$X(r, \theta) = (r \cos\theta, r \sin\theta, u(r)).$$

If $\mathbf{v} = (v_1, v_2, v_3)$, then (4) is

$$\frac{u''}{1 + u'^2} + \frac{u'}{r} = \frac{1}{v_1 r \cos\theta + v_2 r \sin\theta + v_3 u} \tag{18}$$

because

$$N = \frac{X_r \times X_\theta}{|X_r \times X_\theta|} = \frac{1}{\sqrt{1 + u'^2}}(-u' \cos\theta, -u' \sin\theta, 1).$$

Equation (18) can be written as $A(r)\cos\theta + B(r)\sin\theta + C(r) = 0$, where

$$
\begin{aligned}
A(r) &= v_1 r\left(u'(1+u'^2) + ru'' + u'(1+u'^2)\right)\\
B(r) &= v_2 r\left(u'(1+u'^2) + ru'' + u'(1+u'^2)\right)\\
C(r) &= v_3\left(u(u'(1+u'^2) + ru'') - r(1+u'^2)\right).
\end{aligned}
\tag{19}
$$

Since the functions $\{1, \cos\theta, \sin\theta\}$ are linearly independent, all functions A, B, and C must vanish. From $A = 0$ and $B = 0$, we discuss two cases. First, if $v_1 = v_2 = 0$, then necessarily $v_3 \neq 0$ proving that $\mathbf{v} = (0, 0, v_3)$ is parallel to the z-axis, and this is the first part of the theorem. In case that v_1 or v_2 is not 0, from $A = 0$ or $B = 0$, we deduce

$$
u'(1+u'^2) + ru'' + u'(1+u'^2) = 0. \tag{20}
$$

Using now $C = 0$, we have two possibilities:

1. Case $v_3 = 0$. Then L is contained in the vector plane orthogonal to $\mathbf{v} = (v_1, v_2, 0)$, and this proves the second part of the theorem.
2. Case $v_3 \neq 0$. Then the last equation of (19) is $u(u'(1+u'^2)+ru'')-r(1+u'^2) = 0$. Combining with (20), we get $uu' = -r$, obtaining $u(r) = \sqrt{r^2+c}$, for some constant c. However, this function does not satisfy (20), proving that this case is not possible.

□

In consequence, Theorem 5 provides two types of rotational singular minimal surfaces. We analyze the second type of rotational surfaces. In order to follow with our intuition that the direction of the gravity is the z-axis, we rename the coordinates in the proof of the above theorem. Thus we assume that $\mathbf{v} = e_3 = (0, 0, 1)$ and that the rotation axis L is the x-axis. Then the generating curve is $r \mapsto (r, 0, u(r))$ and the parametrization of the surface is $X(r,\theta) = (r, -u(r)\sin\theta, u(r)\cos\theta)$. Now (20) is

$$
\frac{u''}{1+u'^2} = \frac{2}{u}.
$$

Notice that this equation is just the equation of the 2-catenary; see (6). This result is not a coincidence but an example of another more general that we now bring in and whose proof is similar to that of Theorem 5 [37].

Corollary 1 *Let Σ be a surface of revolution about the axis* $(1, 0, 0)$. *Then Σ is an α-singular minimal surface with respect to* $\mathbf{v} = (0, 0, 1)$ *if and only if its generating curve is an* $(\alpha+1)$*-catenary.*

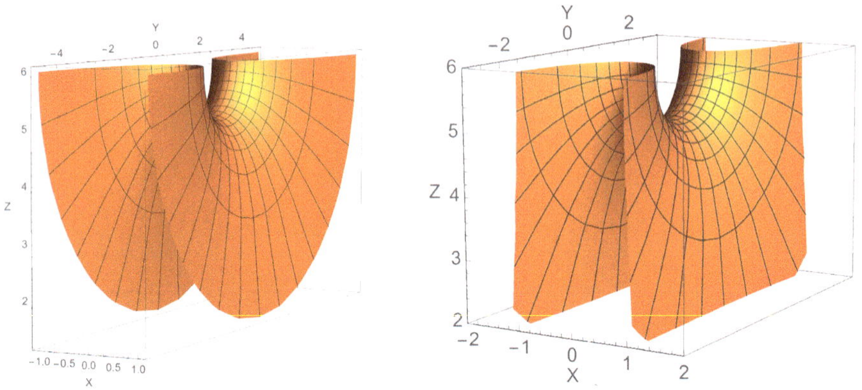

Fig. 4 Left: rotational singular minimal surfaces of the second type of Theorem 5. Right: the same surface once reversed to be the model of a rotational dome

If $\alpha = 0$, then Σ is a rotational minimal surface ($H = 0$) generated by a 1-catenary, but this curve is just the catenary. Indeed, if we place a catenary in the xz-plane and we rotate with respect to x-axis, we obtain a catenoid, which is the only rotational minimal surface. In view of Corollary 1, authors as Leon Simon prefer to name the singular minimal surfaces as *symmetric minimal surfaces* [21].

Although the rotational surfaces of the second type have the rotation axis orthogonal to the direction of the gravity, they follow being models of ideal roofs. If we construct domes using these surfaces, now the circles of the surface of revolution are vertical half-circles whose endpoints are situated in the plane $z = 0$. See Fig. 4. The study of this kind of domes appears in [41].

In the rest of this section, we focus on the first type of surfaces of Theorem 5.

Definition 2 A *rotational tectum* is a rotational singular minimal surface whose rotation axis is parallel to $\mathbf{v}$.

Without loss of generality, we will assume that $\mathbf{v} = e_3 = (0, 0, 1)$. By the proof of Theorem 5, the generating curve $r \mapsto (r, 0, u(r))$ satisfies the third equation of (19), that is,

$$\frac{u''}{1 + u'^2} + \frac{u'}{r} = \frac{1}{u}. \tag{21}$$

A first observation is that there are solutions of (21) that arrive until the plane $z = 0$ at the limit: notice that (21) is degenerated at $u = 0$. This particular solution is $u(r) = r$ defined in $(0, \infty)$ whose surface is a rotational cone whose vertex is the origin. Following with the idea of a roof, this solution, once reversed its position, is a teepee tent of the Native Americans.

The existence of solutions of (21) is assured by standard results once we give initial data at $r_0 > 0$. However, and motivated by the construction of rotational domes, we want that the profile curve of the surface meets orthogonally the rotation axis. This requires that u is defined at $r = 0$ and that $u'(0) = 0$. But just at $r = 0$, Eq. (21) is degenerated. Therefore the existence of solutions of (21) with initial conditions

$$u(0) = u_0 > 0, \quad u'(0) = 0 \tag{22}$$

is a problem that we need to solve.

Proposition 3 *If $u_0 > 0$, then there is a solution $u \in C^2([0, R])$ of (21)–(22) for some $R > 0$.*

Proof We write (21) as

$$\frac{u''}{(1 + u'^2)^{3/2}} + \frac{u'}{r\sqrt{1 + u'^2}} = \frac{1}{u\sqrt{1 + u'^2}}.$$

If we multiply by r, then

$$\left(\frac{ru'(r)}{\sqrt{1 + u'(r)^2}}\right)' = \frac{r}{u(r)\sqrt{1 + u'(r)^2}}. \tag{23}$$

Define the functions

$$g : \mathbb{R}_0^+ \times \mathbb{R} \to \mathbb{R}, \quad g(x, y) = \frac{1}{x\sqrt{1 + y^2}},$$

$$\varphi : \mathbb{R} \to \mathbb{R}, \quad \varphi(y) = \frac{y}{\sqrt{1 + y^2}}.$$

A solution of (23) is just a function $u \in C^2([0, R])$ such that $(r\varphi(u'))' = rg(u, u')$ and $u(0) = u_0$, $u'(0) = 0$. The inverse function $\varphi^{-1} : \mathbb{R} \to (-1, 1)$ is $\varphi^{-1}(x) = x/\sqrt{1 - x^2}$. Fix $R > 0$ that will be determined later and define the operator $\mathsf{T} : C^1([0, R]) \to C^1([0, R])$ by

$$(\mathsf{T}u)(r) = u_0 + \int_0^r \varphi^{-1}\left(\int_0^s \frac{t}{s} g(u, u')dt\right) ds.$$

It is clear that a fixed point of the operator T is a solution of the initial value problem (22)–(23). First, we prove that T is well defined in a closed ball $\overline{B(u_0, \epsilon)}$ of $C^1([0, R])$ for some $\epsilon > 0$, where the space $C^1([0, R])$ is endowed the usual sup-norm $\|u\| = \|u\|_\infty + \|u'\|_\infty$. For this, let $\epsilon > 0$ such that $\epsilon < u_0$ and $R \leq \min\{u_0 - \epsilon, \sqrt{3}\epsilon/2, 2(u_0 - \epsilon)\epsilon/\sqrt{4 + \epsilon^2}\}$. We have

$$\int_0^r \frac{t}{r} g(u, u')\, dt \leq \int_0^r \frac{t}{s} \frac{1}{u_0 - \epsilon}\, dt \leq \frac{R}{2(u_0 - \epsilon)} \leq \frac{1}{2},$$

hence we can use φ^{-1}. We use the Banach fixed point theorem $\mathsf{T} : \overline{B(u_0, \epsilon)} \to \overline{B(u_0, \epsilon)}$. The proof consists of two steps.

1. *The map* T *satisfies* $\mathsf{T}(\overline{B(u_0, \epsilon)}) \subset \overline{B(u_0, \epsilon)}$. If $u \in \overline{B(u_0, \epsilon)}$, and using that φ^{-1} is increasing,

$$|(\mathsf{T}u)(r) - u_0| \leq \int_0^r \varphi^{-1}\left(\int_0^s \frac{t}{s(u_0 - \epsilon)}\, dt\right) ds$$
$$\leq \varphi^{-1}\left(\frac{R}{2(u_0 - \epsilon)}\right) R < \varphi^{-1}\left(\frac{1}{2}\right) R = \frac{R}{\sqrt{3}} \leq \frac{\epsilon}{2}.$$

$$|(\mathsf{T}u)'(r)| \leq \varphi^{-1}\left(\int_0^s \frac{t}{s(u_0 - \epsilon)}\, dt\right) \leq \varphi^{-1}\left(\frac{R}{2(u_0 - \epsilon)}\right)$$
$$= \frac{R}{\sqrt{4(u_0 - \epsilon)^2 - R^2}} \leq \frac{\epsilon}{2},$$

where we have used that $R \leq 2(u_0 - \epsilon)\epsilon/\sqrt{4 + \epsilon^2}$. As a conclusion, $\|\mathsf{T}u\| \leq \epsilon$.

2. *The map* T *is a contraction*. The functions g and φ^{-1} are Lipschitz continuous in $[u_0 - \epsilon, u_0 + \epsilon] \times [-\epsilon, \epsilon]$ and in $[-\epsilon, \epsilon]$, respectively, provided $0 < \epsilon < \min\{u_0, 1\}$. Let $L = \min\{L_g, L_{\varphi^{-1}}\}$, where L_g and $L_{\varphi^{-1}}$ are the Lipschitz constants of g and φ^{-1}, respectively. For all functions u, v in the closed ball $\overline{B(u_0, \epsilon)}$ and for all $r \in [0, R]$,

$$|(\mathsf{T}u)(r) - (\mathsf{T}v)(r)| \leq L \int_0^r \frac{1}{s} \int_0^s t \left|g(u, u') - g(v, v')\right| dt\, ds$$
$$\leq L^2 \int_0^r \frac{1}{s} \int_0^s t \left(\|u - v\|_\infty + \|u' - v'\|_\infty\right) dt\, ds$$
$$= L^2 \|u - v\| \int_0^r \frac{s}{2}\, ds = \frac{r^2 L^2}{4} \|u - v\|.$$

Analogously,

$$|(\mathsf{T}u)'(r) - (\mathsf{T}v)'(r)| \leq \frac{rL^2}{2}.$$

Therefore

$$\|\mathsf{T}u - \mathsf{T}v\| \leq \min\{\frac{R^2 L^2}{4}, \frac{RL^2}{2}\} \|u - v\|.$$

Hence choosing $R > 0$ small enough, we conclude that T is a contraction.

The solution obtained by the Banach fixed point theorem lies in $C^1([0, R]) \cap C^2((0, R])$. Finally, we prove that the solution u extends with C^2-regularity at $r = 0$. This is proved directly with the L'Hôpital rule because from (21),

$$\frac{1}{u_0} = \lim_{r\to 0} u''(r) + \lim_{r\to 0} \frac{u'(r)}{r} = 2 \lim_{r\to 0} u''(r),$$

obtaining

$$\lim_{r\to 0} u''(r) = \frac{1}{2u_0}. \tag{24}$$

□

There are two types of rotational tectums.

1. The existence of surfaces that intersect the rotation axis has been established in Proposition 3. See Fig. 5, left.
2. When surfaces do not intersect with the rotation axis, see Fig. 5, right. The generating curves go back and forth without touching the z-axis and correspond with solutions of (21) with initial conditions $u'(r_0) = 0$, $r_0 > 0$. Indeed, if u attains the rotation axis and by the symmetry of the surface, then $u'(0) = 0$, so u has two critical points. By inspecting (21) together with (24), every critical point $\bar{r}$ of u is a minimum because $u''(\bar{r}) > 0$, obtaining a contradiction. This property is a particular result of other more general cases proving that the solutions of the Dirichlet problem of (3) defined in a convex domain have exactly one critical point [20].

More properties of the rotational tectums were established in the unpublished paper [28]; see also [15]. Focusing on those solutions of the first type, we have the following result.

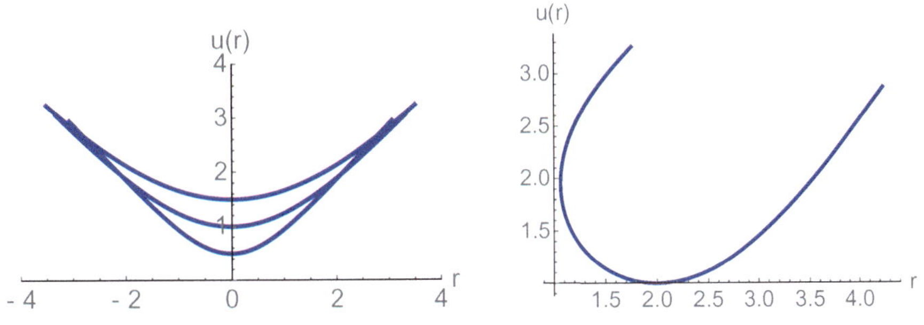

Fig. 5 Left: rotational solutions of (21), where $u'(0) = 0$ for different values of $u(0)$: 0.5, 1, and 1.5. Right: a rotational solution of (21), where $u(2) = 1$ and $u'(2) = 0$

Theorem 6 *Let u be a solution of* (21)–(22). *Then the domain of u is* $[0, \infty)$. *Moreover, u has only one critical point at* $r = 0$*, which it is a global minimum. The function u is asymptotic to* $v(r) = r$.

5 Comparison of Rotational Tectums with Catenary Rotation Surfaces and Paraboloids

We know that singular minimal surfaces are ideal models of domes. If we have in mind a rotational dome whose rotation axis is parallel to the direction of the gravity, its model is a rotational tectum. However, it is natural to think in two surfaces which, not being singular minimal surfaces, could be good "approximations" to the model of a rotational dome. The first surface is the catenary rotation surface that appeared in Sect. 4, and the reason is that its generating curve is the catenary, the model of a hanging chain. The second candidate is the *paraboloid*, the rotational surface obtained by rotating a vertical parabola with respect to its symmetric axis. The motivation has its origin in the well-known dispute between parabolas and catenaries in the construction of arches (see [47]). Furthermore, the paraboloid has a clear advantage from the architectural viewpoint because this surface is defined with polynomials, in contrast to the catenary that uses exponential functions. If we come back to the work of Gaudí, it is clear that he did not think in the catenary rotation surfaces neither in paraboloids as models for an ideal dome.

In this section, we ask if there is a "relevant" difference among the heights of the centers of gravity of rotational tectums, catenary rotation surfaces, and paraboloids; all of them have the same area and the same boundary as prescribed data.

The boundary curve that we prescribed is a horizontal circle of radius $R > 0$ centered at the rotation axis, and, in consequence, the rotational tectum intersects the rotation axis (Theorem 6). Let $u = u(x)$ be a solution of (21)–(22) defined in the interval $[0, R]$. The boundary curve is the horizontal circle

$$\Gamma_R = \{(x, y, u(R)) : x^2 + y^2 = R^2\}, \tag{25}$$

where $u(R) > 0$ is the height of Γ_R. The area A_0 of Σ is

$$A_0 = 2\pi \int_0^R x\sqrt{1 + u'(x)^2}\, dx, \tag{26}$$

and the height h_T of its center of gravity is

$$h_T = \frac{2\pi}{A_0} \int_0^R xu(x)\sqrt{1 + u'(x)^2}\, dx. \tag{27}$$

Let now the catenary rotation surface have the same boundary curve and area. The generating catenary is

$$z_{c,m}(x) = \frac{1}{c}\cosh(cx) + m, \quad x \in [0, R], \tag{28}$$

where $c, m \in \mathbb{R}$ and $c > 0$. Let $C_{c,m}$ denote the corresponding rotational catenary obtained once we rotate $z_{c,m}$ with respect to the z-axis. We will choose the values c and m in (28), so the area of $C_{c,m}$ is A_0, and its boundary is Γ_R. The last condition implies

$$\frac{1}{c}\cosh(cR) + m = u(R). \tag{29}$$

The area of $C_{c,m}$ is obtained by quadratures

$$A = 2\pi \int_0^R x\sqrt{1 + z_{c,m}'(x)^2}\,dx = \frac{2\pi(cR\sinh(cR) - \cosh(cR) + 1)}{c^2}. \tag{30}$$

Finally, the height h_c of the center of gravity of $C_{c,m}$ is

$$\begin{aligned} h_c &= \frac{2\pi}{A_0}\int_0^R x z_{c,m}(x)\sqrt{1 + z_{c,m}'(x)^2}\,dx \\ &= \frac{\pi\left(2c^2R^2 + 2cR\sinh(2cR) - \cosh(2cR) + 1\right)}{4c^3 A_0} + m. \end{aligned} \tag{31}$$

We use *Mathematica* [61] to compute the area (26) and the center of gravity (27) of the rotational tectum. The steps are as follows:

1. Fix $u_0 > 0$ in (22), the lowest point of Σ, and by numerical methods, we solve (21)–(22). We know that the domain of the solution is $[0, \infty)$ by Theorem 6.
2. Fix $R > 0$ and let $[0, R]$ be the domain of u which determines the (compact) rotational tectum Σ. With this value for R, we prescribe Γ_R, the boundary curve of Σ, which is a data in the problem.
3. Compute the area A_0 using (26).
4. Calculate the parameters c and m of $C_{c,m}$ in order that its area is A_0 and its boundary is Γ_R. Here we use (29) and (30).
5. Using (27) and (31), compute the values of the heights h_T and h_c of the centers of gravity of Σ and $C_{c,m}$, respectively.

Since the solutions of (21) (or (4)) are invariant by dilations, we will assume that the value of the lowest point of the rotational tectum is $u_0 = 1$, which will be used as initial condition (22). In Table 1, we show the comparison of the value h_T and h_c for values R between $R = 2$ and $R = 20$. In order to obtain a relation of h_T and h_c, we show the percentage between the difference $h_c - h_T$ with the height $u(R) - u(0)$ of the rotational tectum.

Table 1 Comparison of h_T, h_c, and h_p when the lowest height of Σ is $u_0 = 1$

R	$u(R)$	A_0	h_T	h_c	$\% \frac{hc-hs}{u(R)-u_0}$	h_p	$\% \frac{hc-hp}{(u(R)-u_0)}$
2	1.8854	14.63	1.4773	1.4782	0.0969	1.4778	0.0482
4	3.7295	66.22	2.5786	2.58811	0.3488	2.5843	0.2119
6	5.7681	155.99	3.8651	3.8887	0.4942	3.8805	0.3232
8	7.8292	282.30	5.1997	5.2377	0.5573	5.2253	0.3753
10	9.8837	444.43	6.5477	6.5994	0.5819	6.5830	0.3975
12	11.9278	642.08	7.8989	7.9632	0.5891	7.9430	0.4046
14	13.9628	875.13	9.2498	9.3261	0.5887	9.3023	0.4057
16	15.9903	1143.55	10.5993	10.6870	0.5849	10.6598	0.4033
18	18.0120	1447.30	11.9471	12.0458	0.5798	12.0152	0.4002
20	20.0290	1786.39	13.2932	13.4025	0.5744	13.3686	0.3962

Table 2 Comparison of w_c, w_p, and the lowest height $u_0 = 1$ of Σ

R	$u(R)$	A_0	w_c	$\% \frac{w_c-u_0}{u(R)-u_0}$	w_p	$\% \frac{w_p-u_0}{u(R)-u_0}$
2	1.8854	14.63	1.0452	5.1079	1.0318	3.5900
4	3.7295	66.22	1.3236	11.8550	1.2565	9.3987
6	5.7681	155.99	1.7781	16.3190	1.6501	13.6348
8	7.8292	282.30	2.3128	19.2236	2.1270	16.5027
10	9.8837	444.43	2.8847	21.2156	2.6442	18.5085
12	11.9278	642.08	3.4751	22.6495	3.1822	19.9690
14	13.9628	875.13	4.0751	23.7228	3.7313	21.0701
16	15.9903	1143.55	4.6803	24.5515	4.2866	21.9246
18	18.0120	1447.30	5.2883	25.2075	4.8453	22.6035
20	20.0290	1786.39	5.8976	25.7377	5.4059	23.1536

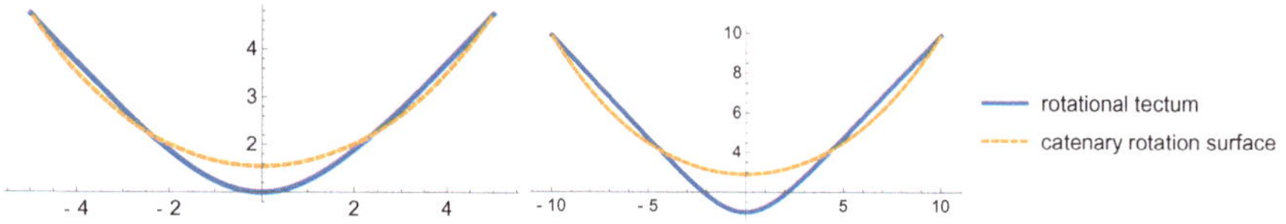

Fig. 6 Comparison between rotational tectums (thick) and catenary rotation surfaces (dashed). Left: $R = 5$ where $h_T = 3.2115$, $h_c = 3.2278$ and $w_c = 1.5364$. Right: $R = 10$, $h_T = 6.5477$ and $h_c = 6.5994$. Here $w_c = 2.8847$

Another useful information is the height of the rotational tectums and of the catenary rotation surfaces which are given by the lowest points denoted by w_T and w_c, respectively. Here $w_T = 1$ in all cases because $u_0 = 1$ in (22). From the architectural viewpoint, this height is that of the dome once reversed the position of the surface. The heights are $u(R) - w_T$ and $u(R) - w_c$, respectively. For $C_{c,m}$, the value of w_c is $1/c + m$. We show again this deviation in relation to the height of the rotational tectum is $(w_c - w_T)/(u(R) - 1)$. This is done in Table 2. In Fig. 6, we show some examples of rotational tectums and catenary rotation surfaces with the same initial data.

In the case of paraboloids, the methodology is the same. Consider the parabola

$$p_{c,m}(x) = cx^2 + m, \quad x \in [0, R],$$

where $c, m \in \mathbb{R}$, $c > 0$, and situated in the coordinate xz-plane. Let $\mathcal{P}_{c,m}$ be the paraboloid obtained by rotating $p_{c,m}$ about the z-axis. Let h_p be the height of the center of gravity. Since the boundary is Γ_R, we have the condition

$$cR^2 + m = u(R).$$

The area of $\mathcal{P}_{c,m}$ is

$$A_0 = 2\pi \int_0^R x\sqrt{1 + p'_{c,m}(x)^2}\, dx = \frac{\pi\left(\left(4c^2R^2 + 1\right)^{3/2} - 1\right)}{6c^2}$$

and the height h_p of its center of gravity is

$$\begin{aligned} h_p &= \frac{2\pi}{A_0}\int_0^R x p_{c,m}(x)\sqrt{1 + p'_{c,m}(x)^2}\, dx \\ &= \frac{\pi\left(\left(6c^2R^2 - 1\right)\left(4c^2R^2 + 1\right)^{3/2} + 1\right)}{60c^3 A_0} + m. \end{aligned}$$

As it is expectable, all expressions for the paraboloid are polynomial in the variables c and R which, from the computational and architectural viewpoint, is much manageable. We show the values of h_p in Table 1 and the values w_p of the lowest point of the paraboloid in Table 2. The value w_p coincides with the parameter m. In Fig. 7 we show some pictures of rotational tectums and paraboloids with the same boundary and area.

We now summarize the conclusions [43].

1. As it is expectable, the heights h_c and h_p are higher than h_T. However the differences $h_c - h_T$ and $h_p - h_T$ are very small. In terms of percentages in relation with the total height of the rotational tectum Σ, the deviation is less than

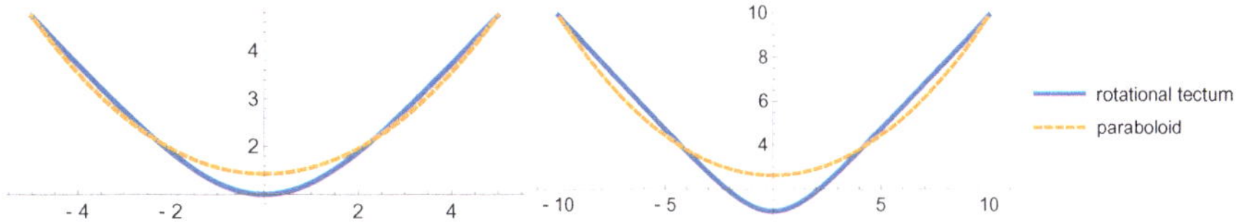

Fig. 7 Comparison between rotational tectums (thick) and paraboloids (dashed). Left: $R = 5$ where $h_T = 3.2115$, $h_p = 3.2219$ and $w_p = 1.4387$. Right: $R = 10$, $h_T = 6.5477$ and $h_p = 6.5994$. Here $w_p = 2.6442$

0.60% for all values of R in the first case and less than 0.41% in the case of paraboloids.

2. Both deviation percentages attain a maximum at $R \approx 12$ and next decreases.
3. The approximation of the paraboloid to the rotational tectum is better than the catenary rotation surface because $h_p < h_c$ for all values of R. This is surprising and contrasts with the reverse property between the centers of gravity of the catenary and the parabola.
4. We have $w_T < w_c$ and $w_T < w_p$. This is expectable although there is no a priori relation between the centers of gravity and the lowest points. Moreover, $w_p < w_c$, showing again that paraboloids adjust better to the rotational tectums than the catenary rotation surfaces.
5. The deviations $w_c - w_T$ and $w_p - w_T$ increase as $R \to \infty$ and are significant. This contrasts with the differences $h_c - h_T$ and $h_p - h_T$. Thus the top of the (inverted) catenary rotation surface or the paraboloid are clearly below that of the rotational tectum.

As far as the author knows, catenary rotation surfaces and paraboloids have not been tested in relation with the centers of gravity nor in comparison with the ideal model of the rotational tectum. It is known that the paraboloid is considered as an example of a quadric employed in architecture, but its usefulness, other than being more tractable in the computational processing of designs in construction, has not been clearly stated: see, for example, [2, 7, 18, 31, 32, 56] where topics concerning to the stress, equilibrium conditions, elasticity, and stress-strain state have been investigated. From our conclusions, we give the following corollary in plain language and easy to understand.

Corollary 2 *Catenary rotation surfaces and paraboloids are surfaces that "adjust extremely well" to the ideal model of a rotational dome, being the paraboloid better than the catenary rotation surface.*

After our results, we can put again in context the work of Gaudí. In the construction of domes, Gaudí used the funicular models explained in Sect. 1 without following mathematical model. Frei Otto reproduced these models in the Institute for Lightweight Structures at the University of Stuttgart [49]. This is the case of the dome of the Palau Güell's Central Hall whose design brings a debate and discussion between architects and engineers as to whether it is actually a paraboloid or not (see [10] and references therein). As a conclusion of our results, we provide mathematical arguments proving that paraboloids can be employed as good approximations of the ideal model of the rotational tectum.

6 Stability Results of Plateau-Rayleigh Type

We study the stability of cylindrical pieces of singular minimal surfaces. We are motivated by the phenomenon of the Plateau-Rayleigh instability of surfaces with constant mean curvature that proves that the length L of stable circular cylinders of radius r must be less than $2\pi r$ [51, 54]. This shows intuitively the known phenomenon that a long circular column of water breaks in droplets after some distance from the source. A singular minimal surface is said to be *stable* if the second derivative of the energy functional (2) is non-negative. In terms of the density e^{ψ} of Sect. 3, Σ is stable if $\mathcal{A}''_{\psi}(0) \geq 0$ for any compactly supported normal variation of Σ. We point out that the stability has received little attention in the literature [13]. The expression of $\mathcal{A}''_{\psi}(0)$ is

$$\mathcal{A}''_{\psi}(0) = -\int_{\Sigma} u\left(\Delta u + \langle \nabla\psi, \nabla u\rangle + (|A|^2 - \overline{\nabla}^2\psi(N, N))u\right) d\mathcal{A}_{\psi},$$

where $u \in C_0^{\infty}(\Sigma)$. Here $|A|^2 = 4H^2 - 2K$ is the norm of the second fundamental form A, ∇ and Δ are the gradient and the Laplacian computed on Σ with the induced Euclidean metric, and $\overline{\nabla}^2$ is the Hessian operator in $\mathbb{R}^3$. The term

$$\Delta_{\psi} u = \operatorname{div}_{\psi} \nabla u = \Delta u + \langle \nabla u, \nabla\psi\rangle$$

is called the ψ*-Laplacian* of a function u and the Jacobi operator is the parenthesis in $\mathcal{A}''_{\psi}(0)$, namely,

$$L[u] = \Delta_{\psi} u + (|A|^2 - \overline{\nabla}^2\psi(N, N))u,$$

acting on the space $C_0^{\infty}(\Sigma)$. Both Δ_{ψ} and L are not self-adjoint with respect to the L^2-inner product, but they are self-adjoint with respect to the weighted inner product $\int_{\Sigma} uv\, d\mathcal{A}_{\psi}$. This allows to define the quadratic form

$$Q[u] = \mathcal{A}''_{\psi}(0) = -\int_{\Sigma} u \cdot L[u]\, d\mathcal{A}_{\psi}, \quad u \in C_0^{\infty}(\Sigma). \tag{32}$$

Coming back to the choice $\psi = \log(z)$, and since

$$\overline{\nabla}^2\psi(N, N) = -\frac{N_3^2}{z^2},$$

we obtain

$$L[u] = \Delta u + \frac{1}{z}\langle \nabla u, e_3\rangle + \left(|A|^2 + \frac{N_3^2}{z^2}\right)u.$$

Once we have the expression of $\mathcal{A}''_{\psi}(0)$, we focus on the case that Σ is a bounded piece of a symmetric cylindrical singular minimal surface. Let

$$\Sigma(a, L) = \{X(s, t) : -a \leq s \leq a,\ 0 \leq t \leq L\}, \quad X(s, t) = (s, t, h(s)),$$

where $h = h(s)$ is the catenary (Theorem 1). After a horizontal translation and a dilation from the origin, we assume that $h(0) = 1$ and $h'(0) = 0$, that is, $h(s) = \cosh(s)$. We call a the *amplitude* and L the *length* of $\Sigma(a, L)$. Motivated by the Plateau-Rayleigh instability in the classical setting, we ask if for a given amplitude $a > 0$, there is a length $L_0 > 0$ such that the surface $\Sigma(a, L)$ is unstable if $L > L_0$. Since we only need to find unstability of the surface, we must find a test function $u \in C_0^\infty(\Sigma(a, L))$ such that, once placed in (32), we have $Q[u] < 0$. The idea is to choose functions u by separation of variables. Similar techniques have been used in the context of surfaces with constant mean curvature [34, 35, 44].

We compute all terms of $L[u]$. Since the coefficients of the first fundamental form are $E = 1 + h'^2 = h^2$, $G = 1$ and $F = 0$, the Laplacian Δ of a function u is

$$\Delta u = \frac{1}{E} u_{ss} + \frac{1}{\sqrt{E}} \left(\frac{1}{\sqrt{E}} \right)_s u_s + u_{tt}.$$

For the computation of $\langle \nabla u, \nabla \psi \rangle$, notice that $\psi = \log h$ does not depend on the variable t, hence

$$\langle \nabla u, \nabla \psi \rangle = \frac{1}{E} \langle \nabla u, \partial_s X \rangle \langle \nabla \psi, \partial_s X \rangle = \frac{\psi_s}{E} u_s = \frac{h'}{h^3} u_s.$$

Since the Gaussian curvature on a cylindrical surface is 0, and by (4),

$$|A|^2 = 4H^2 = (2H)^2 = \frac{N_3^2}{z^2} = \frac{1}{h^4}.$$

Definitively, the expression of the Jacobi operator is

$$L[u] = \frac{1}{E} u_{ss} + \left(\frac{1}{\sqrt{E}} \left(\frac{1}{\sqrt{E}} \right)_s + \frac{h'}{h^3} \right) u_s + u_{tt} + \frac{2}{h^4} u,$$

and using that $h(s) = \cosh(s)$,

$$L[u] = \frac{1}{h^2} u_{ss} + u_{tt} + \frac{2}{h^4} u. \tag{33}$$

Consider now separation of variables for the function u, that is, $u(s, t) = f(s)g(t)$ for functions f and g, $s \in (-a, a)$, $t \in [0, L]$. We can continue with the computation of (33), obtaining

$$L[u] = \frac{1}{h^2} f'' g + f g'' + \frac{2}{h^4} f g.$$

On the other hand, the area element is

$$d\mathcal{A}_\psi = e^\psi d\mathcal{A} = h^2\, ds dt.$$

Since $u(s,t) = f(s)g(t)$ vanishes on the boundary of $[-a,a] \times [0,L]$, then $f(\pm a) = 0$ and $g(0) = g(L) = 0$. For the function g, we choose $g(t) = \sin(\pi t/L)$: notice that $g'' = -\frac{\pi^2}{L^2}g$. Integrating by parts (32),

$$\begin{aligned} Q[u] &= -\int_0^L g(t)^2\,dt \int_{-a}^{a} \left\{ \frac{ff''}{h^2} + \left(\frac{2}{h^4} - \frac{\pi^2}{L^2}\right) f^2 \right\} h^2\,ds \\ &= -\int_0^L g(t)^2\,dt \int_{-a}^{a} \left\{ \left(\frac{2}{h^2} - \frac{\pi^2}{L^2}h^2\right) f^2 - f'^2 \right\} ds. \end{aligned}$$

If we are looking for lengths L where the surface is unstable, we need that

$$I(a,L) = \int_{-a}^{a} \left\{ \left(\frac{2}{h^2} - \frac{\pi^2}{L^2}h^2\right) f^2 - f'^2 \right\} ds \tag{34}$$

is positive. Because we are imposing the condition $f(\pm a) = 0$, consider

$$f(s) = h(s) - h(a).$$

Now (34) is

$$I(a,L) = \int_{-a}^{a} \left\{ \left(\frac{2}{\cosh(s)^2} - \frac{\pi^2\cosh(s)^2}{L^2}\right)(\cosh(s) - \cosh(a))^2 - \sinh(s)^2 \right\} ds. \tag{35}$$

We are in conditions to prove the following Plateau-Rayleigh instability criterion [42].

Theorem 7 *There is a value $a_0 > 0$, $a_0 \approx 1.2391$, such that for all $a > a_0$, there is a critical length $L_0 > 0$, depending only on the amplitude a, such that the singular minimal surfaces $\Sigma(a,L)$ are unstable for all $L > L_0$.*

Proof The integral (35) can be solved explicitly, obtaining

$$I(a,L) = I_1(a) + I_2(a,L),$$

where

$$\begin{aligned} I_1(a) &= 5a + \cosh(a)\left(3\sinh(a) - 16\tan^{-1}\left(\tanh\left(\frac{a}{2}\right)\right)\right), \\ I_2(a,L) &= -\frac{\pi^2}{48L^2}(60a - 44\sinh(2a) + \sinh(4a) + 24\,a\cosh(2a)). \end{aligned}$$

The function $I_1(a)$ is plotted in Fig. 8, left. In particular, there is a unique number a_0 such that $I_1(a) < 0$ if $a < a_0$, $I_1(a_0) = 0$, and $I_1(a) > 0$ if $a > a_0$. The value a_0 can be computed with *Mathematica*, obtaining $a_0 \approx 1.2391$. On the other hand, the

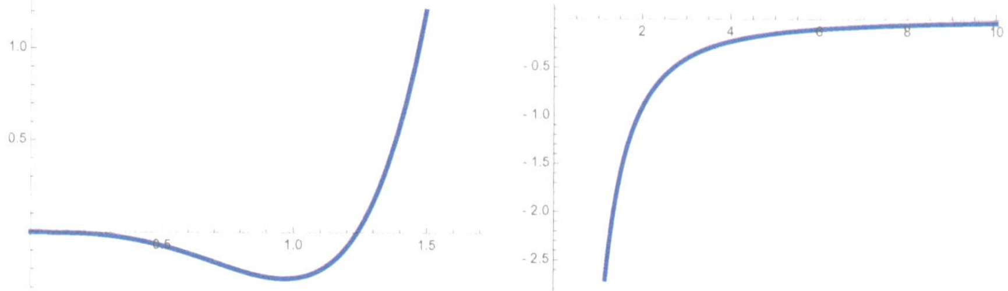

Fig. 8 Left: the function $I_1(a)$. Right: the function $L \mapsto I_2(a, L)$ (here $a = 1$)

parenthesis in the expression of $I_2(a, L)$ is always positive. If we see $I_2(a, L)$ as a function on the variable L fixing a, then

$$\lim_{L \to 0^+} I_2(a, L) = -\infty, \qquad \lim_{L \to \infty} I_2(a, L) = 0 \tag{36}$$

and the function $L \mapsto I_2(a, L)$ is strictly increasing on L. See Fig. 8, right.

It is clear that if $a < a_0$, then $I(a, L) < 0$ for all values of L because is the sum of two negative functions. Let now $a > a_0$, where we know that $I_1(a) > 0$. Letting $L \to \infty$, from (36) and the monotonicity of $L \mapsto I_2(a, L)$, we deduce that there is a unique value L_0 depending on a such that $I(a, L_0) = 0$ and $I(a, L) > 0$ for all $L > L_0$. This proves the result. The value L_0 depends on the amplitude a and it is obtained by solving the equation $I_1(a) + I_2(a, L) = 0$:

$$L_0 = \frac{\pi\sqrt{60a - 44\sinh(2a) + \sinh(4a) + 24a\cosh(2a)}}{4\sqrt{15a + 9\cosh(a)\left(\sinh(a) - 6\tan^{-1}\left(\tanh\left(\frac{a}{2}\right)\right)\right)}}. \tag{37}$$

□

Motivated by the scenario of circular cylinders where, fixed radius r, the instability appears when the length L satisfies $L > 2\pi r$, one expects that increasing the amplitude a, one needs large lengths L to ensure the instability. However, the function $L_0 = L_0(a)$ in (37) is not increasing: there is a value a_1, with $a_1 \approx 1.7964$, such that $L_0(a)$ is decreasing in the interval $[a_0, a_1]$, and then increases for $a > a_1$; see Fig. 9. It is after the amplitude a_1, where if a increases, the critical value L_0 increases as well.

Again, we finish this section with an observation about the works of Gaudí. Recall that his corridors were constructed by repeating inverted catenaries. Theorem 7 establishes that long corridors are unstable. The concepts of mathematical stability and architectural stability are different. In architecture, it is related with the capacity to preserve the design and geometry of the structure under compression or loads. Instability appears after long deformations due to the material used in the construction or a failure in its design. However, Theorem 7 reflects the intuitive idea that corridors with large lengths should collapse with small perturbation.

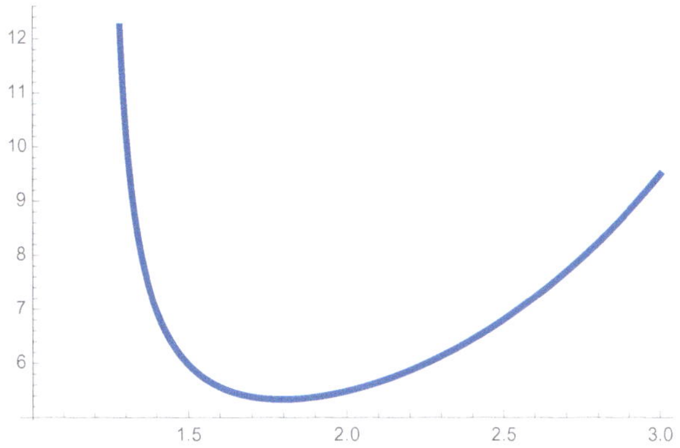

Fig. 9 The function $L_0 = L_0(a)$ given in (37)

7 Compact Singular Minimal Surfaces with Boundary

Coming back to the initial problem of this paper, it is difficult to find general results of existence of singular minimal surfaces for prescribed boundary curve and area. In a similar context, the solvability of the Dirichlet problem associated with (3) is equally difficult, and only partial answers have been obtained [16, 38, 39]. A first question is about the existence of closed singular minimal surfaces, that is, compact surfaces without boundary. Notice that there are not closed examples in the class of surfaces of revolution [15, 28, 37]. In fact, the result holds in general.

Proposition 4 ([37]) *There do not exist closed singular minimal surfaces.*

Proof Let Σ be a closed singular minimal surface and take the height function $h(p) = \langle p, \mathbf{v}\rangle$, $p \in \Sigma$. Then h satisfies

$$\Delta h = 2H\langle N, \mathbf{v}\rangle,$$

where Δ is the Laplace operator on Σ. This equation holds for any surface without any condition on H. Using now (4),

$$\Delta h = \frac{\langle N, \mathbf{v}\rangle^2}{\langle p, \mathbf{v}\rangle}. \tag{38}$$

Integrating on Σ, using the divergence theorem and the fact that Σ is closed, we have

$$0 = \int_\Sigma \frac{\langle N, \mathbf{v}\rangle^2}{\langle p, \mathbf{v}\rangle}\, d\Sigma,$$

obtaining that $\langle N, \mathbf{v}\rangle = 0$ on Σ. This implies that Σ is an open set of a plane, in which it is a contradiction. □

Consequently any compact singular minimal surface has a non-empty boundary, which we denote by $\partial\Sigma$. In order to give the precise terminology, if $\Gamma \subset \mathbb{R}^3_+$ is a closed curve, we say that a compact surface Σ has Γ as its boundary if the restriction of the immersion of Σ to its boundary $\partial\Sigma$ is one-to-one with Γ.

If Γ is a closed curve that spans a surface of minimum area $\bar{A}$, then a necessary condition for the existence of singular minimal surfaces spanning Γ is that the prescribed value for the area A_0 must satisfy $A_0 > \bar{A}$. Thus it is expectable that *the geometry of* Γ *imposes restrictions to the existence of singular minimal surfaces.* For example, given a closed curve Γ invariant by a symmetry $\Phi\colon \mathbb{R}^3 \to \mathbb{R}^3$, it is natural to ask if this symmetry Φ inherits to a singular minimal surface Σ spanning Γ, that is, if $\Phi(\Sigma) = \Sigma$. The typical case is when Γ is a circle where one expects that the surface that spans is a surface of revolution. Along this section, and after a change of coordinates, we will suppose $\mathbf{v} = e_3 = (0, 0, 1)$.

We can follow with the same idea of Proposition 4, obtaining a necessary condition to the existence of singular minimal surfaces.

Proposition 5 *Let* $\Gamma \subset \mathbb{R}^3_+$ *be a closed curve. If*

$$\varpi = \sup\{z(p) : p \in \partial\Sigma\}, \tag{39}$$

then the area $A(\Sigma)$ *of any compact singular minimal surface* Σ *spanning* Γ *satisfies*

$$A(\Sigma) \leq \varpi\, L(\Gamma),$$

where $L(\Gamma)$ *is the length of* Γ.

Proof We now compute Δh^2. The gradient of h is $\nabla h(p) = e_3^T$, where e_3^T is the tangent part on Σ. Then $|\nabla h|^2 = 1 - \langle N, e_3\rangle^2$. From (38),

$$\Delta h^2 = 2h\Delta h + 2|\nabla h|^2 = 2.$$

The divergence theorem implies

$$\int_\Sigma 1\, d\Sigma = A(\Sigma) = \int_{\partial\Sigma} \langle p, e_3\rangle\langle \nu, e_3\rangle ds,$$

where ν is the exterior unit conormal vector along $\partial\Sigma$. By estimating the right-hand side of the above equation,

$$\begin{aligned}\left|\int_{\partial\Sigma} \langle p, e_3\rangle\langle \nu, e_3\rangle ds\right| &\leq \sup\{z(p) : p \in \partial\Sigma\} \int_{\partial\Sigma} |\langle \nu, e_3\rangle|\, ds\\ &\leq \varpi \int_{\partial\Sigma} 1\, ds = \varpi\, L(\Gamma),\end{aligned}$$

obtaining the result. □

Corollary 3 *Let Γ be a Jordan curve contained in the horizontal plane of equation $z = c > 0$ and let Ω be the domain bounded by Γ. Then a necessary condition for the existence of a compact singular minimal surface with boundary Γ is*

$$A(\Omega) \leq c\, L(\Gamma).$$

In particular, if Ω is a round disc of radius r, then $r \leq 2c$.

Notice that the value ϖ in (39) measures the height of the highest point of the boundary curve Γ which is a prescribed data. Proposition 5 and its corollary assert that once we prescribe the boundary curve Γ, the value A_0 of the area is not arbitrary as initial data and depends on the geometry of Γ.

We finish this section answering to the question of how the symmetries of the boundary curve are inherited to the whole surface that spans. This will be carried out thanks to the maximum principle, following the same techniques of the theory of the surfaces with constant mean curvature [36]. By ellipticity, Eq. (3) satisfies a comparison principle as well as a maximum principle [23, Th. 10.1]. This is also a consequence that singular minimal surfaces are minimal surfaces in a space with a conformal metric to the Euclidean or surfaces with zero weighted mean curvature (Theorem 3). The maximum principle establishes that if a singular minimal surface lies in one side of other one around a common tangent point p, then both surfaces coincide in a neighborhood of p. This also holds if p is a boundary point and the tangent lines at p coincide. A first consequence is the following result.

Proposition 6 *If Σ is a compact singular minimal surface, then the height function with respect to a vertical plane does not attain a local maximum at some interior point of Σ.*

Proof Let $\mathbf{w}$ be a horizontal direction (orthogonal to e_3). If $p_0 \in \Sigma$ is an interior point where the function $p \mapsto \langle p, \mathbf{w}\rangle$ attains a local maximum, the tangent plane P at p_0 is also a singular minimal surface, but Σ lies in one side of P which is a contradiction. □

This result implies that a compact singular minimal surface is contained in the vertical solid cylinder $c(\overline{\Omega}) \times \mathbb{R}$, where $c(\overline{\Omega})$ is the convex hull of the planar domain determined by the orthogonal projection of Γ on the plane $z = 0$. With a different argument, we have as follows.

Proposition 7 *Let Σ be a compact singular minimal surface with boundary Γ. Then the height function of Σ attains its maximum at a boundary point. In particular, if Γ is contained in a horizontal plane Π, then Σ lies below Π.*

Proof It is enough to observe that thanks to (38), the function $h(p) = \langle p, e_3\rangle$ satisfies $\Delta h > 0$. □

Finally we answer to the question if the symmetries of Γ are inherited to the whole surface [40].

Theorem 8 *If Σ is an embedded compact singular minimal surface spanning a horizontal circle, then Σ is a rotational tectum.*

Proof The proof follows the standard arguments of the reflection principle of Alexandrov [1]. We only indicate why this technique works in our situation. First, by the above proposition, the surface Σ lies below the boundary plane Π, $\Gamma \subset \Pi$. Next, and using that Σ is embedded, we have determined a bounded 3-domain by $\Sigma \cup \Omega$, where $\Omega \subset \Pi$ is the disc bounded by the circle Γ. Now the Alexandrov method works thanks to the maximum principle, proving that for each horizontal direction $\mathbf{w} \in \mathbb{R}^3$, there exists a plane $P_{\mathbf{w}}$ orthogonal to $\mathbf{w}$ such that Σ is invariant by symmetries about $P_{\mathbf{w}}$. By the compactness of Σ, we finish showing that Σ is axially symmetric about a vertical line. □

Acknowledgments This research has been partially supported by MINECO/MICINN/FEDER grant no. PID2020-117868GB- I00 and by the Maria de Maeztu Excellence Unit IMAG, reference CEX2020-001105- M, funded by MCINN/AEI/10.13039/501100011033/ CEX2020-001105-M.

References

1. Alexandrov, A.D.: Uniqueness theorems for surfaces in the large I. Am. Math. Soc. Transl. **21**, 341–354 (1962)
2. Banerjee, S.P.: Analysis of elliptic-paraboloid shell. J. Struct. Div. **94**, 2213–2230 (1968)
3. Bayle, V.: Propriétés de concavité du profil isopérimétrique et applications. Ph.D. Thesis, Institut Joseph Fourier, Grenoble (2004)
4. Beltrami, E.: Sull equilibrio delle superficie flessibili ed inestensibili. Memorie della Academia delle Scienze dell Istituto di Bologna. Series 4, **3**, 217–265 (1882)
5. Bemelmans, J., Dierkes, U.: On a singular variational integral with linear growth, I: existence and regularity of minimizers. Arch. Ration. Mech. Anal. **100**, 83–103 (1987)
6. Besse, A.L.: Einstein Manifolds. Springer, Berlin (1987)
7. Blaauwendraad, J., Hoefakker, J.H.: Hyperbolic and elliptic-paraboloid roofs. In: Structural Shell Analysis. Solid Mechanics and its Applications, vol. 200. Springer, Dordrecht (2014)
8. Böhme, R., Hildebrandt, S., Taush, E.: The two-dimensional analogue of the catenary. Pac. J. Math. **88**, 247–278 (1980)
9. Castro, K., Rosales, C.: Free boundary stable hypersurfaces in manifolds with density and rigidity results. J. Geom. Phys. **79**, 14–28 (2014)
10. Cortés, A:, Samper, A:, Herrera, B., González, G.: Revisión de la tipología geométrica de la cúpula del Palau Güell. Informes de la Construcción, vol. 72, No. 558 (2020)
11. Darboux, G.: Leçons sur la Théorie Générale des Surfaces et ses Applications Géométriques du Calcul Infinitésimal, vol. 1–4. Chelsea Publ. Co, reprint (1972)
12. Dierkes, U.: A geometric maximum principle, Plateau's problem for surfaces of prescribed mean curvature, and the two-dimensional analogue of the catenary. In: Hildebrandt, S., Leis, R. (eds.) Partial Differential Equations and Calculus of Variations. Springer Lecture Notes in Mathematics, vol. 1357, pp. 116–141 (1988)
13. Dierkes, U.: On the non-existence of energy stable minimal cones. Annales de l'Institut Henri Poincaré. Anal. Non linéaire **7**, 589–601 (1990)
14. Dierkes, U.: Singular minimal surfaces. In: Hildebrandt, S., Karcher, H. (eds.) Geometric Analysis and Nonlinear Partial Differential Equations, pp. 177–193. Springer, Berlin (2003)

15. Dierkes, U., Groh, N.: Symmetric solutions of the singular minimal surface equation. Ann. Global Anal. Geom. **60**, 431–453 (2021)
16. Dierkes, U., Huisken, G.: The n-dimensional analogue of the catenary: existence and nonexistence. Pac. J. Math. 141, 47–54 (1990)
17. Dierkes, U., López, R.: Cylindrical singular minimal surfaces, Atti Accad. Naz. Lincei Cl. Sci. Fis. Mat. Natur. to appear
18. Du, W., Zhu, Z., Zhu, L.: Comparison of four structural schemes for the roof design of Qinyang Stadium. Appl. Mech. Mater. **438/439**, 819–823 (2013)
19. Dunn, W.: The principles of dome construction: I and II. J. R. Inst. Br. Archit. **23**, 401–412 (1908)
20. Enache, C., López, R.: Minimum principles and a priori estimates for some translating soliton type problems. Nonlinear Anal. **187**, 352–364 (2019)
21. Fouladgar, K., Simon, L.: The symmetric minimal surface equation. Indiana Univ. Math. J. **69**, 331–366 (2020)
22. Germain, S.: Recherches sur la Théorie des Surfaces Elastiques. Veuve Courcier, Paris (1821)
23. Gilbarg, D., Trudinger, N.: Elliptic Partial Differential Equations of Second Order, Reprint of the 1998 edition. Classics in Mathematics. Springer, Berlin (2001)
24. Gresy, C. de: Considération sur l'équilibre des surfaces flexibles et inextensibles. Mem. Reale Accad. Sci. Torino **21**, 259–294 (1818)
25. Gromov, M.: Isoperimetry of waists and concentration of maps. Geom. Func. Anal **13**, 178–215 (2003)
26. Heyman, J.: Equilibrium of Shell Structures. Oxford University Press, Oxford (1977)
27. Jellet, J.H.: On the properties of inextensible surfaces. Trans. R. Irish Acad. **22**, 343–378 (1853)
28. Keiper, J.B.: The axially symmetric n-tectum, preprint, Toledo University (1980)
29. Kim, D.-S., Kim, Y.-H, Yoon, D.W.: Some characterizations of catenary rotation surfaces. Kyungpook Math. J. **57**, 667–676 (2017)
30. Kim, D.-S., Kim, Y.-H, Yoon, D.W.: Various centroids and some characterizations of catenary rotation hypersurfaces. Turkish J. Math. **42**, 360–372 (2018)
31. Kollár, L., Tarján, G.: Mechanics of Civil Engineering Structures. Elsevier, Duxford (2021)
32. Krivoshapko, S.N., Gbaguidi-Aisse, G.L.: Geometry, static, vibration and buckling analysis and applications to thin elliptic paraboloid shells. Open Constr. Build. Technol. J. **10**, 576–602 (2016)
33. Lagrange, J.L.: Mécanique Analytique. Tome XI. 4th ed. (1888). Gauthier-Villars, Paris (1788)
34. López, R.: A criterion on instability of cylindrical rotating surfaces. Archiv Math. (Basel) **94**, 91–99 (2010)
35. López, R.: Bifurcation of cylinders for wetting and dewetting models with striped geometry. SIAM J. Math. Anal. **44**, 946–965 (2012)
36. López, R.: Constant Mean Curvature Surfaces with Boundary. Springer Monographs in Mathematics. Springer, Heidelberg (2013)
37. López, R.: Invariant singular minimal surfaces. Ann. Global Anal. Geom. **53**, 521–541 (2018)
38. López, R.: The Dirichlet problem for the α-singular minimal surface equation. Arch. Math. (Basel) **112**, 213–222 (2019)
39. López, R.: Uniqueness of critical points and maximum principles of the singular minimal surface equation. J. Differ. Equ. **266**, 3927–3941 (2019)
40. López, R.: Compact singular minimal surfaces with boundary. Am. J. Math. **142**, 1771–1795 (2020)
41. López, R.: What is the Shape of a Cupola? Am. Math. Monthly **130**, 222–238 (2023).
42. López, R.: Plateau-Rayleigh instability of singular minimal surfaces, Commun. Pure Appl. Anal. **21**, 2981–2997 (2022)

43. López, R.: A dome subjected to compression forces: a comparison study between the mathematical model, the catenary rotation surface and the paraboloid. Chaos Solitons Fractals **161**, 112350 (2022)
44. McCuan, J.: Extremities of stability for pendant drops. In: Ghomi, M., et al. (eds.) Geometric Analysis, Mathematical Relativity, and Nonlinear Partial Differential Equations, Contemporary Mathematics, vol. 599, pp. 157–173. American Mathematical Society, Providence (2013)
45. Nitsche, J.C.C.: A nonexistence theorem for the two-dimensional analogue of the catenary. Analysis **6**, 143–156 (1986)
46. Oppenheim, I.J., Gunaratnam, D.J., Allen, R.H.: Limit state analysis of masonry domes. J. Struct. Eng. **115**, 868–882 (1989)
47. Osserman, R.: Mathematics of the gateway arch. Notices AMS. **57**, 220–229 (2010)
48. Otto, F.: Zugbeanspruchte Konstruktionen. Bd. I, II. Berlin, Frankfurt, Ullstein, Wien (1962/1966)
49. Frei Otto: Spanning the Future: (2020). Documentary film. Dir. Joshua Hassel. https://www.youtube.com/watch?v=P5hKnOyg43k
50. Paradiso, M., Rapallini, M., Tempesta, G.: Masonry domes. Comparison between some solutions under no-tension hypothesis. In: Proceedings of the First International Congress on Construction History, pp. 1571–1581. Instituto Juan de Herrera, Escuela Técnica Superior de Arquitectura, Madrid (2003)
51. Plateau, J.A.F.: Statique Expérimentale et Théorique Des Liquides Soumis Aux Seules Forces Moléculaires, vol. 2. Gauthier-Villars
52. Poisson, S.D.: Mémoire sur les surfaces élastiques. Mémoires de l'Institut de France, **1814/1816**(9), 167–226 (1812)
53. Pottmann, H., Asperl, A., Hofer, M., Kilian, A.: Architectural Geometry. Bentley Institute Press, Exton (2007)
54. Rayleigh, J.W.S.: On the instability of jets. Proc. London Math. Soc. **X**, 4–13 (1879)
55. Schodek, D.L., Bechthold, M.: Structures. Pearson Higher Ed. (2013)
56. Scrivener, J.C.: The analysis of elliptic and hyperbolic paraboloid shell roofs. N. Z. Eng. **19**, 303–310 (1964)
57. Todhunter, I., Pearson, K.: A History of Elasticity and Strength of Materials, vol. 1. Cambridge University Press, Cambridge (1986)
58. Volterra, V.: Sulla deformazione delle superficie flessibili ed inestensibili. Atti della R. Accad. Dei Lincei, Rendiconti, Series 4, vol. 1, pp. 274–278 (1884/1885)
59. Wikipedia contributors. The Free Encyclopedia. Colegio Teresiano de Barcelona. https://es.wikipedia.org/w/index.php?title=Colegio_Teresiano_de_Barcelona&oldid=134544852
60. Wikipedia contributors. The Free Encyclopedia. Maqueta funicular de la cripta de la colonia Güell. https://es.wikipedia.org/w/index.php?title=Maqueta_funicular_de_la_cripta_de_la_colonia_G%C3%BCell&oldid=134388617
61. Wolfram Research, Inc.: Mathematica, Version 10.0.0, Champaign (2014)

Geometry of $[\varphi, \mathbf{e}_3]$-Minimal Surfaces in $\mathbb{R}^3$

Antonio Martínez and A. L. Martínez-Triviño

Abstract In this survey we report a general and systematic approach to study $[\varphi, \mathbf{e}_3]$-minimal surfaces in $\mathbb{R}^3$ from a geometric viewpoint and show some fundamental results obtained in the recent development of this theory.

Keywords $[\varphi, \mathbf{e}_3]$-minimal surface · Weighted area functional · Weierstrass representation · Cauchy problem · Calabi's correspondence · spacelike $[\varphi, \mathbf{e}_3]$-maximal surface

1 Introduction

In this survey, we would like to report some recent results about critical points of the weighted area functional

$$\mathcal{A}^{\varphi}(\Sigma) = \int_{\Sigma} e^{\varphi}\, d\Sigma, \tag{1}$$

on surfaces Σ in a domain $\mathfrak{D}^3 \subset \mathbb{R}^3$ when φ is the restriction on Σ of a smooth function depending only on the last coordinate of $\mathfrak{D}^3$ and where $d\Sigma$ denotes the volume element induced by the Euclidean metric $\langle \cdot, \cdot \rangle$ in $\mathbb{R}^3$.

The Euler-Lagrange equation of (1) is given in terms of the mean curvature vector $\mathbf{H}$ of Σ as follows:

$$\mathbf{H} = (\overline{\nabla}\varphi)^{\perp} = \dot{\varphi}\, \mathbf{e}_3^{\perp}, \tag{2}$$

where $\perp$ denotes the projection on the normal bundle, $\overline{\nabla}$ stands the usual gradient operator in $\mathbb{R}^3$, and $(\dot{\ })$ denotes derivate with respect to the third coordinate. Ilmanen in [21] proved that (2) means also that Σ is a *minimal* immersion in the so-called

A. Martínez (✉) · A. L. Martínez-Triviño
University of Granada, Granada, Spain
e-mail: amartine@ugr.es; aluismartinez@ugr.es

A. Alarcón et al. (eds.), *New Trends in Geometric Analysis*,
RSME Springer Series 10, https://doi.org/10.1007/978-3-031-39916-9_12

Ilmanen's space, $(\mathfrak{D}^3, g_\varphi)$, that is, $\mathfrak{D}^3$ endowed with the conformally Euclidean changed metric

$$g_\varphi := e^\varphi \langle \cdot, \cdot \rangle. \tag{3}$$

From a physical point of view (see [42, pp. 173–187]), Eq. (2) gives the equilibrium condition of a flexible inextensible surface in the absence of intrinsic forces under the gravitational force field

$$\mathcal{F} := \overline{\nabla} \mathrm{e}^\varphi = (0, 0, \dot{\varphi}\, \mathrm{e}^\varphi).$$

Any surface satisfying (2) will be called $[\varphi, \mathbf{e}_3]$-*minimal* and if Σ is the vertical graph of a function $u : \Omega \subseteq \mathbb{R}^2 \longrightarrow \mathbb{R}$, we also refer to u as $[\varphi, \mathbf{e}_3]$-minimal. Hence, u is $[\varphi, \mathbf{e}_3]$-minimal if and only if it solves the so-called $[\varphi, \mathbf{e}_3]$-minimal equation,

$$(1 + u_x^2)u_{yy} + (1 + u_y^2)u_{xx} - 2u_y u_x u_{xy} = \dot{\varphi}(u) W^2, \quad (x, y) \in \Omega, \tag{4}$$

where $W = \sqrt{1 + u_x^2 + u_y^2}$. If Ω is simply connected then, the Poincare's Lemma gives that Eq. (4) is equivalent to integrability of the following differential system:

$$\phi_{xx} = e^{\varphi(u)} \frac{1 + u_x^2}{W}, \quad \phi_{xy} = e^{\varphi(u)} \frac{u_x u_y}{W}, \quad \phi_{yy} = e^{\varphi(u)} \frac{1 + u_y^2}{W}, \tag{5}$$

for a convex function $\phi : \Omega \to \mathbb{R}$ unique, up to linear polynomials.

This kind of surfaces has been widely studied specially from the viewpoint of calculus of variations. Classical results about the Euler equation and the existence and regularity for the solutions of the Plateau problem for (1) can be found in [4, 15–17, 48]. But contributions from a more geometric viewpoint only have been given for some particular functions φ, namely,

- for translating solitons: if φ is just the height function, $\varphi(z) = z$, that is, surfaces such that

$$t \to \Sigma + t\mathbf{e}_3$$

  is a mean curvature flow, i.e., the normal component of the velocity at each point is equal to the mean curvature at that point. Recent advances in the understanding of their local and global geometry can be found in [8, 18–20, 32, 33, 47, 50]
- for singular α-minimal surfaces: if $\varphi(z) = \alpha \log z$, $z > 0$, α=const. (when $\alpha = -2$, Σ is a minimal surface in the Poincaré upper half hyperbolic space model and when $\alpha = 1$, Σ describes the shape of a "hanging roof," i.e., a heavy surface in a gravitational field that, according to the architect F. Otto [40, p. 290], are of importance for the construction of perfect domes). We refer to [4, 11, 12, 14, 28–31, 34, 39] for some progress in this family.

The aim of this paper is to develop a systematic and general approach to the study of $[\varphi, \mathbf{e}_3]$-minimal surfaces from a geometric point of view. Since this class of surfaces is too large, actually, we could find almost any geometric asymptotic behavior; it will be necessary to give some conditions on the function φ. Here, as a general assumption, we will always consider φ strictly monotone, that is,

$$\varphi :]a, b[\subseteq \mathbb{R} \to \mathbb{R} \text{ is a strictly increasing (or decreasing) function} \tag{6}$$

and $\Sigma \subset \mathfrak{D}^3 = \mathbb{R}^2 \times]a, b[$.

2 The Most Symmetric Examples

The first section consists in the description of the examples of $[\varphi, \mathbf{e}_3]$-minimal surfaces which are invariant by either horizontal translations or vertical rotations.

2.1 *The One-dimensional Variational Problem*

Let us consider the one-dimensional variational integral

$$\mathcal{I}_\varphi(u) = \int_I e^\varphi \sqrt{1 + u'(x)^2} dx, \tag{7}$$

with $u : I \subseteq \mathbb{R} \to]a, b[$ a differentiable function. It is easy to check that extremals for $\mathcal{I}_\varphi$ must be solutions of the following ODE:

$$u''(x) = \dot{\varphi}(u)(1 + u'(x)^2). \tag{8}$$

Looking for complete examples, we will assume that $\varphi :]a, +\infty[\to \mathbb{R}$, $a \in \mathbb{R} \cup \{-\infty\}$ is strictly monotone. Then, taking $z = \varphi(u)$ and $u' = \tan(v)$, we get that (8) is equivalent to

$$\left.\begin{aligned} v' &= \lambda(z), \\ z' &= \lambda(z) \tan(v) \end{aligned}\right\} \tag{9}$$

where λ is the function defined by $\lambda(z) = \dot{\varphi}(\varphi^{-1}(z))$ on $\varphi(]a, +\infty[)$.

Thus, $e^z \cos(v)$ is constant along the solutions of (9) and from the Phase portrait of (9) (see Fig. 1); for any maximal solution u of (8), there exists a unique $x_0 \in I$ such that $v(x_0) = 0$. It is not a restriction to consider $x_0 = 0$ and u satisfying

$$u(0) = u_0, \quad u'(0) = 0. \tag{10}$$

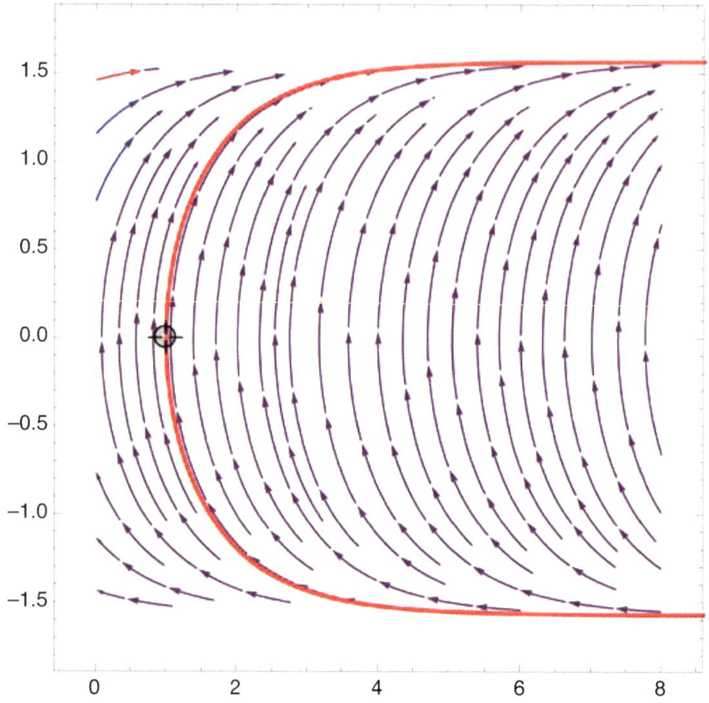

Fig. 1 Phase portrait of (9)

In this case, the solution to (8)–(10) is given by

$$u(x) = (\mathcal{X} \circ \varphi)^{-1}(x), \text{ with } \mathcal{X}(z) = \int_{z_0}^{z} \frac{d\tau}{|h(\tau)|\sqrt{e^{2(\tau - z_0)} - 1}}, \tag{11}$$

where $z_0 = \varphi(u_0)$. Thus, we obtain that u is even, and it is defined in the interval $]-\Lambda_{u_0}, \Lambda_{u_0}[$, where

$$\Lambda_{u_0} = \lim_{u \to +\infty} \int_{\varphi(u_0)}^{\varphi(u)} \frac{d\tau}{|h(\tau)|\sqrt{e^{2(\tau - z_0)} - 1}}. \tag{12}$$

Theorem 2.1 ([34, Theorem 3.2 and Theorem 3.3]) *Let $\varphi :]a, +\infty[\to]b, c[$, $a, b \in \mathbb{R} \cup \{-\infty\}$, $c \in \mathbb{R} \cup \{+\infty\}$ be a strictly increasing diffeomorphism. Then the solution u of* (8)–(10) *is defined in* $]-\Lambda_{u_0}, \Lambda_{u_0}[$, $\Lambda_{u_0} \in \mathbb{R}^+ \cup \{+\infty\}$*; it is convex and symmetric about the y-axis and has a minimum at $x = 0$. Moreover,*

- *if $c < +\infty$, then $\Lambda_{u_0} = +\infty$ and,* $\begin{cases} \lim_{x \to \pm\infty} u(x) = +\infty, \\ \lim_{x \to \pm\infty} u'(x) = \pm\sqrt{e^{2(c - \varphi(u_0))} - 1}. \end{cases}$

- *if* $c = +\infty$, $\lim_{x\to\pm\Lambda_{u_0}} u(x) = +\infty$, $\lim_{x\to\pm\Lambda_{u_0}} u'(x) = \pm\infty$.
  In particular, if $\Lambda_{u_0} < +\infty$, *the graph of* u *is asymptotic to two vertical lines. Moreover,*
- $\Lambda_{u_0} < +\infty$ *if and only if* $e^{-\varphi} \in L^1(]u_0, +\infty[)$, $\left(i.e., \int_{u_0}^{\infty} e^{-\varphi(\tau)} d\tau < \infty\right)$.
- *If* $\Lambda_\tau < +\infty$ *and* $\dot{\varphi}$ *is increasing (respectively, decreasing), then* Λ_τ *is decreasing (respectively, increasing) in* τ.

Theorem 2.2 ([34, Theorem 3.4]) *Let* $\varphi :]a, +\infty[\longrightarrow]b, c[$, $a, b \in \{\mathbb{R}, -\infty\}$, $c \in \{\mathbb{R}, +\infty\}$ *be a strictly decreasing diffeomorphism, then the solution* u *of* (8)–(10) *is defined in* $]-\Lambda_{u_0}, \Lambda_{u_0}[$, $\Lambda_{u_0} \in \mathbb{R}^+ \cup \{+\infty\}$, *it is concave, symmetric about the y-axis and has a maximum at* $x = 0$. *Moreover,*

- *if* $c < +\infty$, *then* $\Lambda_{u_0} < +\infty$ *and,* $\begin{cases} \lim_{x\to\pm\Lambda_{u_0}} u(x) = a, \\ \lim_{x\to\pm\Lambda_{u_0}} u'(x) = \pm\sqrt{e^{2(c-\varphi(u_0))} - 1}. \end{cases}$
- *if* $c = +\infty$, *then* $\Lambda_{u_0} < +\infty$ *if and only if* $\int_a^{u_0} e^{-\varphi(\tau)} d\tau < \infty$, *and,*

$$\lim_{x\to\pm\Lambda_{u_0}} u(x) = a, \qquad \lim_{x\to\pm\Lambda_{u_0}} u'(x) = \pm\infty.$$

Motivated by their physical interpretation, for each solution u of (8)–(10), we will refer $\mathcal{G}^{u_0} := \{(x, y, u(x)) \mid (x, y) \in I \times \mathbb{R}\}$ as a $[\varphi, \mathbf{e}_3]$-*catenary cylinder* surface (see Fig. 2 (left)).

Remark 1 I would like to conclude this section mentioning a special example only for translating solitons. If we rotate a grim reaper cylinder translating soliton an angle $\theta \in]0, \pi/2[$ about the x-axis and dilate with factor $1/\cos\theta$, the resulting surface is again a translating soliton and is called *tilted grim reaper cylinder* (see Fig. 2 (right)).

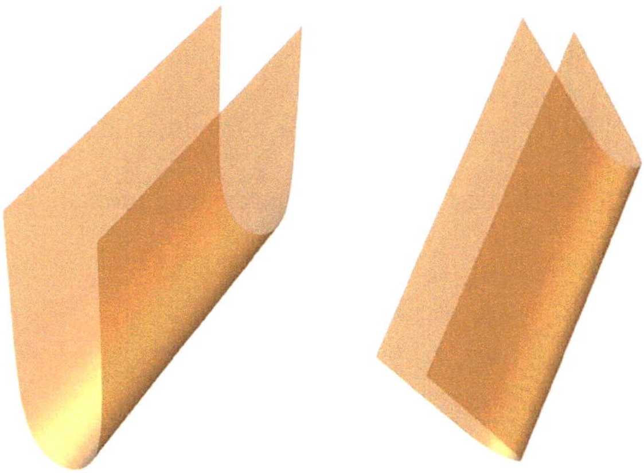

Fig. 2 The grim reaper cylinder and its corresponding tilted grim reaper cylinder, respectively

2.2 Rotationally Symmetric Solutions

In the rotationally symmetric case, Eq. (4) for $[\varphi, \mathbf{e}_3]$-minimal graphs $u = u(r)$ with $r = \sqrt{x^2 + y^2}$ reduces to the following ODE:

$$u''(r) = (1 + u'(r)^2)\left(\dot{\varphi}(u) - \frac{u'(r)}{r}\right), \tag{13}$$

where $(')$ denotes the derivative with respect to r. Notice that (13) is degenerated, and then, the existence and uniqueness of solution around $r = 0$ is not guaranteed by standard theory of ordinary differential equations. Moreover, by applying [46, Theorem 2], solutions of (4) do not have isolated singularities. Consequently, it is not a restriction to look for solutions of (13) with the following initial data:

$$u(0) = u_0 \ , \quad u'(0) = 0. \tag{14}$$

In this case, we can assert (see [34, Proposition 4.1]) that the Initial Value Problem (13)–(14) has a unique solution $u \in C^2([0, R])$ for some $R > 0$, which depends continuously on the initial data.

Now, once the existence of solution is guaranteed, we want to describe $[\varphi, \mathbf{e}_3]$-minimal surfaces that are invariant under the one-parameter group of rotations that fix the $\mathbf{e}_3$ direction. A such surface with generating curve the arc-length parametrized curve

$$\gamma(s) = (x(s), 0, z(s)), \qquad s \in I \subset \mathbb{R}$$

is given by

$$\psi(s, t) = (x(s)\cos(t), x(s)\sin(t), z(s)) \, , \ (s, t) \in I \times \mathbb{R}. \tag{15}$$

The inner normal of ψ writes as

$$N(s, t) = \left(-z'(s)\cos(t), -z'(s)\sin(t), x'(s)\right), \tag{16}$$

and the coefficients of the first and second fundamental form are given by

$$\begin{aligned} &\langle \psi_s, \psi_s \rangle = 1, \quad \langle \psi_s, N_s \rangle = -\kappa, \\ &\langle \psi_t, \psi_t \rangle = x^2, \ \langle \psi_t, N_t \rangle = -x\, z', \\ &\langle \psi_s, \psi_t \rangle = 0, \quad \langle \psi_s, N_t \rangle = 0, \end{aligned} \tag{17}$$

where κ is the curvature of γ and by $'$ we denote derivative respect to s. Consequently, from (2), (15), (16), and (17), the surface ψ is a $[\varphi, \mathbf{e}_3]$-minimal surface if and only if

$$\begin{cases} x' = \cos(\theta) \\ z' = \sin(\theta), \\ \theta' = \dot{\varphi}(z)\cos(\theta) - \frac{\sin(\theta)}{x}, \end{cases} \tag{18}$$

where

$$\theta(s) = \int_0^s \kappa(t)dt.$$

Along this section, we will consider that $\varphi :]a, +\infty[\to \mathbb{R}$ is a strictly increasing and convex function, that is,

$$\dot{\varphi} > 0, \quad \ddot{\varphi} \geq 0, \qquad \text{on }]a, +\infty[. \tag{19}$$

2.3 Globally Convex Examples

Let us consider the solutions of (18) with the following initial conditions:

$$x(0) = 0, \qquad z(0) = z_0 \in]a, \infty[, \qquad \theta(0) = 0. \tag{20}$$

Then, the surface ψ intersects orthogonally the rotation axis and it is globally convex. In fact, by application of L'Hôpital's rule, we have that $2\theta'(0) = \dot{\varphi}(z_0) > 0$ and γ is a strictly locally convex planar curve around $s = 0$. We assert that $\theta'(s) > 0$ for $s \geq 0$; otherwise from (20), there exists a first value $s_0 > 0$ such that $\theta'(s_0) = 0$ and $\theta''(s_0) \leq 0$. As $\theta' > 0$ on $[0, s_0[$, from (18), we have that $0 < 2\theta(s_0) < \pi$, and by differentiation of (18), we get

$$\theta''(s_0) = \frac{\sin(2\theta(s_0))}{2}\left(\ddot{\varphi}(z(s_0)) + \frac{1}{x(s_0)^2}\right) > 0,$$

getting to contradiction.

Theorem 2.3 ([34, Theorem 4.5]) *Under the conditions* (20), *the curve γ is the graph of a strictly convex symmetric function $u(x)$ defined on a maximal interval $]-\omega_+, \omega_+[$ which has a minimum at* 0 *and*

$$\lim_{x \to \pm\omega_+} u(x) = +\infty.$$

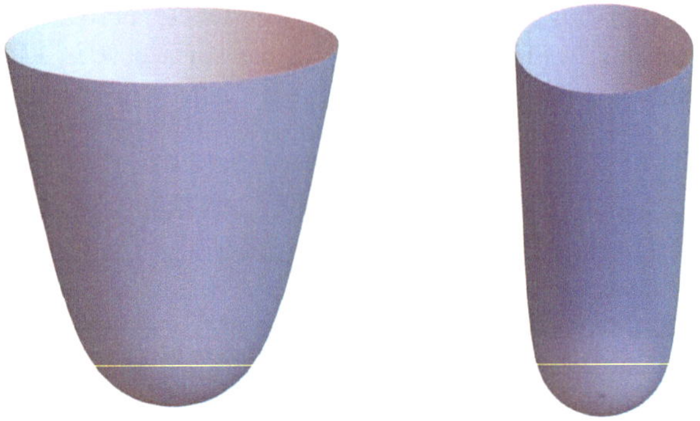

Fig. 3 $[\varphi, \mathbf{e}_3]$-minimal bowls for $\dot{\varphi}(u) = e^{-1/u}$ and $\dot{\varphi}(u) = u^2$, respectively

If γ is a graph as in Theorem 2.3, we are going to say that the revolution surface with generating curve γ is a $[\varphi, \mathbf{e}_3]$-*minimal bowl* (or a $[\varphi, \mathbf{e}_3]$-*bowl* in short) (Fig. 3).

2.4 Non-convex Examples

Now, we consider the solutions of (18) with the following initial conditions:

$$x(0) = x_0 > 0, \qquad z(0) = z_0 \in]a, +\infty[, \qquad \theta(0) = 0. \tag{21}$$

From standard theory, the existence and uniqueness of solution to the problem (18)–(21) are guaranteed. Let $]-s_-, s_+[$ be the maximal interval of existence, and consider $\gamma^+ := \gamma\big|_{[0,s_+[}$ the right branch of γ. As in Theorem 2.3, γ^+ is the graph of a convex function $u = u(x)$ defined on a maximal interval $]x_0, \omega_+[$, such that

$$\lim_{x \to \omega_+} u(x) = +\infty.$$

For studying the left branch of γ, we are going to consider $\gamma^-(s) = \gamma(-s)$ for $s \in [0, s_-[$. Then, by taking $\overline{x}(s) = x(-s)$, $\overline{z}(s) = z(-s)$ and $\overline{\theta}(s) = \theta(-s) + \pi$ for $s \in [0, s_-[$, we have that $\{\overline{x}, \overline{z}, \overline{\theta}\}$ is a solution of (18) on $[0, s_-[$ satisfying

$$\overline{x}(0) = x_0 > 0, \qquad \overline{z}(0) = z_0 \in]a, +\infty[, \qquad \overline{\theta}(0) = \pi.$$

Then, we have as follows.

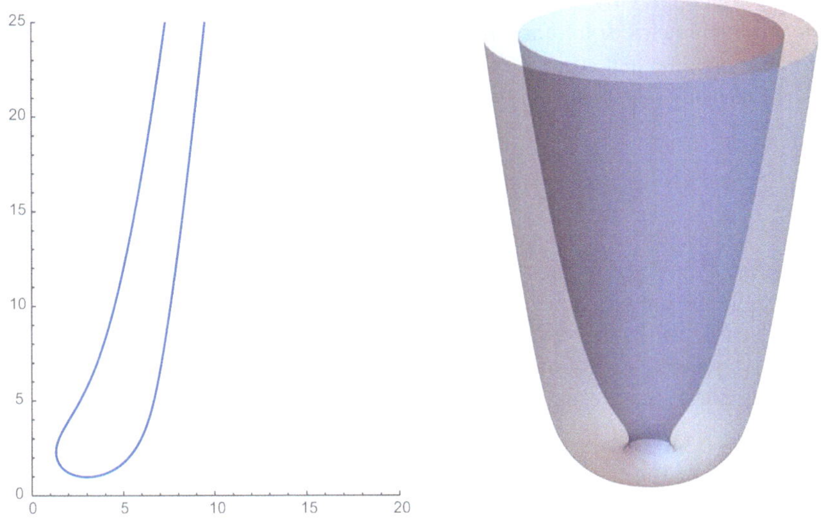

Fig. 4 $[\varphi, \mathbf{e}_3]$-minimal catenoid with $\dot{\varphi}(u) = e^{-1/u}$

Lemma 1 ([34, Section 4.2.2]) *The following statements hold:*

- *There exists* $s_0 \in]0, s_-[$ *such that* $2\overline{\theta}(s_0) = \pi$.
- *If* $s \in]s_0, s_-[$, *then* $0 < 2\overline{\theta}(s) < \pi$.
- $\overline{\theta}$ *has a minimum at a point* $s_1 \in]s_0, s_-[$ *and* $\overline{\theta}' > 0$ *on* $]s_1, s_-[$.
- *The profile curve* γ *is embedded.*

Theorem 2.4 ([34, Theorem 4.11]) *For every* $x_0 > 0$, *there exists a complete embedded rotational* $[\varphi, \mathbf{e}_3]$*-minimal surface; see Fig. 4 (right) with the annulus topology whose distance to axis of revolution is* x_0, *and whose generating curve* γ *is of winglike type, see Fig. 4 (left). These examples will be called* $[\varphi, \mathbf{e}_3]$*-minimal catenoids.*

2.5 The Asymptotic Behavior

One of the questions that we ask ourselves is to know if the rotationally symmetric examples are cylindrically asymptotic or not.

Indeed, we may prove the following.

Lemma 2 ([34, Proposition 4.12]) *Depending on the asymptotic behavior of* φ, *we have:*

- *If* $\dot{\varphi}$ *has at most a linear growth, then* $\omega_+ = +\infty$ *and* $\overline{x}_- = +\infty$.
- *If* $\dot{\varphi}$ *growths as* u^α *for some* $\alpha > 1$, *then* $\omega_+, \overline{x}_- < +\infty$.

Motivated by the previous result, we get to control the asymptotic behavior of the rotationally symmetric solutions when φ has at most quadratic growth. Hence, we are going to assume that $\varphi :]a, +\infty[\to \mathbb{R}$ is a smooth function satisfying (19) and with the following expansion at infinity:

$$\dot{\varphi}(u) = \Lambda u + \beta + \sum_{n=1}^{\infty} \frac{a_n}{u^n}, \quad a_n \in \mathbb{R}, \tag{22}$$

where either $\Lambda > 0$ and the first non-vanishing a_k is positive or $\Lambda = 0$, $\beta > 0$ and the first non-vanishing a_k is negative.

In this case, we can give explicit formulas for the asymptotic behavior of a rotationally symmetric solutions of (13) that generalize the result of J. Clutterbuck, O. Schnüre, and F. Schulze in [8] for translating solitons.

Theorem 2.5 ([34, Theorem A]) *If $\dot{\varphi}$ satisfies* (22), *then any rotationally symmetric solution u of* (4) *has the following asymptotic behavior:*

- *If $\Lambda > 0$,*

$$\varphi(u)(r) = C\, e^{\alpha\, r^2} + O(r^2), \quad C > 0, \tag{23}$$

- *If $\Lambda = 0$ and up to a constant, we have:*

$$\mathcal{G}(u)(r) = \frac{r^2}{2} - \frac{1}{\beta^2} \log(r) + O(r^{-2}), \tag{24}$$

where $\mathcal{G}$ is the strictly increasing function given by $\mathcal{G}(u) = \int_{u_0}^{u} \frac{d\xi}{\dot{\varphi}(\xi)}$.

3 Flat $[\varphi, \mathbf{e}_3]$-Minimal Surfaces

Bearing in mind the $[\varphi, \mathbf{e}_3]$-catenary cylinders that we have already described, we mention the following classification of complete flat $[\varphi, \mathbf{e}_3]$-minimal surfaces.

Let $\Sigma = \gamma \times \Pi^{\perp}$ be a complete ruled surface where γ is a complete regular curve in a plane $\Pi \subset \mathbb{R}^3$. Then, Σ can be parametrized by $\psi(s,t) = \gamma(s) + t\mathbf{v}$ with $\mathbf{v}$ a unit vector orthogonal to Π. We may assume that $(s,t) \in \mathbb{R}^2$ and $|\gamma'| = 1$; then, the Gauss map N of ψ and its mean curvature H are given by

$$N(s,t) = \gamma'(s) \wedge \mathbf{v} \ , \quad H(s,t) = \kappa_\gamma(s),$$

where κ_γ if the curvature of γ. Hence, Σ is $[\varphi, \mathbf{e}_3]$-minimal if and only if

$$\kappa_\gamma(s) = -\dot{\varphi}(\langle \gamma(s) + t\mathbf{v}, \mathbf{e}_3\rangle)\langle \gamma'(s) \wedge \mathbf{v}, \mathbf{e}_3\rangle.$$

Differentiating with respect to t in the above expression, we have the following classification result.

Theorem 3.6 ([34, Theorem 3.7]) *Let $\varphi : \mathbb{R} \to \mathbb{R}$ be an increasing diffeomorphism and Σ be a complete flat $[\varphi, \mathbf{e}_3]$-minimal surface in $\mathbb{R}^3$. Then, one of the following statements holds:*

- *Σ is a vertical plane.*
- *Σ is a grim reaper cylinder (maybe tilted).*
- *Σ is a $[\varphi, \mathbf{e}_3]$-catenary cylinder.*

Remark 2 An analogous classification was proved for translating solitons by F. Martín, A. Savas-Halilas, and K. Smoczyk in [33] and for singular α-minimal surfaces by R. López in [28].

4 Mean Convex $[\varphi, \mathbf{e}_3]$-Minimal Surfaces

In this section, we show several results concerning area and curvature estimates for properly embedded $[\varphi, \mathbf{e}_3]$-minimal surfaces with mean curvature $H \leq 0$ together with a convexity result for the same.

Previously, in 1983, R. Schoen [45] obtained an estimate for the length of the second fundamental form S of a stable minimal surface Σ in a 3-manifold. In particular, in $\mathbb{R}^3$, he proved the existence of a constant C such that

$$|S(p)| \leq \frac{C}{d_\Sigma(p, \partial\Sigma)}, \quad p \in \Sigma,$$

where d_Σ stands for the intrinsic distance of Σ. Later, in 2010, H. Rosenberg, R. Souam, and E. Toubiana [44] obtained an estimate for the length of the second fundamental form, depending on the distance to the boundary, for any stable H-surface Σ in a complete Riemannian 3-manifold of bounded sectional curvature $|\mathbb{K}| \leq \beta < +\infty$. They proved the existence of a constant $C > 0$ such that

$$|S(p)| \leq \frac{C}{\min\{d_\Sigma(p, \partial\Sigma), \pi/2\sqrt{\beta}\}}, \quad p \in \Sigma.$$

More recently, in 2016, B. White [51] obtained an estimate for the length of the second fundamental form for minimal surfaces with finite total absolute curvature less than 4π in 3-manifolds, depending on the distance to the boundary, on the sectional curvature, and on the gradient of the sectional curvature of the ambient space, and following C. H. Colding and W. P. Minicozzi method, [9, 10], J. Spruck and L. Xiao [47] have also obtained area and curvature bounds for complete mean convex translating solitons in $\mathbb{R}^3$. As application and using the Omori-Yau maximum principle (see, for example, [2]), they have proved one of the fundamental results in the recent development of translating solitons theory conjectured by X. Wang in [49].

Theorem 4.7 ([47, Theorem 1.1]) *Let $\Sigma \subset \mathbb{R}^3$ be a complete immersed two-sided translating soliton with non-negative mean curvature. Then Σ is convex.*

In a very recent study, we have extended the results of [47] to mean convex $[\varphi, \mathbf{e}_3]$-minimal surfaces. In our case, by mean convex surfaces, we will refer to those surfaces with $H \leq 0$ everywhere. More precisely, we will consider mean convex $[\varphi, \mathbf{e}_3]$-minimal oriented surfaces Σ with empty boundary in $\mathbb{R}^3_\alpha = \{p \in \mathbb{R}^3 \mid \langle p, \mathbf{e}_3 \rangle > \alpha\}$, where $\varphi : \mathbb{R} \longrightarrow \mathbb{R}$ is a smooth function satisfying

$$\dot{\varphi} > 0, \quad \ddot{\varphi} \geq 0 \quad \text{on }]\alpha, +\infty[. \tag{25}$$

In order to describe our results, let us consider F_t the normal variation associated with a compactly supported variational vector field on the normal bundle of Σ. Then, see [6, Appendix], the second variation of $\mathcal{A}^\varphi$ is given by

$$\left.\frac{d^2}{dt^2}\right|_{t=0} \mathcal{A}^\varphi(F_t(\Sigma)) = -\int_\Sigma \varrho \mathcal{L}_\varphi(\varrho)\, d\Sigma \quad \text{for any } \varrho \in C_0^\infty(\Sigma),$$

where $\mathcal{L}_\varphi$ is a gradient Schrödinger's type operator defined on $C^2(\Sigma)$ and given by

$$\mathcal{L}_\varphi(\cdot) = \Delta^\varphi(\cdot) + (|S|^2 - \ddot{\varphi}\eta_3^2)(\cdot)$$

where $\Delta^\varphi(\cdot) = \Delta(\cdot) + \langle \nabla\varphi, \nabla(\cdot)\rangle$ and $\eta_3 = \langle N, \mathbf{e}_3\rangle$.

As usual, we will say that Σ is stable if and only if for any compactly supported smooth function ϱ, it holds that

$$-\int_\Sigma \varrho \mathcal{L}_\varphi(\varrho) e^\varphi\, d\Sigma \geq 0. \tag{26}$$

It is interesting to mention the following fact.

Proposition 1 ([38, Proposition 4.4]) *Let $\varphi :]\alpha, +\infty[\rightarrow \mathbb{R}$ be a regular function satisfying* (25) *and Σ be an oriented $[\varphi, \mathbf{e}_3]$-minimal immersion in $\mathbb{R}^3$ with $H \leq 0$. Then, Σ is stable.*

Remark 3 The existence of stable surfaces is not guaranteed for any function φ. X. Cheng, T. Mejia, and D. Zhou [6] proved that if the Ilmanen's space is complete and $\ddot{\varphi} \leq -\epsilon < 0$, for some $\epsilon > 0$, then there are not stable surfaces without boundary and with finite weighted area.

From Proposition 1 and following the same method as in [47], we get the following area estimate.

Theorem 4.8 ([38, Proposition 4.10]) *Let Σ be a $[\varphi, \mathbf{e}_3]$-minimal immersion in $\mathbb{R}^3_\alpha$ with $H \leq 0$ and φ satisfying (25) and*

$$\Gamma := \sup_{]\alpha,+\infty[} (2\ddot{\varphi} - \dot{\varphi}^2) < +\infty.. \tag{27}$$

If $2\rho\,\dot{\varphi}(\rho + \mu(p)) < \log(2)$ and $\sqrt{|\Gamma|}\,\rho < 1$, then the geodesic disk $\mathcal{D}_\rho(p)$ of radius ρ centered at p is disjoint from the cut locus of p and

$$\mathcal{A}(\mathcal{D}_\rho(p)) < 4\pi\rho^2, \tag{28}$$

where $\mathcal{A}(\cdot)$ is the intrinsic area of Σ in $\mathbb{R}^3$.

But, for obtaining curvature bounds, we need a better control at infinity of the function φ . To be more precise, we are going to consider that $z \mapsto \frac{\dot{\varphi}(z)}{z}$ is analytic at $+\infty$; i.e., $\dot{\varphi}$ has the following series expansion at $+\infty$:

$$\dot{\varphi}(u) = \Lambda\, u + \beta + \sum_{i=1}^{\infty} \frac{c_i}{u^i}, \quad u \text{ large enough}, \tag{29}$$

with $\Lambda \geq 0$ and $\beta > 0$ if $\Lambda = 0$.

Remark 4 It is worth to note that condition (29) implies (27). Besides a natural extension of the best known examples, conditions (25) and (29) are interesting because under these assumptions, it is possible to know explicitly the asymptotic behavior of rotational and translational invariant examples (see §2).

Bearing in mind Proposition 1, Theorem 4.8 and the compactness Theorem 2.1 of B. White [52] for minimal surfaces together with the classification of all complete translating soliton graphs in $\mathbb{R}^3$ (see [19]), we can prove the following Blow-up result.

Theorem 4.9 ([38, Theorem 4.13]) *Let Σ be a properly embedded $[\varphi, \mathbf{e}_3]$-minimal surface in $\mathbb{R}^3_\alpha$ with $H \leq 0$, locally bounded genus and φ satisfying (25) and (29). Consider any sequence $\{\lambda_n\} \to +\infty$ and suppose that there exists a sequence $\{p_n\}$ in Σ such that $\{\dot{\varphi}(\mu(p_n))/\lambda_n\} \to C$ for some constant $C \geq 0$. Then, after passing to a subsequence, $\Sigma_n = \lambda_n(\Sigma - p_n)$ converge smoothly to*

1. *a plane when $C = 0$,*
2. *one of the following translating solitons when $C > 0$:*
   - *vertical plane,*
   - *grim reaper surface,*
   - *titled grim reaper surface,*
   - *bowl soliton,*
   - *Δ-Wing translating soliton.*

From Theorem 4.9 and by combining the methods of H. Rosenberg, R. Souam, and E. Toubiana [44] and J. Spruck and L. Xiao [47], we have the following curvature estimates.

Theorem 4.10 ([38, Theorem A]) *Let Σ be a properly embedded $[\varphi, \mathbf{e}_3]$-minimal surface in $\mathbb{R}^3_\alpha$ with non-positive mean curvature, locally bounded genus and $\varphi : \mathbb{R} \to \mathbb{R}$ satisfying* (25) *and* (29). *Then $|S|/\dot{\varphi}$ is bounded on Σ. In particular, if $\Lambda = 0$, $|S|$ is bounded, and if $\Lambda \neq 0$, $|S|$ may go to infinity but with at most a linear growth in height.*

As we have already mentioned, J. Spruck and L. Xiao [47] proved that any complete translating soliton in $\mathbb{R}^3$ with $H \leq 0$ is convex. Later, D. Hoffman, T. Ilmanen, F. Martín, and B. White in [20], by using the Omori-Yau's maximum principle for the Laplacian Δ [2, Theorem 3.2], have obtained a more simplified proof of it.

As an extension of this result, we have the following.

Theorem 4.11 ([38, Theorem B]) *Let Σ be a properly embedded $[\varphi, \mathbf{e}_3]$-minimal surface in $\mathbb{R}^3_a$ with non-positive mean curvature, locally bounded genus, and $\varphi : \mathbb{R} \to \mathbb{R}$ satisfying* (25), (29), *and $\dddot{\varphi} \leq 0$ on $]a, +\infty[$. Then Σ is convex if and only if the function ΛK is bounded from below, where K is the Gauss curvature.*

Remark 5 The main tool in the proof of Theorem 4.11 is the Omori-Yau's maximum principle for the drift Laplacian Δ^φ [2, Theorem 3.2], and it is remarkable that the condition (29) on φ is essential for proving that this maximum principle can be applied (see [38, Theorem 5.1]).

5 Uniqueness of Dirichlet's Problems at Infinity

Despite this large family of surfaces and the very general conditions on φ, it is possible to prove uniqueness of bowls and $[\varphi, \mathbf{e}_3]$-catenary cylinders from their asymptotic behavior.

5.1 Uniqueness of $[\varphi, \mathbf{e}_3]$-Bowls

Let $\varphi :]a, +\infty[\to \mathbb{R}$, $a \in \mathbb{R} \cup \{-\infty\}$ be a strictly increasing convex smooth function satisfying (22) and let Σ be a complete connected properly embedded $[\varphi, \mathbf{e}_3]$-minimal surface and $\mathfrak{D}^3 = \mathbb{R}^2 \times]a, +\infty[$. From Theorem 2.5, it is natural to say that an end of Σ is *smoothly asymptotic* to a rotational-type example if Σ can be expressed outside a Euclidean ball as a vertical graph of a function u_Σ so that

$$\varphi(u_\Sigma)(x) = C\, e^{\Lambda\, |x|^2} + O\left(|x|^2\right), \quad \text{if} \quad \Lambda > 0, \tag{30}$$

where C is a positive constant and, up to a constant,

$$\mathcal{G}(u_\Sigma)(x) = \frac{|x|^2}{2} - \frac{1}{\beta^2}\log(|x|) + O\left(|x|^{-2}\right), \text{ if } \Lambda = 0 \text{ and } \beta > 0. \tag{31}$$

where $\mathcal{G}$ is the strictly increasing function given by $\mathcal{G}(u) = \int_{u_0}^{u} \frac{d\xi}{\dot{\varphi}(\xi)}$.

Under these conditions, the following result holds.

Theorem 5.12 ([34, Theorem B]) *Let $\varphi :]a, +\infty[\to \mathbb{R}$, $a \in \mathbb{R} \cup \{-\infty\}$ be a strictly increasing convex smooth function satisfying* (22) *and Σ be a complete properly embedded $[\varphi, \mathbf{e}_3]$-minimal surface in $\mathbb{R}^3$ with a single end that is smoothly asymptotic to a $[\varphi, \mathbf{e}_3]$-minimal bowl. Then the surface Σ is a $[\varphi, \mathbf{e}_3]$-minimal bowl.*

Remark 6 The proof of Theorem 5.12 is based on the use of the Alexandrov reflection principle (see [1]) to prove that the surface is symmetric with respect to any vertical plane through the origin. Although this principle is applied in a standard way, it is crucial in the proof (see [34, Lemmas 6.3 and 6.4]) to show that it is possible to start the reflection with respect to any vertical plane far enough from the origin.

Remark 7 In the particular case of translating solitons, Theorem 5.12 was proved in [33, Theorem A].

5.2 Uniqueness of $[\varphi, \mathbf{e}_3]$-Catenary Cylinders

Let $\varphi :]a, +\infty[\to]b, c[$ $a, b \in \mathbb{R} \cup \{-\infty\}$, $c \in \mathbb{R} \cup \{+\infty\}$ be a strictly increasing diffeomorphism such that $e^{-\varphi} \in L^1(]a, +\infty[)$ and let us consider $\mathcal{G}^{u_0}$, $u_0 \in]a, +\infty[$ one of the $[\varphi, \mathbf{e}_3]$-catenary cylinders described in §2.

If Σ is a complete connected $[\varphi, \mathbf{e}_3]$-minimal graph and $\Pi(0)$ is the vertical plane through the origin orthogonal to $\mathbf{e}_1$, we will say that a smooth surface Σ is C^k-*asymptotic* to the *right part*

$$\mathcal{G}^{u_0}_+(0) = \{p \in \Sigma \, : \, \langle p, \mathbf{e}_1 \rangle \geq 0\}$$

of $\mathcal{G}^{u_0}$ if for any $\varepsilon > 0$, there exists $\delta > 0$ such that Σ can parametrized as a graph over $\mathcal{G}^{u_0}$ as follows:

$$\widetilde{F} : T^+_{\delta, u_0} \subset \mathbb{R}^2 \to \mathbb{R}^3 \quad \widetilde{F} = F + \overline{u}\, N_F, \tag{32}$$

where $T^+_{\delta,u_0} :=]\Lambda_{u_0} - \delta, \Lambda_{u_0}[\times\mathbb{R}$, $F(x_1, x_2) = (x_1, x_2, u(x_1))$ parametrizes $\mathcal{G}^{u_0}$ on T^+_{δ,u_0}, u is a solution of (8) with $u(0) = u_0$, $\overline{u} : T^+_{\delta,u_0} \to \mathbb{R}$ is a function in $C^k(T_{\delta,u_0}(+))$ such that

$$\sup_{T^+_{\delta,u_0}} |\overline{u}| < \varepsilon \ , \quad \sup_{T^+_{\delta,u_0}} |D^j\overline{u}| < \varepsilon, \text{ for any } j \in \{1, \cdots, k\}. \tag{33}$$

and N_F is the downwards unit normal of $\mathcal{G}^{u_0}$.

Analogously, we will say that a smooth surface Σ is C^k-*asymptotic* to left *left part*

$$\mathcal{G}^{u_0}_-(0) = \{p \in \Sigma \ : \ \langle p, \mathbf{e}_1\rangle \ \leq 0\}$$

of $\mathcal{G}^{u_0}$ if for any $\varepsilon > 0$ there exists $\delta > 0$ such that Σ can parametrized as a graph over $\mathcal{G}^{u_0}$ as follows

$$\widetilde{F} : T^-_{\delta,u_0} \subset \mathbb{R}^2 \to \mathbb{R}^3 \quad \widetilde{F} = F + \overline{u}\, N_F, \tag{34}$$

where $T^-_{\delta,u_0} :=] - \Lambda_{u_0}, -\Lambda_{u_0} + \delta[\times\mathbb{R}$, $F(x_1, x_2) = (x_1, x_2, u(x_1))$ parametrizes $\mathcal{G}^{u_0}$ on T^-_{δ,u_0}, u is a solution of (8) with $u(0) = u_0$, $\overline{u} : T^-_{\delta,u_0} \to \mathbb{R}$ is a function in $C^k(T^-_{\delta,u_0})$ such that

$$\sup_{T^-_{\delta,u_0}} |\overline{u}| < \varepsilon \ , \quad \sup_{T^-_{\delta,u_0}} |D^j\overline{u}| < \varepsilon, \text{ for any } j \in \{1, \cdots, k\}. \tag{35}$$

In particular, we say that Σ is C^k-*asymptotic* to $\mathcal{G}^{u_0}$ if and only if Σ is C^k-asymptotic to the both branches $\mathcal{G}^{u_0}_+(0)$ and $\mathcal{G}^{u_0}_-(0)$. Moreover, a smooth surface Σ is called C^k-*asymptotic to* $\mathcal{G}^{u_0}$, *outside a cylinder*, if there exists a solid cylinder $\mathfrak{c}$ whose axis is $\mathcal{G}^{u_0} \cap \Pi(0)$ and the set $\Sigma - \mathfrak{c}$ consists of two connected components Σ_1 and Σ_2 which are C^k-asymptotic to $\mathcal{G}^{u_0}_+(0)$ and $\mathcal{G}^{u_0}_-(0)$, respectively.

As a consequence of the compactness result [37, Theorem 3.4] and [37, Lemma 4.3], we have the following.

Proposition 2 ([37, Proposition 4.5]) *Let* $\varphi :]a, +\infty[\to]b, c[$, $a, b \in \mathbb{R} \cup \{-\infty\}$, $c \in \mathbb{R} \cup \{+\infty\}$ *be a convex strictly increasing diffeomorphism with* $e^{-\varphi} \in L^1(]a, +\infty[)$ *and* Σ *be a connected* $[\varphi, \mathbf{e}_3]$*-minimal immersion* C^∞*-asymptotic to* $[\varphi, \mathbf{e}_3]$*-catenary cylinder* $\mathcal{G}^h$*, outside a cylinder, for some* $h \in]a, +\infty[$*. For any sequence of points* $\{(p_{1,n}, p_{2,n}, p_{3,n})\}$ *of* Σ *such that* $\{p_{2,n}\}$ *diverges and* $\{p_{3,n}\}$ *is bounded, the sequence* $\{\Sigma_n = \Sigma - (0, p_{2,n}, 0)\}_{n\in\mathbb{N}}$ *converges smoothly, after subsequence, to some* $[\varphi, \mathbf{e}_3]$*-catenary cylinder with the same asymptotic behavior that* $\mathcal{G}^h$.

From Proposition 2, there exists n_0 large enough such that for any $n \geq n_0$, each Σ_n can be parametrized as a graph over some $\mathcal{G}^{u_0'}$ in $\mathcal{T}^+_{\delta,u_0,n}$ (respectively, $\mathcal{T}^-_{\delta,u_0,n}$) by

$$\widetilde{F}_n : \mathcal{T}^+_{\delta,u_0,n} \to \mathbb{R}^3 \quad \widetilde{F}_n = F + \overline{u_n} N_F, \tag{36}$$

where $\mathcal{T}^+_{\delta,u_0,n} =]-\Lambda_{u_0'} + \delta, \Lambda_{u_0'} - \delta[\times]m_{1,n}, m_{2,n}[\to \mathbb{R}$ (respectively,$\mathcal{T}^-_{\delta,u_0,n} =]-\Lambda_{u_0'} + \delta, \Lambda_{u_0'} - \delta[\times]-m_{2,n}, -m_{1,n}[$) with $\Lambda_{u_0'} = \Lambda_{u_0}$ due to the asymptotic behavior, $\delta > 0$ only depends on n, $\{m_{1,n}\}_{n\in\mathbb{N}}$, $\{m_{2,n}\}_{n\in\mathbb{N}}$ are strictly monotonous sequences with $m_{1,n} < m_{2,n}$, and each $\overline{u_n} : \mathcal{T}^+_{\delta,u_0,n} \to \mathbb{R}$ (respectively, $\overline{u_n} : \mathcal{T}^-_{\delta,u_0,n} \to \mathbb{R}$) is a smooth function satisfying the following inequalities:

$$\sup_{\mathcal{T}^+_{\delta,u_0,n}} |\overline{u_n}| < \varepsilon, \quad \sup_{\mathcal{T}^+_{\delta,u_0,n}} |D^j \overline{u_n}| < \varepsilon, \quad \text{for any } j \in \mathbb{N}. \tag{37}$$

By using the above inequalities, it is possible to prove that the function

$$\frac{\eta_2}{\eta_3} := \frac{\langle N, \mathbf{e}_2 \rangle}{\langle N, \mathbf{e}_3 \rangle}$$

goes to zero at infinity, and then, there exists an interior point where the function η_2/η_3 attains either a local minimum in $\{p \in \Sigma : \eta_2(p) < 0\}$ or a local maximum in $\{p \in \Sigma : \eta_2(p) > 0\}$. But then, it is possible to deduce that η_2 vanishes everywhere and so Σ is invariant under translations in the direction $\mathbf{e}_2$ which gives the following.

Theorem 5.13 ([37, Theorem 1.4]) *Let $\varphi :]a, +\infty[\to]b, c[$, $a, b \in \mathbb{R} \cup \{-\infty\}$ and $c \in \mathbb{R} \cup \{+\infty\}$ be a strictly increasing convex diffeomorphism such that $e^{-\varphi} \in L^1(]a, +\infty[)$ and bounded quotient $\ddot{\varphi}/\dot{\varphi}$. If Σ is a complete connected $[\varphi, \mathbf{e}_3]$-minimal graph C^∞-asymptotic to $[\varphi, \mathbf{e}_3]$-catenary cylinder $\mathcal{G}^{u_0}$, outside a cylinder, for some $h \in]a, +\infty[$, then Σ coincides with some $[\varphi, \mathbf{e}_3]$-catenary cylinder with the same behavior that $\mathcal{G}^{u_0}$.*

Remark 8 F. Martín, J. Pérez-García, A. Savas-Halilaj, and K. Smoczyk [32] proved that, if Σ is a properly embedded translating soliton with locally bounded genus and C^∞-topology to two vertical planes outside a cylinder, then Σ must coincide with some grim reaper translating soliton.

6 Weierstrass' Type Representation

Thanks to the property of to being "minimal" in the Ilmanen's space, we obtain a Weierstrass's type formula for translating soliton and singular minimal surfaces in $\mathbb{R}^3$ from its normal Gauss map.

Let $\varphi : I \to \mathbb{R}$ be a smooth function on a real open interval $I \subseteq \mathbb{R}$ and $\psi : \Sigma \to \mathfrak{D}^3 = \mathbb{R}^2 \times I \subseteq \mathbb{R}^3$ be an immersion. Consider a local conformal parameter

$\zeta = u + iv$ of Σ on an open simply connected domain $\mathcal{U} \subset \mathbb{C}$ such that the induced metric ds^2 writes

$$ds^2 := \lambda^2(du^2 + dv^2) = \lambda^2|d\zeta|^2 \tag{38}$$

and set, as usual, the Wirtinger's operators by

$$\partial_\zeta = \frac{1}{2}(\partial_u - i\partial_v), \quad \partial_{\overline{\zeta}} = \frac{1}{2}(\partial_u + i\partial_v).$$

If $\psi_\zeta = e^{\frac{1}{2}\varphi}(f, g, h)$ then, the conformality conditions write

$$\lambda^2 = 2|\psi_\zeta|^2 = 2\,e^\varphi\left(|f|^2 + |g|^2 + |h|^2\right) \tag{39}$$

$$f^2 + g^2 + h^2 = 0, \tag{40}$$

and by an straightforward computation, we obtain that ψ is a $[\varphi, \mathbf{e}_3]$-minimal immersion if and only if

$$e^{\frac{1}{2}\varphi} h_{\overline{\zeta}} = \frac{1}{2}\dot{\varphi}\left(|f|^2 + |g|^2\right), \quad e^{\frac{1}{2}\varphi} f_{\overline{\zeta}} = -\frac{1}{2}\dot{\varphi}\overline{f}h, \quad e^{\frac{1}{2}\varphi} g_{\overline{\zeta}} = -\frac{1}{2}\dot{\varphi}\overline{g}h \tag{41}$$

Now, if we introduce the complex functions

$$F = f - i\,g \quad \text{and} \quad G = \frac{h}{F}, \tag{42}$$

from (39) and (40), we have that G is a smooth map into the Riemann sphere and if G is not constant

$$h = FG\ ,\ f = \frac{1}{2}F(1 - G^2)\ ,\ g = \frac{i}{2}F(1 + G^2). \tag{43}$$

Moreover, the Gauss map N of ψ in the Euclidean space $\mathbb{R}^3$ is given in terms of G as

$$N := \left(\frac{2G}{1 + |G|^2}, \frac{1 - |G|^2}{1 + |G|^2}\right).$$

We are going to say also that G is the *Gauss map* of ψ.

By using (39) and (40), we obtain that (41) is equivalent to

$$\begin{aligned} 2e^{\frac{1}{2}\varphi} F_{\overline{\zeta}} &= \dot{\varphi}\,|F|^2|G|^2\overline{G}, \\ 4e^{\frac{1}{2}\varphi} G_{\overline{\zeta}} &= \dot{\varphi}\,\overline{F}(1 - |G|^4) \\ e^{\frac{1}{2}\varphi}\langle\psi, \mathbf{e}_3\rangle_\zeta &= FG. \end{aligned} \tag{44}$$

Remark 9 Observe that from (44), G is holomorphic if and only if $|G| \equiv 1$, and, in this case, it is clear that G must be constant and $\psi(\Sigma)$ lies on a vertical plane in $\mathbb{R}^3$.

Let us consider ψ_k a $[\varphi_k, \mathbf{e}_3]$- minimal surface where

$$\varphi_k(z) = z, \quad z \in \mathbb{R}, \quad \text{if } k = 1, \tag{45}$$

$$\varphi_k(z) = \frac{2}{k-1}\log(z) \quad z > 0, \quad \text{if } k \neq 1 \tag{46}$$

Then, from (39), (40), (44), (45), and (46), the Gauss map G of ψ_k satisfies the following complex equation:

$$G_{\zeta\overline{\zeta}} + 2\frac{|G|^2}{1-|G|^4}\overline{G}G_\zeta G_{\overline{\zeta}} + 2k\frac{|G_{\overline{\zeta}}|^2}{1-|G|^4}G = 0. \tag{47}$$

and

$$\frac{G_{\overline{\zeta}}}{1-|G|^4}, \quad \frac{\overline{G}G_{\overline{\zeta}}}{1-|G|^4}, \quad \frac{\overline{G}^2 G_{\overline{\zeta}}}{1-|G|^4}$$

are smooth functions on Σ. In this case, Eq. (47) gives the integrability conditions of the system (44), and we have as follows.

Theorem 6.14 ([36, Theorem 3.2]) *Let G be a not holomorphic solution of* (47) *defined on a simply connected domain $\mathcal{U} \subset \mathbb{C}$. Then the map $\psi_1 : \mathcal{U} \to \mathbb{R}^3$ given by*

$$\psi_1 = 4\,\Re\left(\int_{\zeta_0}^{\zeta}\frac{\overline{G}_\zeta(1-G^2)}{1-|G|^4}d\zeta, \int_{\zeta_0}^{\zeta} i\,\frac{\overline{G}_\zeta(1+G^2)}{1-|G|^4}d\zeta, 2\int_{\zeta_0}^{\zeta}\frac{\overline{G}_\zeta G}{1-|G|^4}d\zeta\right) \tag{48}$$

is a conformal translating soliton in $\mathbb{R}^3$ with Gauss map G. Conversely, any translating soliton which is not on a vertical plane can be locally represented in this way.

Theorem 6.15 ([36, Theorem 3.3]) *Let G be a not holomorphic solution of* (47) *defined on a simply connected domain $\mathcal{U} \subset \mathbb{C}$. Then for any $k \neq 0, 1$, the map $\psi_k : \mathcal{U} \to \mathbb{R}^3$ given by*

$$\psi_k = \left(4k\Re\int_{\zeta_0}^{\zeta}\frac{\overline{G}_\zeta(1-G^2)}{1-|G|^4}\Gamma\, d\zeta, 4k\Re\int_{\zeta_0}^{\zeta} i\,\frac{\overline{G}_\zeta(1+G^2)}{1-|G|^4}\Gamma\, d\zeta, \frac{2k}{k-1}\Gamma\right), \tag{49}$$

where

$$\Gamma = \mathrm{e}^{4(k-1)\Re\int_{\zeta_0}^{\zeta}\frac{\overline{G}_\zeta G}{1-|G|^4}d\zeta}$$

is a conformal $\dfrac{2}{k-1}$*-singular minimal surface in* $\mathbb{R}^3_+$ *with Gauss map* G. *Conversely, any singular minimal surface in* $\mathbb{R}^3_+$ *which is not on a vertical plane can be locally represented in this way.*

Remark 10 We would like to point out that it came to our knowledge that Theorem 6.14 was also proved in [25, Theorem 4].

The case $k = 0$ (i.e., of minimal surfaces in Hyperbolic space) has been studied in [22].

7 The Cauchy Problem

The Weierstrass representation described in the above Section can be applied to solve the following general Cauchy problem:

> Let $\beta = (\beta_1, \beta_2, \beta_3) : I \to \mathbb{R}^3$ be a regular analytic curve and let $V : I \to \mathbb{S}^2$ be an analytic vector field along β such that $\langle \beta', V \rangle = 0$, $|\Pi \circ V| < 1$, and $\beta_3 > 0$ if $k \neq 1$, where Π denotes the stereographic projection from the south pole. Find $[\varphi_k, \mathbf{e}_3]$-minimal surfaces containing β with unit normal in $\mathbb{R}^3$ along β given by V.

This problem has been inspired by the classical Björling problem for minimal surfaces in $\mathbb{R}^3$, proposed by E. G. Björling in 1844 and solved by H.A. Schwarz in 1890. Any pair β, V in the conditions of that problem *a pair of Björling data.*

Theorem 7.16 ([36, Theorem 4.3]) *For any* $k \in \mathbb{R}$, $k \neq 0$, *there exists a unique* $[\varphi_k, \mathbf{e}_3]$*-minimal which is a solution to the Cauchy problem with Björling data* $\beta = (\beta_1, \beta_2, \beta_3)$, $V = (V_1, V_2, V_3)$. *This solution*

$$\psi : \mathcal{U} = I \times]-\epsilon, \epsilon[\subseteq \mathbb{C} \longrightarrow \mathbb{R}^3,$$

can be constructed in a neighborhood of β *as follows: let* $G : \mathcal{U} \to \mathbb{C}$ *be the unique solution to the following system of Cauchy-Kowalewski's type, [43]:*

$$\begin{cases} G_{\zeta\overline{\zeta}} + 2\dfrac{|G|^2 G_\zeta G_{\overline{\zeta}}}{1-|G|^4}\overline{G} + 2k\dfrac{|G_{\overline{\zeta}}|^2}{1-|G|^4}G = 0, \\[2ex] G(s,0) = \dfrac{\phi_3(s)}{\phi_1(s) - i\phi_2(s)} = -\dfrac{\phi_1(s) + i\phi_2(s)}{\phi_3(s)}, \\[2ex] G_{\overline{\zeta}}(s,0) = \begin{cases} \dfrac{1-|G(s,0)|^4}{4}\left(\overline{\phi}_1(s) + i\overline{\phi}_2(s)\right), & \text{if } k = 1, \\ \dfrac{1-|G(s,0)|^4}{2(k-1)\beta_3}\left(\overline{\phi}_1(s) + i\overline{\phi}_2(s)\right), & \text{if } k \neq 1, \end{cases} \end{cases} \tag{50}$$

where

$$\phi(s) = (\phi_1(s), \phi_2(s), \phi_3(s)) = \frac{1}{2}(\beta'(s) - i\beta'(s) \wedge V(s)), \quad u \in I. \tag{51}$$

Then ψ is given, up to an appropriate translation, by (48) *if $k = 1$ and by* (49) *if $k \neq 1$, and G is its Gauss map.*

8 A Calabi's Type of Correspondence

In this last section, we establish a bijection between $[\varphi, \mathbf{e}_3]$-minimal surfaces in $\mathbb{R}^3$ and its corresponding spacelike $[\varphi, \mathbf{e}_3]$-maximal surfaces in the Lorentz-Minkowski space $\mathbb{L}^3$ and show how their geometric properties are related.

Let us consider $\mathbb{L}^3$ the Minkowski space $\mathbb{R}^3$ with the Lorentz metric

$$\langle\langle \cdot, \cdot \rangle\rangle = dx^2 + dy^2 - dz^2, \tag{52}$$

A surface in $\mathbb{L}^3$ is called *spacelike* if the induced metric on the surface is a positive definite Riemannian metric. This kind of surfaces has played a major role in Lorentzian geometry; for a survey of some results, we refer to [3].

Let $\varphi : I \to \mathbb{R}$ be a smooth function on a real open interval $I \subseteq \mathbb{R}$. A spacelike surface $\widetilde{\Sigma}$ in $\mathfrak{D}^3 = \mathbb{R}^2 \times I \subseteq \mathbb{L}^3$ is called $[\varphi, \mathbf{e}_3]$-*maximal* if its mean curvature vector $\widetilde{\mathbf{H}}$ satisfies

$$\widetilde{\mathbf{H}} = \left(\overline{\nabla}^{\mathbb{L}^3} \varphi\right)^{\perp} = -\dot{\varphi}\, \mathbf{e}_3, \tag{53}$$

where $\overline{\nabla}^{\mathbb{L}^3}$ denotes the gradient operator in $\mathbb{L}^3$.

As in the Euclidean case, a $[\varphi, \mathbf{e}_3]$-maximal spacelike surface can be also viewed either a critical point of the weighted volume functional

$$\widetilde{V}_{\varphi}(\widetilde{\Sigma}) := \int_{\widetilde{\Sigma}} e^{\varphi}\, dA_{\widetilde{\Sigma}}, \tag{54}$$

or a maximal (zero mean curvature) spacelike surface in the conformally changed metric

$$\widetilde{g}_{\varphi} := e^{\varphi} \langle\langle \cdot, \cdot \rangle\rangle. \tag{55}$$

Well-known examples of spacelike $[\varphi, \mathbf{e}_3]$-maximal are the spacelike maximal surfaces and the spacelike translating solitons, whose study is an exciting and already classical mathematical research field; see [7, 13] for some results. As in the

Euclidean case, a spacelike $[\varphi, \mathbf{e}_3]$-maximal with $\varphi(p) = \alpha \log |\langle\langle p, \mathbf{e}_3 \rangle\rangle|$, $\alpha \in \mathbb{R}$, $\alpha \neq 0$ will be called *singular* α*-maximal* surface.

If $\widetilde{\Omega}$ is a simply connected planar domain, the vertical graph in $\mathbb{L}^3$ of a function $\tilde{u} : \widetilde{\Omega} \to \mathbb{R}$ is a $[\varphi, \mathbf{e}_3]$-maximal spacelike if and only if $\tilde{u}$ is a solution of the following elliptic partial differential equation:

$$(1 - \tilde{u}_{\tilde{x}}^2)\tilde{u}_{\tilde{y}\tilde{y}} + (1 - \tilde{u}_{\tilde{y}}^2)\tilde{u}_{\tilde{x}\tilde{x}} + 2\tilde{u}_{\tilde{x}}\tilde{u}_{\tilde{y}}\tilde{u}_{\tilde{x}\tilde{y}} + \dot{\varphi}(\tilde{u})\widetilde{W}^2 = 0, \quad (\tilde{x}, \tilde{y}) \in \widetilde{\Omega}, \tag{56}$$

where $\widetilde{W} = \sqrt{1 - \tilde{u}_{\tilde{x}}^2 - \tilde{u}_{\tilde{y}}^2}$.

Equation (56) is equivalent to the integrability of the following differential system:

$$\widetilde{\phi}_{\tilde{x}\tilde{x}} = \frac{1 - \tilde{u}_{\tilde{x}}^2}{\widetilde{W}} \mathrm{e}^{\varphi(\tilde{u})}, \quad \widetilde{\phi}_{\tilde{x}\tilde{y}} = -\frac{\tilde{u}_{\tilde{x}}\tilde{u}_{\tilde{y}}}{\widetilde{W}} \mathrm{e}^{\varphi(\tilde{u})}, \quad \widetilde{\phi}_{\tilde{y}\tilde{y}} = \frac{1 - \tilde{u}_{\tilde{y}}^2}{\widetilde{W}} \mathrm{e}^{\varphi(\tilde{u})} \tag{57}$$

for a convex function $\widetilde{\phi} : \widetilde{\Omega} \longrightarrow \mathbb{R}$ (unique, modulo linear polynomials).

In [5], E. Calabi observed that there is a natural (local) connection between Euclidean minimal graphs and Lorentzian spacelike maximal graphs which is useful for describing examples and for applying similar methods to the study of their geometrical and topological properties. Recent advances of this correspondence in other ambient spaces could be found in [23, 24, 26, 27, 41].

Now, we show how Calabi's correspondence can be extended to one between the family of $[\varphi, \mathbf{e}_3]$-minimal surfaces in $\mathbb{R}^3$ and the family of spacelike $[\varphi, \mathbf{e}_3]$-maximal surfaces in $\mathbb{L}^3$.

Theorem 8.17 ([35, Theorem 4.1]) *Let Ω be a simply connected planar domain, $\psi : \Omega \to \mathbb{R}^3$, $\psi(x, y) = (x, y, u)$ be a vertical $[\varphi, \mathbf{e}_3]$-minimal graph in $\mathbb{R}^3$, ϕ be a solution to the system* (5) *and ϑ be a primitive function of* e^{φ} *(that is, $\dot{\vartheta} = \mathrm{e}^{\varphi}$). Then $\widetilde{\psi} : \Omega_1 \to \mathbb{L}^3$ given by*

$$\widetilde{\psi} := (\phi_x, \phi_y, \vartheta(u)), \tag{58}$$

is a $[-\varphi \circ \vartheta^{-1}, \mathbf{e}_3]$-maximal spacelike graph in the Lorentz-Minkowski space whose Gauss map $\widetilde{N}$ writes

$$\widetilde{N} = (u_x, u_y, W), \tag{59}$$

The induced metrics g and $\widetilde{g}$ of ψ and $\widetilde{\psi}$, respectively, are conformal, and the mean curvature H $(\widetilde{H})$ and Gauss curvature K $(\widetilde{K})$ of ψ $(\widetilde{\psi})$ satisfy

$$\widetilde{H} + W^2 \mathrm{e}^{-\varphi(u)} H = 0, \tag{60}$$

$$\widetilde{K} + W^4 \mathrm{e}^{-2\varphi(u)} K = 0. \tag{61}$$

Remark 11 Observe that on a simply connected $[\varphi, \mathbf{e}_3]$-minimal immersion ψ in $\mathbb{R}^3$ the correspondence (58) writes as follows:

$$\widetilde{\psi} = \int e^{\varphi(\psi_3)}(\mathbf{e}_3 \wedge (d\psi \wedge N) + \langle d\psi, \mathbf{e}_3\rangle \mathbf{e}_3), \tag{62}$$

where $\wedge$ denotes the cross product in $\mathbb{R}^3$. Moreover, the singularities of $\widetilde{\psi}$ hold where the angle function $\langle \mathbf{e}_3, N\rangle$ vanishes.

Theorem 8.18 ([35, Theorem 4.4]) *Let $\widetilde{\Omega}$ be a simply connected planar domain, $\widetilde{\psi} : \widetilde{\Omega} \to \mathbb{L}^3$, $\widetilde{\psi}(\tilde{x}, \tilde{y}) = (\tilde{x}, \tilde{y}, \tilde{u})$ be a vertical $[\varphi, \mathbf{e}_3]$-maximal graph in $\mathbb{L}^3$, $\widetilde{\phi}$ be a solution to the system (57), and $\tilde{\vartheta}$ be a primitive of e^{φ}. Then the immersion given by*

$$\psi := (\widetilde{\phi}_{\tilde{x}}, \widetilde{\phi}_{\tilde{y}}, \tilde{\vartheta}(\tilde{u})), \tag{63}$$

is a $[-\varphi \circ \vartheta^{-1}, \mathbf{e}_3]$-minimal graph in $\mathbb{R}^3$ whose induce metric, mean curvature H, and Gauss curvature K satisfy

$$g := \frac{e^{2\varphi(\tilde{u})}}{\widetilde{W}^2} \widetilde{g}, \tag{64}$$

$$H + e^{-\varphi(\tilde{u})}\, \widetilde{W}^2\, \widetilde{H} = 0, \tag{65}$$

$$K + e^{-2\varphi(\tilde{u})}\, \widetilde{W}^4\, \widetilde{K} = 0, \tag{66}$$

where $\widetilde{g}$, $\widetilde{H}$, and $\widetilde{K}$ are the induced metric, the mean curvature, and the Gauss curvature of the spacelike graph of $\tilde{u}$.

Remark 12 The correspondence (63) also writes as follows:

$$\psi = \int e^{\varphi(\widetilde{\psi}_3)} \left(\mathbf{e}_3 \wedge_{\mathbb{L}^3} (d\widetilde{\psi} \wedge_{\mathbb{L}^3} \widetilde{N}) + \langle\!\langle d\widetilde{\psi}, \mathbf{e}_3\rangle\!\rangle \mathbf{e}_3\right), \tag{67}$$

where $\wedge_{\mathbb{L}^3}$ denotes the cross product in $\mathbb{L}^3$ and $\widetilde{N}$ is the Gauss map of $\widetilde{\psi}$. The singular points of ψ hold where the angle function $\langle\!\langle \mathbf{e}_3, \widetilde{N}\rangle\!\rangle$ vanishes.

8.1 Some Consequences and Applications

Using the above correspondence, we can prove the following statements; see [35, Section 5].

- If ψ is a translating soliton in $\mathbb{R}^3$, then $\widetilde{\psi}$ is a singular (-1)-maximal spacelike surface in $\mathbb{L}^3$; see Fig. 5.
- If $\widetilde{\psi}$ is a translating soliton in $\mathbb{L}^3$ then ψ is a singular (-1)-minimal surface in $\mathbb{R}^3$; see Fig. 6.

Fig. 5 Soliton in $\mathbb{R}^3$ and its corresponding spacelike singular (-1)-maximal surface in $\mathbb{L}^3$

Fig. 6 Singular 1-minimal surface in $\mathbb{R}^3$ and its corresponding translating soliton in $\mathbb{L}^3$

Fig. 7 Translating soliton and singular (-2)-maximal bowl of elliptic type in $\mathbb{L}^3$

- Be a revolution surface with vertical rotational axis is a preserving property.
- There exists a rotationally symmetric, entire, smooth, strictly convex spacelike translating soliton (unique up to translation) and of linear growth, Fig. 7 left.
- For any $\alpha < -1$, there exists a rotationally symmetric, entire, smooth, strictly convex spacelike singular α-maximal graph (unique up to homothety) and with a linear growth, Fig. 7 right.
- For any $\alpha < -1$, there exist, up to an homothety (translation), two entire spacelike singular α-maximal graphs (spacelike translating solitons) in $\mathbb{L}^3$ with linear growth and with an isolated singularity at the origin which are asymptotic to the light cone. This kind of examples are called either *winglike solitons* or *winglike singular α-maximal surfaces*, respectively; see Fig. 8.

Fig. 8 Winglike soliton and singular (-3)-maximal winglike in $\mathbb{L}^3$

Acknowledgments Research partially supported by Ministerio de Economía y Competitividad Grant No: MTM2016-80313-P, Junta de Andalucía Grant No. A-FQM-139-UGR18, P18-FR-4049 and the 'Maria de Maeztu" Excellence Unit IMAG, reference CEX2020-001105-M, funded by MCIN/AEI/10.13039/501100011033.

References

1. Alexandrov, A.D.: Uniqueness theorems for surfaces in the large. Vestnik. Leninger. Univ. Math. **11**(5), 17 (1956)
2. Alías, L., Mastrolia, P., Rigoli, M.: Maximum Principles and Geometric Applications. Springer Monographs in Mathematics. Springer, Cham (2016)
3. Bartnik, R.: Maximal surfaces and general relativity. In: Proc. Miniconf. on Anal. Geom. (Austral. Nat. Univ.), (1986), (J.Hutchinson, ed.) Centre for Math. Anal. (ANU)
4. Böme, R., Hildebrant, S., Tausch, E.: The two-dimensional analogue of the catenary. Pac. J. Math. No. 2. **88**, 247–278 (1980)
5. Calabi, E.: Examples of Bernstein problems for some non-linear equations. Proc. Sympos. Pure Math. **15**, 222–230 (1970)
6. Cheng X., Mejia, T., Zhou, D.: Stability and compactness for complete f-minimal surfaces. Trans. Am. Math. Soc. **367**(6), 4041–4059 (2015)
7. Cheng, S.Y., Yau, S.T.: Maximal spacelike surfaces in Lorentz-Minkowski spaces. Ann. Math. **104**, 407–419 (1976)
8. Clutterbuck, J., Schnüre, O., Schulze, F.: Stability of translating solitons to mean curvature flow. Calc. Var. **29**, 281–293 (2007)
9. Colding, T.H., Minicozzi II, W.P.: Estimates for parametric elliptic integrands. Int. Math. Res. Not. **6**, 291–297 (2002)
10. Colding, T.H., Minicozzi II, W.P.: A course in minimal surfaces. Graduate studies in Mathematics, vol. 121, American Mathematical Society, Providence (2011)
11. Dierkes, U.: Singular minimal surfaces. In: Hildebrandt, S., et. al (eds.) Geometric Analysis and Nonlinear Partial Differential Equations, pp. 177–193 (2003)
12. Dierkes, U., Huisken, G.: The N-dimensional analogue of the catenary: prescribed area. In: Jost, J. (ed) Calculus of Variations and Geometric Analysis, pp. 1–13. International Press (1996)
13. Ding, Q.: Entire spacelike translating solitons in Minkowski space. J. Funct. Anal **265**, 3133–3162 (2013)
14. Dunn, W: The principles of dome construction: I and II. J. R. Inst. Br. Archit. **23**, 401–412 (1908)
15. Hildebrant, S.: On the regularity of solutions of two-dimensional variational problems with obstructions. Commun. Pure Appl. Math. **25**, 479–496 (1972)
16. Hildebrant, S.: Interior $C^{1+\alpha}$-regularity of solutions of two-dimensional variational problems with obstacles. Math. Z. **131**, 233–240 (1973)

17. Hildebrant, S., Kaul, H.: Two-dimensional variational problems with obstructions, and Plateau's problem for H-surfaces in a Riemannian manifold. Commun. Pure Appl. Math. **25**, 187–223 (1972)
18. Hoffman, D., Martín, F., White, B.: Scherk-like Translators for Mean Curvature Flow. J. Differ. Geom. **122**(3), 421–465 (2022).
19. Hoffman, D., Ilmanen, T., Martín, F., White, B.: Graphical translators for mean curvature flow. Calc. Var. PDE's **58**, 117 (2019)
20. Hoffman, D., Ilmanen, T., Martín, F., White, B.: Notes on translating solitons for mean curvature flow. In: Minimal Surfaces: Integrable Systems and Visualisation. Springer Proceedings in Mathematics & Statistics, vol. 349, pp. 147–168. Springer, Cham (2021)
21. Ilmanen, T.: Elliptic regularization and partial regularity for motion by mean curvature. Mem. Am. Math. Soc. **108**(520), x+90pp (1994)
22. Kokubo, M.: Weierstrass representation for minimal surfaces in hyperbolic space. Tôhoku Math. J. **49**, 367–377 (1997)
23. Lee, H.: Extensions of the duality between minimal surfaces and maximal surfaces. Geom. Dedicata **151**, 373–386 (2011)
24. Lee, H.: Calabi's correspondence and its generalization. In: Proceedings of The Sixteenth International Workshop on Differential Geometry, vol. 16, pp. 155–160 (2012)
25. Lee, H.: The H-flow translating solitons in $\mathbb{R}^3$ and $\mathbb{R}^4$. https://arxiv.org/abs/1204.0243
26. Lee, H.: Minimal surface systems, maximal surface systems and special Lagrangian equations. Trans. Am. Math. Soc. **365**(7), 3775–3797 (2013)
27. Lee, H., Manzano, J.M.: Generalized Calabi correspondence and complete spacelike surfaces. Asian J. Math. **23**(1), 35–48 (2019)
28. López, R.: Invariant singular minimal surfaces. Ann. Glob. Anal. Geom. **53**, 521–541 (2018)
29. López, R.: The Dirichlet problem for the α-singular minimal surface equation. Arch. Math. (Basel) **112**, 213–222 (2019)
30. López, R.: Uniqueness of critical points and maximum principles of the singular minimal surface equation. J. Differ. Equ. **266**, 3927–3941 (2019)
31. López, R.: Compact singular minimal surfaces with boundary. Am. J. Math. **142**, 521–541 (2020)
32. Martín, F., Pérez-García, J., Savas-Halilaj, A., Smoczyk, K.: A characterization of the grim reaper cylinder. J. Reine Angew. Math. **746**, 209–234 (2019)
33. Martín, F., Savas-Halilaj, A., Smoczyk, K.: On the topology of translating solitons of the mean curvature flow. Cal. Var. **54**, 2853–2882 (2015)
34. Martínez, A., Martínez-Triviño, A.L.: Equilibrium of surfaces in a vertical force field. Mediterr. J. Math. **19**(1), 3, 28pp. (2022)
35. Martínez, A., Martínez-Triviño, A.L.: A Calabi's type correspondence. Nonlinear Anal. **191** 111637 17pp. (2020)
36. Martínez, A., Martínez-Triviño, A.L.: A Weierstrass type representation for translating solitons and singular minimal surfaces. J. Math. Anal. Appl. **516**, 126528 (2022)
37. Martínez-Triviño, A.L., dos Santos, J.P.: Uniqueness of the $[\varphi, \mathbf{e}_3]$-catenary cylinders by their asymptotic behavior. J. Math. Anal. Appl. **514**, 126347 (2022)
38. Martínez, A., Martínez-Triviño, A.L., dos Santos, J.P.: Mean convex properly embedded $[\varphi, \mathbf{e}_3]$-minimal surfaces in $\mathbb{R}^3$. Rev. Mat. Iberoam. https://doi.org/10.4171/RMI/1352
39. Nitsche, J.C.C.:A nonexistence theorem for the two-dimensional analogue of the catenary. Analysis **6**, 143–156 (1986)
40. Otto, F.: Zugbeanspruchte Konstruktionen. Bd. I, II. Berlín, Frankfurt/M. Ullstein, Wien (1962, 1966)
41. Pelegrín, J.A., Romero, A., Rubio, R.M.: An extension of Calabi's correspondence between the solutions of two Bernstein problems to more general elliptic nonlinear equations Math. Notes. **105**(1–2), 280–284 (2019)
42. Poisson, S.D.: Sur les surfaces elastique, pp. 167–225. Men. CL. Sci. Math. Phys. Inst. Frace, deux (1975)

43. Petrovsky, I.G.: Lectures on Partial Differential Equations. Interscience Publishers. New York (1954)
44. Rosenberg, H., Souam, R., Toubiana, E.: General curvature estimates for stable H-surfaces in 3-manifolds and applications. J. Differ. Geom. **84**(3), 623–648 (2010)
45. Schoen, R.: Uniqueness, symmetry and embeddedness of minimal surfaces. J. Differ. Geom. **18**, 791–809 (1984)
46. Serrin, J.: Removable singularities of solutions of elliptic equations II. Arch. Ration. Mech. Anal. **20**, 163–169 (1965)
47. Spruck, J., Xiao, L.: Complete translating solitons to the mean curvature flow in $\mathbb{R}^3$ with nonnegative mean curvature. Am. J. Math. **142**(3), 993–1015 (2020)
48. Tausch, E.: A class of variational problems with linear growth. Math. Z. **164**, 159–178 (1978)
49. Wang, X.: Convex solutions to the mean curvature flow of mean-convex sets. J. Am. Math. Soc. **16**, 123–138 (2003)
50. White, B.: Curvature estimates and compactness theorems in 3-manifolds for surfaces that are stationary for parametric elliptic functionals. Invent. Math **88**(2), 243–256 (1987)
51. White, B.: Lectures on minimal surfaces theory (2016). prePrint. arXiv: 1308.3325v4
52. White, B.: On the compactness theorem for embedded minimal surfaces in 3-manifolds with locally bounded area and genus. Commun. Anal. Geom. **26**(3), 659–678 (2018)

Uniqueness of Constant Mean Curvature Spheres

Pablo Mira and Joaquín Pérez

Abstract We survey on the problem of finding all constant mean curvature two-spheres inside a given ambient Riemannian three-manifold. While this task is seemingly hopeless in such generality, the authors recently solved (with Meeks and Ros) this classification problem when the ambient metric is homogeneous. This classification implies, in particular, that two compact surfaces of genus zero with the same constant mean curvature in a Riemannian homogeneous three-manifold must coincide, up to ambient isometry. The result generalizes, among others, two fundamental uniqueness theorems by Hopf and Abresch-Rosenberg. The proof of our uniqueness theorem is extremely long and technical. In these notes we present a detailed sketch of our proof of the classification of constant mean curvature spheres in Riemannian three-manifolds and discuss some relevant open problems of the theory.

Keywords Constant mean curvature · Hopf theorem · Minimal spheres · Homogeneous three-manifolds

1 Introduction

The simplest examples of compact, constant mean curvature (CMC) surfaces are the round, totally umbilic spheres in the Euclidean three-space $\mathbb{R}^3$. In 1951, H. Hopf [21] proved the following remarkable uniqueness theorem for them: *any compact CMC surface of genus zero immersed in* $\mathbb{R}^3$ *is a round sphere*. The proof

P. Mira (✉)
Departamento de Matemática Aplicada y Estadística, Universidad Politécnica de Cartagena, Cartagena, Murcia, Spain
e-mail: pablo.mira@upct.es

J. Pérez
Department of Geometry and Topology and Institute of Mathematics (IMAG), University of Granada, Granada, Spain
e-mail: jperez@ugr.es

A. Alarcón et al. (eds.), *New Trends in Geometric Analysis*,
RSME Springer Series 10, https://doi.org/10.1007/978-3-031-39916-9_13

by Hopf introduced a holomorphic quadratic differential for CMC surfaces in $\mathbb{R}^3$, the so-called *Hopf differential*, which has been a key ingredient in the subsequent development of CMC surface theory. The proof by Hopf was later on extended to the remaining *three*-dimensional space forms $\mathbb{S}^3$ and $\mathbb{H}^3$; see [1, 7]. However, for a long time, the extension of these classification results of immersed CMC spheres to other Riemannian ambient three-manifolds remained elusive.

This situation only changed less than 20 years ago. In 2004, Abresch and Rosenberg [3] discovered that CMC surfaces in the homogeneous Riemannian product spaces $\mathbb{S}^2 \times \mathbb{R}$ and $\mathbb{H}^2 \times \mathbb{R}$ also admit a holomorphic quadratic differential. They used this differential to extend Hopf theorem to these spaces, by proving that *any compact CMC surface of genus zero immersed in $\mathbb{S}^2 \times \mathbb{R}$ or $\mathbb{H}^2 \times \mathbb{R}$ is a sphere of revolution*. In [4], Abresch and Rosenberg extended their holomorphic differential as well as their classification of CMC spheres to any rotationally invariant, simply connected Riemannian homogeneous three-manifold. Again, immersed CMC spheres in these spaces turned out to be rotational spheres. These results were the starting point of a great development of CMC surface theory in homogeneous three-manifolds in the following decade; see the expository works [11, 17] for an account of the beginnings of the theory.

The results by Abresch and Rosenberg raised the following fundamental question: *can one classify all immersed CMC spheres in any Riemannian homogeneous three-manifold?* One of the substantial difficulties in this problem was that the approach by Abresch and Rosenberg did not work in this more general, non-rotational setting.

In 2013, Daniel and Mira [12] introduced a new method for studying constant mean curvature spheres in the homogeneous Thurston *three*-dimensional manifold Sol_3. Using this method, they classified constant mean curvature spheres in Sol_3 for values H of the mean curvature greater than $\frac{1}{\sqrt{3}}$ and reduced the general classification problem to the obtention of area estimates for the family of spheres of constant mean curvature greater than any given positive number. These crucial area estimates were subsequently proved by Meeks [25]. This completed the classification of constant mean curvature spheres in Sol_3: *for any $H > 0$ there is a unique*[1] *constant mean curvature sphere S_H in* Sol_3 *with mean curvature H; moreover, S_H is maximally symmetric and embedded* [12, 25].

The method by Daniel and Mira could be used in principle in any Riemannian homogeneous three-manifold, to obtain a classification of CMC spheres for large enough values of the mean curvature H. However, in order to obtain a full classification of CMC spheres in any homogeneous three-manifold, it is necessary to understand the limits of the moduli space of such CMC spheres, and this turned out to be a very delicate issue.

Following this path, in the works [28, 29], the authors were able to obtain, jointly with Meeks and Ros, a classification of all immersed CMC spheres in arbitrary Riemannian homogeneous three-manifolds. For instance, they showed that if two

[1] Uniqueness is meant up to ambient isometries of Sol_3.

immersed spheres in a simply connected homogeneous three-manifold have the same constant mean curvature H, then they differ by an ambient isometry. They also obtained the exact values of the mean curvature H for which such spheres exist, and they described their basic geometry regarding stability index, embeddedness, maximal symmetry, and moduli space parametrization.

When the ambient homogeneous manifold M is diffeomorphic to $\mathbb{S}^3$, the arguments of the proof are not too lengthy; see [28]. However, when M is diffeomorphic to $\mathbb{R}^3$, the proof is extremely technical. For instance, it needs to take into account that these homogeneous manifolds M can have very different geometric properties, and it relies on a case-by-case study depending on the ambient manifold, in order to prove a crucial *area estimate* for the family of spheres of constant mean curvature greater than a geometrically relevant number, namely, the critical mean curvature of M.

However, except for the proof of this area estimate, the basic argument for the classification of CMC spheres in any homogeneous three-manifold can be explained in a quite intuitive, non-technical way. Our objective in this work is to explain the strategy behind the proof of the classification in [28, 29] and to place it into a more general context of the uniqueness of immersed spheres in Riemannian three-manifolds, homogeneous or not, by means of topological index arguments.

We next explain the organization of the work. In Sect. 2 we describe the classification of simply connected Riemannian homogeneous three-manifolds, following the notes [30] by Meeks and Pérez. In Sect. 3 we present the proofs by Hopf and Abresch-Rosenberg of the classification of CMC spheres in space forms and rotationally symmetric homogeneous three-manifolds, respectively. Both rely on the existence of a holomorphic quadratic differential, an object not available in general homogeneous three-manifolds.

In Sect. 4 we will depart from the argumentative path followed in [28, 29] and, instead, review the Gálvez-Mira uniqueness theory for immersed spheres in three-manifolds in terms of elliptic equations and index theory, [18]. Their result ensures, in particular, uniqueness of immersed CMC spheres provided one can foliate the unit tangent bundle of the ambient Riemannian three-manifold by Legendrian lifts of CMC surfaces.

In Sect. 5 we discuss the problem of classifying minimal two-spheres in Riemannian three-spheres. We review the classification of minimal two-spheres in homogeneous three-spheres by Meeks, Ros, and the authors [28], and we view it in the light of the Gálvez-Mira theory and some recent existence and uniqueness theorems for minimal two-spheres by Ambrozio, Marques, and Neves [2].

In Sect. 6 we go back to the classification of CMC spheres in homogeneous three-manifolds, by reviewing some basic properties regarding the stability operator of such CMC spheres. In Sect. 7 we state the classification theorems in [28, 29] that describe all CMC spheres in an arbitrary homogeneous three-manifold and discuss the scope, limits, and sharpness of the result. In Sect. 8 we give a brief outline of the proof of this classification. We conclude the paper by presenting some open problems on the uniqueness of CMC spheres in Sect. 9.

2 Homogeneous Three-Manifolds

The space of simply connected Riemannian homogeneous three-manifolds M can be classified from several viewpoints; a detailed presentation of this fact can be found in the notes [30] by Meeks and Pérez. In this section we will discuss briefly this classification.

Let M be a simply connected homogeneous three-manifold. First, from a topological viewpoint, M is homeomorphic to either $\mathbb{R}^3$, $\mathbb{S}^3$, or $\mathbb{S}^2 \times \mathbb{R}$. Moreover, if M has the topology of $\mathbb{S}^2 \times \mathbb{R}$, then it is actually (isometric to) a Riemannian product space $\mathbb{S}^2(c) \times \mathbb{R}$, where $c > 0$ is the Gaussian curvature of $\mathbb{S}^2(c)$.

The simplest examples of homogeneous three-manifolds are the Riemannian space forms, that is, the Euclidean space $\mathbb{R}^3$, the sphere $\mathbb{S}^3(c)$ of constant curvature $c > 0$, and the hyperbolic three-space $\mathbb{H}^3(c)$ of constant negative curvature $c < 0$. We denote such spaces by $\mathbb{Q}^3(c)$. Their isometry group has dimension 6. Very simple examples of homogeneous three-manifolds are also the product spaces $\mathbb{H}^2(c) \times \mathbb{R}$ and $\mathbb{S}^2(c) \times \mathbb{R}$. Their isometry group is *four*-dimensional.

In addition to these examples, there is a natural way of creating a homogeneous three-manifold, as follows. Let X be a simply connected *three*-dimensional Lie group, and fix $\langle , \rangle$ an inner product on its Lie algebra $T_e X$. Then, $\langle , \rangle$ can be extended by left translations of X to define a global, left-invariant Riemannian metric g on X. We say in these conditions that $X \equiv (X, g)$ is a *metric Lie group*. Generically (but not always), a metric Lie group has a *three*-dimensional isometry group.

The most relevant example of homogeneous three-manifold with a *three*-dimensional isometry group is the Thurston geometry Sol_3. It can be seen as the Riemannian manifold $(\mathbb{R}^3, g_{\mathrm{Sol}})$, with

$$g_{\mathrm{Sol}} = e^{2x_3}\, x_1^2 + e^{-2x_3}\, x_2^2 + x_3^2,$$

where (x_1, x_2, x_3) are canonical coordinates of $\mathbb{R}^3$. The space Sol_3 has a Lie group structure with respect to which the above metric is left-invariant, given by

$$(x_1, x_2, x_3) \cdot (y_1, y_2, y_3) = (x_1 + e^{-x_3}\, y_1, x_2 + e^{x_3}\, y_2, x_3 + y_3).$$

The isometry group of Sol_3 is of dimension 3, and the connected component of the identity is generated by the left translations with respect to the previous group product.

By construction, any metric Lie group is a simply connected homogeneous three-manifold. Conversely we have:

Fact Any simply connected homogeneous three-manifold that is not isometric to $\mathbb{S}^2(c) \times \mathbb{R}$ for some $c > 0$ is isometric to a metric Lie group X.

This fact essentially reduces the classification of homogeneous three-manifolds to the algebraic classification of metric Lie groups. It is worth mentioning that a simply connected homogeneous three-manifold could be isometric to two non-isomorphic metric Lie groups.

The following basic fact is a consequence of the left invariance of the metric:

Fact On any metric Lie group X, there exist three linearly independent Killing vector fields F_1, F_2, F_3, which are right-invariant for the group product of X.

Let n denote the dimension of the isometry group of the homogeneous three-manifold M. Then, $n = 6, 4$, or 3. If $n = 6$, then M is a Riemannian space form $\mathbb{Q}^3(c)$. We explain next the geometric properties of M when $n = 4$ or $n = 3$.

2.1 Four-Dimensional Isometry Group: $\mathbb{E}(\kappa, \tau)$-Spaces

Let M be a simply connected homogeneous three-manifold with $\dim(\mathrm{Iso}(M)) = 4$. Then, M admits a specially nice structure: it is an $\mathbb{E}(\kappa, \tau)$-space for some $(\kappa, \tau) \in \mathbb{R}^2$ with $\kappa \neq 4\tau^2$ (see [9]); we next make a quick review of some standard aspects of these geometries.

Every $\mathbb{E}(\kappa, \tau)$-space admits a canonical Riemannian fibration $\pi : \mathbb{E}(\kappa, \tau) \to \mathbb{M}^2(\kappa)$ over the simply connected *two*-dimensional surface $\mathbb{M}^2(\kappa)$ of constant curvature κ. After choosing orientations, we can define the vertical unit field ξ associated with this fibration, which is a Killing field on $\mathbb{E}(\kappa, \tau)$ (since *vertical* translations along the fibers are isometries of $\mathbb{E}(\kappa, \tau)$). The number $\tau \in \mathbb{R}$ is determined by the equation

$$\nabla_X \xi = \tau X \times \xi,$$

for any vector field on $\mathbb{E}(\kappa, \tau)$, where $\times$ stands for the vector product with respect to the chosen orientation.

If $\tau = 0$ we get the Riemannian product spaces $\mathbb{H}^2(\kappa) \times \mathbb{R}$ and $\mathbb{S}^2(\kappa) \times \mathbb{R}$. When $\tau \neq 0$, we obtain the Riemannian Heisenberg space Nil_3 for $\kappa = 0$, the so-called Berger spheres for $\kappa > 0$, and a rotationally symmetric left-invariant metric on the universal cover of $\mathrm{PSL}(2, \mathbb{R})$ if $\kappa < 0$.

When $\kappa > 0$, the space $\mathbb{E}(\kappa, \tau)$ is homeomorphic to $\mathbb{S}^3$ if $\tau \neq 0$ and to $\mathbb{S}^2 \times \mathbb{R}$ if $\tau = 0$. When $\kappa \leq 0$, $\mathbb{E}(\kappa, \tau)$ is homeomorphic to $\mathbb{R}^3$. All spaces $\mathbb{E}(\kappa, \tau)$ are rotationally invariant, i.e., for each $p \in \mathbb{E}(\kappa, \tau)$, there is a continuous $\mathbb{S}^1$-family of orientation-preserving isometries of $\mathbb{E}(\kappa, \tau)$ leaving pointwise fixed the fiber $\pi^{-1}(\pi(p))$.

There is a useful *coordinate model* $\mathcal{R}^3(\kappa, \tau)$ for $\mathbb{E}(\kappa, \tau)$. Specifically, let $\mathcal{R}^3(\kappa, \tau)$ be $\mathbb{R}^3$ if $\kappa \geq 0$, or $\mathbb{D}\left(2/\sqrt{-\kappa}\right) \times \mathbb{R}$, if $\kappa < 0$, where $\mathbb{D}(\rho) = \{(x_1, x_2) \in \mathbb{R}^2 ;\, x_1^2 + x_2^2 < \rho^2\}$, endowed in any case with the Riemannian metric

$$ds^2 = \lambda^2(dx_1^2 + dx_2^2) + \left(\tau\lambda(x_2 dx_1 - x_1 dx_2) + dx_3\right)^2, \qquad \lambda = \frac{1}{1 + \frac{\kappa}{4}(x_1^2 + x_2^2)}. \tag{1}$$

When $\kappa \leq 0$, $(\mathcal{R}^3(\kappa, \tau), ds^2)$ is globally isometric to $\mathbb{E}(\kappa, \tau)$. When $\kappa > 0$, $(\mathcal{R}^3(\kappa, \tau), ds^2)$ is isometric to $(\mathbb{S}^2(\kappa) \setminus \{p\}) \times \mathbb{R}$ if $\tau = 0$ and to the Riemannian universal cover of the Berger sphere $\mathbb{E}(\kappa, \tau)$ minus one fiber if $\tau \neq 0$.

In the coordinate model $\mathcal{R}^3(\kappa, \tau)$, the vertical unit Killing field ξ of $\mathbb{E}(\kappa, \tau)$ and the projection $\pi : \mathbb{E}(\kappa, \tau) \to \mathbb{M}^2(\kappa)$ are represented by $\frac{\partial}{\partial x_3}$ and $(x_1, x_2, x_3) \mapsto (x_1, x_2)$, respectively.

2.2 *Metric Lie Groups*

As explained previously, the case of a *three*-dimensional isometry group corresponds to the situation in which M is a metric Lie group X, with the property that the set of isometries of X that fix the identity element $e \in X$ is a finite group. This is the generic situation and in general the hardest to deal with. An important fact in this situation is that the metric on X determines the algebraic Lie group structure, i.e., two isometric homogeneous three-manifolds with a *three*-dimensional isometry group always have isomorphic underlying Lie group structures. This is not true for isometry groups of dimension ≥ 4.

In general, one can divide metric Lie groups into three different classes.

2.2.1 X Is Isomorphic to SU(2)

The Lie group SU(2) is diffeomorphic to the three-sphere $\mathbb{S}^3$. Thus, by considering an inner product on the Lie algebra of SU(2) and extending it to a left-invariant Riemannian metric on SU(2) by left translations, we obtain a homogeneous three-sphere. Conversely, any homogeneous three-sphere can be seen as SU(2) endowed with a left-invariant metric. This is true in particular for spheres of constant curvature $\mathbb{S}^3(c)$ and Berger three-spheres $\mathbb{E}(\kappa, \tau)$, with $\kappa > 0$ and $\tau \neq 0$.

The Lie group SU(2) is unimodular. This property and the previous discussion implies that if $X = (\mathrm{SU}(2), \langle , \rangle)$ is a homogeneous three-sphere; then there exists an orthonormal left-invariant frame $\{E_1, E_2, E_3\}$ on X such that

$$[E_2, E_3] = c_1 E_1, \quad [E_3, E_1] = c_2 E_2, \quad [E_1, E_2] = c_3 E_3, \tag{2}$$

for certain positive constants $c_1, c_2, c_3 > 0$. The Ricci tensor of X diagonalizes in the basis $\{E_1, E_2, E_3\}$ with eigenvalues

$$\mathrm{Ric}(E_1) = 2\mu_2\mu_3, \quad \mathrm{Ric}(E_2) = 2\mu_1\mu_3, \quad \mathrm{Ric}(E_3) = 2\mu_1\mu_2,$$

where

$$\mu_1 = \frac{1}{2}(-c_1 + c_2 + c_3), \quad \mu_2 = \frac{1}{2}(c_1 - c_2 + c_3), \quad \mu_3 = \frac{1}{2}(c_1 + c_2 - c_3).$$

The space of homogeneous three-spheres is parametrized by the structure constants (c_1, c_2, c_3). If the three constants c_i coincide (resp. two or none coincide), then X is a round sphere (resp. a Berger sphere or X has a three-dimensional isometry group).

2.2.2 X Is Isomorphic to a Semidirect Product

Given a matrix $A \in \mathcal{M}_2(\mathbb{R})$, the semidirect product $\mathbb{R}^2 \rtimes_A \mathbb{R}$ is the Lie group $(\mathbb{R}^3 \equiv \mathbb{R}^2 \times \mathbb{R}, *)$ endowed with the group operation

$$(\mathbf{p}_1, z_1) * (\mathbf{p}_2, z_2) = (\mathbf{p}_1 + e^{z_1 A}\, \mathbf{p}_2, z_1 + z_2); \tag{3}$$

here e^B is the usual exponential of $B \in \mathcal{M}_2(\mathbb{R})$. Let

$$A = \begin{pmatrix} a & b \\ c & d \end{pmatrix}, \qquad e^{zA} = \begin{pmatrix} a_{11}(z) & a_{12}(z) \\ a_{21}(z) & a_{22}(z) \end{pmatrix}, \tag{4}$$

for each $z \in \mathbb{R}$. Then, a left-invariant frame $\{E_1, E_2, E_3\}$ of $X = \mathbb{R}^2 \rtimes_A \mathbb{R}$ is given by

$$E_1(x, y, z) = a_{11}(z)\partial_x + a_{21}(z)\partial_y, \quad E_2(x, y, z) = a_{12}(z)\partial_x + a_{22}(z)\partial_y, \quad E_3 = \partial_z. \tag{5}$$

Observe that $\{E_1, E_2, E_3\}$ is the left-invariant extension with respect to the group structure (3) of the canonical basis $(\partial_x)_e, (\partial_y)_e, (\partial_z)_e$ of the tangent space T_eX at the identity element $e = (0, 0, 0)$. The right-invariant extensions on X of the same vectors of T_eX define the frame $\{F_1, F_2, F_3\}$ where

$$F_1 = \partial_x, \quad F_2 = \partial_y, \quad F_3(x, y, z) = (ax + by)\partial_x + (cx + dy)\partial_y + \partial_z. \tag{6}$$

Definition 1 We define the *canonical left-invariant metric* $\langle , \rangle$ on $\mathbb{R}^2 \rtimes_A \mathbb{R}$ to be that one for which the left-invariant frame $\{E_1, E_2, E_3\}$ given by (5) is orthonormal.

Some basic properties of the canonical left invariant metric $\langle , \rangle$ on $\mathbb{R}^2 \rtimes_A \mathbb{R}$ are:

1. The right-invariant vector fields F_1, F_2, F_3 are Killing.
2. The mean curvature of each leaf of the foliation $\mathcal{F} = \{\mathbb{R}^2 \rtimes_A \{z\} \mid z \in \mathbb{R}\}$ with respect to the unit normal vector field E_3 is the constant $H = \mathrm{trace}(A)/2$.
3. The expression of $\langle , \rangle$ in the x, y, z coordinates of X is

$$\begin{aligned} \langle , \rangle = {} & \left[a_{11}(-z)^2 + a_{21}(-z)^2\right] dx^2 + \left[a_{12}(-z)^2 + a_{22}(-z)^2\right] dy^2 + dz^2 \\ & + \left[a_{11}(-z)a_{12}(-z) + a_{21}(-z)a_{22}(-z)\right] (dx \otimes dy + dy \otimes dx)\,. \end{aligned} \tag{7}$$

Different choices of the matrix A give rise then to different metric Lie groups, non-isomorphic in general. We remark that if A is a singular matrix, then $\mathbb{R}^2 \rtimes_A \mathbb{R}$ with its canonical metric has an isometry group of dimension ≥ 4. Thus, for the case where X has a *three*-dimensional isometry group, the matrix A is non-singular.

2.2.3 X Is Isomorphic to $\widetilde{\mathrm{SL}}(2, \mathbb{R})$

The Lie group $\widetilde{\mathrm{SL}}(2, \mathbb{R})$ is the universal cover of the special linear group $\mathrm{SL}(2, \mathbb{R}) = \{A \in \mathcal{M}_2(\mathbb{R}) \mid \det A = 1\}$ and of the projective special linear group $\mathrm{PSL}(2, \mathbb{R}) = \mathrm{SL}(2, \mathbb{R})/\{\pm I_2\}$. The Lie algebra of any of the groups $\widetilde{\mathrm{SL}}(2, \mathbb{R})$, $\mathrm{SL}(2, \mathbb{R})$, and $\mathrm{PSL}(2, \mathbb{R})$ is $\mathfrak{sl}(2, \mathbb{R}) = \{B \in \mathcal{M}_2(\mathbb{R}) \mid \mathrm{trace}(B) = 0\}$. It is worth recalling that $\mathrm{PSL}(2, \mathbb{R})$ has several isomorphic models: for instance, the group of orientation-preserving isometries of the hyperbolic plane or the group of conformal automorphisms of the unit disk.

The matrices in the Lie algebra $\mathfrak{sl}(2, \mathbb{R})$ given by

$$(E_1)_e = \begin{pmatrix} 1 & 0 \\ 0 & -1 \end{pmatrix}, \quad (E_2)_e = \begin{pmatrix} 0 & 1 \\ 1 & 0 \end{pmatrix}, \quad (E_3)_e = \begin{pmatrix} 0 & -1 \\ 1 & 0 \end{pmatrix} \tag{8}$$

define a left-invariant frame $\{E_1, E_2, E_3\}$ on $\widetilde{\mathrm{SL}}(2, \mathbb{R})$ with the property that

$$[E_1, E_2] = -2E_3, \qquad [E_2, E_3] = 2E_1, \qquad [E_3, E_1] = 2E_2.$$

Every left-invariant metric in $\widetilde{\mathrm{SL}}(2, \mathbb{R})$ can be obtained by choosing numbers $\lambda_1, \lambda_2, \lambda_3 > 0$ and declaring the length of the left-invariant vector field E_i to be λ_i, 1, 2, 3 while keeping them orthogonal. For instance, by declaring the left-invariant frame $\{E_1, E_2, E_3\}$ to be orthonormal, we obtain a left-invariant metric $\langle , \rangle$ on $\widetilde{\mathrm{SL}}(2, \mathbb{R})$ such that the metric Lie group $(\widetilde{\mathrm{SL}}(2, \mathbb{R}), \langle , \rangle)$ is isometric to the $\mathbb{E}(\kappa, \tau)$-space with base curvature $\kappa = -4$ and bundle curvature $\tau^2 = 1$.

3 Holomorphic Quadratic Differentials and CMC Spheres

3.1 CMC Spheres in Space Forms $\mathbb{Q}^3(c)$

Constant mean curvature spheres in $\mathbb{R}^3$ were classified by Hopf; see [21, 22]. Specifically, he proved that any CMC sphere in $\mathbb{R}^3$ must be a round, totally umbilic sphere. The proof by Hopf also works for the case where the ambient manifold is the sphere $\mathbb{S}^3(c)$ or the hyperbolic space $\mathbb{H}^3(c)$, a theorem commonly associated with the names of Chern [7] and Almgren [1]. Thus:

Theorem 3 *Any CMC sphere in $\mathbb{Q}^3(c)$ is a round totally umbilic sphere.*

The proof by Hopf introduced an object that has played a fundamental role in the development of CMC surface theory: a holomorphic quadratic differential for any CMC surface in $\mathbb{R}^3$. We give a brief sketch of the proof.

Let $\psi : \Sigma \looparrowright \mathbb{Q}^3(c)$ be a conformally immersed surface in $\mathbb{Q}^3(c)$, and let z denote a local conformal parameter for Σ. Let λ, H, Q denote, respectively, the conformal factor of the metric of ψ, the mean curvature function of ψ, and the quadratic differential

$$Q\,dz^2 := -\langle \psi_z, \eta_z \rangle dz^2,$$

where $\eta : \Sigma \to TU(\mathbb{Q}^3(c))$ is the unit normal of ψ. Observe that the zeros of Q agree with the umbilic points of Σ. The data (λ, H, Q) determine the surface up to rigid motions in $\mathbb{Q}^3(c)$ and satisfy the Codazzi equation

$$2Q_{\bar{z}} = \lambda H_z.$$

In particular, if H is constant, $Q\,dz^2$ is a holomorphic quadratic differential on Σ.

Now, assume that Σ is homeomorphic to $\mathbb{S}^2$. By uniformization, $Q\,dz^2$ defines a holomorphic quadratic differential on the Riemann sphere $\overline{\mathbb{C}}$, and so $Q \equiv 0$ on Σ. Therefore, Σ is a totally umbilic sphere, hence round.

Remark 1 In $\mathbb{R}^3$, round CMC spheres exist for all non-zero values of the mean curvature H. In $\mathbb{S}^3$, they exist for all values of H; for instance, when $H = 0$, such a sphere is a totally geodesic *equatorial sphere* of $\mathbb{S}^3$. In $\mathbb{H}^3(c)$, round CMC spheres exist only for the values $|H| > \sqrt{-c}$.

One should note that the value $\sqrt{-c}$ appearing in Remark 1 is the mean curvature of horospheres in $\mathbb{H}^3(c)$. An easy application of the maximum principle with respect to these horospheres implies that *there are no compact surfaces in* $\mathbb{H}^3(c)$ *with* $|H| \leq \sqrt{-c}$ *at every point.*

3.2 CMC Spheres in Homogeneous $\mathbb{E}(\kappa, \tau)$-Spaces

The extension of Theorem 3 to other Riemannian ambient three-manifolds (M^3, g) of non-constant sectional curvature remained unanswered for a long time, until Abresch and Rosenberg [3, 4] proved the striking fact that, on any homogeneous $\mathbb{E}(\kappa, \tau)$-space, there exists a holomorphic quadratic differential for CMC surfaces. This allowed them to classify all CMC spheres in $\mathbb{E}(\kappa, \tau)$-spaces. We next explain their theorem. For a more detailed approach, see [11, 17].

Let $\psi : \Sigma \to \mathbb{E}(\kappa, \tau)$ be a conformal immersion with unit normal map N, and consider on Σ a local conformal parameter $z = s + it$. Recall that ξ denotes the vertical Killing field of $\mathbb{E}(\kappa, \tau)$.

In these conditions, we will call the *fundamental data* of ψ to

$$(\lambda|dz|^2, \nu, H, p\,dz^2, A\,dz),$$

where λ, H, Q are defined as in $\mathbb{Q}^3(c)$, while $\nu = \langle N, \xi\rangle$ is the *angle function* of Σ and $A := \langle \xi, \psi_z\rangle$.

Once here, a set of necessary and sufficient conditions for the integrability of CMC surfaces in $\mathbb{E}(\kappa, \tau)$ can be written in terms of these fundamental data. This is a key influential result by B. Daniel [9]. The Daniel integrability conditions can be rewritten in terms of a conformal parameter z as follows (see Fernández-Mira [15]):

$$\begin{cases} \textbf{(C.1)} \quad Q_{\bar z} &= \dfrac{\lambda}{2}(H_z + \nu A(\kappa - 4\tau^2)). \\ \textbf{(C.2)} \quad A_{\bar z} &= \dfrac{\nu\lambda}{2}(H + i\tau). \\ \textbf{(C.3)} \quad \nu_z &= -(H - i\tau)A - \dfrac{2Q}{\lambda}\bar{A}. \\ \textbf{(C.4)} \quad \dfrac{4|A|^2}{\lambda} &= 1 - \nu^2. \end{cases} \tag{9}$$

The fundamental data $(\lambda|dz|^2, \nu, H, p\,dz^2, A\,dz)$ satisfying (9) determine the immersion ψ uniquely up to ambient isometries preserving the orientations of base and fiber of $\mathbb{E}(\kappa, \tau)$. We note that **(C.1)** is the Codazzi equation, while the rest of equations express the geometry of the surface with respect to the special Killing vertical direction ξ.

A simple computation from the integrability conditions proves:

Theorem 4 ([3, 4]) *The quadratic differential*

$$Q_{AR}\,dz^2 := \left(2(H + i\tau)Q - (\kappa - 4\tau^2)A^2\right) dz^2$$

is holomorphic for any CMC surface in $\mathbb{E}(\kappa, \tau)$.

As a consequence, one can classify CMC spheres in $\mathbb{E}(\kappa, \tau)$, as follows:

Theorem 5 (Abresch-Rosenberg) *Any CMC sphere in* $\mathbb{E}(\kappa, \tau)$ *is a rotational sphere.*

Proof Arguing as in Hopf's proof for $\mathbb{R}^3$, we have from Theorem 4 that $Q_{AR} \equiv 0$ on any CMC sphere in $\mathbb{E}(\kappa, \tau)$. So, we just need to show that CMC spheres with $Q_{AR} \equiv 0$ are rotational.

On any CMC surface Σ in $\mathbb{E}(\kappa, \tau)$, the equation $Q_{AR} = 0$ together with the integrability conditions (9) implies that $w := \operatorname{arctanh}(\nu)$ is a harmonic function on Σ (here ν is the angle function of the surface). Define ζ as a local conformal parameter on Σ with $\operatorname{Re}\zeta = w$. Using this parameter and $Q_{AR} = 0$ in (9), one can show that all fundamental data of the surface depend only on w (and not on $\operatorname{Im}\zeta$).

This implies that the surface is a local piece of some CMC surface invariant by a continuous 1-parameter subgroup of ambient isometries of $\mathbb{E}(\kappa, \tau)$.

If the surface is diffeomorphic to $\mathbb{S}^2$, this 1-parameter subgroup can only be the group of rotations around a vertical fiber of $\mathbb{E}(\kappa, \tau)$. Hence, CMC spheres in $\mathbb{E}(\kappa, \tau)$ are rotational. □

Remark 2 For each $H \in \mathbb{R}$ there is at most one rotational CMC sphere in $\mathbb{E}(\kappa, \tau)$ with mean curvature H, up to ambient isometry. The values of the mean curvature H for which there exist rotational CMC spheres in $\mathbb{E}(\kappa, \tau)$ are:

1. If $\kappa > 0$, there exist CMC spheres for every $H \in \mathbb{R}$. When $H = 0$ in $\mathbb{S}^2(c) \times \mathbb{R}$, these are the totally geodesic slices $\mathbb{S}^2(c) \times \{t_0\}$.
2. If $\kappa \leq 0$, then there exist CMC spheres only if $|H| > \frac{\sqrt{-\kappa}}{2}$.

As a matter of fact, if $\kappa \leq 0$, there exist *entire graphs* in $\mathbb{E}(\kappa, \tau)$ of constant mean curvature $H = \frac{\sqrt{-\kappa}}{2}$. Following [17], we call this value the critical mean curvature of $\mathbb{E}(\kappa, \tau)$. By the maximum principle, there are no compact CMC surfaces in $\mathbb{E}(\kappa, \tau)$ with $|H| \leq \frac{\sqrt{-\kappa}}{2}$. The class of entire graphs of critical constant mean curvature in $\mathbb{E}(\kappa, \tau)$ was classified in the works [10, 14, 16, 20]; see [11, 17].

Remark 3

1. There exist some rotational CMC spheres in some Berger spheres $\mathbb{E}(\kappa, \tau)$ that are not embedded, i.e., they have self-intersections. These examples were found and classified by Torralbo in [33].
2. The space of minimal spheres inside a Berger sphere $\mathbb{E}(\kappa, \tau)$ agrees with the space of minimal spheres in the unit sphere $\mathbb{S}^3$, i.e., with the space of equators of $\mathbb{S}^3$, when both $\mathbb{E}(\kappa, \tau)$, $\mathbb{S}^3$ are seen as the Lie group SU(2) with two different Riemannian metrics. This was observed by Torralbo in [34]. See Sect. 5 for more details.

Motivated by the Abresch-Rosenberg holomorphic differential, the existence of a holomorphic quadratic differential for CMC surfaces has also been intensively sought in the case that the ambient manifold is a homogeneous three-manifold with a *three*-dimensional isometry group. This holomorphic differential does not seem to exist; however, there is a particular situation in which it does. In [12], Daniel and Mira found a holomorphic quadratic differential defined on any minimal surface in the Thurston geometry Sol_3. Their construction does not work in other homogeneous manifolds or for non-zero values of the mean curvature in Sol_3. The existence of this holomorphic differential is yet to be exploited.

4 General Uniqueness of Immersed Spheres

In this section we present a theory due to Gálvez and Mira [18] that allows to prove uniqueness results for immersed spheres in three-manifolds without having a holomorphic quadratic differential. It is based on elliptic PDE theory and the

Poincaré-Hopf index theorem, and we will use it in the classification of CMC spheres of homogeneous three-manifolds.

Let $M = (M^3, g)$ be an oriented Riemannian three-manifold, and let $TU(M) = \{(p, v) : p \in M, v \in T_pM, |v| = 1\}$ be its unit tangent bundle. Consider a general Weingarten equation for immersed oriented surfaces Σ in M, of the form

$$W(\kappa_1, \kappa_2, x, \eta) = 0, \tag{10}$$

where W is smooth on $\mathbb{R}^2 \times TU(M)$, symmetric in (κ_1, κ_2), and satisfies the ellipticity condition

$$\frac{\partial W}{\partial \kappa_1}\frac{\partial W}{\partial \kappa_2} > 0 \quad \text{in } W^{-1}(0). \tag{11}$$

Here, $\kappa_1, \kappa_2, x, \eta$ are, respectively, the principal curvatures, the position vector, and the unit normal of Σ. Of course, the CMC condition H = constant is a very particular example of (10). In the sequel, we will call H-surface to any surface in M with constant mean curvature $H \in \mathbb{R}$.

Definition 2 We say that a family $\mathcal{F}$ of immersed surfaces in M *foliates* $TU(M)$ if:

1. For each $S \in \mathcal{F}$, the map $q \in S \mapsto (q, \eta(q)) \in TU(M)$ is one-to-one.
2. For every $(p, v) \in TU(M)$ there exists a unique $S \in \mathcal{F}$ with $p \in S$ and $v = \eta(p)$.
3. The family $\mathcal{F}$ depends smoothly on the initial conditions $(p, v) \in TU(M)$.

Theorem 6 shows that one can classify the immersed spheres in M that satisfy (10) as long as there exists a family of surfaces in M satisfying (10) that foliates the unit tangent bundle of M:

Theorem 6 ([18]) *Assume that, on a Riemannian three-manifold M, there exists a family of surfaces $\mathcal{F}$ that foliates $TU(M)$ and whose elements satisfy an elliptic Weingarten equation* (10).

Then, any immersed sphere in M that satisfies (10) *is an element of the family $\mathcal{F}$.*

As an example, given $H > 0$, the family of round spheres in $\mathbb{R}^3$ of radius $1/H$ foliates $TU(\mathbb{R}^3)$. Since all such spheres have constant mean curvature H, we can deduce Hopf theorem from Theorem 6: *any immersed H-sphere in $\mathbb{R}^3$ is a round sphere of radius $1/H$.*

More generally, one can check by ODE analysis that if S_H is a rotational H-sphere in a homogeneous space $\mathbb{E}(\kappa, \tau)$, then the family of all surfaces of the form $\Phi(S_H)$, where Φ is an orientation-preserving isometry of $\mathbb{E}(\kappa, \tau)$, actually foliates the unit tangent bundle of $\mathbb{E}(\kappa, \tau)$. As a consequence, we recover the Abresch-Rosenberg Theorem 5 without using the existence of a holomorphic quadratic differential.

We next explain the idea of the proof of Theorem 6. Let Σ be an immersed sphere in M that satisfies (10). For any $q \in \Sigma$, let S_q denote the element of the family $\mathcal{F}$

that passes through $(q, \eta(q))$. We define the smooth quadratic form Λ on Σ given for any $q \in \Sigma$ by

$$\Lambda(q) := \sigma(q) - \sigma_{S_q}(q), \tag{12}$$

where σ, σ_{S_q} denote, respectively, the second fundamental forms of Σ and of S_q. Thus, Λ compares the second fundamental forms of a *fixed* surface Σ with a *moving* surface S_q that depends on q. The quadratic form Λ vanishes on Σ if and only if Σ is an element of the family $\mathcal{F}$.

Consider local coordinates (x, y, z) on M around q, so that $q \equiv (0, 0, 0)$ and Σ (resp. S_q) is a local graph $z = u(x, y)$ (resp. $z = u_0(x, y)$) in these coordinates. Note that $u = u^0$ and $Du = Du^0$ at the origin. Since both u, u^0 satisfy the same elliptic equation, coming from (10), one can show that $h := u - u_0$ has at the origin a zero of finite order, and the nodal structure of h around the origin is, up to affine transformation, the one of a harmonic function.

On the other hand, if one rewrites (12) for any $p \in \Sigma$ close to q as

$$\Lambda(p) = \sigma(p) - \sigma_{S_q}(p) + \sigma_{S_q}(p) - \sigma_{S_p}(p)$$

and uses the above properties of the function h, a computation shows that, in the (x, y, z) coordinates, we have $\Lambda = D^2 h + \cdots$ (here, the dots mean higher order terms which vanish at q). By our previous control on h, we can then deduce that, if Λ is not identically zero on Σ, then:

1. The zeros of Λ are isolated.
2. Λ is a Lorentzian metric on Σ away from these zeros.
3. Around any zero of Λ, the null line fields L_1, L_2 of the Lorentzian metric Λ have negative (Poincaré-Hopf) index.

On the other hand, since Σ has genus zero, the Poincaré-Hopf theorem implies that the sum of all indices of a continuous line field with isolated singularities in Σ (in particular, both L_1, L_2 are in these conditions) is equal to 2. This is a contradiction. Thus $\Lambda \equiv 0$ on Σ, i.e., Σ is an element of $\mathcal{F}$. This proves Theorem 6.

Remark 4 Note that Theorem 6 reduces the uniqueness of immersed elliptic Weingarten spheres in a given (M^3, g) to the existence of a family $\mathcal{F}$ of such surfaces foliating $TU(M)$.

5 Minimal Spheres in Riemannian Three-Spheres

If we consider the three-sphere $\mathbb{S}^3$ endowed with some Riemannian metric g, it is an important open problem to understand the space of minimal spheres in $(\mathbb{S}^3, g)$. There is a very general existence theorem, due to Simon and Smith [32], showing that this space is always non-empty:

Theorem 7 (Simon-Smith) *On any Riemannian three-sphere* $(\mathbb{S}^3, g)$*, there exists at least one embedded minimal sphere.*

As a matter of fact, it is widely believed that any Riemannian three-sphere $(\mathbb{S}^3, g)$ should admit at least *four* embedded minimal spheres. This statement is a natural analogous of the Lusternik-Schnirelmann theorem, according to which any Riemannian two-sphere $(\mathbb{S}^2, g)$ admits at least three closed embedded geodesics. The number *four* cannot be sharpened, since B. White [37] has proved that there exist some Riemannian three-spheres $(\mathbb{S}^3, g)$ that contain exactly four embedded minimal two-spheres. White's Riemannian metrics are suitable small perturbations of the round metric g_{can} on $\mathbb{S}^3$. For some contributions to this problem, see [6, 13, 19, 23, 36, 37].

Besides existence, another important question is uniqueness. As we already explained in Sect. 3.1, by a theorem of Almgren [1], any immersed minimal sphere in the round unit sphere $\mathbb{S}^3 \subset \mathbb{R}^4$ is an *equatorial* totally geodesic sphere, i.e., the intersection of $\mathbb{S}^3$ with a hyperplane of $\mathbb{R}^4$ containing the origin. See Theorem 3.

The notion of *equatorial sphere* has a natural meaning if we consider a homogeneous three-sphere, i.e. $(\mathbb{S}^3, g_h)$, where g_h is a Riemannian homogeneous metric. In that situation, by classification, $(\mathbb{S}^3, g_h)$ can be seen as the Lie group SU(2) endowed with a left-invariant metric. Now, we can view the Lie group structure SU(2) as the group of unit quaternions with its natural product operation. Specifically, identify $\mathbb{R}^4$ with the space of quaternions $\mathbf{H} := \text{span}\{1, \mathbf{i}, \mathbf{j}, \mathbf{k}\}$ in the usual way. Then, the quaternionic product acts naturally on the unit quaternions $\mathbb{S}^3 \subset \mathbf{H}$, defining a Lie group structure isomorphic to SU(2). The space of unit quaternions with zero real part is then an equatorial sphere $\mathbf{S}$ of $\mathbb{S}^3$, and any other equatorial sphere can be seen as a left translation of $\mathbf{S}$. This allows for a definition of equatorial sphere only in terms of the Lie group structure on $\mathbb{S}^3$, independent of the chosen Riemannian metric. Note, in any case, that two equatorial spheres in a homogeneous three-sphere $(\mathbb{S}^3, g_h)$ are always congruent, since they differ by a left translation of their underlying Lie group structure, and these translations are isometries of the ambient space.

In general, equatorial spheres $\mathbf{S}$ are not totally geodesic in the homogeneous three-sphere $(\mathbb{S}^3, g_h)$. However, we have:

Lemma 1 ([2, 33]) *Any equatorial sphere* $\mathbf{S}$ *in a homogeneous three-sphere* $(\mathbb{S}^3, g_h)$ *is minimal.*

This statement was proved by Torralbo [34] when $(\mathbb{S}^3, g_h)$ is a Berger three-sphere, i.e., when g_h is rotationally invariant. For the general case of an arbitrary homogeneous metric g_h on $\mathbb{S}^3$, the result was recently proved by Ambrozio, Marques, and Neves [2].

The class of immersed minimal spheres in a homogeneous three-sphere $(\mathbb{S}^3, g_h)$ was classified by Meeks, Ros, and the authors in [28]. Specifically, it was shown in [28] that on any homogeneous three-sphere $(\mathbb{S}^3, g_h)$, there exists an embedded minimal two-sphere S_0 and that any other immersed minimal sphere in $(\mathbb{S}^3, g_h)$ is a left translation of S_0 with respect to the underlying Lie group structure of $(\mathbb{S}^3, g_h)$. In particular, up to ambient isometry, there exists a unique minimal sphere in $(\mathbb{S}^3, g_h)$.

Now, using Lemma 1 and Theorem 6, we can make this uniqueness statement more explicit:

Theorem 8 *Any (immersed) minimal sphere in a homogeneous three-sphere* $(\mathbb{S}^3, g_h)$ *is an equatorial sphere* **S**.

To see this, it suffices to make the trivial observation that the family $\mathcal{F}$ of equators in $\mathbb{S}^3$ foliates the unit tangent bundle of $\mathbb{S}^3$; see Definition 2.

Outside the realm of homogeneous three-spheres, the only known classification theorem for immersed minimal spheres in Riemannian three-spheres $(\mathbb{S}^3, g)$ has been very recently obtained by Ambrozio, Marques, and Neves in [2]. In that work, they constructed Riemannian metrics g on $\mathbb{S}^3$ with the following property:

> There exists a smooth family $\mathcal{F} = \{\Sigma_\sigma\}_{\sigma\in\mathbb{RP}^3}$ of embedded minimal two-spheres in $(\mathbb{S}^3, g)$, with the property that for each $x \in \mathbb{S}^3$ and each *two*-dimensional plane $\Pi \subset T_x\mathbb{S}^3$, there is a unique element $\sigma \in \mathbb{RP}^3$ such that $T_x\Sigma_\sigma = \Pi$. Ambrozio, Marques, and Neves called such $\mathcal{F} = \{\Sigma_\sigma\}_\sigma$ a *Zoll family* of minimal spheres.

Besides the clear independent interest of the existence of these Riemannian three-spheres with Zoll families, the result is specially interesting from the viewpoint of uniqueness of minimal spheres. Indeed, it is immediate that the conditions imposed on the definition of a Zoll family $\mathcal{F}$ automatically imply that $\mathcal{F}$ foliates the unit tangent bundle of $\mathbb{S}^3$, following Definition 2. In particular, the Gálvez-Mira uniqueness Theorem 6 applies in this context and allows to prove a full classification of immersed minimal two-spheres in any Riemannian three-sphere with a Zoll family $\mathcal{F}$.

For example, we can extract from [2] the following existence and uniqueness theorem for minimal spheres in Riemannian three-spheres. We should note that the results in [2] are actually more general; for instance, they hold for arbitrary dimension and for more general perturbations of the canonical metric g_{can} in $\mathbb{S}^3$ that the one given in the theorem below.

Theorem 9 ([2]) *Let* $\dot{\rho}$ *denote a smooth odd function on* $\mathbb{S}^3$. *Then, there exists a smooth* 1*-parameter family of smooth functions* $\rho(t)$ *in* $\mathbb{S}^3$, *with* $t \in (-\delta, \delta)$, $\rho(0) = 0$ *and* $\rho'(0) = \dot{\rho}$, *such that* $(\mathbb{S}^3, e^{2\rho(t)}g_{\text{can}})$ *admits a Zoll family* $\mathcal{F}$ *of embedded minimal two-spheres.*

Thus, by Theorem 6, any immersed minimal two-sphere in any such Riemannian three-sphere $(\mathbb{S}^3, e^{2\rho(t)}g_{\text{can}})$ *is one of the embedded minimal spheres of the family* $\mathcal{F}$.

As explained above, any homogeneous three-sphere $(\mathbb{S}^3, g_h)$ has the property that all equatorial spheres are minimal. In [2], Ambrozio, Marques, and Neves showed that there also exist non-homogeneous, antipodally symmetric Riemannian three-spheres $(\mathbb{S}^3, g)$ with this property. Then, by the Gálvez-Mira uniqueness theorem, equators are the only minimal spheres in any such $(\mathbb{S}^3, g)$:

Theorem 10 ([2]) *There exist Riemannian metrics* g *on* $\mathbb{S}^3$ *such that* $(\mathbb{S}^3, g)$ *has a discrete isometry group, and any equatorial sphere of* $\mathbb{S}^3$ *is minimal in* $(\mathbb{S}^3, g)$.

By Theorem 6, any immersed minimal two-sphere in any such $(\mathbb{S}^3, g)$ *is an equatorial sphere.*

6 Uniqueness of Index-One CMC Spheres in Metric Lie Groups

6.1 Basic Stability Properties

We start with some basic definitions regarding stability and index of CMC surfaces. We refer the reader to the handbook [31] by Meeks, Pérez, and Ros for a more detailed approach to this topic.

Let M be a Riemannian manifold and let Σ be a immersed, two-sided[2] hypersurface in M. Then, the *Jacobi operator (stability operator)* of Σ is defined as

$$\mathcal{L} = \Delta + |\sigma|^2 + \mathrm{Ric}(N). \tag{13}$$

Here, Δ is the Laplacian on Σ, $|\sigma|^2$ denotes the square of the norm of the second fundamental form of Σ, and N is the unit normal of Σ. A smooth function on Σ is called a *Jacobi function* if it lies in the kernel of $\mathcal{L}$, i.e., $\mathcal{L}u = 0$.

A domain $\Omega \subset \Sigma$ with compact closure is *stable* if $-\int_\Sigma u\mathcal{L}u \geq 0$ for all compactly supported smooth functions $u \in C_0^\infty(\Omega)$. The *index* of a domain $\Omega \subset \Sigma$ with compact closure is the number of negative eigenvalues of $-\mathcal{L}$ on Ω; thus, Ω is stable if and only if its index is zero. Since the index of stability is non-decreasing with the respect to the inclusion of subdomains of Σ with compact closure, one can define the *index of stability* of Σ as the supremum of the indices over any increasing sequence of subdomains $\Omega_i \subset \Sigma$ with compact closure and $\cup_i \Omega_i = \Sigma$. Σ is called *stable* if its index is zero.

Assume next that the ambient manifold M is a metric Lie group X and $\Sigma = S_H$ is an H-sphere in X. Then, the index of S_H is at least 1. This is due to the fact that on X, there exist three linearly independent, right-invariant Killing vector fields F_1, F_2, F_3 (see Fact 2) and that the functions $u_i := \langle F_i, N\rangle$, $i = 1, 2, 3$ are Jacobi functions on S_H. Since right-invariant vector fields on X are either non-vanishing or identically zero, and tangent vector fields on spheres always vanish somewhere, we deduce that the functions $u_i = \langle F_i, N\rangle$, $i = 1, 2, 3$, are linearly independent, i.e., $\mathrm{Ker}(\mathcal{L})$ has dimension ≥ 3. Thus, 0 cannot be the first eigenvalue of $-\mathcal{L}$, and so the index of S_H is at least one.

On the other hand, it follows from Theorem 3.4 in Cheng [8] that if S_H is an index-one H-sphere, then $\mathrm{Ker}(\mathcal{L})$ always has dimension *at most* three. We can deduce then that $\mathrm{Ker}(\mathcal{L})$ is actually three-dimensional, and it is spanned by the

[2] Σ is two-sided if there exists a globally defined unit normal vector field N to Σ.

geometric Jacobi functions u_i defined above. We say then that the index-one H-sphere S_H has *nullity three.*

The index-one condition is related to (weak) stability of CMC surfaces. Here, we say that a compact surface S is *weakly stable* if $-\int_S u\mathcal{L}u \geq 0$ for any $u \in C^\infty(S)$ satisfying $\int_S u = 0$. Solutions to the isoperimetric problem are compact, embedded, weakly stable CMC surfaces. Also, weakly stable surfaces have index 0 or 1. In particular, when X is a metric Lie group, any compact, weakly stable CMC surface (in particular, any solution of the isoperimetric problem) has index one. This follows from the more general fact that there are no compact CMC surfaces of index 0 in X. Indeed, if Σ was such a surface, $\mathrm{Ker}(\mathcal{L})$ would be *one*-dimensional. In particular, there would exist two linearly independent right invariant vector fields F_1, F_2 on X for which $\langle F_i, N\rangle = 0, i = 1, 2$. By left translating Σ so that it contains the identity element e of X, this means that Σ would be a compact *two*-dimensional subgroup of X. But such subgroups do not exist.

The geometric properties of index-one H-spheres in a metric Lie group X will be of crucial importance for classifying all H-spheres in X. We actually seek to prove that any H-sphere in X has index one. However, the approach to prove this result is indirect. We will first show a uniqueness property for index-one H-spheres via their Gauss map.

6.2 Gauss Map and Uniqueness of Index-One Spheres

An important tool for the study of CMC surfaces in metric Lie groups is the left-invariant Gauss map that we define below. When X is $\mathbb{R}^3$ with its usual abelian Lie group structure, this definition coincides with the usual Gauss map for surfaces in $\mathbb{R}^3$.

Definition 3 Let Σ be an immersed oriented surface in a metric Lie group X. The *left-invariant Gauss map* of Σ is the map $g : \Sigma \to \mathbb{S}^2$ taking values in the unit sphere of the tangent space T_eX at the identity element $e \in X$, obtained by left-translating the unit normal of Σ at each $p \in \Sigma$ to T_eX.

A connected oriented surface[3] in a metric Lie group has constant left-invariant Gauss map if and only if it is a left coset of some two-dimensional subgroup; see Lemma 3.9 in [30].

Contrary to what happens in Euclidean three-space, this left-invariant Gauss map $g : \Sigma \to \mathbb{S}^2$ is not a harmonic map for CMC surfaces in a general metric Lie group X. Still, it satisfies a nice complex PDE, of the form

$$g_{z\bar{z}} = A(g, \bar{g})g_z g_{\bar{z}} + B(g, \bar{g})|g_z|^2. \tag{14}$$

[3] This property holds true for connected oriented hypersurfaces in an $(n + 1)$-dimensional metric Lie group.

Here, g is viewed as a map $g : \Sigma \to \mathbb{C} \cup \{\infty\}$ after stereographic projection on T_eX with respect to an adequately chosen orthonormal basis $\{E_1, E_2, E_3\}$ of T_eX. The functions A, B are rational functions on $g, \overline{g}$, that depend on the algebraic structure constants of the metric Lie group X and on the constant value H of the mean curvature; see [28] for the specific form of these coefficients. For the case where $X = \text{Sol}_3$, this equation was obtained by Daniel and Mira [12].

One also has a converse to this result, in the form of a representation theorem. This result roughly says that any solution g to (14) on a simply connected Riemann surface Σ is the left-invariant Gauss map of an H-surface in the metric Lie group X. See [28, Theorem 3.7].

The next result is key for the classification of CMC spheres in metric Lie groups:

Theorem 11 ([28]) *If S_H is an index-one H-sphere in a metric Lie group X, then the left-invariant Gauss map of S_H is an orientation preserving diffeomorphism onto $\mathbb{S}^2$.*

We give a sketch of the proof. Assume that the left-invariant Gauss map g of the index-one H-sphere S_H has a point z_0 where dg has rank ≤ 1. Then, g has a first-order contact at z_0 with another solution $\widehat{g}$ to (14) that depends only on one variable. Using the representation theorem, $\widehat{g}$ generates an H-surface Σ_0 in Σ that has at z_0 a second-order contact with S_H and which is invariant by a 1-parameter group of left translations of X (since $\widehat{g}$ only depends on one variable). In particular, there exists a right-invariant Killing vector field F in X that is everywhere tangent to Σ_0. Therefore, F is tangent to S_H at z_0. This implies that the Jacobi function $u := \langle F, N\rangle$ on S_H satisfies $u(z_0) = (\nabla u)_{z_0} = 0$. In particular, the nodal set $u^{-1}(0)$ of u around z_0 consists of an equiangular system of $n \geq 2$ arcs intersecting at z_0. On the other hand, as a consequence of Courant's nodal domain theorem, a second eigenfunction of $-\mathcal{L}$ on a two-sphere must have a regular, embedded curve as a nodal set; see Corollary 3.5 in [8]. Since S_H has index one, 0 is a second eigenvalue of $-\mathcal{L}$, and so the above behavior for the Jacobi function u is not possible. This contradiction proves that $g \colon S_H \to \mathbb{S}^2$ is a local diffeomorphism, from where Theorem 11 follows.

Theorem 11 implies that if we consider the family $\mathcal{F}$ of all possible left translations in X of the index-one H-sphere S_H, then $\mathcal{F}$ foliates the unit tangent bundle of X, as in Definition 2. In particular, we have directly from Theorem 6 and Theorem 11:

Corollary 1 ([28]) *If S_H is an index-one sphere in a metric Lie group X for some value $H \in \mathbb{R}$, then any other immersed H-sphere Σ in X is a left-translate of S_H.*

We remark that Theorem 11 and Corollary 1 were proved by Daniel-Mira [12] in the case that X is the Thurston geometry Sol_3.

7 Classification of CMC Spheres in Homogeneous Three-Manifolds

In this section, we will state the results that classify all immersed CMC spheres in any simply connected homogeneous three-manifold M. As explained in Sect. 2, M must be homeomorphic to either $\mathbb{S}^3$, $\mathbb{R}^3$, or $\mathbb{S}^2 \times \mathbb{R}$. We discuss each of these topological cases separately.

First of all, recall that if M is homeomorphic to $\mathbb{S}^2 \times \mathbb{R}$, then M is actually isometric to a Riemannian product $\mathbb{S}^2(c) \times \mathbb{R}$. The classification of CMC spheres in $\mathbb{S}^2(c) \times \mathbb{R}$ was obtained by Abresch-Rosenberg [3]: for any $H \in \mathbb{R}$ there exists a unique immersed H-sphere S_H in $\mathbb{S}^2(c) \times \mathbb{R}$. This sphere is actually embedded and rotationally symmetric. If $H \neq 0$, S_H has index-one. The minimal spheres in $\mathbb{S}^2(c) \times \mathbb{R}$ are the totally geodesic equators $\mathbb{S}^2(c) \times \{t_0\}$, whose index is zero.

In the next two sections, we describe the classification for the cases that M is diffeomorphic to $\mathbb{S}^3$ and to $\mathbb{R}^3$, respectively. One should observe that, by Fact 2, any such homogeneous manifold M can be seen as a metric Lie group X.

7.1 CMC Spheres in Homogeneous Three-Spheres

In Theorem 12, X denotes a homogeneous three-sphere, i.e., $X = (\mathbb{S}^3, g)$ where g is a homogeneous metric on $\mathbb{S}^3$. By classification, X can be seen as the Lie group SU(2) endowed with a left-invariant metric.

Theorem 12 ([28]) *Let $X = (\mathbb{S}^3, g)$ be a homogeneous three-sphere. Then:*

1. *For every $H \in \mathbb{R}$, there exists an H-sphere S_H in X.*
2. *Up to ambient isometry, S_H is the unique H-sphere in X.*
3. *There exist two points in X, called the centers of symmetry of S_H, such that the isometries of X that fix any of these points also leave S_H invariant.*
4. *S_H is* Alexandrov embedded, *i.e., the immersion $f : S_H \looparrowright X$ of S_H in X can be extended to an isometric immersion $F : B \to X$ of a Riemannian three-ball such that $\partial B = S_H$ is mean convex.*
5. *S_H has index one and nullity three for the Jacobi operator.*

Moreover, let $\mathcal{M}_X$ be the set of oriented immersed spheres of constant mean curvature in X whose center of symmetry is a given point $e \in X$. Then, $\mathcal{M}_X$ is an analytic family $\{S(t)\}_{t\in\mathbb{R}}$ parameterized by the mean curvature value t of $S(t)$.

Let us make some comments regarding this theorem.

1. Clearly, allowing the values of the mean curvature H to move in $[0, \infty)$ is just a matter of choosing an orientation on the H-surfaces; after doing this, the center of symmetry is also well-defined if we choose it in the mean convex side.
2. The uniqueness of each S_H among H-spheres is not only up to ambient isometry of X, but up to *left translation* of X. That is, any H-sphere of X is a left

translation of S_H. This is the key aspect in proving the existence of the center of symmetry of S_H. The fact that S_H is invariant by all isometries of X that fix its center of symmetry imply that S_H is round if X is the standard round sphere (Theorem 3) and that S_H is rotational if X is a Berger three-sphere $\mathbb{E}(\kappa, \tau)$, for $\kappa > 0, \tau \neq 0$ (Theorem 5).
3. One cannot claim that S_H is embedded in general. As explained in Sect. 3.2, Torralbo found examples of non-embedded CMC spheres in Berger three-spheres.
4. Recall from Sect. 6.1 that any compact, weakly stable H-surface in a metric Lie group X has index one. The H-spheres S_H of Theorem 12 have index one. However, Torralbo and Urbano [35] found some examples of such H-spheres S_H in Berger three-spheres that are not weakly stable. In other words, the Alexandrov embeddedness and index-properties of S_H in Theorem 12 cannot be improved to embeddedness or weak stability.
5. Recall from Sect. 5 that there exists a unique minimal sphere S_0 in $X = (\mathbb{S}^3, g)$. This sphere S_0 is embedded, and it is actually explicit, as it is an equatorial sphere of $\mathbb{S}^3$.
6. By Theorem 11, the left-invariant Gauss map of S_H is a diffeomorphism.

7.2 CMC Spheres in Metric Lie Groups Diffeomorphic to $\mathbb{R}^3$

In Theorem 13, X denotes a homogeneous three-manifold diffeomorphic to $\mathbb{R}^3$. By Fact 2, X can be seen as a metric Lie group. In this situation, there are no compact minimal surfaces in X; this follows by a simple application of the maximum principle and the fact that there exist entire graphs in X with constant mean curvature. In particular, the existence part of Theorem 12 does not hold in this non-compact context, and in order to classify CMC spheres, one needs to determine the values of the mean curvature H for which such spheres exist. This will be done in terms of the *Cheeger constant* of X.

The *Cheeger constant* $\mathrm{Ch}(Y)$ of a Riemannian manifold Y with infinite volume is defined as the infimum of the quotients

$$\frac{\mathrm{Area}(\partial D)}{\mathrm{Volume}(D)}$$

where D is any smooth compact domain in Y. If $Y = X$ is a homogeneous manifold diffeomorphic to $\mathbb{R}^3$, the value of $\mathrm{Ch}(X)$ is explicitly known, except when X is isometric to the universal cover of the special linear group $\widetilde{\mathrm{SL}}(2, \mathbb{R})$ endowed with a left-invariant metric with a three-dimensional isometry group. For such a space $X = (\widetilde{\mathrm{SL}}(2, \mathbb{R}), \langle , \rangle)$, it is still known that $\mathrm{Ch}(X)$ is positive; see [27]. In general, for any homogeneous manifold X diffeomorphic to $\mathbb{R}^3$, the value $\mathrm{Ch}(X)/2$ agrees with the so-called critical mean curvature $H(X)$ of the space X, defined by

$$H(X) := \inf\{\max_\Sigma |H_\Sigma| : \Sigma \in \mathcal{A}\},$$

where $\mathcal{A}$ is the space of all compact orientable surfaces immersed in X and H_Σ denotes the mean curvature function of any $\Sigma \in \mathcal{A}$. See Theorem 1.4 in [27].

For instance, $H(X) = 0 = \mathrm{Ch}(X)/2$ if X is the Euclidean three-space, and $H(X) = 1 = \mathrm{Ch}(X)/2$ if X is the hyperbolic three-space $\mathbb{H}^3$. For a metric semidirect product $X = \mathbb{R}^2 \rtimes_A \mathbb{R}$, it holds $2H(X) = \mathrm{Ch}(X) = \mathrm{trace}(A)$.

Theorem 13 ([29]) *Let X be a homogeneous manifold diffeomorphic to $\mathbb{R}^3$, and let* $\mathrm{Ch}(X)$ *denote the Cheeger constant of X. Then:*

1. *For any $H \in \mathbb{R}$ with $|H| > \mathrm{Ch}(X)/2$, there exists an H-sphere S_H in X. Conversely, if $|H| \leq \mathrm{Ch}(X)/2$, there are no compact H-surfaces in X.*
2. *Up to ambient isometry, S_H is the unique H-sphere in X.*
3. *There exists a well-defined point in X called the* center of symmetry *of S_H such that the isometries of X that fix this point also leave S_H invariant.*
4. *S_H is* Alexandrov embedded.
5. *S_H has index one and nullity three for the Jacobi operator.*

Moreover, if $\mathcal{M}_X(p)$ denotes the space of spheres of positive constant mean curvature in X that have a base point $p \in M$ as a center of symmetry, then the map $S_H \in \mathcal{M}_X(p) \mapsto H \in (\mathrm{Ch}(X)/2, \infty)$ that assigns to each sphere S_H its mean curvature H is a real analytic homeomorphism between $\mathcal{M}_X(p)$ and $(\mathrm{Ch}(X)/2, \infty)$.

As we already discussed in connection with Theorem 12, S_H is actually unique up to left translations in X, and the left-invariant Gauss map of S_H is a diffeomorphism. Also, the invariance of S_H by isometries of X fixing its center of symmetry implies that S_H is as symmetric as the space X allows. That is, S_H is round if X has constant curvature, it is rotationally symmetric if X is an $\mathbb{E}(\kappa, \tau)$-space, and it inherits all symmetries of the (discrete) isotopy group of X if the isometry group of X is three-dimensional.

We do not know if all the spheres S_H are embedded. In relation with this, it is an open problem to determine if every sphere S_H of X bounds an isoperimetric region of X; see Sect. 9.

8 A Sketch of Proof of the Classification

Let X denote a metric Lie group. The first observation is that, in X, one can solve the isoperimetric problem for any small enough volume $V > 0$ (in fact, we can solve this problem for *any* $V > 0$, since the quotient of X by its isometry group is compact). These small isoperimetric regions of X are bounded by embedded, almost round, H-spheres S_H with $H > 0$ very large. By our discussion in Sect. 6.1, these isoperimetric spheres S_H have index one, and the kernel of their Jacobi operator $\mathcal{L}$ satisfies

$$\mathrm{Ker}(\mathcal{L}) = \mathrm{Span}\{u_1, u_2, u_3\} \tag{15}$$

for the Jacobi functions $u_i := \langle N, F_i \rangle$, where $\{F_1, F_2, F_3\}$ is a basis of right-invariant Killing vector fields of X.

Because of this control on the kernel of $\mathcal{L}$, one can use the implicit function theorem to deform analytically any index-one H-sphere S_H of X into a 1-parameter family of CMC spheres $\{S_{H'}\}_{H'}$, with $H' \in (H - \varepsilon, H + \varepsilon)$ (here $\varepsilon > 0$ is a small number depending on H). Moreover, all these spheres $S_{H'}$ also have index one and nullity three, by continuity of eigenvalues in the deformation and (15). Therefore, the set of values $H \geq 0$ for which there exists an index-one H-sphere in X is open in $[0, \infty)$, and it contains some interval of the form (a, ∞). Let $\mathcal{J} \subset [0, \infty)$ be the largest such interval. Then, either $\mathcal{J} = [0, \infty)$ or $\mathcal{J} = (h_0(X), \infty)$, for some $h_0(X) \geq 0$.

One should observe here that the properties of H-spheres in X stated in Theorems 12 and 13 hold for every value $H \in \mathcal{J}$. Indeed, for any $H \in \mathcal{J}$, there exists an index-one H-sphere S_H, which by uniqueness (Corollary 1) comes from the above deformation from an isoperimetric sphere. S_H is the unique immersed H-sphere in X (again by Corollary 1), and it is Alexandrov embedded since that property is preserved in the deformation process. The fact that S_H is unique not only up to ambient isometry, but up to left translation, can be used to show the existence of a center of symmetry for S_H (see Section 6 in [28]).

Thus, in order to prove Theorems 12 and 13, we only need to show existence of index-one H-spheres for the expected range of mean curvatures, that is,

(i) If X is a homogeneous three-sphere, then $\mathcal{J} = [0, \infty)$, and
(ii) if X is diffeomorphic to $\mathbb{R}^3$, then $\mathcal{J} = (H(X), \infty)$, where $H(X) = \mathrm{Ch}(X)/2$ is the critical mean curvature of X.

We will discuss these two cases separately. We note however that some of the arguments in the case that X is a three-sphere are also used when X is diffeomorphic to $\mathbb{R}^3$.

8.1 The Case Where X Is a Homogeneous Three-Sphere

Let X be a homogeneous three-sphere. Arguing by contradiction, assume that $\mathcal{J} \neq [0, \infty)$, i.e., it is of the form $\mathcal{J} = (h_0(X), \infty)$ for some $h_0(X) \geq 0$. Thus, for any $H > h_0(X)$ there exists an index-one H-sphere S_H, which is unique up to left translation in X due to Corollary 1, and there are no index-one spheres of constant mean curvature $h_0(X)$ in X. By elliptic theory and openness, this means that for any sequence of values $(H_n)_n \in \mathcal{J}$ decreasing to $h_0(X)$, one of the following two obstructions must appear:

1. The norms of the second fundamental forms of the spheres S_{H_n} are not bounded.
2. The areas of the spheres S_{H_n} are not bounded.

The first problem can be ruled out by a blow-up analysis. Roughly, if there is a sequence of index-one spheres S_{H_n} whose maxima of their norms of the second fundamental form diverge, then after a sequence of left translations, a standard blow-up, and passing to a subsequence, they converge to a complete minimal surface Σ_∞ immersed in $\mathbb{R}^3$ with a non-planar point at the origin. Thus, the Gauss map of Σ_∞ is a local diffeomorphism around the origin which locally reverses orientation. However, the left-invariant Gauss maps of the spheres S_{H_n} are orientation preserving diffeomorphisms, by Theorem 11. This gives a contradiction, which proves a uniform *curvature estimate* for index-one spheres S_H, when H varies in a bounded interval of $\mathcal{J}$.

Thus, we only need to rule out the second possible obstruction, i.e., we need to obtain an *area estimate* for the spheres S_H in $\mathcal{J}$.

Arguing by contradiction, assume that $(S_n)_n$ is a sequence of index-one spheres with mean curvatures $H_n \in \mathcal{J}$ converging to $h_0(X)$. We assume up to left translation that all of them pass through the identity element e of X and have the same unit normal vector at e. Moreover, assume that $\text{Area}(S_n) \to \infty$. Since we have proved that the norms of the second fundamental forms of the spheres S_n is uniformly bounded, we can use elliptic theory to show that, up to subsequence, the S_n converge uniformly on compact sets to a complete, non-compact limit immersion $\Sigma \subset X$ of constant mean curvature $h_0(X)$ that passes through e. Take now a divergent sequence $(q_n)_n \in \Sigma$, and let $(\Sigma_n)_n$ be the surfaces obtained from Σ by left translation, $\Sigma_n := (q_n)^{-1}\Sigma$. By the previous argument, these surfaces $(\Sigma_n)_n$ converge again (up to extracting a subsequence) to a complete surface Σ_∞ immersed in X of constant mean curvature $h_0(X)$ that passes through the origin. Moreover, Σ_∞ has the following fundamental properties:

(a) *The left-invariant Gauss map of* Σ_∞ *is singular at every point*, i.e., there exists a right-invariant vector field $F \in \mathfrak{X}(X)$ everywhere tangent to Σ_∞. Indeed, assume that there is some point of Σ_∞ around which the Gauss map of Σ_∞ is a local diffeomorphism. Then, there exists a sequence of disjoint compact domains $\Omega_n \subset \Sigma$ whose spherical Gauss map areas are bounded from below by some constant $\varepsilon > 0$. Thus, the spherical areas of the $(S_n)_n$ should be infinite, what is impossible, as all these spherical areas are actually 4π, since the Gauss map of S_n is a diffeomorphism.
(b) Σ_∞ *is stable*. Indeed, if Σ_∞ had a strictly unstable compact domain, there would exist a sequence of disjoint strictly unstable compact domains in Σ. This contradicts that all the $(S_n)_n$ have index one.

We finally prove that such a surface Σ_∞ cannot exist. First, since X is isomorphic to SU(2), the fact that Σ_∞ is everywhere tangent to a right-invariant $F \in \mathfrak{X}(M)$ implies that Σ_∞ is actually a *Hopf cylinder*, i.e., the lift via the Hopf fibration $\pi : \mathbb{S}^3 \to \mathbb{S}^2$ of a complete curve in $\mathbb{S}^2$. Thus, Σ_∞ is topologically either a torus or a cylinder.

The case that Σ_∞ is a torus can be easily ruled out as follows. If V denotes a right-invariant Killing field in X linearly independent from F that is tangent to Σ_∞ at e, then the Jacobi function $u := \langle V, N\rangle$ vanishes at e. If $u \equiv 0$ on Σ, then both F, V are everywhere tangent to Σ, and so, the Lie Bracket $[F, V]$ is everywhere tangent to Σ; this is impossible, since $F, V, [F, V]$ are linearly independent everywhere in X (because SU(2) has no *two*-dimensional subgroups). Therefore, u must change sign on the torus Σ_∞; this contradicts stability.

Assume next that Σ_∞ is a cylinder. Then, one can show that Σ_∞ has at most linear area growth. This follows from two facts: firstly, that for every $R_0 > 0$, there exists a constant $C = C(R_0) > 0$ such that every intrinsic ball in Σ_∞ of radius R_0 has area less than C (this follows from Bishop's second theorem, since the Gaussian curvature of Σ_∞ is bounded from below), and secondly, that the fibers $\pi^{-1}(p)$ that foliate Σ_∞ are embedded closed curves in X of uniformly bounded length. The *at most linear* area growth property implies that Σ_∞ is conformally parabolic. Also, by the arguments in the previous paragraph, we can create a bounded Jacobi function on Σ_∞ that changes sign; this contradicts stability in this parabolic setting by a theorem of Manzano, Pérez, and Rodríguez [24]. This final contradiction proves Theorem 12.

8.2 The Case Where X Is Diffeomorphic to $\mathbb{R}^3$

We now let X be a metric Lie group diffeomorphic to $\mathbb{R}^3$. Take $\mathcal{J} = (h_0(X), \infty)$, and let $(S_n)_n$ be a sequence of index-one spheres with mean curvatures $H_n \in \mathcal{J}$ converging to $h_0(X)$. Following the discussion in Sect. 8.1, there exists a complete, stable limit immersion $f : \Sigma_\infty \looparrowright X$ of constant mean curvature $h_0(X)$ of adequate left translations of the H-spheres S_n, such that $e \in f(\Sigma_\infty)$. Moreover, there exists a right-invariant Killing vector field $F \in \mathfrak{X}(X)$ everywhere tangent to $f(\Sigma_\infty)$. Besides, Σ_∞ can be proven to have, at most, quadratic area growth (see Corollary 6.7 in [29]).

We want to show that $h_0(X) = H(X)$, the critical mean curvature of X. For that, it suffices to prove:

Theorem 14 *The limit surface $f(\Sigma_\infty)$ can be chosen to be an entire Killing graph in X.*

Here, we say that a surface $S \subset X$ is an entire Killing graph if there exists a right-invariant vector field $V \in \mathfrak{X}(X)$ such that every orbit of V intersects S exactly once, transversally. Note that if $S \subset X$ is an entire Killing H-graph, then $H \leq H(X)$; this follows by the mean curvature comparison principle, the definition of $H(X)$, and the fact that X is diffeomorphic to $\mathbb{R}^3$. In particular, Theorem 14 indeed implies that $h_0(X) = H(X)$, and this would complete the proof of Theorem 13.

We now explain some of the main ideas involved in the proof of Theorem 14. First, since the Gauss map of $f : \Sigma_\infty \looparrowright X$ is singular, it is either constant (in which case $f(\Sigma_\infty)$ is a two-dimensional subgroup of X as mentioned in Sect. 6.2, and thus, Theorem 14 holds), or its Gauss map image $\gamma := g(\Sigma_\infty)$ into $\mathbb{S}^2 \subset T_eX$ is a complete regular curve. In the sequel, we will assume this last case holds. One of the key steps for proving Theorem 14 is:

Claim The limit immersion $f : \Sigma_\infty \looparrowright X$ can be chosen so that its Gauss map image γ is a closed, regular embedded curve in $\mathbb{S}^2$.

To prove Claim 8.2, one first ensures that γ is embedded. Indeed, γ cannot have tangential self-intersections by ODE uniqueness and the Gauss map equation (14). Besides, if γ had a non-tangential self-intersection, we would contradict the fact that $f : \Sigma_\infty \looparrowright X$ is a limit of the spheres $(S_n)_n$, all of which satisfy that their Gauss maps are diffeomorphisms. In addition, by the curvature estimate for the spheres S_n proved in Sect. 8.1, the curve γ has bounded geodesic curvature in $\mathbb{S}^2$.

These properties imply that the closure $\overline{\gamma}$ has the structure of a one-dimensional lamination of $\mathbb{S}^2$, and it can be proven that the leaves of this lamination are precisely the Gauss map images of limits of subsequences of the spheres $(S_n)_n$ (Lemma 5.10 in [29]). This indicates that there are two possibilities: either γ is a closed curve (as we want to prove), or, by taking a sublamination of $\overline{\gamma}$ with no proper sublaminations and applying Zorn's lemma, γ must be *quasiperiodic*, in the sense that given any compact arc $\sigma \subset \gamma$ there must exist a sequence of disjoint compact arcs $\sigma_n \subset \gamma$ with $(\sigma_n)_n \to \sigma$.

The argument to rule out this second case is quite involved and will not be described here. In it, one needs to distinguish different cases depending on the algebraic structure of the metric Lie group X and the right-invariant vector field F on X that is everywhere tangent to $f(\Sigma_\infty)$.

Once Claim 8.2 is proved, it follows that there are three possibilities for the complete, stable, $h_0(X)$-immersion $f : \Sigma_\infty \looparrowright X$:

(L1) Σ_∞ is diffeomorphic to an annulus, and it has linear area growth.
(L2) Σ_∞ is simply connected, $f(\Sigma_\infty)$ is an immersed annulus in X, and Σ_∞ has quadratic area growth.
(L3) Σ_∞ is simply connected and there exists $a \in X - \Gamma$ such that the left translation by a in X leaves $f(\Sigma_\infty)$ invariant. Here, Γ is the 1-parameter subgroup of X that generates the right-invariant vector field F.

Once here, we again need to split the proof of Theorem 14 into separate cases. We give a very brief outline of the argument.

First, assume that $X = \mathbb{R}^2 \rtimes_A \mathbb{R}$. If case (L3) above holds, then we showed in [29, Theorem 7.2] that the right-invariant Killing field F is *horizontal*, i.e., it is a linear combination of the vector fields F_1, F_2 in (6). Once we know this, we showed in [29, Lemma 7.3] that $f(\Sigma_\infty)$ is an entire graph with respect to any horizontal Killing field V linearly independent from F. Therefore, Theorem 14 holds in case (L.3) above. So, it remains to rule out cases (L1) and (L2) above. The argument to rule out (L2) is based on the construction of a sequence of geodesic balls of a certain fixed radius $R^* > 0$ in the abstract Riemannian three-balls B_n that the Alexandrov-embedded index-one spheres S_n bound and whose volumes tend to infinity as $n \to \infty$. This unbounded volume result eventually provides a contradiction with Bishop's theorem; see Section 7.3 in [29]. Finally, the argument to rule out (L1) is based on the construction of an abstract three-dimensional Riemannian cylinder bounded by Σ_∞ that submerses isometrically into X with boundary the cylinder $f(\Sigma_\infty)$ and then proving that a certain CMC flux of Σ_∞ in this abstract cylinder is different from zero. This gives a contradiction with the homological invariance of the CMC flux and the fact that $f(\Sigma_\infty)$ is a limit of the (homologically trivial) Alexandrov-embedded constant mean curvature spheres S_n; see Section 7.4 in [29].

The argument and basic strategy of the proof of Theorem 14 when X is isomorphic to $\widetilde{\mathrm{SL}}(2,\mathbb{R})$ follow closely those from the case $X = \mathbb{R}^2 \rtimes_A \mathbb{R}$. However, several of these arguments are by necessity different, as many geometric properties of metric semidirect products do not have analogous counterparts in $\widetilde{\mathrm{SL}}(2,\mathbb{R})$. The details are quite technical, so they will not be discussed here; the interested reader can find them in Section 8 of [29].

9 Open Problems

We start by considering the case of constant mean curvature spheres in homogeneous three-manifolds. The most ambitious theorem here would probably be to prove the following conjecture by Meeks, Ros, and the authors.

Conjecture 1 ([29]) Let M be a homogeneous manifold diffeomorphic to $\mathbb{R}^3$. Then, any isoperimetric region of M is bounded by a CMC sphere, and conversely, any CMC sphere of M bounds an isoperimetric region.

One should recall here that any isoperimetric region of a homogeneous three-manifold is bounded by a compact, embedded, weakly stable CMC surface.

The conjecture is known to hold only in the Euclidean plane $\mathbb{R}^3$, the hyperbolic space $\mathbb{H}^3(c)$, and the product space $\mathbb{H}^2(c) \times \mathbb{R}$. It is interesting to comment what is known about it in other spaces, since some partial results toward Conjecture 1 would also be of great interest.

To start, consider the Heisenberg space Nil_3. By the Abresch-Rosenberg theorem, all CMC spheres in Nil_3 are rotational. They are also weakly stable, a necessary condition for them bounding isoperimetric regions. However, it is not known if isoperimetric regions in Nil_3 are topological balls. The problem here is that Nil_3 does not admit reflections, i.e., all its isometries are orientation preserving. In particular, one cannot apply the Alexandrov reflection technique (see [5]) (or other methods) to deduce that isoperimetric regions are simply connected. In this respect, the following well-known conjecture would actually prove Conjecture 1 for $X = \mathrm{Nil}_3$.

Conjecture 2 Any compact, embedded CMC surface in Nil_3 is a CMC sphere.

We now consider the case $M = \mathrm{Sol}_3$. This time, Sol_3 admits two linearly independent families of planes of reflective symmetry. By applying the Alexandrov reflection technique with respect to them, Rosenberg proved that any compact embedded CMC surface in Sol_3 is, topologically, a sphere; see [12]. In particular, isoperimetric regions of Sol_3 are bounded by CMC spheres S_H, and these spheres were classified by Daniel-Mira and Meeks in [12, 25]. All these spheres are embedded and have index one. However, it is not known if they are weakly stable. In this respect, the following conjecture was proposed by Daniel and Mira:

Conjecture 3 ([12]) Any CMC sphere in Sol_3 is weakly stable.

A positive answer to this problem can be used to prove that all CMC spheres in Sol_3 bound isoperimetric regions.

For the general case of homogeneous three-manifolds, the following result is expected:

Conjecture 4 Let M be a homogeneous manifold diffeomorphic to $\mathbb{R}^3$. Then, any CMC sphere in M is embedded.

Conjecture 4 is obviously a particular case of Conjecture 1. By Theorem 13, H-spheres in such a homogeneous manifold M are embedded for large values of H, and they constitute a real analytic family as H decreases. A possible way to prove embeddedness would then be to consider the first value $H_0 > 0$ where S_{H_0} fails to be embedded and look for a contradiction.

The are some cases where Conjecture 4 has been proved. For the case where the homogeneous manifold M admits an *algebraic open book structure*, Meeks and the authors showed that CMC spheres in M are embedded; see [26]. For the case where M is Sol_3, the embeddedness of CMC spheres was proved by Daniel and Mira [12].

We close the survey with a uniqueness problem regarding the results on minimal spheres in Riemannian three-spheres $(\mathbb{S}^3, g)$ described in Sect. 5. Let $E \subset \mathbb{R}^4$ be a *three*-dimensional ellipsoid

$$E = \{x = (x_1, \ldots, x_4) \in \mathbb{R}^4 : \sum a_i^2 x_i^2 = c^2\},$$

for some $a_1, \ldots, a_4, c > 0$. The intersection of $S_i := E \cap \{x_i = 0\}$ with each of the coordinate hyperplanes $x_i = 0$, $i \in \{1, \ldots, 4\}$, is an embedded minimal two-sphere in E. Yau asked in 1987 if these four *equatorial* spheres S_i are the only embedded minimal two-spheres in E.

In [19], Haslhofer and Ketover gave a negative answer to Yau's question, in the case of ellipsoids which are very elongated in one direction. In [6], Bettiol and Piccione constructed embedded minimal two-spheres in sufficiently elongated ellipsoids of revolution. These minimal spheres are rotationally invariant. An interesting problem related to Yau's question is to understand whether any immersed minimal two-sphere in an ellipsoid of revolution is either a rotational sphere or one of the equatorial spheres of the ellipsoid.

Acknowledgments This research has been financially supported by Projects PID2020-118137GB-I00 and PID2020-117868GB-I00 by MINECO/MICINN/FEDER; the "Maria de Maeztu" Excellence Unit IMAG, funded by MCINN/AEI/10.13039/ 501100011033/CEX2020-001105-M; Junta de Andalucia grants no. P18-FR-4049 and A-FQM-139-UGR18; and CARM, Programa Regional de Fomento de la Investigación, Fundación Séneca-Agencia de Ciencia y Tecnología Región de Murcia, reference 21937/PI/22.

References

1. Almgren, F.J.: Some interior regularity theorems for minimal surfaces and an extension of Bernstein's theorem. Ann. Math. **84**, 277–292 (1966)
2. Ambrozio, L., Marques, F.C., Neves, A.: Riemannian metrics on the sphere with Zoll families of minimal hypersurfaces. Preprint, arXiv:2112.01448v1
3. Abresch, U., Rosenberg, H.: A Hopf differential for constant mean curvature surfaces in $\mathbb{S}^2 \times \mathbb{R}$ and $\mathbb{H}^2 \times \mathbb{R}$. Acta Math. **193**, 141–174 (2004)
4. Abresch, U., Rosenberg, H.: Generalized Hopf differentials. Mat. Contemp. **28**, 1–28 (2005)
5. Alexandrov, A.D.: Uniqueness theorems for surfaces in the large, I. Vestnik Leningrad Univ. **11**, 5–17 (1956). (English translation: Am. Math. Soc. Transl. **21**, 341–354 (1962)).
6. Bettiol, R.G., Piccione, P.: Nonplanar minimal spheres in ellipsoids of revolution. Preprint, arXiv:2111.14995
7. Chern, S.S.: On surfaces of constant mean curvature in a three-dimensional space of constant curvature. In Palis, J. (ed.) Geometric Dynamics. Lecture Notes in Mathematics, vol. 1007, pp. 104–108. Springer, Berlin (1983)
8. Cheng, S.Y.: Eigenfunctions and nodal sets. Comment. Math. Helv. **51**, 43–55 (1976)
9. Daniel, B.: Isometric immersions into 3-dimensional homogeneous manifolds. Comment. Math. Helv. **82**, 87–131 (2007)
10. Daniel, B., Hauswirth, L.: Half-space theorem, embedded minimal annuli and minimal graphs in the Heisenberg group. Proc. Lond. Math. Soc. (3) **98**(2), 445–470 (2009)
11. Daniel, B., Hauswirth, L., Mira, P.: Constant Mean Curvature Surfaces in Homogeneous Manifolds. Korea Institute for Advanced Study, Seoul, Korea (2009)
12. Daniel, B., Mira, P.: Existence and uniqueness of constant mean curvature spheres in Sol_3. J. Reine Angew. Math. **685**, 1–32 (2013)
13. Deaibes, S.: Minimal 2-spheres and optimal foliations in 3-spheres with arbitrary metric. Preprint, arXiv:2001.07695v2
14. Fernández, I., Mira, P.: Harmonic maps and constant mean curvature surfaces in $\mathbb{H}^2 \times \mathbb{R}$. Am. J. Math. **129**, 1145–1181 (2007)
15. Fernández, I., Mira, P.: A characterization of constant mean curvature surfaces in homogeneous 3-manifolds. Differ. Geom. Appl. **24**, 281–289 (2007)
16. Fernández, I., Mira, P.: Holomorphic quadratic differentials and the Bernstein problem in Heisenberg space. Trans. Am. Math. Soc. **361**(11), 5737–5752 (2009)
17. Fernández, I., Mira, P.: Constant mean curvature surfaces in 3-dimensional Thurston geometries. In: Proceedings of the International Congress of Mathematicians, Volume II (Invited Conferences), pp. 830–861. Hindustan Book Agency, New Delhi (2010)
18. Gálvez, J.A., Mira, P.: Uniqueness of immersed spheres in three-manifolds. J. Differ. Geom. **116**, 459–480 (2020)
19. Haslhofer, R., Ketover, D.: Minimal 2-spheres in three-spheres. Duke Math. J. **168**, 1929–1975 (2019)
20. Hauswirth, L., Rosenberg, H., Spruck, J.: On complete mean curvature $\frac{1}{2}$ surfaces in $\mathbb{H}^2 \times \mathbb{R}$. Commun. Anal. Geom. **16**(5), 989–1005 (2008)
21. Hopf, H.: Uber Flachen mit einer Relation zwischen den Hauptkrummungen, Math. Nachr. **4**, 232–249 (1951)
22. Hopf, H.: Differential Geometry in the Large, volume 1000 of Lecture Notes in Mathematics. Springer, Berlin (1989)
23. Jost, J.: Embedded minimal surfaces in manifolds diffeomorphic to the three-dimensional ball or sphere. J. Differ. Geom. **30**, 555–577 (1989)
24. Manzano, M., Pérez, J., Rodríguez, M.: Parabolic stable surfaces with constant mean curvature. Calc. Var. Partial Differ. Equ. **42**, 137–152 (2011)
25. Meeks, W.H.: Constant mean curvature spheres in Sol_3. Am. J. Math. **135**, 763–775 (2013)
26. Meeks, W.H., Mira, P., Pérez, J.: Embeddedness of spheres in homogeneous three-manifolds. Int. Math. Res. Not. **2017**, 4796–4813 (2017)

27. Meeks, W.H., Mira, P., Pérez, J., Ros, A.: Isoperimetric domains of large volume in homogeneous three-manifolds. Adv. Math. **264**, 546–592 (2014)
28. Meeks, W.H., Mira, P., Pérez, J., Ros, A.: Constant mean curvature spheres in homogeneous three-spheres, J. Differ. Geom. **120**, 307–343 (2022)
29. Meeks, W.H., Mira, P., Pérez, J., Ros, A.: Constant mean curvature spheres in homogeneous three-manifolds. Invent. Math. **224**, 147–244 (2021)
30. Meeks, W.H., Pérez, J.: Constant mean curvature surfaces in metric Lie groups. In: Geometric Analysis, vol. 570, pp. 25–110. Contemporary Mathematics (2012)
31. Meeks, W.H., Pérez, J., Ros, A.: Stable constant mean curvature surfaces. In: Ji, L., Li, P., Schoen, R., Simon, L. (eds.) Handbook of Geometrical Analysis, vol. 1, pp. 301–380. International Press (2008)
32. Smith, F.: On the existence of embedded minimal 2-spheres in the 3-sphere. Endowed with an Arbitrary Riemannian Metric. Ph.D. Thesis, University of Melbourne, 1982. Supervisor: L. Simon
33. Torralbo, F.: Rotationally invariant constant mean curvature surfaces in homogeneous 3-manifolds. Differ. Geom. Appl. **28**, 593–607 (2010)
34. Torralbo, F.: Compact minimal surfaces in the Berger spheres. Ann. Global Anal. Geom. **41**, 391–405 (2012)
35. Torralbo, F., Urbano, F.: Compact stable constant mean curvature surfaces in homogeneous 3-manifolds. Indiana Univ. Math. J. **61**, 1129–1156 (2012)
36. Wang, Z., Zhou, X.: Existence of four minimal spheres in S^3 with a bumpy metric. Preprint, arXiv:2305.08755
37. White, B.: The space of minimal submanifolds for varying Riemannian metrics. Indiana Univ. Math. J. **40**, 161–200 (1991)

GPSR Compliance

The European Union's (EU) General Product Safety Regulation (GPSR) is a set of rules that requires consumer products to be safe and our obligations to ensure this.

If you have any concerns about our products, you can contact us on ProductSafety@springernature.com

In case Publisher is established outside the EU, the EU authorized representative is:

Springer Nature Customer Service Center GmbH
Europaplatz 3
69115 Heidelberg, Germany

Batch number: 10732076

Printed by Printforce, the Netherlands